T0390299

# New Trends in Algebras and Combinatorics

Proceedings of the 3rd International Congress in Algebras and Combinatorics (ICAC2017)

In Honour of Professor Leonid Bokut
on the Occasion of His 80th Birthday

# New Trends in Algebras and Combinatorics

Proceedings of the 3rd International Congress in Algebras and Combinatorics (ICAC2017)

In Honour of Professor Leonid Bokut
on the Occasion of His 80th Birthday

Hong Kong, China    25 – 28 August 2017

**Editors**

## K P Shum
*Yunnan University, China*

## E Zelmanov
*University of California, San Diego, USA*

## P Kolesnikov
*Sobolev Institute of Mathematics, Russia*

## S M Anita Wong
*The Open University of Hong Kong, Hong Kong*

 **World Scientific**

NEW JERSEY · LONDON · SINGAPORE · BEIJING · SHANGHAI · HONG KONG · TAIPEI · CHENNAI · TOKYO

*Published by*

World Scientific Publishing Co. Pte. Ltd.
5 Toh Tuck Link, Singapore 596224
*USA office:* 27 Warren Street, Suite 401-402, Hackensack, NJ 07601
*UK office:* 57 Shelton Street, Covent Garden, London WC2H 9HE

**British Library Cataloguing-in-Publication Data**
A catalogue record for this book is available from the British Library.

**NEW TRENDS IN ALGEBRAS AND COMBINATORICS**
**Proceedings of the 3rd International Congress in Algebras and Combinatorics (ICAC2017)**

ISBN 978-981-121-546-9

For any available supplementary material, please visit
https://www.worldscientific.com/worldscibooks/10.1142/11694#t=suppl

# Preface

This volume composed of twenty four research articles which are selected from the keynotes speakers and invited lectures presented in the $3^{rd}$ International Congress in Algebra and Combinatorics (ICAC2017) held on 25-28 August 2017 in Hong Kong and one additional invited article. This congress was especially dedicated to *Professor Leonid Bokut* on the occasion of his $80^{th}$ birthday. Leonid Bokut is known for his solution of the Malcev problem on semigroup algebras and for his contribution to the degrees of unsolvability of the combinatorial problems for groups. He initiated a broad program of study in combinatorial algebra based on the method of Gröbner–Shirshov bases.

This Congress has been organized every 10 years since 1997. The ICAC2017 was hosted by the Open University of Hong Kong and Southeast Asian Mathematical Society (SEAMS). It was organized by *Kar Ping Shum* (Yunnan University of China), *Leonid Bokut* (Sobolev Institute of Mathematics), *Efim Zelmanov* (University of California, San Diego), *Kin Chung Ho* (Open University of Hong Kong), *Li Guo* (Rutgers University, Newark), *Xiuyun Guo* (Shanghai University), *Xueming Ren* (Xian University of Architecture and Technology), *Yuqun Chen* (South China Normal University), *J.M.P. Balmaceda* (University of the Philippines), *Tuan Hoa Le* (Vietnam Institute for Advance Study in Mathematics), *Van Sanh Nguyen* (Mahidol University of Bangkok), *San Ling* (Nanyang Technological University, Singapore), *Intan Muchtadi-Alamsyah* (Institute Teknologi Bandung, Indonesia), *Wahyuni Sri* (Gadjah Mada University, Indonesia). ICAC2017 was regarded as a great success event and around 500 participants attended from over 24 countries. There were 21 plenary speakers and 80 invited talks in this congress. Our high appreciation is to all the plenary speakers and reviewers: Leonid Bokut, Efim Zelmanov, Bern Wegner, Alexander Olshanskiy, Alexander A Ivanov, Pavel Kolesnikov, Yuqun Chen, Anatoly Yakovlev, Viacheslav Artamonov, Laszlo Marki, John Meakin, Vassily Manturov, Michel Jambu, Andrey Vasil'ev, Luchezar Avramov, Vsevolod Gubarev, JiPing Zhang, Alexander Guterman, Gyula Katona, Tuan Hoa Le, Qaiser Mushtaq, and to all invited and contributed speakers as well as to all the participants for making this conference successful and enjoyable. Moreover, we are very grateful for the financial support by Mr. Roland Wong, the director of the International Mathematical Olympiad Hong Kong Committee Limited.

**Editors**

# A Short Biography of Professor Leonid Bokut

P. Kolesnikov

*Sobolev Institute of Mathematics,*
*Akad. Koptyug prosp., 4,*
*Novosibirsk 630090, Russia*

*pavelsk@math.nsc.ru*

Professor Leonid Bokut was born in 1937 in a small town Krupki located in the Republic of Belarus which had been a part of the Soviet Union at that time. In 1953, he attended Moscow State University (MSU), Department of Mechanics and Mathematics. His supervisor at MSU was A.I. Shirshov, who later became one of the founders of the Siberian School of Algebra and Logics together with A.I. Malcev and M.I. Kargapolov.

Leonid Bokut graduated from MSU in 1959 with Honors diploma, his Master thesis was related to the theory of Lie algebras. At that time, Akademgorodok of Novosibirsk was established, S.L. Sobolev founded the Institute of Mathematics (now called Sobolev Institute of Mathematics), and many leading mathematicians of Soviet Union were invited to join that institute. Among others, A.I. Malcev and A.I. Shirshov decided to switch Moscow to Novosibirsk. Shirshov invited him to attend graduate study at the new Institute of Mathematics, within the Malcev's laboratory "Algebra and Logic", where Leonid was enrolled on March 1, 1960. He had been developing as a mathematician under the direct personal influence of the world-wide famous researchers A.G. Kurosh, A.I. Malcev and A.I. Shirshov.

Since October 1, 1962 to the present time he has been working in Sobolev Institute of Mathematics. He defended Ph.D. Thesis entitled "Some embedding theorems for rings" and obtained the Candidate of Science degree in January 1963. One of the main results of the Thesis was the proof that every Lie (nonassociative, commutative, or anti-commutative) algebra is embeddable into an algebraically closed Lie (resp., nonassociative, commutative, or anti-commutative) algebra. Here the term "algebraically closed" means the strongest possible sense: an algebra is algebraically closed if it contains a solution for every non-trivial equation.

In May 1969, he received the Doctor of Science degree (Habilitation), with the Thesis entitled "Some problems of the combinatorial theory of groups and rings". The main result of the Thesis presented complete solution of the problem stated by A.I. Malcev in 1937. There was constructed a semigroup $Q$ such that its semigroup algebra $kQ$ has no zero divisors but it is not embeddable into a skew field whereas its multiplicative semigroup is embeddable into a group. By now, the semigroup $Q$ constructed in that work is the only known semigroup with the above properties. It was also proved that, given a Turing degree $\alpha$ (i.e., complexity) of insolubility there exists a group in which the conjugacy problem is of degree of insolubility $\alpha$ (it was a solution of another Malcev's problem posed in 1956). To prove the above two results, he used the idea of "groups with standard bases" inspired by the ideas of A.I. Shirshov.

Later on, he solved some algorithmic problem for algebras. In particular, he proved algorithmic insolubility of the word problem for finitely presented Lie algebras in the spirit of weak form of the Higman Embedding Theorem for groups (1972): some recursively presented Lie algebras are embeddable into finitely presented Lie algebras. Then he showed that every associative (Lie) algebra can be embedded into a simple associative (resp., algebraically closed Lie) algebra which is a sum of four given associative (resp., Lie) algebras with some cardinality condition (1976, 1978). It was a linear version of Ph. Hall's theorem (1974) on the embedding of a group into a union of given groups. He posed a problem of validity of the Higman's Theorem for associative algebras which was solved positively by V. Belyaev (1976). A similar problem for Lie algebras remains open.

An essential part of the scientific life of Prof. Leonid Bokut is devoted to the theory of what is now called Gröbner–Shirshov bases. It was initiated by A.I. Shirshov for non-associative and Lie algebras, as well as (implicitly) for associative algebras in 1962. H. Hironaka in 1964 applied the same ideas for local commutative algebras, and B. Buchberger in 1965—for commutative algebras.

He worked in the theory of Gröbner–Shirshov bases with many colleagues, including A. Klein, P. Malcolmson, S.-J. Kang, K.-H. Lee, L.-S. Shiao, Y. Fong, W.-F. Ke, V. Chainikov, K.P. Shum. He inspired creation of a research group in South China Normal University (SCNU), Guangzhou, China, with Y. Chen, Z. Zhang, W. Chen, Y. Li, Q. Mo, Y.-S. Chen, J. Huang, J. Li, X. Deng, et al. A remarkable series of applications of the technique inspired by A.I. Shirshov has been reflected in the change of terminology throughout the algebra world: people are now often using the term "Gröbner–Shirshov bases" rather than "non-commutative

Gröbner bases".

Prof. Leonid Bokut worked at visiting positions in USA, Israel, Korea, Taiwan, Hong Kong. In 2006–2016 he had been a Professor of SCNU. With Y. Chen, K.P. Shum, and students of SCNU he published more than 50 papers on different classes of universal algebras. Several Ph.D. theses have been prepared by his students in SCNU.

We know Prof. Leonid Bokut not just as a researcher, but also as a teacher and organizer. In the period 1961–1981 and from 1988 until the present he worked at the Novosibirsk State University. He has 36 successful Ph.D. Students in Novosibirsk and Guangzhou. The most celebrated student of him is Efim Zelmanov (Fields Medal winner in 1994), and there are many other resounding names in the list of his students: V. Gerasimov, R. Gonchigdorzh, A. Grishkov, A. Kemer, V. Kharchenko, Yu. Maltsev, et al. The total number of his research descendants is over 60.

He initiated and organized a number of international scientific events, even during the Soviet period when it was a hard task to invite a foreign researcher to a conference. Among other, let us mention the series of Siberian Schools on varieties of algebraic systems (10 events in 1979–1990), Siberian Schools on algebra and analysis (10 events in 1987–1996), International Conferences on algebra dedicated to the memory of A.I. Malcev (1989) and A.I. Shirshov (1991).

In the period 1994–2000, he was a Vice-Chairman of the Program Committee of 4 Siberian Congresses on Industrial and Applied mathematics (INPRIM), M.M. Lavrentiev was the Chairman. INPRIM events were devoted to the memory of L.V. Kantorovich (Nobel prize winner, one of the organizers of the Sobolev Institute of Mathematics), S.L. Sobolev (inventor of the theory of generalized functions), A.A. Lyapunov, A.P. Ershov, I.A. Poletaev (pioneers of cybernetics), and M.A. Lavrentiev (founder of the Siberian Branch of the Academy of Science).

He initiated establishing the Siberian Society for Promotion of Science and Education (SIBOS), headed by Yu.G. Reshetnyak (1991–2000) and S.V. Goldin (1997–2000). In 1994–1996, SIBOS distributed USD 500,000 granted by the G. Soros Fund into about 1500 small group grants for researchers of Novosibirsk Akademgorodok. SIBOS also organized 4 Conferences for Soros High School Teachers of Novosibirsk Region. Prof. Leonid Bokut had been the secretary of the Malcev Seminar "Algebra and Logic" in 1961–1970.

We would like to congratulate Professor Leonid Bokut on his 80th birthday and take this opportunity to thank him for his invaluable contribution to the mathematical community. His experience, gained over many years of work in mathematics, is truly an invaluable asset of the algebra people over the world; energy and efficiency, goodwill, willingness to learn new topics are an encouraging example for all of us, especially for young people making their first steps in research. We are always inspired by his inexhaustible sense of humor which, together with a firm belief in the best, helps us to overcome all obstructions.

The International Congress in Algebra and Combinatorics (ICAC 2017) is an occasion for us to celebrate the remarkable results of his research, teaching, and organizing activity in more than 60 years. We would like to acknowledge the invaluable contribution of his wife Svetlana Bokut who supported him a lot in his life. We wish him to remain healthy, active, and producing fruitful ideas inspiring the young generation of mathematicians. We hope he will celebrate more anniversaries.

# Contents

1

# My life in mathematics, 60 years

Leonid A. Bokut

*School of Mathematical Sciences,*
*South China Normal University,*
*Guangzhou 510631, P. R. China;*
*Sobolev Institute of mathematics,*
*Akad. Koptyug prosp., 4,*
*Novosibirsk 630090, Russia*

*bokut@math.nsc.ru*

This paper is about my life in mathematics, including my supervisors, my students and my main results. I will introduce them mainly in chronological order.

*2010 Mathematics subject classification:* 17B63, 16S15, 13P10.

## 1. My education and positions

During 1953–1959, I had been an undergraduate student at Moscow State University, where A.I. Shirshov was my supervisor (Honors diploma). During September 1959–February 1960, I had been a postgraduate student of the Moscow State University, A. G. Kurosh's Algebra Department(supervisor A. I. Shirshov). Then during March 1960 - October 1962 I had been a postgraduate student of the (now Sobolev) Institute of mathematics, Novosibirsk, A.I. Malcev Laboratory "Algebra and Logic" (though by family reasons I was at Kurosh's department untill September 1961). Again, A.I. Shirshov was my supervisor.

In January 1963 I got my Ph.D. (in Russia: Candidate of Sciences) degree. The topic of my thesis was "Some embedding theorems for rings", United Math.-Phys. Scientific Council of the Siberian Branch of the Academy of Sciences of USSR, supervisor A.I. Shirshov, referees P.G. Kontorovich, A.T. Gainov, official organization Moscow State University, A.G. Kurosh. In May 1969 I became a Doctor of Sciences, the title of my thesis was "Some problems of the combinatorial theory of groups and rings" (600 pages), Sobolev Institute of Mathematics, Novosibirsk. The referees were S.I. Adyan, V.A. Andrunakievich, M.D. Grindlinger, official organization was the Moscow State Pedagogical Institute, L.Ya. Kulikov, A.L. Shmelkin.

From 1962 to the present time I have been working in Sobolev Institute of Mathematics. At first in A.I. Malcev's Laboratory of Algebra and Logic, till Malcev's death in 1967. Then I worked in A.I. Shirshov's Laboratory of Ring Theory till Shirshov's death in 1981. After that, I become the head of the Laboratory "Associative and Lie rings" (1981–1995), a leading researcher of the Laboratory of Ring Theory (1996–p.t.). V.Filippov, V.Zhelyabin and P.Kolesnikov are the heads of the Laboratory during these years.

During the years 1961–1981 and the years 1988–p.t. I also worked at the Novosibirsk State University and became a full professor in 1972. I was also the secretary of the Malcev Seminar "Algebra and Logic" during 1961–1970.

In the recent 30 years I also worked abroad. I worked in USA, Israel, Korea, Taiwan, Hong Kong. In 2006–2016 I was a professor of South China Normal University, Guangzhou, P.R. China.

## 2. My teachers

The teacher I would like to introduce first is A.I.Shirshov. He was not only my teacher, but also my supporter, mentor, scientific father since 1953 until his death in the year 1981. His influence on me is so profound that I can never forget him. I am always proud of being a student of him. Under his guidance, I prepared my Master Thesis on Lie algebras and my PhD Thesis on Lie and associative algebras, and semigroups. He prompted me to study P.S.Novikov's book "Insolubility of the word problem for groups", and it was important for my future Doctor of Sciences Thesis. I have been using his Gröbner-Shirshov basis method since the 1960th.

I would like to introduce next A.G.Kurosh, who gave the first introductory lecture for 400 freshmen of the Department of Mathematics of MSU in 1953. In the same year, he delivered lectures on algebra for 200 first year undergraduate students, and I was fortunate to be one of the students. He was a professor of the Higher Algebra chair during my study at MSU in 1953-1959. I participated in the Schmidt-Kurosh's algebra seminar at MSU and the optional courses taught by the latter. A.I.Shirshov asked A.G. Kurosh to co-advise me during my PhD study in Moscow, 1960-1961 (because A.I.Shirshov was a deputy-director of (Sobolev) Institute of mathematics, Novosibirsk, since 1959).

The last but not least, I would like to mention A.I.Malcev. I was a member of his laboratory and the secretary of his famous seminar at Novosibirsk. He and Shirshov were my advisers for the (second) Doctor of Sciences degree. My second Doctor degree Thesis was devoted to the solution of two Malcev's problems.

## 3. Main results

Concerning embedding of algebras, I proved several results. In 1962, I proved that every Lie (nonassociative, (anti-)commutative) algebra is embeddable into a strong algebraically closed Lie (nonassociative, (anti-) commutative) algebra $L$ (any equation $f(x_1, \ldots, x_n) = 0$, $f \in L * Lie(X)$, $Lie(X)$ is a free Lie algebra, $f \notin L$, has a solution in $L$). My problems (see ref. 111) in the list of my papers below) on existence strong algebraically closed associative algebras and groups were solved positively by S.Brodsky (1974) for groups and L.Makar-Limanov (1981) for associative algebras. Then I constructed an associative algebra without zero-divisors (semigroup algebra k(Q)), that is not embeddable into skew field whereas the multiplicative semigroup is embeddable into a group (Malcev's problem, 1937) (1967, 1969). Up to now, the semigroup $Q$ is the only known semigroup with the above property. I found a large class of semigroups algebras $k(S)$ without zero divisors whereas multiplicative semigroups $k(S)^*$ imbeddable into groups without torsions. I also proved that for every Turing degree (complexity) of insolubility there exists a group with the given degree of insolubility of the conjugacy problem (a solution of Malcev's problem, 1956) (1968). To prove the above three results, I used the idea of "groups with standard bases" (1968). E.Chibrikov (2012) had a progress toward a positive solution of the last problem.

I also proved some algorithmic problem for certain algebras. I proved the algorithmic insolubility of the word problem for Lie algebras in the weak form of Higman embedding theorem for groups (some recursively presented Lie algebras are embeddable into finitely presented Lie algebras) (1972). After that I showed that every associative (Lie) algebra is embeddable into a simple associative (algebraically closed Lie) algebra which is a sum of 4 given associative (Lie) algebras with some cardinality condition (1976, 1978), which is a linear version of Ph.Hall's last theorem on the embedding of a group into a join of given groups (1974). My problem (see ref. 95) of validity of Higman theorem for associative algebras was solved positively by V.Belyaev (1976). A similar problem (see ref. 95) for Lie algebras is still open.

What is now called Gröbner and Gröbner-Shirshov bases theory was initiated by A. I. Shirshov for non-associative and Lie algebras, and implicitely for associative algebras in 1962, by H. Hironaka in 1964 for local commutative algebras, and, by B. Buchberger in 1965 for commutative algebras. I worked in the theory of Gröbner-Shirshov bases with many colleagues. We found Gröbner-Shirshov bases of simple Lie algebras (four joint papers with Abraham Klein (Israel), 1993-1997). It served as an alternative proof of Serre's theorem for simple Lie algebras. We also found Gröbner-Shirshov bases of Drinfeld—Jimbo quantum groups (joint paper with Peter Malcolmson (USA), 1995). It offered an alternative proof of Roso—Yamane theorem for linear bases of Drinfeld—Jimbo quantum groups.

We proved PBW theorem in the Shirshov's form: Let $L = \mathsf{Lie}(X|S)$ be a Lie algebra over a field $k$ with a product $[xy]$ and let $U(L) = k\langle X|S\rangle$ be its universal enveloping associative algebra, where $[xy] = xy - yx$ in $S$. Then $S$ is a Lie Gröbner-Shirshov basis iff $S$ is an associative Gröbner-Shirshov basis (joint paper with Peter Malcolmson, ICAC1997). We found Gröbner-Shirshov basis of classical Lie superalgebras (joint paper with Seok-Jin Kang, Kyu-Hwan Lee(South Korea), and Peter Malcolmson, 1999). It offered an alternative proof of the Serre type theorem for classical Lie superalgebras. We found Gröbner-Shirshov bases of finite Coxeter groups (joint paper with L.-S.Shiao (Taiwan), 2000). It served for an alternative proof of the Coxeter theorem for finite Coxeter groups. We established a theory of Gröbner-Shirshov bases for associative conformal algebras (joint paper with Yuen Fong and Fong Ke (Taiwan), 2004). We also constructed Gröbner - Shirshov bases of braids groups in Burau generators and Adyan extension of Novikov group (joint works with V.Chainikov and K.P.Shum, 2007, and with V.Chainikov, 2008,correspondingly). It gives alternative proofs of Artin–Markov and Adyan–Rabin theorems. We proved a CD-lemma for free $G$-associative algebras, where $G$ is a group (joint paper with K.P.Shum, 2008). It had applications to the Malcev's problem above and to Bruhat normal forms for algebraic groups.

We also found Gröbner-Shirshov bases for braid groups in Artin–Garside and Birman–Ko-Lee–Garsaid generators (2008, 2009) It offered an alternative proofs of Garside and Birman–Ko-Lee–Garside normal form theorems for braid groups. With Yuqun Chen, K.P.Shum and students of SCNU we published more than 50 papers during the time of my visiting Chinese University of Hong Kong (2000–05) and working at South China Normal University (2006–16). The papers were on different classes of universal algebras. Several PhD theses have been prepared by my Chinese students: Chanyan Zhong, Gröbner-Shirshov basis theory and its applications for some groups and modules (2010); Yongshan Chen, Topics on Gröbner-Shirshov Bases for Lie algebras (2011); Qiuhui Mo, The Development and Applications of Gröbner-Shirshov Bases Theories for Semirings and Lie Algebras (2013); Yu Li, Gröbner-Shirshov Bases for Categories, pre-Lie and Non-Associative Algebras (2013); Lili Ni, New Composition-Diamond lemmas for conformal algebras and their modules (2016); Jianjun Qiu, The applications and generalizations of the Gröbner-Shirshov bases theory for Lie algebras; Guangliang Zhang, Gröbner-Shirshov bases theory for dialgebras and commutative dialgebras and its applications. Zerui Zhang, On Novikov superalgebras, Novikov - Poisson algebras, Poisson algebras and metabelian Poisson algebras. Also Rabigul Tuniyaz (2018) defended PhD Thesis on Groebner - Shirshov bases for some associative and nonassociative algebras (supervisor Abdukadir Obul, I was a co-supervisor).

## 4. My activities in mathematical life

Starting from 1979 I initiated four series of conferences, and I attended many important conferences.

(1) Siberian Schools on varieties of algebraic systems (10 Schools, Barnaul, Omsk, Magnitogorsk, 1979–1990).

(2) Siberian Schools on algebra and analysis (10 Schools, Kemerovo, Tomsk, Lake Baikal (7), Novosibirsk, 1987–1996.

(3) First Siberian Winter School "Algebra and Analysis" (1987, Kemerovo), AMS, 150 pp (A.D.Aleksandrov, O.V.,Belegradek, L. A.Bokut, Yu.L.Ershov, Eds).

(4) Second Siberian Winter School "Algebra and Analysis" (1988, Tomsk), AMS, 100 pp (I.A.Aleksandrov, L. A.Bokut, Yu.G.Reshetnyak, Eds).

(5) Third Siberian School: Algebra and Analysis (1989, Irkutsk, Lake Baikal), AMS, 150 pp (L. A.Bokut, M.Hazewinkel, Yu.G.Reshetnyak, Eds).

(6) International Conferences on algebra dedicated to the memories of A.I.Malcev, A.I.Shirshov and M.I.Kargapolov.

(7) Malcev Conference, Novosibirsk, 1989, Proceedings, AMS, 3 volumes, 2100 pp. (L. A.Bokut, Yu.L.Ershov, A.I.Kostrikin, Eds).

(8) Shirshov Conference, Barnaul, 1991, Proceedings, AMS, 400 pp, (L. A.Bokut, A.I.Kostrikin, S.S.Kutateladze, Eds).

(9) Kargapolov Conference, Krasnoyarsk, 1993, Proceedings, Walter de Gruyter, 1996, 300 pp, (Yu.L.Ershov, E.I.Khukhro, V.M.Levchuk,

N.D.Podufalov, Eds).

(10) Novosibirsk Congresses on Industrial and Applied mathematics (IN-PRIM) dedicated to: L.V.Kantorovich (1994); A.A.Lyapunov - A.P.Ershov - I.A.Poletaev (1996); S.L.Sobolev (1998); M.A.Lavrentiev (2000). Program Committee included M.M.Lavrentiev (Chairman), L. A.Bokut (Deputy-Chairman), Yu.G.Reshetnyak (Deputy-Chairman), Yu.V.Merekin (Organizer Committee), E.P.Volokitin (Secretary of Kantorovich INPRIM), S.A.Treskov (Secretary of Lyapunov - Ershov - Poletaev INPRIM), V.L.Vaskevich (Secretary of Sobolev INPRIM and Lavrentiev INPRIM), A.I.Rylov (Secretary of Sobolev INPRIM and Lavrentiev INPRIM). There were up to 25 Sections on pure, applied and industrial mathematics. We published up to two thousand abstracts of participants of the Congresses.

I initiated AMS publications of 4 Dissertations of my students:

(1) Kemer, Identities of associative algebras, AMS, 200 pp, 1994.

(2) V.Gerasimov, N.Nesterenko, A.Valitskas, AMS, 200 pp, 1991.

(3) I initiated Siberian Society for Promotion of Science and Education (SIBOS), 1991–2000, Yu.G.Reshetnyak (President, 1991–1996), S.V. Goldin (President, 1997–2000).

(4) SIBOS distributed a Soros grant of 500,000 USD for our Novosibirsk Akademgorodok (SB RAS), and one thousand five hundred small group grants. On behalf of SIBOS, we, with Yu.V.Merekin, organized 4 Conferences for Soros High School Teachers of Novosibirsk Region (1994-1996).

## 5. List of my students

(1) Kukin, Georgii, 1973 (2 PhD students);

(2) Maltsev, Yurii, 1973 (3 PhD students);

(3) Kharchenko, Vladislav, 1975 (3 PhD students);

(4) Lvov, Igor, 1975 (1 PhD student);

(5) Grishkov, Alexander, 1979;

(6) Gerasimov, Victor, 1980;

(7) Yagzhev, Alexander, 1980;

(8) Zelmanov, Efim, 1981 (13 PhD students);

(9) Kemer, Alexander, 1981 (3 PhD students);

(10) Gonchigdorj, Radnaasumberel, 1982;

(11) Kolotov, Alexander, 1982;

(12) Tarasov, Boris, 1982;

(13) Urman, Alexander, 1982;

(14) Talapov, Vladimir, 1983;

(15) Chekhonadskikh, Alexander, 1983;
(16) Bobkov, Oleg, 1985;
(17) Ananin, Alexander, 1987 (1 PhD student);
(18) Dashdorj, Tserendoj, 1987;
(19) Nesterenko, Nikolai, 1987;
(20) Vdovin, Victor, 1987;
(21) Kryazhovskikh, Galina, 1988;
(22) Stern, Alexander, 1988;
(23) Vais, Arkadii, 1988;
(24) Valitskas, Alexei, 1988;
(25) Sidorov, Alexander, 1989;
(26) Petrov, Evgenii, 1996;
(27) Kulchinovskii, Vladislav, 2001;
(28) Poroshenko, Evgenii, 2002;
(29) Kolesnikov, Pavel, 2003 (2 PhD students);
(30) Chibrikov, Evgenii, 2004;
(31) Firdman, Ilya, 2007;
(32) Dolguntseva, Irina, 2008;
(33) Zhong, Chanyan, 2010;
(34) Chen, Yongshan, 2011;
(35) Li, Yu, 2013;
(36) Mo, Qiuhui, 2013.

Totally, I have 36 PhD students and 63 PhD descendants. There are also members of my seminar "Associative and Lie rings" who got PhD abroad: Rumynin Dmitrii, USA; Kapovich Ilya, USA; Krushkal Vladislav, USA; Mineev Igor, USA; Chainikov Vladimir, USA; Ulyanov Alexander, USA; Movlyutov Anvar, USA.

## 6. Main publications

### 6.1. *Books*

(1) Associative rings I, Novosibirsk, State University, 1977.
(2) Associative rings II, Novosibirsk, State University, 1979.
(3) (with V. Kharchenko and I. L'vov), Non commutative rings, Springer, Algebra-2, 1990.
(4) (with G. Kukin), Algorithmic and Combinatorial algebra, Kluwer, Dordrecht, 1994, 384 p.
(5) (with A.D. Aleksandrov, O.Belegradek, Yu.Ershov, Eds.), Algebra and Analysis, Proceedings of the First Siberian School, AMS, 1991,

112.

(6) (with I. Aleksandrov, Yu.Reshetnyak, Eds.), Algebra and Analysis, Proceedings of the Second Siberian School, AMS, 1992.

(7) (with M. Hazewinkel, Yu. Reshetnyak, Eds.), Algebra and Analysis, Proceedings of the Third Siberian School, AMS, 1995.

(8) (with Yu. Ershov, A. Kostrikin, Eds), Proceedings of the International Conference in Algebra in the Memory of A.I.Malcev, AMS, Contemp. Math., 131, part 1, part 2, part 3, AMS, 1992.

(9) (with A. Kostrikin, S. Kutateladze, Eds), Proceedings of the Second International Conference on Algebra in the Memory of A.I.Shirshov, AMS, Contemp. Math., 184, AMS, 1995.

(10) (with S.Kutateladze, M.Lavrent'ev, Eds), Proceedings of the Siberian Conference in Applied and Industrial Mathematics in the Memory of L.V.Kantorovich, Novosibirsk, Institute of Mathem., Vol. 1, Vol. 2, 1997.

(11) (with Yu.G.Reshetnyak, I.A.Taimanov, S.K.Vodopyanov, Eds.), Algebra, Geometry, Analysis and Mathematical Physics, 10-th Siberian school, Novosibirsk, Institute of Mathematics, 1997.

(12) (with Yuqun Chen), Gröbner-Shirshov Bases and Shirshov Algorithm, Educational tutorial lecture notes, Novosibirsk State University, 2014.

### 6.2. *Papers*

(13) (with Y.Q. Chen and A. Obul) Some new results on Grobner-Shirshov bases for Lie algebras and around, International Journal of Algebra and Computation, 28 (2018), no. 8, 1403-1423.

(14) (with R. Tuniyaz, M. Xiryazidin, and A. Obul) Gröbner-Shirshov bases for free Gelfand-Dorfman-Novokov algebras and for right ideals of free right Leibniz algebras, Comm. Algebra, 46 (2018), no. 10, 4392-4402.

(15) Early history of the theory of rings in Novosibirsk, Bul. Acad. Stiinte Repub. Mold. Mat., 2 (2017), p. 5-23.

(16) (with Y.Q. Chen and Zerui Zhang) Gröbner–Shirshov bases method for Gelfand–Dorfman–Novikov algebras, Journal of Algebra and Its Applications, 16 (2017), no. 1, 22 p.

(17) (with Y.Q. Chen and Zerui Zhang) On free Gelfand–Dorfman–Novikov–Poisson algebras and a PBW theorem. J. Algebra, 500 (2018), 153-170.

(18) (with Y.Q. Chen, Weiping Chen, Jing Li) New approaches to plactic monoid via Groebner-Shirshov bases, Journal of Algebra, 423(2015), 301-317.

(19) (with Y.Q. Chen), Gröbner-Shirshov bases and their calculation, Bull. Math. Sci., 4(2014), 325-395.

(20) (with Y.Q. Chen), Gröbner-Shirshov bases for universal algebra, Journal of SCNU, 46(6)(2014), 1-9.

(21) (with Y.Q. Chen), Gröbner-Shirshov and PBW theorems, Journal of Siberian Federal University. Mathematics & Physics, 6(4)(2013), 417-C427.

(22) (with Y.Q. Chen, Qiuhui Mo), Gröbner-Shirshov bases for semirings, Journal of Algebra, 385(2013) 47-63.

(23) (with Y.Q. Chen, Y. Li), Lyndon-Shirshov basis and anticommutative algebras, Journal of Algebra, 378(2013), 173-183.

(24) (with Y.Q. Chen, K.P. Shum), Some new results on Groebner-Shirshov bases, in: Proceedings of International Conference on Algebra 2010, Advances in Algebraic Structures, 2012, pp.53-102.

(25) (with Y.Q. Chen, Y. Li), Groebner-Shirshov bases for categories, Nankai Series in Pure, Applied Mathematics and Theoretical Physics, Operads and Universal Algebra, Vol.9(2012), 1-23.

(26) (with Y.Q. Chen, Jiapeng Huang), Groebner-Shirshov bases for L-algebras, International Journal of Algebra and Computation, 23 (2013), no. 3, 547-571.

(27) (with Y.Q. Chen, Yongshan Chen), Groebner-Shirshov bases for Lie algebras over a commutative algebra, Journal of Algebra, 337(2011), 82-102.

(28) (with Y.Q. Chen, Qiuhui Mo), Groebner-Shirshov bases and embeddings of algebras, International Journal of Algebra and Computation, 20(2010), 875-900.

(29) (with Y.Q. Chen, Xueming Deng), Groebner-Shirshov bases for Rota-Baxter algebras, Siberian Mathematical Journal, 51(6)(2010), 978-988.

(30) (with Y.Q. Chen, Cihua Liu), Groebner-Shirshov bases for dialgebras, International Journal of Algebra and Computation, 20(3)(2010), 391-415.

(31) (with Y.Q. Chen, Yongshan Chen), Composition-Diamond lemma for tensor product of free algebras, Journal of Algebra, 323(2010), 2520-2537.

(32) (with Y.Q. Chen, Jianjun Qiu), Groebner-Shirshov bases for

associative algebras with multiple operations and free Rota-Baxter algebras, Journal of Pure and Applied Algebra, 214(2010), 89-100.

(33) (with Y.Q. Chen, Xiangui Zhao), Groebner-Shirshov bases for a free inverse semigroup, International Journal of Algebra and Computation, 19(2)(2009), 129-143.

(34) (with Y.Q. Chen, Y. Li), Anti-commutative Groebner-Shirshov bases of a free Lie algebra, Science in China, 52(2)(2009), 244-253.

(35) (with Y.Q. Chen), Groebner-Shirshov bases: some new results, Advances in Algebra and Combinatorics, World Scientific, 2008, 35-56.

(36) (with Y.Q. Chen, Y. Li), Groebner-Shirshov bases for Vinberg-Koszul-Gerstenhaber right-symmetric algebras, Fundamental and Applied Mathematics, 14(8)(2008), 55-67. (in Russian) Journal of Mathematical Sciences, 166(2010), 603-612.

(37) Gröbner-Shirshov bases for braid groups in Artin-Garside generators. Journal of Symbolic Computation, 43(2008), 397-405.

(38) Gr"obner–Shirshov bases for the braid group in the Birman-Ko-Lee generators, Journal of Algebra, 321(2009), 361-376.

(39) (with V. Chainikov), Gröbner-Shirshov bases of Adjan extension of the Novikov group, Discrete Mathematics, 2008.

(40) (with V.V. Chainikov, K.P. Shum), Markov and Artin normal form theorem for braid groups, Comm. Algebra, 35 (2007), 2105-2115.

(41) (with Yuqun Chen), Grobner-Shirshov bases for Lie algebras: after A.I. Shirshov, Southeast Asian Bull. Math., 31, 811-831(2007).

(42) (with D. Lee) Gröbner-Shirshov basis for Temperley-Lieb-Kaufman monoid, Izvestia Ural State University, 36 (2005), 7, 49-66.

(43) (with V. Bardakov and A. Vesnin), On the conjugacy problem for cyclic extensions of free groups, Comm. Algebra, 33(6), 2005, 1979-1996.

(44) (with E. Chibrikov), Lyndon-Shirshov words, Gröbner-Shirshov bases, and free Lie algebras, Non-associative algebras and its applications, Taylor&Francis Group, Boca Raton, FL, 2005, p. 17-34.

(45) (with P. Kolesnikov), Gröbner-Shirshov bases, conformal algebras and pseudo-algebras, Journal of Mathematical Sciences,Journal of Mathematical Sciences, vol. 131, N. 5 (2005),5962–6003.

(46) (with A. Vesnin), Gröbner-Shirshov bases for some braid groups, Journal for Symbolic Computations, 41 (2005), 3-4, 357-371.

(47) (with V.Bardakov, A.Vesnin), Twisted conjugacy problem in free groups and Makanin's question, Southeast Asian Bulletin of Math-

ematics, 29 (2005), 2, 209-226.

(48) (with K.P. Shum), Gröbner and Gröbner-Shirshov bases in algebra: an elementary approach, Southeast Asian Bulletin of Mathematics, 29 (2005), 2, 227-252.

(49) (with Y. Fong, W.-F.Ke), Composition - Diamond lemma for associative conformal algebras, J. of Algebra, 272 (2004), 2, 739-774.

(50) (with V.Bardakov, A.Vesnin), Twisted conjugacy problem in free groups and Makanin's question, RIMS-GARC Preprint Series 03-13, Seoul National University, 2003, 19 pp.

(51) (with Y. Fong, W.-F. Ke, L.-S. Shiao), Gröbner-Shirshov bases for semigroups of positive braids. Advances in Algebra and related topics. Proceedings of the ICM2002 satellite conference on algebra, Hong Kong (K.P.Shum, Ed), River Edge, World Scientific, 2003, pp 14-25.

(52) (with Y. Fong, L.S. Shiao), Gröbner-Shirshiv bases for algebras, groups and semigroups. Proceedings of the Third International Algebra Conference, Tainan, Taiwan (Y. Fong, L.S. Shiao, E. Zelmanov, Eds), Kluwer, 2003, pp 17-32.

(53) (with P. Kolesnikov), Gröbner-Shirshov bases: from their incipiency to the present, Journal of Mathematical Sciences, 116 (2003), 1, 2894-2916.

(54) (with A. Vesnin), Rewriting systems for the braid groups $B_3$ and $B_4$ with Gorin-Lin generators, Kolmogorov Conference, Kolmogorov and Modern Mathematics, Moscow, 2003, 793.

(55) Gröbner bases: the non-commutative case. The Concise Handbook in Algebra (A.V. Mikhalev, G. Pilz, Eds), Kluwer, 2002, 265-267.

(56) Word problems and rewriting systems. The Concise Handbook in Algebra (A.V. Mikhalev, G. Pilz, Eds), Kluwer, 2002, 550-553.

(57) (with A. Vesnin), New rewriting system for the braid group $B_4$, Proceedings of the Buchberger Conference, Logic, Mathematics and Computer Science: Interaction, (LMCS, 2002), Linz, Austria, pp 48-60.

(58) (with A.Vesnin), New rewriting system for the braid group $B_4$, RIMS-GARC Preprint series 02-12, Seoul National University, 2002, pp 11.

(59) (with L.-S. Shiao), Gröbner-Shirshov bases for Coxeter groups, Comm. in Algebra, 29 (2001), N9, 4305-4319.

(60) (with P. Kolesnikov), Gröbner—Shirshov bases: from incipient to nowdays, Proceedings of the POMI, 272 (2000), 26-67.

(61) (with Y. Fong, W.-F. Ke, P. Kolesnikov), Gröbner and Gröbner—Shirshov bases in Algebra and Conformal algebras, Fundamental and Applied Mathematics, 6 (2000), N3, 669-706.

(62) (with Y. Fong, W.-F. Ke), Free associative conformal algebras, Proceedings of the 2nd Tainan-Moscow Algebra and Combinatorics Workshop, Tainan.– Springer, Hong Kong, 2000. p.13-25.

(63) (with Y. Fong, W.-F. Ke), Gröbner–Shirshov bases and the composition lemma for associative conformal algebras: an example. Contemporary Mathematics, N 264, AMS, 2000, p.63-91.

(64) (with K.I.Beidar and Y.Fong), Prime rings with semigroup generalized identity, Comm. in Algebra, 28 (2000), N3, 1497-1501.

(65) (with S.-J. Kang, K.-H. Lee, P. Malcolmson), Gröbner - Shirshov bases for Lie superalgebras and their universal enveloping algebras, J.Algebra, 217(1999), 461-495.

(66) (with P. Malcolmson), Gröbner — Shirshov bases for relations of a Lie algebra and its enveloping algebra, Algebra and Combinatorics(Hong Kong), 47-54, Springer, Singapore, 1999.

(67) Gröbner — Shirshov bases for Lie and associative algebras, Collection of Abstracts, ICAC 97, Hong Kong, 1997, 139-142.

(68) (with A.A. Klein), Gröbner — Shirshov bases for exceptional Lie algebras. I, J. Pure Applied Algebra, 133, 1998, 51-57.

(69) (with A.A. Klein), Gröbner — Shirshov bases for exceptional Lie algebras E6, E7, E8. Algebra and Combinatorics (Hong Kong), 37-46, Springer, Singapore, 1999.

(70) The method of Gröbner and Shirshov bases, in: Algebra, Geometry, Analysis and Mathematical Physics, Novosibirsk, 1997, 30-39.

(71) (with A.A. Klein), Serre relations and Gröbner — Shirshov bases for simple Lie algebras I, Internat. J. Algebra Comput. 6, 1996, N 4, 389-400.

(72) (with A.A. Klein), Serre relations and Gröbner – Shirshov bases for simple Lie algebras II, Internat. J. Algebra Comput. 6, 1996, N 4, 401-412.

(73) (with P. Malcolmson), Gröbner–Shirshov bases for quantum enveloping algebras, Israel Math. J., 96, 1996, 97-113.

(74) Abstract semigroups and groups of formal series on dependent variables, Siberian Advances in Math., 6(1996), no.3, 1-27.

(75) (with I.Shestakov), Some results by A.I.Shirshov and his school, Contemp. Math., 184 (1995), 1-12.

(76) (with L.Makar–Limanov), Basis of a free metabelian associative

algebra, Siberian Math.J., 1991, 32, 6, 35-46.

(77) A fragment of a group constructed by P.S.Novikov in 1952, Siberian Math.J., 1989, 30, 6, 42-51.

(78) Some new results in the combinatorial theory of rings and groups, Lecture Notes in Math., 1352, 34-43, Springer, 1988.

(79) (with G.Kukin), Undecidable algorithmic problems for semigroups, groups and rings, Algebra, Topology and Geometry, Vol. 25, p.3-66, VINITI, Moscow, 1987.

(80) Imbedding of rings, Russian Mathem. Series, 1987, 42, 4, 89-111.

(81) On the Novikov and Boone – Borisov groups, Illinois J. Math., 1986, 30, 2, 355-359.

(82) A remark on the Borisov–Boone group, Sibir. Math. J., 26 (1985), 5, 43-46.

(83) The centrally symmetric Novikov group, Sibir. Math. J., 1985, 26, 2, 18-28.

(84) New examples of groups with the standard normal forms, Doklady AN USSR, 1984, 77, 2, 277-280.

(85) Some problems of combinatorial algebra, Serdica, 9 (1983), 4, 387-395.

(86) Algorithmic problems and imbedding theorems: some open problems for rings, groups and semigroups, Mathematika, Izvestija Vuzov, 1982, 11, 3-11.

(87) New results of the theory of associative and Lie rings, Algebra Logika, 20 (1981), 5, 531-545.

(88) On a condition for solvability of Lie algebras, Sibir. Math. J., 1981, 22, 3, 15-20.

(89) (with the collaboration of D.Collins) Malcev problem and groups with a normal form, Word problems, II, G.Higman, W.W.Boone, S.I. Adjan , Eds., North Holland, 1980, p.29-54.

(90) Decision problems for ring theory, the same Word problems, II, p.55-70.

(91) Some questions in the theory of rings, Serdica, 3 (1977), 4, 299-308.

(92) Imbedding into algebraically closed and simple Lie algebras, Trudy Steklov Institute AN USSR (Leningradskoe Otdelenie), 1978, p.30-42.

(93) The introduction to the Russian translation of P.Cohn "Free rings and there relations", Moscow, Mir, 1977.

(94) Imbedding into simple associative algebras, Algebra and Logics, 1976, 15, 2, 117-142.

(95) Insolvability of the word problem for Lie algebras, and subalgebras of finitely presented Lie algebras, Izvestija AN USSR (mathem.), 1972, 36, 6, 1173-1219.

(96) Insolvability of the word problem for Lie algebras, Doklady AN USSR, 1972, 206, 6, 1288-1391.

(97) The nilpotent Lie algebras, Algebra and Logics, 1971,10, 2, 135-168.

(98) Insolvability of certain algorithmic problems in the class of associative rings, Algebra i Logika, 9 (1970), 137-144.

(99) Some questions of the combinatorial theory of groups and rings, Dissertation, Institute of Mathematics, Novosibirsk, 1969, pp.650.

(100) The problem of Malcev, Sibir. Math.J., 1969, 10, 5, 965-1005.

(101) Groups of fractions of multiplication semigroups of certain rings. III, Sibir. Math.J., 1969, 10, 4, 800-819.

(102) Groups of fractions of multiplication semigroups of certain rings. II, Sibir. Math.J., 1969, 10, 4, 744-799.

(103) Groups of fractions of multiplication semigroups of certain rings. I, Sibir. Math.J., 1969, 10, 2, 246-286.

(104) Degrees of insolvability of the conjugacy problem for finitely presented groups. I, II, Algebra i Logica, 1968, 7, 5, 4–70; 1968, 7, 6, 4-52.

(105) Groups with a relative standard basis, Sibir. Math.J., 1968, 9, 3, 499-521.

(106) The extensions of isomorphisms of rings, Algebra i Logika, 1968, 7,1, 15-24.

(107) The imbedding of rings in skew fields, Doklady AN USSR, 1967, 175, 4, 755-758.

(108) On some classes of rings without zero divisors, International Math. Congress, Moscow, 1966.

(109) On the Novikov groups, Algebra i Logika, 1967, 6, 1, 25-38.

(110) On one property of the Boone groups, Algebra i Logika, 1966, 5,5, 5-23; II, 1967, 6, 1, 25-38.

(111) Theorems of imbedding in the theory of algebras, Colloq. Math., 1966, 14, 349-353.

(112) Factorization theorems for certain classes of rings without zero divisors. I, Algebra i Logika, 1965, 4, 4, 25-52.

(113) Factorization theorems for certain classes of rings without zero divisors. II, Algebra i Logika, 1965, 4, 5, 17-46.

(114) Some examples of rings without zero divisors, Algebra i Logika,

1964, 3, 5-6, 5-28.

(115) A basis for free polynilpotent Lie algebras, Algebra i Logika, 1963, 2, 3, 13-20.

(116) Some imbedding theorems for rings and semigroups, Siberian Math.J., 1963, 3, 4, 500–518; II, 1963, 3, 4, 729-743.

(117) On a problem of Kaplansky, Sibir. Math. J., 1963, 4, 1184-1195.

(118) Embedding of Lie algebras into the algebraically closed Lie algebras, Algebra i Logika, 1962, 1, 2, 47-53.

(119) Embedding algebras into algebraically closed algebras, Doklady AN USSR, 154 (1962), 963-964.

## Acknowledgments

The author is grateful to his wife, who supported his mathematical life a lot. The author also thanks his teachers for guidance in mathematics and in the everyday life. Finally, the author thanks all the people who helped him kindly.

# The incipience of Gröbner–Shirshov bases*

Leonid A. Bokut

*Sobolev Institute of mathematics,
Akad. Koptyug prosp., 4,
Novosibirsk 630090, Russia*

*bokut@math.nsc.ru*

"This short paper [A. I. Shirshov, Some algorithmic problems for Lie algebras. Sib. Mat. Zh. 3, 292-296 (1962) – L.B.] describes work similar to that appearing in B. Buchbergers 1965 thesis inventing Gröbner bases, but in the context of Lie algebras. Preceding Buchberger by only three years, this paper, along with the two cited references, are the original papers defining what has become known as Gröbner–Shirshov bases." This is a quotation by Michael Abramson, together with Rebecca Abramson the translator of the first English translation of the paper under review, cf. [*A.I. Shirshov*, "Certain algorithmic problems for Lie algebras," SIGSAM Bull. 33, No. 2, 3-6 (1999; Zbl 1097.17502)]. Incidentally, there is another translation available in [*L. A. Bokut, V. Latyshev, I. Shestakov, E. Zelmanov* (editors), Selected works of A.I. Shirshov. Translated by Murray Bremner and Mikhail V. Kotchetov. Contemporary Mathematicians. Basel: Birkhäuser (2009; Zbl 1188.01028)].

The Shirshov paper was in line with A.G. Kurosh's program to study (relatively) free nonassociative algebras initiated in [*A. Kurosh*, "Nonassociative free algebras and free products of algebras" (Russian), Mat. Sb., N. Ser. 20 (62), 239–262 (1947; Zbl 0041.16803)]. Kurosh proved that any subalgebra of a free nonassociative algebra (over a field) is again free (an analogue of the Nilsen–Schreier theorem for free groups). For Lie algebras this was proved by Shirshov [Zbl 0052.03004, see below]. Also Kurosh proved the subalgebra theorem for free products of nonassociative algebras (an analogue of the Kurosh theorem for free products of groups). Later Shirshov [Zbl 0104.26003, see below] used his Gröbner–Shirshov basis theory to

---

*This paper is the review Zbl 0104.26004 on the paper by Shirshov, Anatoli Illarionovich, "Some algorithmic problems for Lie algebras", Sib. Mat. Zh. 3, 292-296 (1962).

give a negative answer to the conjecture that the same theorem is valid for free products of Lie algebras.

A.I. Zhukov, a student of A.G. Kurosh (1908–1972), solved the membership problem for finitely generated ideals of free nonassociative algebras (magma algebras) in 1950 [*A.I. Zhukov*, "Reduced systems of defining relations in nonassociative algebras" (Russian), Mat. Sb., N. Ser. 27 (69), 267–280 (1950; Zbl 0038.17001)]. In a sense, this was the beginning of the Gröbner–Shirshov basis theory for nonassociative algebras. The main difference with the later Shirshov approach was that Zhukov did not use any linear ordering of nonassociative monomials. Instead, he chose any monomial of maximal degree as a "leading" monomial of a polynomial. Also for nonassociative algebras there is no "composition of intersection" ("*s*-polynomial"). In this sense it cannot be a model for associative or Lie algebras.

Shirshov, another student of Kurosh's, defended his PhD thesis at Moscow University in 1953 that can be viewed as a background of his later Gröbner–Shirshov basis theory. He proved the free subalgebra theorem for free Lie algebras [*A.I. Shirshov*, "Subalgebras of free Lie algebras" (Russian), Mat. Sb. N. Ser. 33 (75), 441–452 (1953; Zbl 0052.03004)] (now known as Shirshov–Witt theorem), using the elimination process rediscovered by *M. Lazard* ["Anneaux de Lie et problème de Burnside," C.I.M.E., Gruppi, Anelli di Lie e Theoria della Coomologia (1960; Zbl 0134.26003)] (now known as Lazard or Lazard–Shirshov elimination process). Shirshov used the elimination process later, cf. Lemma 1 in [*A.I. Shirshov*, "On free Lie rings" (Russian), Mat. Sb., N. Ser. 45 (87), 113–122 (1958; Zbl 0080.25503)] to prove the property of Lyndon–Shirshov bases of free Lie algebras (viz., the unique bracketing of regular (Lyndon–Shirshov) words) that was important for his Gröbner–Shirshov basis theory.

Also he proved the free subalgebra theorem for (anti) commutative nonassociative algebras [*A.I. Shirshov*, "Subalgebras of free commutative and free anti-commutative algebras" (Russian), Mat. Sb., N. Ser. 34 (76), 81–88 (1954; Zbl 0055.02703)]. He used the paper later [*A.I. Shirshov*, "Some algorithmic problems for ∈-algebras" (Russian), Sib. Mat. Zh. 3, 132–137 (1962; Zbl 0143.25602)] in his Gröbner–Shirshov basis theory for (anti) commutative nonassociative algebras.

Last but not least, he found a series of generalized Hall bases (now known as Hall bases or Hall–Shirshov bases) of a free Lie algebra published 10 years later [*A.I. Shirshov*, "On bases of a free Lie algebra" (Russian), Algebra Logika 1, No. 1, 14–19 (1962; Zbl 0145.25803)] (rediscovered by

*G. Viennot* ["Une généralisation des ensembles de Hall," C.R. Acad. Sci. Paris, Sér. A 276, 599–602 (1973; Zbl 0252.17002)]). In the definition of his series of bases, Shirshov used the Hall standard Lie commutator construction together with any ordering of nonassociative monomials such that $uv > v$. Actually it is the real background of his definition of a Lyndon–Shirshov basis below.

The Gröbner–Shirshov basis theory for Lie algebras is based on the example of Hall–Shirshov bases. This is the basis of regular (now called Lyndon or Lyndon–Shirshov) Lie monomials invented in the paper by Shirshov from 1958 mentioned above [Zbl 0080.25503] and independently by Chen, Fox and Lyndon [*K.-T. Chen, R.H. Fox, R.C. Lyndon*, "Free differential calculus, IV. The quotient groups of the lower central series," Ann. Math. (2) 63, 81–95 (1958; Zbl 0083.01403)].

Shirshov's definition was as follows. Let $X = \{x_i \colon i = 1, 2, \ldots\}$ be a totally ordered alphabet, and let $x_i > x_j$ if $i > j$. A non-empty word $u \in X^*$ is called regular if $u = vw > wv$ for any non-empty words $v, w$. Let $>$ be the lexicographical ordering of regular words. A nonassociative monomial $[u]$ is called regular if (1) the associative word $u$ is regular, (2) if $[u] = [[v][w]]$, then $[v], [w]$ are regular words, (3) if $[v] = [[v_1][v_2]]$, then $v_2 \leq w$.

It is easy to see that regular words can be defined inductively starting with $x_i$. From condition (2) it follows that $v > w$. From (1)–(3) it follows easily that $w$ is the longest proper regular suffix of $u$, cf. [*A.I. Shirshov*, "Some problems in the theory of rings which are related to associative rings" (Russian), Usp. Mat. Nauk 13, No. 6 (84), 3–20 (1958; Zbl 0088.03202)].

In view of the third part of his 1953 thesis [see Zbl 0145.25803 above] the definitions of regular words and regular monomials were very natural for Shirshov after he decided to use the lexicographical ordering of monomials and their underlying words.

Shirshov culminated his study of free Lie algebras and free (anti) commutative algebras in 1962 in the paper under review and in [*A.I. Shirshov*, "Some algorithmic problems for $\epsilon$-algebras" (Russian), Sib. Mat. Zh. 3, 132–137 (1962; Zbl 0143.25602)] by inventing what is now called Gröbner–Shirshov basis theory for Lie algebras and (anti) commutative nonassociative algebras explicitly, and for associative algebras and nonassociative algebras implicitly.

Let us sketch Shirshov's approach for Lie algebras. First of all he views a free Lie algebra $\mathrm{Lie}(X)$ as a Lie subalgebra of the free associative algebra $k\langle X\rangle$. His decisive idea is to define a composition $[f, g]_w$ of two Lie polynomials $f, g$ relative to some associative word $w$. He defines first the associative composition $(f, g)_w$. Let us fix the degree-lexicographical (deg-lex for short) ordering on the set of associative words $X^*$ (to compare two words first by degree and then lexicographically using $x_1 < x_2 < \ldots < x_n < \ldots$). Let $\bar{f}$ be the maximal word of $f$ as an associative non-commutative polynomial. A polynomial $f$ is called monic if the coefficient of $\bar{f}$ is equal to 1. Without loss of generality one may assume that $f, g$ are monic. Let $w = \bar{f}b = a\bar{g}$, where $a, b \in X^*$, $\deg(\bar{f}) + \deg(\bar{g}) > \deg(w)$, (i.e., $\bar{f}$ and $\bar{g}$ have non-empty intersection as subwords of $w$). Then the associative composition is $(f, g)_w = fb - ag$. To define a Lie composition $[f, g]_w$ Shirshov puts some Lie bracketing on $fb$ and $ag$. He uses his Special Bracketing Lemma, Lemma 4 from the above-mentioned paper [Zbl 0080.25503].

First of all, the words $\bar{f}$, $\bar{g}$ and $w$ are associative regular (i.e., Lyndon–Shirshov) words. According to the Special Bracketing Lemma, any regular word $w$ with a fixed regular subword $u$, $w = aub$, has a special bracketing $[w]_u = (a[u]b)$, where $[u]$ is a standard Lyndon–Shirshov bracketing, and $\overline{[w]_u} = w$. The proof of this lemma relies on the so-called principal property of regular (Lyndon–Shirshov) words that any word $c$ is a non-strictly increasing product of regular (Lyndon–Shirshov) words. Namely, in the standard bracketing $[aub]$ of the regular word $aub$ one pair of brackets has a position $[a[uc]d]$, where $cd = b$, $c$ may be empty. If it is empty, we are at home. Otherwise, $c$ has a presentation $c = c_1 c_2 \ldots c_k$, where the $c_i$ are regular words and $c_1 \leq c_2 \leq \ldots \leq c_k$. Then by definition $[w]_u = [a[[[u][c_1]] \ldots [c_k]]d]$.

Then Shirshov defines the Lie composition as $[f, g]_w = [fb]_{\bar{f}} - [ag]_{\bar{g}}$. Here, for example, $[fb]_{\bar{f}}$ is the result of substitution $[\bar{f}] \mapsto f$ into the special bracketing monomial $[w]_{\bar{f}}$.

To formulate the main Composition Lemma (Lemma 3 of the paper under review), we need to define the Shirshov reduction algorithm for Lie polynomials. Let $f, g$ be monic Lie polynomials and $\bar{f} = a\bar{g}b$, where $a, b \in X^*$. Then the transformation $f \mapsto h = (f - [agb]_{\bar{g}})$ is called the elimination of the leading word (ELW) of $g$ into $f$. By the Special Bracketing Lemma, $h$ is a Lie polynomial with the leading associative monomial less than $\bar{f}$. The ELW is the reduction algorithm for Lie polynomials.

A monic set $S$ of Lie polynomials is called an irreducible (or reduced) set if the leading associative monomials of $S$ do not contain each other as subwords (so they are irreducible as associative polynomials). In particular, the leading associative monomials of two different polynomials of $S$ are not equal.

In view of the Special Bracketing Lemma, it is clear that for any finite set $S$, in a finite number of steps we can find a finite irreducible set $R$ that generates the same Lie ideal as $S$ (for a formal proof see Lemma 1 in the paper [Zbl 0143.25602] above).

Now we are ready to formulate Shirshov's main Composition Lemma: Let $S$ be an irreducible set of Lie polynomials in $\mathrm{Lie}(X)$, and let $S^c$ be an irreducible set that is the result of the process of joining to $S$ a multiple composition and subsequent reduction. Then if a polynomial $f$ belongs to the Lie ideal generated by $S$, then the leading associative monomial $\bar{f}$ contains $\bar{r}$ for some $r \in S^c$.

Shirshov assumed that $S$ is a so-called stable set but he did not use this condition in the proof. On the other hand, the stability condition is the essentially only known sufficient condition for $S^c$ to be a recursive set for a finite set $S$.

He constantly uses the simple corollary of the lemma that the set of regular (Lyndon–Shirshov) Lie monomials $[u]$ such that the corresponding regular (Lyndon–Shirshov) associative words $u$ do not contain $\bar{s}$, $s \in S^c$, is a $k$-basis of the Lie algebra $\mathrm{Lie}(X)/\mathrm{Id}(S) = \mathrm{Lie}(X \mid S)$ with generators $X$ and defining relations $S$.

The set $S^c$ is now called a Gröbner–Shirshov basis of the ideal generated by $S$. The process of adding a composition and subsequent reduction in the Shirshov algorithm is in general infinite in the same way as later discovered for the Knuth–Bendix algorithm [*D. Knuth, E. Bendix*, "Simple word problems in universal algebras," Comput. Problems Abstract Algebra, Proc. Conf. Oxford 1967, 263–297 (1970; Zbl 0188.04902)].

Shirshov's Composition Lemma was formulated somewhat later in the following form, see [*L. A. Bokut*, "Unsolvability of the equality problem and subalgebras of finitely presented Lie algebras," Math. USSR, Izv. 6 (1972), 1153–1199 (1974; Zbl 0275.02041)]. The above Lie composition $[f, g]_w = [fa]_{\bar{f}} - [bg]_{\bar{g}}$, $w = acb$, $\bar{f} = ac$, $\bar{g} = cb$, $c \neq 1$ is called the composition of intersections. Now we define the composition of inclusions of monic Lie polynomials, $[f, g]_w = f - [agb]_{\bar{g}}$, $w = \bar{f} = a\bar{g}b$. A composition $[f, g]_w$ is called trivial $\mathrm{mod}(S, w)$ if it has a presentation $[f, g]_w = \sum \alpha_i[a_is_ib_i]$, $\alpha_i \in k$, $s_i \in S$, $a_i, b_i \in X^*$, $\overline{[a_is_ib_i]} = a_i\bar{s}_ib_i < w$. By the Special Bracketing Lemma and Composition Lemma above, this is the same as to say that the composition goes to zero by elimination of leading words of $S$ (for a Gröbner–Shirshov basis $S$).

A monic set $S$ of $\mathrm{Lie}(X)$ is called closed under compositions (i.e., a Gröbner–Shirshov basis) if for every $f, g \in S$ every composition $[f, g]_w$ of $f$

and $g$ is trivial $\mod(S, w)$.

Shirshov Composition Lemma: Let $S$ be a monic subset of $\mathrm{Lie}(X)$ that is closed under compositions. If $f \in \mathrm{Id}(S)$, then $\bar{f}$ contains $\bar{s}$ for some $s \in S$.

As a corollary, the set of all irreducible Lyndon–Shirshov monomials constitutes a $k$-linear basis of $\mathrm{Lie}(X|S)$.

A modern name of the Shirshov Composition Lemma is Composition-Diamond Lemma because of the Bergman Diamond Lemma (see below) for associative algebras that is equivalent to the Composition Lemma for associative algebras.

Shirshov Composition-Diamond Lemma for Lie algebras: Let $S$ be a monic subset of $\mathrm{Lie}(X)$. Then the following conditions are equivalent: (1) $S$ is a Gröbner–Shirshov basis, (2) if $f \in \mathrm{Id}(S)$, then $\bar{f}$ contains $\bar{s}$ for some $s \in S$, (3) $\mathrm{Irr}(S) = \{[u] : [u] \text{ is an } S\text{-irreducible Lyndon–Shirshov Lie word}\}$ is a $k$-basis of $\mathrm{Lie}(X \mid S)$.

The main application of Shirshov's 1962 paper is the Shirshov algorithm to find a Gröbner–Shirshov basis for finitely presented Lie algebras. As we mentioned above, it has two essential parts. One is the Lie reduction algorithm, i.e., the elimination of leading words of one Lie polynomial into another. The second is to join a nontrivial composition to a set of Lie polynomials.

Shirshov's paper contains three specific applications of his algorithm and Composition Lemma:

1) The Shirshov algorithm gives a solution of the word problem for homogeneous finitely presented Lie algebras. It is much faster than an algorithm based on solutions of systems of linear equations.

2) The Shirshov algorithm gives a solution of the word problem for one-relator Lie algebras since there is no composition $[f, f]_w$ for a Lie polynomial $f$.

3) The Freiheitssatz (freeness theorem) is valid for one-relator Lie algebras. This is so since for any Lie polynomial $f$ and any fixed letter $x$ involved in $f$ (presented in the regular basis), one may assume that $x$ is involved in the leading monomial of $f$ (one may use a new Deg-function such that $\mathrm{Deg}(x) > \deg(f)$, $\mathrm{Deg}(y) = 1$, $y \in X \smallsetminus \{x\}$). Then all regular monomials in $X \smallsetminus \{x\}$ are irreducible, i.e., linearly independent.

In his subsequent paper [*A.I. Shirshov*, "On a conjecture in the theory of Lie algebras" (Russian), Sib. Mat. Zh. 3, 297–301 (1962; Zbl 0104.26003)] he found a linear basis of the free product of Lie algebras with amalgamated subalgebras as an application of his Composition Lemma for Lie algebras. Also he answered negatively the conjecture that Kurosh's theorem on subalgebras of free products of nonassociative algebras is valid for Lie algebras.

From the previous observations it is clear how to proceed in the associative case. The reviewer is a witness that Shirshov understood the associative case as a simple version of the Lie algebra case. Explicitly this was formulated in [*L. A. Bokut*, "Embedding into simple associative algebras," Algebra Logic 15 (1976), 73–90 (1977; Zbl 0355.16)] and [*G.M. Bergman*, "The Diamond lemma for ring theory," Adv. Math. 29, 178–218 (1978; Zbl 0326.16019)]. The current formulation of the Shirshov lemma is as follows.

Shirshov Composition-Diamond Lemma for associative algebras: Let $S$ be a monic subset of $k\langle X \rangle$. Then the following conditions are equivalent: (1) $S$ is a Gröbner–Shirshov basis, (2) If $f \in \mathrm{Id}(S)$, then $\bar{f}$ contains $\bar{s}$ for some $s \in S$, (3) $\mathrm{Irr}(S) = \{u \colon u$ is a reduced word in $X$ relative to $S\}$ is a $k$-basis of $k\langle X \mid S \rangle$.

# Some new results on cluttered ordering on several special bipartite graphs

Tomoko Adachi

*Department of Information Sciences, Toho University*
*2-2-1 Miyama, Funabashi, Chiba, 274-8510, Japan*

*adachi@is.sci.toho-u.ac.jp*

A cluttered ordering is a type of a cyclic ordering of a graph. It can provide minimize the number of disk operations. Mueller et al. (2005) decomposed the complete bipartite graph into isomorphic copies of the special bipartite graph $H(h;t)$, where $h$ and $t$ are positive integers. They presented some edge labeling of the infinite families of $H(1;t)$, $H(2;t)$. Adachi (2017) defined the special bipartite graph $H(h,k;t)$, which is extension of $H(h;t)$, and presented some edge labeling of the infinite families of $H(1,2;t)$. In this paper, we investigate the case of $H(1,k;t)$, and obtain the corresponding cluttered orderings.

*Keywords*: Cluttered ordering; cyclic ordering; edge labeling; RAID

## 1. Introduction

Some studies replaced the problem of the RAID (redundant arrays of independent disks) system in computer science with the problem of cyclic orderings in design theory. (see Ref. 6). We refer as RAID system to Ref. 5,7.

A cluttered ordering is a type of a cyclic ordering of the complete graph and the complete bipartite graph. It is introduced by Cohen et. al.[4], and can provide minimize the number of disk operations in RAID systems. Mueller et al.[8] decomposed the complete bipartite graph into isomorphic copies of the special bipartite graph $H(h;t)$, where $h$ and $t$ are positive integers. They presented some edge labeling of the infinite families of $H(1;t)$, $H(2;t)$, and constructed cluttered ordering for the complete bipartite graph $K_{\ell,\ell}$ where $\ell$ is $3t$ and $10t$. Also, the cases of $H(3;t)$ and $H(4;t)$ are known, that is, the case of $\ell = 21t, 36t$ (see Ref. 2,3).

Adachi[1] developed the concept of $H(h;t)$, and defined the special bipartite graph $H(h,k;t)$ for positive integers $h, k, t$ and $h \neq k$. $H(h;t)$ has each $h(t+1)$ vertices as upper vertex set and lower vertex set. While,

$H(h, k; t)$ has $h(t + 1)$ vertices as upper vertex set and $k(t + 1)$ vertices as lower vertex set. Adachi[1] presented some edge labeling in the most simple case $H(1, 2; t)$. In this paper, we investigate the more general case $H(1, k; t)$ which includes above result.

## 2. Known Results

First, we define a cluttered ordering. Let $G = (V, E)$ be a graph with $n = |V|$ and $E = \{e_0, e_1, \cdots, e_{m-1}\}$. Let $d \le m$ be a positive integer, called a *window* of $G$, and $\pi$ a permutation on $\{0, 1, \cdots, m - 1\}$, called an *edge ordering* of $G$. Then, given a graph $G$ with edge ordering $\pi$ and window $d$, we define $V_i^{\pi, d}$ to be the set of vertices which are connected by an edge of $\{e_{\pi(i)}, e_{\pi(i+1)}, \cdots, e_{\pi(i+d-1)}\}$, $0 \le i \le m - 1$, where indices are considered modulo $m$. The cost of accessing a subgraph of $d$ consecutive edges is measured by the number of its vertices. An upper bound of this cost is given by the *d-maximum access cost* of $G$ defined as $\max_i |V_i^{\pi, d}|$. An ordering $\pi$ is a $(d, f)$-*cluttered ordering*, if it has $d$-maximum access cost equal to $f$. We are interested in minimizing the parameter $f$.

Second, we define the special bipartite graph $H(h, k, t)$. Let $h, k, t$ be positive integers. For each parameter $h$, $k$, $t$, we define a bipartite graph denoted by $H(h, k; t) = (U, E)$. Its vertex set $U$ is partitioned into $U = V \cup W$ and consists of the following $2(h + k)t$ vertices:

$$V := \{v_i | 0 \le i < (h + k)t\},$$
$$W := \{w_i | 0 \le i < (h + k)t\}.$$

In order to cyclic, we consider that vertices $v_0 = v_{(h+k)(t-1)}$, $v_1 = v_{(h+k)(t-1)+1}$, $\cdots$, $v_{h+k-1} = v_{(h+k)t-1}$, $w_0 = w_{(h+k)(t-1)}$, $w_1 = w_{(h+k)(t-1)+1}$, $\cdots$, $w_{h+k-1} = w_{(h+k)t-1}$. The edge set $E$ is partitioned into subsets $E_s$, $0 \le s < 2t$. defined by

$$E_s := E'_s \cup E''_s \cup E'''_s, \quad \text{for } 0 \le s < 2t,$$
$$E := \bigcup_{s=0}^{2t-1} E_s.$$

In the case $s$ is even (i.e. $s = 0, 2, 4, \cdots, 2t - 2$ ), we define as follows:

$$E'_s := \{\{v_i, w_j\} | s \cdot (h + k) \le i < s \cdot (h + k) + h,$$
$$s \cdot (h + k) \le j < s \cdot (h + k) + k\},$$
$$E''_s := \{\{v_i, w_{k+j}\} | s \cdot (h + k) \le j \le i < s \cdot (h + k) + h\},$$
$$E'''_s := \{\{v_{h+i}, w_j\} | s \cdot (h + k) \le i \le j < s \cdot (h + k) + k\}.$$

In the case $s$ is odd ( i.e. $s = 1, 3, 5, \cdots, 2t-1$ ), we define as follows:

$$E'_s := \{\{v_i, w_j\} | s \cdot (h+k) + h \le i < (s+1) \cdot (h+k),$$
$$s \cdot (h+k) + k \le j < (s+1) \cdot (h+k)\},$$
$$E''_s := \{\{v_i, w_{k+j}\} | s \cdot (h+k) + h \le j \le i < (s+1) \cdot (h+k)\},$$
$$E'''_s := \{\{v_{h+i}, w_j\} | s \cdot (h+k) + k \le i \le j < (s+1) \cdot (h+k)\}.$$

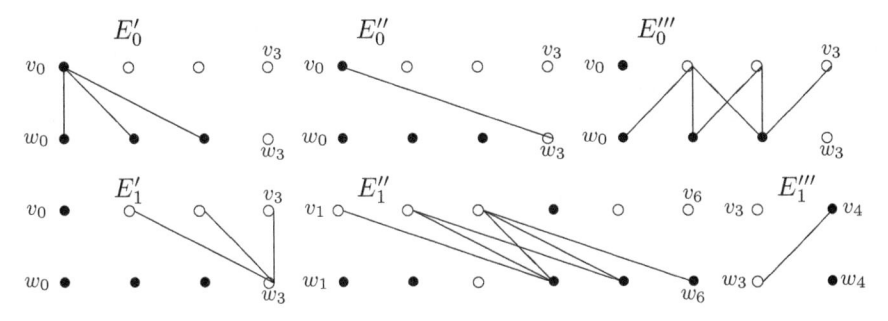

Fig. 1.  A part of partition of the edge set of $H(1, 3; 1)$.

Figure 1 shows a part of the edge partition of $H(1, 3; 1)$, that is, $E'_s$, $E''_s$ and $E'''_s$ in the case of $s = 0, 1$.

For the number of edges, it holds $|E| = 2t \cdot (h \cdot k + \frac{h(h+1)}{2} + \frac{k(k+1)}{2}) = t(h+k)(h+k+1)$. In this paper, we deal with the case of $h = 1$. Then a bipartite graph $H(1, k; t)$ has $2t(k+1)$ vertices and $t(k+1)(k+2)$ edges.

The following Proposition related to the composite subgraphs of a bipartite graphs can be found in Ref. 1.

**Proposition 2.1.** [1] *Let $G(E)$ be an induced graph defined by an edge-set $E$. Let $s$ be a non-negative integer and even. Let $G_{s,s+1}$ be a composite graph of six subgraphs $G(E'_s)$, $G(E''_s)$, $G(E'''_s)$, $G(E'_{s+1})$, $G(E''_{s+1})$, and $G(E'''_{s+1})$. Then, for any even integer $s$ $(s = 0, 2, 4, \cdots)$, bipartite graphs $G_{s,s+1}$ are isomorphic to $G_{0,1}$.*

A bipartite graph $G_{0,1}$ is equal to a bipartite graph $H(h, k; 1)$. Intuitively speaking to Proposition 2.1, the bipartite graph $H(h, k; t)$ consists of $t$ consecutive copies of $G_{0,1}$, where the last $(h+k)$ vertices of $V$ and $W$ respectively of one copy are identified with the first $(h+k)$ vertices of $V$ and $W$ respectively of the next copy.

Known results [2,3,8] studied orderings of the complete bipartite graph in the case of $h = k$. In this paper, we consider the case of $h \neq k$, in which it is not the complete bipartite graph.

## 3. The initial stages for a bipartite graph $H(1, k; t)$

First, we consider a bipartite graph $H(1, k; 1)$, that is, the case of $t = 1$. For example, the case of $k = 3$, i.e., a bipartite graph $H(1, 3; 1)$ is illustrated as Figure 2. A bipartite graph $H(1, k; 1)$ has an edge $\{ v_0, w_k \}$ in the set $E_0'$. On the other hand, a bipartite graph $H(1, k; 1)$ has an edge $\{ v_{k+1}, w_k \}$ in the set $E_1'''$. In order to cyclic, we consider that vertex $v_0 = v_{k+1}$. Then, the bipartite graph $H(1, k; 1)$ has multi-edges such as $\{ v_0, w_k \}$ and $\{ v_{k+1}, w_k \}$. The bipartite graph with multi-edges such is not suitable for mathematical modeling of RAID system. Hence, we do not consider this case furthermore.

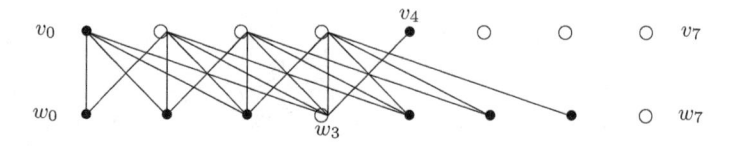

Fig. 2.   A bipartite graph $H(1, 3; 1)$

Second, we consider a bipartite graph $H(1, k; 2)$, that is, the case of $t = 2$. For example, the case of $k = 3$, i.e., a bipartite graph $H(1, 3; 2)$ is illustrated as Figure 3. In order to cyclic, we consider that vertices $v_0 = v_8$, $v_1 = v_9$, $v_2 = v_{10}$, $v_3 = v_{11}$, $w_0 = w_8$, $w_1 = w_9$, $w_2 = w_{10}$, $w_3 = w_{11}$. Then, the bipartite graph $H(1, 3; 2)$ becomes $K_{8,8} - 8K_{1,3}$. Hence, generally, we get the following result.

**Lemma 3.1.** *Let $h, k, t$ be positive integers. In the case of $h = 1, t = 2$, a bipartite graph $H(h, k; t)$ is a graph $K_{2(k+1),2(k+1)} - 2(k + 1)K_{1,k}$.*

**Proof.**    In order to cyclic, we consider that vertices $v_0 = v_{2(k+1)}$, $v_1 = v_{2(k+1)+1}$, $v_2 = v_{2(k+1)+2}$, $v_3 = v_{2(k+1)+3}$, $w_0 = w_{2(k+1)}$, $w_1 = w_{2(k+1)+1}$, $w_2 = w_{2(k+1)+2}$, $w_3 = w_{2(k+1)+3}$. Then, the bipartite graph $H(1, k; 2)$ becomes $K_{2(k+1),2(k+1)} - 2(k + 1)K_{1,k}$.

(Q.E.D.)

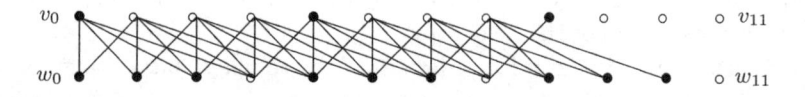

Fig. 3.   A bipartite graph $H(1, 3; 2)$

The bipartite graph $K_{2(k+1),2(k+1)} - 2(k+1)K_{1,k}$ does not have multi-edges. This is suitable for mathematical modeling of RAID system. Hence, we consider an ordering of this case.

Similarly, $H(1,k;3)$ has multi-edges, and it is not suitable. $H(1,k;4)$ does not have multi-edges, and it is suitable. By the mathematical induction, we obtain that $H(1,k;t)$ has multi-edges when $t$ is odd, and it does not when $t$ is even.

## 4. A Cluttered Ordering of $H(1,k;2t)$

In previous section, we know that the bipartite graph $H(1,k;2)$ becomes $K_{2(k+1),2(k+1)} - 2(k+1)K_{1,k}$. In this section, first, we consider an ordering of $K_{2(k+1),2(k+1)} - 2(k+1)K_{1,k}$. Next, we will give result for an ordering of $K_{2(k+1)t,2(k+1)t} - 2(k+1)tK_{1,k}$ corresponding to $H(1,k;2t)$.

The following Lemma is to describe the properties and edge ordering of some composite subgraphs of a special bipartite graph $H(1,k;2)$.

**Lemma 4.1.** *Let $G^{(k)}(E)$ be an induced graph defined by an edge-set $E$ of a bipartite graph $H(1,k;2)$. Then, a bipartite graph $K_{2(k+1),2(k+1)} - 2(k+1)K_{1,k}$ is decomposed to twelve subgraphs $G^{(k)}(E'_0)$, $G^{(k)}(E''_0)$, $G^{(k)}(E'''_0)$, $G^{(k)}(E'_1)$, $G^{(k)}(E''_1)$, $G^{(k)}(E'''_1)$, $G^{(k)}(E'_2)$, $G^{(k)}(E''_2)$, $G^{(k)}(E'''_2)$, $G^{(k)}(E'_3)$, $G^{(k)}(E''_3)$ and $G^{(k)}(E'''_3)$.*

**Proof.** By Lemma 3.1, the graph $K_{2(k+1),2(k+1)} - 2(k+1)K_{1,k}$ is a bipartite graph $H(1,k;2)$. From the definition of a bipartite graph $H(h,k;t)$, we easily see that the bipartite graph $K_{2(k+1),2(k+1)} - 2(k+1)K_{1,k}$ is decomposed to twelve subgraphs $G^{(k)}(E'_0)$, $G^{(k)}(E''_0)$, $G^{(k)}(E'''_0)$, $G^{(k)}(E'_1)$, $G^{(k)}(E''_1)$, $G^{(k)}(E'''_1)$, $G^{(k)}(E'_2)$, $G^{(k)}(E''_2)$, $G^{(k)}(E'''_2)$, $G^{(k)}(E'_3)$, $G^{(k)}(E''_3)$ and $G^{(k)}(E'''_3)$.

$$(Q.E.D.)$$

**Lemma 4.2.** *Let $G^{(k)}(E)$ be an induced graph defined by an edge-set $E$. Let $s$ be a non-negative integer and even. Let $G^{(k)}_{s,s+1}$ be a composite graph of six subgraphs $G^{(k)}(E'_s)$, $G^{(k)}(E''_s)$, $G^{(k)}(E'''_s)$, $G^{(k)}(E'_{s+1})$, $G^{(k)}(E''_{s+1})$, and $G^{(k)}(E'''_{s+1})$. Then, the bipartite graph $K_{2(k+1),2(k+1)} - 2(k+1)K_{1,k}$ is decomposed to two isomorphic subgraphs $G^{(k)}_{0,1}$ and $G^{(k)}_{2,3}$.*

**Proof.** By Lemma 3.1, Then, the bipartite graph $K_{2(k+1),2(k+1)} - 2(k+1)K_{1,k}$ is decomposed to twelve subgraphs $G^{(k)}(E'_0)$, $G^{(k)}(E''_0)$, $G^{(k)}(E'''_0)$,

$G^{(k)}(E'_1)$, $G^{(k)}(E''_1)$, $G^{(k)}(E'''_1)$, $G^{(k)}(E'_2)$, $G^{(k)}(E''_2)$, $G^{(k)}(E'''_2)$, $G^{(k)}(E'_3)$, $G^{(k)}(E''_3)$ and $G^{(k)}(E'''_3)$. For $s = 0, 2$, every bipartite graph $G^{(k)}_{s,s+1}$ is a composite graph of six subgraphs $G^{(k)}(E'_s)$, $G^{(k)}(E''_s)$, $G^{(k)}(E'''_s)$, $G^{(k)}(E'_{s+1})$, $G^{(k)}(E''_{s+1})$, and $G^{(k)}(E'''_{s+1})$. By Proposition 4.2, for any even integer $s$ ($s = 0, 2$), bipartite graphs $G^{(k)}_{s,s+1}$ are isomorphic to $G^{(k)}_{0,1}$. Then, we easily see.

$$(\text{Q.E.D.})$$

From Lemma 3.1, we may obtain an edge-ordering of a bipartite graph $G^{(k)}_{2,3}$ by adding the total sum 12 of edges of $G^{(k)}_{0,1}$ to an edge-ordering of a bipartite graph $G^{(k)}_{0,1}$.

Here, we consider an ordering of a bipartite graph $G^{(k)}_{0,1}$. We investigate edge-labeling such that we minimize the number $f$ of virtices connecting $d$ edges when we fix the number $d$ of edge-windows. We try and error many times. At last, we can obtain edge-labeling of $G^{(k)}_{0,1}$ as the following Figure 4. Here, we illustrate the labeling of the edge set in $E'''_0$ and $E''_1$ as the following Figure 5 and Figure 6, respectively. For example, we give an edge-labeling in the case of $k = 3$ as the following Figure 7. Therefore, we obtain the following result.

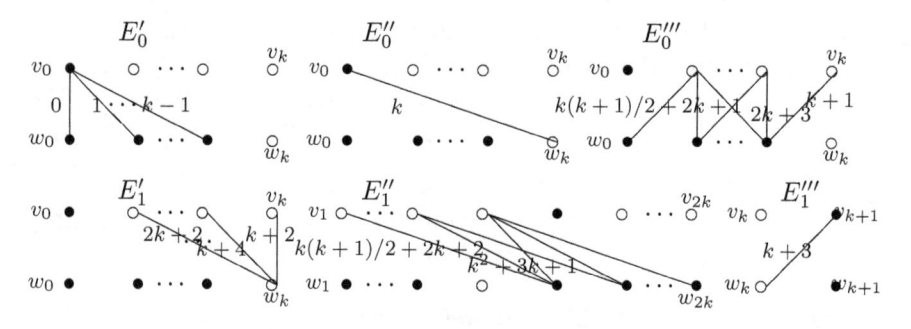

Fig. 4.    Labeling of the edge set of $G^{(k)}_{0,1}$.

**Proposition 4.1.** *A bipartite graph $G^{(k)}_{0,1}$ has a $(d, f)$-ordering with $d = k$ and $f = k + 1$. Moreover, a bipartite graph $G^{(k)}_{0,1}$ has a $(d, f)$-ordering with $d = k \times s + r$ and $f = (k + 1)s + r$, $s > 0$, $r = 0, 1, 2, \cdots, k - 1$.*

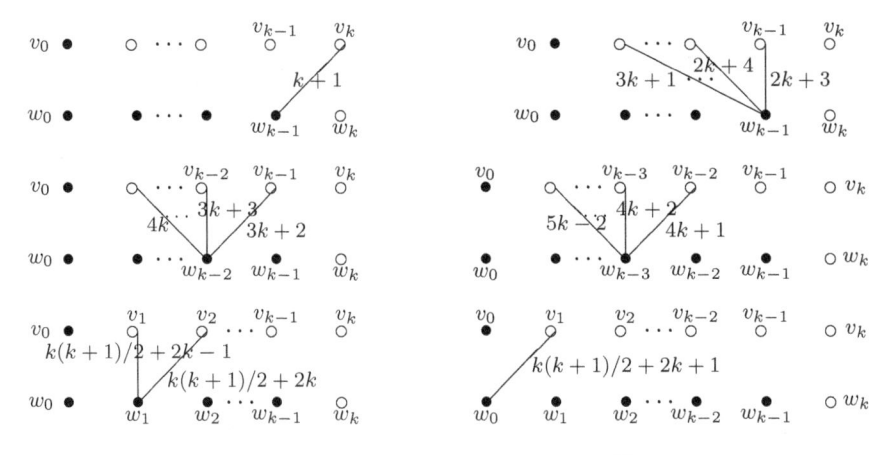

Fig. 5.   Labeling of the edge set of $E_0'''$.

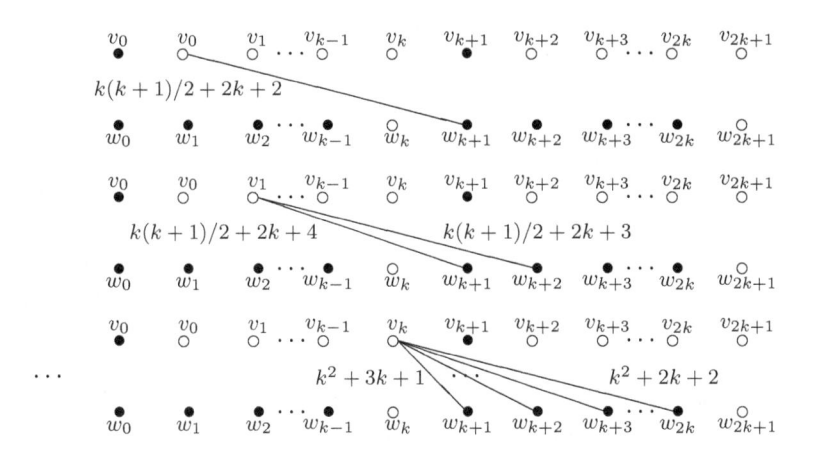

Fig. 6.   Labeling of the edge set of $E_1''$.

**Proof.**    From $E_0'$ in Figure 4, $k$ edges labeled $\{\ 0, 1, 2, \cdots, k - 1,\ \}$ connected with $k + 1$ vertices $\{\ v_0, w_0, w_1, w_2, \cdots, w_{k-1}\ \}$. From $E_0'$ and $E_0''$ in Figure 4, $k$ edges labeled $\{\ 1, 2, \cdots, k - 1, k\ \}$ connected with $k + 1$ vertices $\{\ v_0, w_1, w_2, \cdots, w_{k-1}, w_k\ \}$. From $E_0'$, $E_0''$ and $E_0'''$ in Figure 4, $k$ edges labeled $\{\ 2, 3, \cdots, k - 1, k, k + 1\ \}$ connect with $k + 1$ vertices $\{\ v_0, v_k, w_2, w_3, \cdots, w_{k-1}, w_k\ \}$. From $E_0'$, $E_0''$, $E_0'''$ and $E_1'$ in Figure 4,

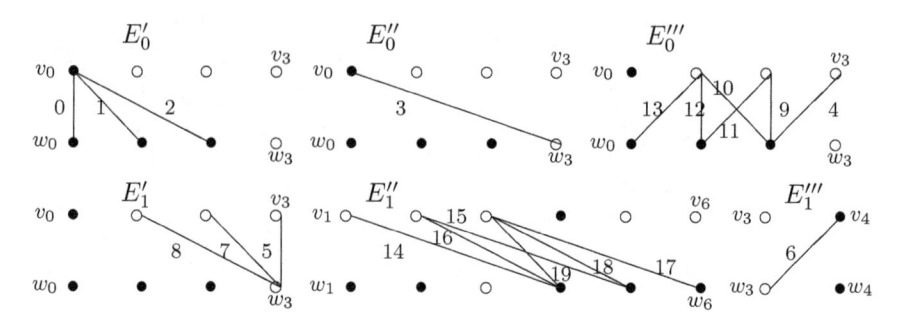

Fig. 7.   Labeling of the edge set of $G_{0,1}^{(3)}$.

$k$ edges labeled $\{\ 3, 4, \cdots, k-1, k, k+1, k+2\ \}$ connect with $k$ vertices $\{\ v_0, v_k, w_3, w_4, \cdots, w_{k-1}, w_k\ \}$. From $E_0'$, $E_0''$, $E_0'''$, $E_1'$ and $E_1'''$ in Figure 4, $k$ edges labeled $\{\ 4, 5, \cdots, k-1, k, k+1, k+2, k+3\ \}$ connect with $k$ vertices $\{\ v_0, v_k, v_{k+1}, w_4, w_5, \cdots, w_{k-1}, w_k\ \}$. From $E_0'$, $E_0''$, $E_0'''$, $E_1'$ and $E_1'''$ in Figure 4, $k$ edges labeled $\{\ 5, 6, \cdots, k-1, k, k+1, k+2, k+3, k+4\ \}$ connect with $k$ vertices $\{\ v_0, v_{k-1}, v_k, v_{k+1}, w_5, w_6, \cdots, w_{k-1}, w_k\ \}$. Similarly, when we repeat till $k$ edges labeled $\{\ k, k+1, \cdots, 2k-1\ \}$, we see that these edges connect with $k$ vertices $\{\ v_0, v_4, \cdots, v_k, v_{k+1}, w_k\ \}$, since edges labeled $\{\ k+4, k+5, \cdots, 2k\ \}$ exist in $E_1'$.

From $E_0''$, $E_0'''$, $E_1'$ and $E_1'''$ in Figure 4, $k$ edges labeled $\{\ k+1, k+2, \cdots, 2k\ \}$, connect with $k+1$ vertices $\{\ v_3, v_4, \cdots, v_k, v_{k+1}, w_{k-1}, w_k\ \}$. From $E_0'''$, $E_1'$ and $E_1'''$ in Figure 4, $k$ edges labeled $\{\ k+2, k+3, \cdots, 2k+1\ \}$ connect with $k+1$ vertices $\{\ v_2, v_3, \cdots, v_k, v_{k+1}, w_k\ \}$. From $E_1'$ and $E_1'''$ in Figure 4, $k$ edges labeled $\{\ k+3, k+4, \cdots, 2k+2\ \}$ connect with $k+1$ vertices $\{\ v_1, v_2, \cdots, v_{k-1}, v_{k+1}, w_k\ \}$.

From $E_1'$ and $E_1'''$ in Figure 4 and Figure 5, $k$ edges labeled $\{\ k+4, k+5, \cdots, 2k+3\ \}$ connect with $k+1$ vertices $\{\ v_1, v_2, \cdots, v_{k-1}, w_{k-1}, w_k\ \}$. From $E_1'$ in Figure 4 and Figure 5, $k$ edges labeled $\{\ k+5, k+6, \cdots, 2k+4\ \}$ connect with $k$ vertices $\{\ v_1, v_2, \cdots, v_{k-2}, w_{k-1}, w_k\ \}$. Similarly, when we repeat till $k$ edges labeled $\{\ 2k+2, 2k+3, \cdots, 3k+1\ \}$, we see that these edges connect with $k+1$ vertices $\{\ v_1, v_2, \cdots, v_{k-1}, w_{k-1}, w_k\ \}$.

From $E_1'$ in Figure 4 and Figure 5, $k$ edges labeled $\{\ 2k+3, 2k+4, \cdots, 3k+2\ \}$ connect with $k+1$ vertices $\{\ v_1, v_2, \cdots, v_{k-1}, w_{k-2}, w_{k-1}\ \}$. From Figure 5, $k$ edges labeled $\{\ 2k+4, 2k+5, \cdots, 3k+3\ \}$ connect with $k+1$ vertices $\{\ v_1, v_2, \cdots, v_{k-1}, w_{k-2}, w_{k-1}\ \}$. Similarly, when we repeat till $k$ edges labeled $\{\ 3k+1, 3k+2, \cdots, 4k\ \}$, we see that these edges connect with $k+1$ vertices $\{\ v_1, v_2, \cdots, v_{k-1}, w_{k-2}, w_{k-1}\ \}$.

From Figure 5, $k$ edges labeled $\{ 3k+2, 3k+3, \cdots, 4k+1 \}$ connect with $k+1$ vertices $\{ v_1, v_2, \cdots, v_{k-1}, w_{k-3}, w_{k-2} \}$. Similarly, when we repeat till $k$ edges labeled $\{ k(k+1)/2+k+2, k(k+1)/2+k+3, \cdots, k(k+1)/2+2k+1 \}$, we see that the number of vertices with which these edges connect is equal or less than $k + 1$. Since the number of edges which connect with $2x$ vertices $\{ v_1, v_2, \cdots, v_x, w_0, w_1, \cdots, w_{x-1} \}$ is $x(x+1)/2$, and it holds $2x = x(x+1)/2+1$ in the case of $x = 1, 2$, $2x = x(x+1)/2$ in the case of $x = 3$, $2x < x(x+1)/2$ in the case of $x > 3$.

From $E_1''$ in Figure 4 and Figure 5, we see that the number of vertices with which $k$ edges labeled $\{ k(k+1)/2+k+3, k(k+1)/2+k+4, \cdots, k(k+1)/2 + 2k + 2 \}$ connect is equal or less than $k + 1$. Since, we remove one vertex $v_x$ and add one vertex $w_{k+1}$. Similarly, when we repeat till $k$ edges labeled $\{ k^2 + 2k + 2, k^2 + 2k + 3, \cdots, k^2 + 3k + 1 \}$, we see that the number of vertices with which these edges connect is equal or less than $k + 1$.

For every edge-window which has $k$ edges, the maximum $f$ of the number of vertices for these edge-windows is equal to $k + 1$ when a window (that is, the number of edges for these edge-windows) $d$ is fixed to $k$. Hence, we see that a bipartite graph $G_{0,1}^{(k)}$ has a $(k, k + 1)$-ordering. Similarly, we can easily see that a bipartite graph $G_{0,1}^{(k)}$ has a $(d, f)$-ordering with $d = k \times s + r$ and $f = (k + 1)s + r$, $s > 0$, $r = 0, 1, 2, \cdots, k - 1$.

$$(Q.E.D.)$$

In the following Theorem, we consider and describe the cluttered ordering of the graph $K_{2(k+1),2(k+1)} - 2(k + 1)K_{1,k}$ corresponding to $H(1, k; 2)$.

**Theorem 4.1.** *Let $k$ be a positive integer. For all $k$, there is a $(d, f)$-cluttered ordering of a bipartite graph $K_{2(k+1),2(k+1)} - 2(k+1)K_{1,k}$ with $d = k$ and $f = k+1$. Moreover, a bipartite graph $K_{2(k+1),2(k+1)} - 2(k+1)K_{1,k}$ has a $(d, f)$-cluttered ordering with $d = k \times s + r$ and $f = (k + 1)s + 1 + r$, $s > 0$, $r = 0, 1, 2, \cdots, k - 1$.*

**Proof.** From Lemma 4.2, a bipartite graph $K_{2(k+1),2(k+1)} - 2(k+1)K_{1,k}$ is decomposed to two isomorphic subgraphs $G_{0,1}^{(k)}$ and $G_{2,3}^{(k)}$. Since the number of edges of $G_{0,1}^{(k)}$ and $G_{2,3}^{(k)}$ are $k^2 + 3k + 2$ respectively, we start at the labels of the edges $0 - (k - 1) \equiv 2(k^2 + 3k + 2) - (k - 1) = 2k^2 + 5k + 5 \pmod{2(k^2 + 3k + 2)}$.

Similarly to the proof of Proposition 4.1, we consider edge-windows from $\{ 2k^2+5k+5, 2k^2+5k+6, \cdots, 2k^2+6k+3, 0 \}$ to $\{ 2k^2+6k+3, 0, 1, \cdots, k-2 \}$.

Here, we begin at the wrong end, that is, we consider edge-windows from $\{\, 2k^2+6k+3, 0, 1, \cdots, k-2 \,\}$ to $\{\, 2k^2+5k+5, 2k^2+5k+6, \cdots, k^2+3k+2, 0 \,\}$.

From $E_0'$ and $E_1''$ in Figure 4, $k$ edges labeled $\{\, 2k^2+6k+3, 0, 1, \cdots, k-2 \,\}$ connected with $k+1$ vertices $\{\, v_0, v_{2k}, w_0, w_1, \cdots, w_{k-2} \,\}$. From $E_0'$ and $E_1''$ in Figure 4, $k$ edges labeled $\{\, 2k^2+6k+2, 2k^2+6k+3, 0, 1, \cdots, k-3 \,\}$ connect with $k+1$ vertices $\{\, v_0, v_{2k}, w_0, w_1, \cdots, w_{k-2} \,\}$. Similarly, when we repeat till $k$ edges labeled $\{\, 2k^2+5k+5, 2k^2+5k+6, \cdots, 2k^2+6k+3, 0 \,\}$, from $E_0'$ in Figure 4 and Figure 6 these edges connected with $k+1$ vertices $\{\, v_0, v_{2k+1}, w_0, w_1, \cdots, w_{k-2} \,\}$.

For every edge-window which has $k$ edges, the maximum $f$ of the number of vertices for these edge-windows is equal to $k+1$ when a window (that is, the number of edges for these edge-windows) $d$ is fixed to $k$. Hence, we see that a bipartite graph $K_{2(k+1),2(k+1)} - 2(k+1)K_{1,k}$ has a $(k, k+1)$-ordering.

Similarly, we can easily see that a bipartite graph $K_{2(k+1),2(k+1)} - 2(k+1)K_{1,k}$ has a $(d, f)$-cluttered ordering with $d = k \times s + r$ and $f = (k+1)s + r$, $s > 0$, $r = 0, 1, 2, \cdots, k-1$.

$$\text{(Q.E.D.)}$$

Finally, for any positive integers $k, t$, we give a theorem concerning the cluttered ordering of the graph $K_{2(k+1)t,2(k+1)t} - 2(k+1)tK_{1,k}$ corresponding to $H(1, k; 2t)$.

**Theorem 4.2.** *Let $k, t$ be positive integers. For all $k, t$, there is a $(d, f)$-cluttered ordering of a bipartite graph $K_{2(k+1)t,2(k+1)t} - 2(k+1)tK_{1,k}$ with $d = k$ and $f = k+1$. Moreover, for all $t$, there is a $(d, f)$-cluttered ordering of a bipartite graph $K_{2(k+1)t,2(k+1)t} - 2(k+1)tK_{1,k}$ with $d = k \times s + r$ and $f = (k+1)s + r$, $s > 0$, $r = 0, 1, 2, \cdots, k-1$.*

**Proof.**    Similarly to the proof of Lemma 4.2, a bipartite graph $K_{2(k+1)t,2(k+1)t} - 2(k+1)tK_{1,k}$ is decomposed to $t$ isomorphic subgraphs $G_{0,1}$ and $t$ isomorphic subgraphs $G_{2,3}$. That is, a bipartite graph $K_{2(k+1)t,2(k+1)t} - 2(k+1)tK_{1,k}$ is decomposed to $t$ isomorphic subgraphs $K_{2(k+1),2(k+1)} - 2(k+1)K_{1,k}$. Similarly to the proof of Theorem 4.1, we can easily see that a bipartite graph $K_{2(k+1)t,2(k+1)t} - 2(k+1)tK_{1,k}$ has a $(k, k+1)$-cluttered ordering. Similarly, we can easily see that a bipartite graph $K_{2(k+1)t,2(k+1)t} - 2(k+1)tK_{1,k}$ has a $(d, f)$-cluttered ordering with $d = k \times s + r$ and $f = (k+1)s + r$, $s > 0$, $r = 0, 1, 2, \cdots, k-1$.

$$\text{(Q.E.D.)}$$

## Acknowledgement

The author gratefully acknowledgments the support from Initiative for Realizing Diversity in the Research Environment.

## References

1. T. Adachi, A cyclic ordering of composition from bipartite graphs and its application to RAID, *Congressus Numerantium*, **229**(2017), 295–311.
2. T. Adachi, Some construction of cluttered orderings for the complete bipartite graph, *Congressus Numerantium*, **227**(2016), 309–323.
3. T. Adachi, D. Kikuchi, Some sequence of wrapped $\Delta$-labellings for the complete bipartite graph, *Applied Mathematics*, **5(1)**(2015), 195–205.
4. M. Cohen, C. Colbourn, D. Froncek, Cluttered orderings for the complete graph, *Computing and Combinatorics: Proc. 7th annual international conference, COCOON 2001*, Lect. Notes Comp. Sci. 2108, Springer-Verlag, 2001, 420–431.
5. P. Chen, E. Lee, G. Gibson, R. Katz, D. Ptterson, RAID: High-performance, reliable secondary storage, *ACM Computing Surveys*, **26**(1994), 145–185.
6. M. Dewar and B. Stevens, *Ordering Block Designs – Gray Codes, Universal Cycles and Configuration Orderings*, Springer, (2012).
7. L. Hellerstein, G. Gibson, R. Karp, R. Katz, D. Patterson, Coding techniques for handling failures in large disk arrays, *Algorithmica*, **12**(1994), 182–208.
8. M. Mueller, T. Adachi, M. Jimbo, Cluttered orderings for the complete bipartite graph, *Discrete Applied Mathematics*, **152**(2005), 213–228.

# On $n$-ary polynomially complete quasigroups*

V. A. Artamonov

*Department of Algebra, Faculty of Mechanics and Mathematics, Moscow State University and Department of Informatics and Mathematics, VAVT and RANEPA*

*artamon@mech.math.msu.su*

To Prof. L.A. Bokut on the occasion of his 80th birthday

The paper presents a method for recognition of polynomial completeness of $n$-ary finite quasigroup given by its Latin square. The method uses properties of the group $G(Q)$. There is also given a construction of ternary finite quasigroups whose all binary deducts are polynomially complete.

## 1. Introduction

Quasigroups are groupoids with multiplication in which the operators of left and right multiplications are invertible. It means that quasigroups are good choice of algebraic structures for cryptographic purpose[1-4,6,8]. Here we give an example of cryptographic transformations based on of quasigroup multiplication.

Let $Q = \{x_1, \ldots, x_n\}$ be a finite set (alphabet) and denote by $Q^\dagger = \{x_1 x_2 \cdots x_t \mid x_i \in Q, t \geqslant 1\}$ the set of all finite strings over $Q$. Consider a quasigroup $(Q, \cdot)$ and message space $\mathcal{M} = $ crypt space $\mathcal{C} = Q^\dagger$. For each fixed permutation $g$ of $Q$, define an encryption transformation $E_g : \mathcal{M} \to \mathcal{C}$ as follows: if $M = x_1 x_2 \cdots x_t \in \mathcal{M} = Q^\dagger$ then

$$E_g(M) = y_1 y_2 \cdots y_t,$$

$$\text{where } y_i = \begin{cases} g(x_1), & i = 1, \\ y_{i-1} \cdot x_i, & 2 \leqslant i \leqslant t. \end{cases} \tag{1}$$

Each transformation $E_g$ is a bijection on the set $Q^\dagger = \mathcal{M}$ preserving lengths of words. So $E_g$ generate a subgroup $\mathcal{E}$ in the permutation group of the set

---

*The research is partially supported by NIR AAAA-A16-116070810025-5

$Q^\dagger$. Each element of $\mathcal{E}$ can be used as an encryption transformation. It is common to choose $g$ as a left multiplication $L_l$ for some element $l \in Q$.

The choice of $Q$ and of an element from $\mathcal{E}$ can be considered as a choice of a key. Since encryption transformations are based on quasigroup operations the security depends on its algebraic properties. For example high non-associative is one of the significant algebraic properties for cryptographic suitable choice of quasigroup[4].

High non-associativity is defined in terms of multiplication group $\mathrm{Mult}\,Q$. But this group is not stable under isotopy. Instead of $\mathrm{Mult}\,Q$ we suggest to consider the group $G(Q)$ which maps to its conjugate under an isotopy. So algebraic properties of $G(Q)$ remain almost the same under isotopy. For example we consider the property of double transitive action of $G(Q)$. This property is stable under isotopy and in this case the quasigroup is polynomially complete. According to Ref. 7 the problem of solutions of systems of equations over these quasigroup is NP-complete.

In the present paper we extend the results of Ref. 1–3 from binary to $n$-ary case. We define an analogy of the group $G(Q)$ and show that in $n$-ary case if $G(Q)$ acts doubly transitive in $Q$, then $Q$ is polynomially complete. In the final section we construct ternary quasigroup in which any its binary deduct is polynomially complete. $n$-ary case is more interesting because the number of these quasigroups increases very high and we can add as keys not only quasigroups and also a choice of binary sections.

All necessary definition concerning quasigroups can be found in Ref. 5.

## 2. Isotopies of $n$ quasigroups

Let $n \geqslant 2$. By $n$-ary quasigroup we shall mean a finite set $Q$ with $n$-ary operation of multiplication $x_1 \ldots x_n$ such that for any $i = 1, \ldots, n$ and any elements $a = (a_1, \ldots, a_{i-1}, a_{i+1}, \ldots, a_n) \in Q^{n-1}$ the map

$$M_{i,a}x = a_1 \ldots a_{i-1}xa_{i+1}a_n \tag{2}$$

is a bijection. We shall consider $n$-ary quasigroup $Q$ as a general algebra with $n$-ary operations of multiplications and with all operations

$$M_i^{-1}(a_1, \ldots, a_{i-1}, x, a_{i+1}, \ldots, a_n) = M_{i,a}^{-1}x, \quad i = 1, \ldots, n.$$

Denote by $G(Q)$ the subgroup of permutations on $Q$ generated by all maps $M_{i,a}M_{i,b}^{-1}$ for all $i$ and all $a, b \in Q^{n-1}$.

Each $n$-ary quasigroup can be presented by its Latin $n$-cube (Cayley table). Let $Q = \{x_1, \ldots, x_m\}$. Then the entry of the cube located at the point $(i_1, \ldots, i_n)$ is equal to the product $x_1 \ldots x_n$.

Let

$$x_1 \ldots x_n, \quad [x_1 \ldots x_n] \tag{3}$$

be two $n$-quasigroup multiplications on a set $Q$. We say that the first of them is an *isotope* of the second one if there exist permutations $\pi, \pi_1, \ldots, \pi_n$ on $Q$ such that

$$[x_1 \ldots x_n] = \pi \left( \pi_1^{-1}(x_1) \ldots \pi_2^{-1}(x_n) \right) \tag{4}$$

for all $x_i \in Q$.

**Theorem 2.1.** *Let $G(Q)$ and $G_{[]}(Q)$ be permutation group as above associated with multiplications (3). If these multiplication are isotopic with an isotopy (4) then $G(Q_{[]}) = \pi G(Q) \pi^{-1}$.*

**Proof.** Let

$$x = (x_1, \ldots, x_{i-1}, x_{i+1}, \ldots, x_n) \in Q^{n-1} \tag{5}$$

and

$$\overline{x} = (\pi_1(x_1), \ldots, \pi_{i-1}(x_{i-1}), \pi_{i+1}(x_{i+1}), \ldots, \pi_n(x_n)) \in Q^{n-1}.$$

Then

$$M_{i,\overline{x}} \left( \pi_i(x_i) \right) = \pi_1(x_1) \ldots \pi_n(x_n) = \pi[x_1 \ldots x_n] = \pi M'_{i,x}(x_i),$$

where $M'_{i,x}$ is defined as in (2) with respect to the second multiplication in (3). Hence $M_{i,\overline{x}}\pi_i = \pi M'_{i,x}$, It follows that

$$M_{i,\overline{y}} M_{i,\overline{x}}^{-1} = \pi M'_{i,y} \pi_i^{-1} \pi_i \left( M'_{i,x} \right)^{-1} \pi^{-1} = \pi M'_{i,y} \left( M'_{i,x} \right)^{-1} \pi^{-1}.$$

Since the map $x \mapsto \overline{x}$ is a bijection we complete the proof. $\qquad\square$

It is interesting to observe the way of changing of the encryption transformation $E_g$ from (1) under an isotopy $(\pi, \pi_1, \pi_2)$ from (4) in the binary case. Take $M = x_1 x_2 \cdots x_t \in \mathcal{M} = Q^\dagger$ then under the new multiplication (4) we get

$$E_g(M) = z_1 z_2 \cdots z_t,$$

$$\text{where } z_i = \begin{cases} g(x_1), & i = 1, \\ \pi \left( \pi_1^{-1}(y_{i-1}) \pi_2^{-1}(x_i) \right), & 2 \leqslant i \leqslant t. \end{cases}$$

## 3. Affine $n$-quasgroups

A $n$-quasigroups $Q$ is *affine* if there exists a structure of an additive abelian group in $Q$ such that

$$x_1 \ldots x_n = a_0 + \alpha_1(x_1) + \cdots + \alpha_n(x_n) \tag{6}$$

for all $x_i \in Q$. Here $a_0 \in Q$ and $\alpha_1, \ldots, \alpha_n$ are group automorphisms of $(Q, +)$.

Calculate the group $G(Q)$ for an affine $n$-quasigroups $Q$. By (6) for $x$ from (5) we have

$$M_{i,x}(x_i) = \alpha_i(x_i) + \sum_{j \neq i} \alpha_j(x_j) + a_0.$$

Hence

$$M_{i,x}^{-1}(x_i) = \alpha_i^{-1}(x_i) - \alpha_i^{-1} \left( \sum_{j \neq i} \alpha_j(x_j) + a_0 \right),$$

and therefore for any $y = (y_1, \ldots, y_{i-1}, x_{i+1}, \ldots, y_n) \in Q^{n-1}$ we obtain

$$M_{i,x} M_{i,y}^{-1}(x_i) = x_i - \left( \sum_{j \neq i} \alpha_j(x_j) + a_0 \right) + \sum_{j \neq i} \alpha_j(x_j) + a_0$$

$$= x_i + \sum_{j \neq i} \alpha_j(x_j - y_j).$$

So the map $M_{i,x} M_{i,y}^{-1}$ is a transfer in the abelian group $(Q, +)$. All transfers form an abelian group isomorphic to $(Q, +)$. Thus we have proved

**Theorem 3.1.** *The group $G(Q)$ in the affine $n$-quasigroup $Q$ is abelian and it is isomorphic to the group $(Q, +)$. In particular if $Q$ is isotopic to an affine quasigroup then $G(Q)$ is abelian.*

## 4. Polynomial completeness

Let $\wp$ be a congruence in $n$-quasigroup $Q$. If $(x, y) \in \wp$ where $x, y \in Q$, then $(g(x), g(y)) \in \wp$ for any $g \in G(Q)$. Hence $G(Q)$ is not a primitive permutation group. Thus we have

**Theorem 4.1.** *Let $G(Q)$ act primitively in $Q$. Then $Q$ is a simple $n$-quasigroup.*

Also we have

**Theorem 4.2.** *Suppose that $|Q| \geqslant 2$ and $G(Q)$ is 2-transitive permutation group. Then $Q$ is polynomially complete.*

**Proof.** Suppose that $G(Q)$ is abelian. Since $G(Q)$ is transitive the stabilizer group of any point is the same and therefore it is the identity group.

Let $z \neq x \neq y \in Q$. By 2-transitivity there exists an element $f \in G(Q)$ such that $f(x) = x$ and $f(y) = z$. As it was noticed in this case $f$ is the identity map, a contradiction.

Hence $G(Q)$ is not abelian and therefore by Theorem 3.1 the quasigroup $Q$ is not affine.

Also any doubly transitive permutation group is primitive. Hence $Q$ is simple by Theorem 4.1. $\qquad\square$

**Theorem 4.3.** *Let $1 \leqslant i < j \leqslant n$ and*

$$x = (x_1, \ldots, x_{i-1}, x_{i+1}, \ldots, x_{j-1}, x_{j+1}, \ldots, x_n).$$

*Denote by $Q_{ij,x}$ the quasigroup with multiplication*

$$ab = x_1 \ldots x_{i-1} a x_{i+1} \ldots x_{j-1} b x_{j+1} \ldots x_n$$

*If $G_{ij,x} = G(Q_{ij,x})$ then $G_{ij,x} \subseteq G(Q)$. In particular if $G_{ij,x}$ acts doubly transitive in $Q$ for some $i, j, x$ then $Q$ is polynomially complete.*

**Proof.** Put

$$x'(a) = (x_1, \ldots, x_{i-1}, x_{i+1}, \ldots, x_{j-1}, b, x_{j+1}, \ldots, x_n) \in Q^{n-1},$$
$$x''(b) = (x_1, \ldots, x_{i-1}, a, x_{i+1}, \ldots, x_{j-1}, x_{j+1}, \ldots, x_n) \in Q^{n-1}.$$

It follows from (2) that the left and right multiplication maps in $Q_{ij,x}$ have the forms $L_a(b) = M_{j,x'(a)}(b)$, $R_b(a) = M_{i,x''(b)}(a)$. So the generators

$$L_a L_c^{-1} = M_{j,x'(a)} M_{j,x'(c)}^{-1}, \quad R_b R_d^{-1} = M_{i,x''(b)} M_{i,x''(d)}^{-1}$$

of $G_{ij,x}$ belong to $G(Q)$. $\qquad\square$

## 5. Construction of ternary polynomially complete quasigroups

We are going to construct ternary polynomially complete quasigroups $Q$ in which each group $G(Q_{ij,x})$ acts doubly transitive in $Q$. So any binary deduct of $Q$ is polynomially complete by Theorem 4.3.

Take two Latin squares $L, T$ of order $n$ with row permutations $\sigma_1 = 1, \sigma_2, \ldots, \sigma_n$ and $\xi_1 = 1, \xi_2, \ldots, \xi_n$, respectively. The case $L = T$ is allowed. Assume that the subgroup generated by $\sigma_2 = \sigma_2 \sigma_1^{-1}, \ldots, \sigma_n = \sigma_n \sigma_1^{-1}$ contains the alternative group $\mathbf{A}_n$. Similarly row permutations $\xi_1 = 1, \xi_2, \ldots, \xi_n$ generate a subgroup which contains $\mathbf{A}_n$.

The following Latin squares satisfy this property:

|       | $x_1$ | $x_2$ | $x_3$ | $x_4$ | $x_5$ | $x_6$ | $x_7$ | $x_8$ |
|-------|-------|-------|-------|-------|-------|-------|-------|-------|
| $x_1$ | $x_1$ | $x_2$ | $x_3$ | $x_4$ | $x_5$ | $x_6$ | $x_7$ | $x_8$ |
| $x_2$ | $x_2$ | $x_3$ | $x_4$ | $x_5$ | $x_6$ | $x_7$ | $x_8$ | $x_1$ |
| $x_3$ | $x_3$ | $x_1$ | $x_5$ | $x_8$ | $x_7$ | $x_2$ | $x_4$ | $x_6$ |
| $x_4$ | $x_4$ | $x_5$ | $x_2$ | $x_1$ | $x_3$ | $x_8$ | $x_6$ | $x_7$ |
| $x_5$ | $x_5$ | $x_6$ | $x_7$ | $x_3$ | $x_8$ | $x_4$ | $x_1$ | $x_2$ |
| $x_6$ | $x_6$ | $x_7$ | $x_8$ | $x_2$ | $x_1$ | $x_3$ | $x_5$ | $x_4$ |
| $x_7$ | $x_7$ | $x_8$ | $x_1$ | $x_6$ | $x_4$ | $x_5$ | $x_2$ | $x_3$ |
| $x_8$ | $x_8$ | $x_4$ | $x_6$ | $x_7$ | $x_2$ | $x_1$ | $x_3$ | $x_5$ |

;

and

|       | $x_1$ | $x_2$ | $x_3$ | $x_4$ | $x_5$ |
|-------|-------|-------|-------|-------|-------|
| $x_1$ | $x_1$ | $x_2$ | $x_3$ | $x_4$ | $x_5$ |
| $x_2$ | $x_2$ | $x_3$ | $x_4$ | $x_5$ | $x_6$ |
| $x_3$ | $x_3$ | $x_1$ | $x_5$ | $x_2$ | $x_4$ |
| $x_4$ | $x_4$ | $x_5$ | $x_1$ | $x_3$ | $x_2$ |
| $x_5$ | $x_5$ | $x_4$ | $x_2$ | $x_1$ | $x_3$ |

.

Now we take the Latin cube with vertical sections $S_1 = \xi_1 L = L$, $S_2 = \xi_2 L, \ldots, S_n = \xi_n L$. Each of these sections corresponds to a quasigroup $Q_{12,x}$ for some $x \in Q$. Its row permutations are $\xi_i \sigma_1, \ldots, \xi_i \sigma_n$. Hence the group $G_{12,x}$ contains a subgroup $\xi_i G(Q) \xi_i^{-1} \supseteq \mathbf{A}_n$.

Now we can consider another vertical (or horizontal) section passing through a row or a column of $L$. Then other rows (columns) of the section will be obtained by applying to them $\xi_1 = 1, \xi_2, \ldots, \xi_n$. These sections correspond to quasigroups $Q_{13,x}, Q_{23,x}$ whose groups $G_{13,x}, G_{23,x}$ contain $\mathbf{A}_n$.

By this construction one can obtain a ternary quasigroup $Q$ whose all binary deducts $Q_{ij,x}$ are polynomially complete.

## References

1. V.A. Artamonov, S. Chakrabarti, S. Gangopadhyay and S.K. Pal, On Latin squares of polynomially complete quasigroups and quasigroups generated by shifts, Quasigroups and Related Systems 21 (2013), 201–214.

2. V.A. Artamonov, S. Chakrabarti and S.K. Pal, Characterization of Polynomially Complete Quasigroups based on Latin Squares for Cryptographic Transformations, J. Discrete Applied Mathematics, doi:10.1016/j.dam.2015.06.033.

3. V.A. Artamonov, S. Chakrabarti and S.K. Pal, Characterizations of highly non-associative quasigroups and associative triples. Quasigroups and Related Systems 25 (2017), 1–19.

4. J. Dénes & A.D. Keedwell, "Latin Squares and their Applications". Akadémiai Kiadó, Budapest; Academic Press, New York; English Universities Press, London, 1974.

5. J. Dénes & A.D. Keedwell, "Latin squares. New developments in the theory and applications". Annals of Discrete Mathematics, 46. North-Holland, Amsterdam, 1991.

6. V. Dimitrova, S. Markovski, D. Gligoroski, Classification of Quasi-groups as Boolean Functions, their Algebraic Complexity and Application of Gröbner Bases in Solving Systems of Quasi-group Equations.

7. G. Horvth, C. L. Nehaniv, Cs. Szab. An assertion concerning functionally complete algebras and NP-completeness. Theoret. Comput. Sci., 407:591–595, 2008.

8. Guohao Liu, Yunqing Xu, Cryptographic classification of quasigroups of order 4, International Workshop on Cloud Computing and Information Security (CCIS 2013), 278–281.

# Matrix rings as one sided $\sigma$-$(S,1)$ rings

V. K. Bhat

*School of Mathematics, SMVD University,*
*P/o SMVD University, Katra, J and K, India-182320*

*vijaykumarbhat2000@yahoo.com*

Pradeep Singh

*School of Mathematics, SMVD University,*
*P/o SMVD University, Katra, J and K, India-182320*

*pradeep333singh@gmail.com*

Let $R$ be a ring and $\sigma$ an endomorphism of $R$. We recall that $R$ is called an $(S,1)$-ring if for $a$, $b \in R$, $ab = 0$ implies $aRb = 0$. We involve $\sigma$ to generalize this notion. We say that $R$ is a left $\sigma$-$(S,1)$ ring if for $a$, $b \in R$, $ab = 0$ implies $aRb = 0$ and $\sigma(a)Rb = 0$. We say that $R$ is a right $\sigma$-$(S,1)$ ring if for $a$, $b \in R$, $ab = 0$ implies $aRb = 0$ and $aR\sigma(b) = 0$. $R$ is called a $\sigma$-$(S,1)$ ring if it is both right and left $\sigma$-$(S,1)$ ring. In this paper we give examples of such rings and a relation between $\sigma$-$(S,1)$ rings, 2-primal rings, and $\sigma(*)$-rings.

We show that a certain class of matrix rings, with suitable endomorphisms $\sigma$ are left $\sigma$-$(S,1)$ but not right $\sigma$-$(S,1)$, and vice versa.

*2010 Mathematics subject classification:* 16-XX, 16S36, 16N40, 16P40, 16W20.

*Keywords:* Endomorphism, matrix ring, left $\sigma$-$(S,1)$ ring.

## 1. Introduction and Preliminaries

**Notation:** All rings are associative with $1 \neq 0$. The ring of integers, the field of rational numbers and the field of real numbers are denoted by $\mathbb{Z}$, $\mathbb{Q}$ and $\mathbb{R}$ respectively, unless otherwise stated. $Spec(R)$ denotes the set of prime ideals of $R$. $MinSpec(R)$ denotes the set of minimal prime ideals of $R$. The Prime radical and the set of nilpotent elements of $R$ are denoted by $P(R)$ and $N(R)$ respectively. Zero matrix and identity matrix are denoted by $O$ and $I$ respectively.

**Definition 1.1.** (Kim and Lee[2]). Let $R$ be a ring. Then $R$ is called an $(S,1)$-ring if for $a$, $b \in R$, $ab = 0$ implies $aRb = 0$.

This notion was actually introduced by Shin (i.e., a ring satisfying $SI$ property, Lemma 1.2 of Ref. 3). Reduced rings (i.e., rings without nonzero nilpotent elements) are obviously $(S,1)$-rings, right (left) duo rings are $(S,1)$-rings (Ref. 3, Lemma 1.2). Shin (Ref. 3, Theorem 1.5) showed that an $(S,1)$-ring $R$ is 2-primal (i.e., $N(R) = P(R)$).

**Example 1.1.** Consider the ring of real quaternions

$$\mathbb{H} = \{a + bi + cj + dk : a, b, c, d \in \mathbb{R}\},$$

where $i^2 = j^2 = k^2 = ijk = -1$ and $ij = -ji$, $ik = -ki$, $jk = -kj$.

(1) Let $R = \left\{ \begin{pmatrix} a & 0 \\ 0 & b \end{pmatrix}; a, b \in \mathbb{H} \right\}$. The only matrices $A \in R$ and $B \in R$ satisfying $AB = O$ are of the type $A = \begin{pmatrix} a & 0 \\ 0 & 0 \end{pmatrix}$ and $B = \begin{pmatrix} 0 & 0 \\ 0 & b \end{pmatrix}$ where $a, b \in \mathbb{H}$.

Now for all $K = \begin{pmatrix} c & 0 \\ 0 & d \end{pmatrix} \in R$,

$$AKB = \begin{pmatrix} a & 0 \\ 0 & 0 \end{pmatrix} \begin{pmatrix} c & 0 \\ 0 & d \end{pmatrix} \begin{pmatrix} 0 & 0 \\ 0 & b \end{pmatrix} = \begin{pmatrix} 0 & 0 \\ 0 & 0 \end{pmatrix}.$$

This implies that $R$ is an $(S,1)$-ring.

(2) Let $R = \left\{ \begin{pmatrix} a & b \\ 0 & c \end{pmatrix}; a, b, c \in \mathbb{H} \right\}$. Let $A = \begin{pmatrix} a & 0 \\ 0 & 0 \end{pmatrix}$ and $B = \begin{pmatrix} 0 & 0 \\ 0 & b \end{pmatrix}$ where $0 \neq a, 0 \neq b \in \mathbb{H}$. Then $AB = 0$. Now let $K = \begin{pmatrix} c & d \\ 0 & 0 \end{pmatrix} \in R$, with $d \neq 0$. Then

$$AKB = \begin{pmatrix} a & 0 \\ 0 & 0 \end{pmatrix} \begin{pmatrix} c & d \\ 0 & 0 \end{pmatrix} \begin{pmatrix} 0 & 0 \\ 0 & b \end{pmatrix}$$
$$= \begin{pmatrix} 0 & adb \\ 0 & 0 \end{pmatrix} \neq \begin{pmatrix} 0 & 0 \\ 0 & 0 \end{pmatrix}$$

Thus $R$ is not an $(S,1)$-ring.

We note that $R$ is not an $(S,1)$-ring even if $\mathbb{H}$ is replaced by $\mathbb{Z}$ or a field $F$.

**Example 1.2.** (Example (5.3) of Ref. 3). Let $F = \mathbb{Z}_2(y)$ be the field of rational functions over $\mathbb{Z}_2$ with $y$ an indeterminate. Consider the ring $R = \{f(x) \in F[x] \mid xy + yx = 1\}$. Then clearly $R$ is a domain, so it is reduced and hence an $(S, 1)$-ring.

## 2. σ-(S, 1) rings

In this article we generalize the notion of $(S, 1)$-rings by involving an endomorphism $\sigma$ of $R$ as follows:

**Definition 2.1.** Let $R$ be a ring and $\sigma$ an endomorphism of $R$. We call $R$ a left $\sigma$-$(S, 1)$ ring if for $a, b \in R$, $ab = 0$ implies $aRb = 0$ and $\sigma(a)Rb = 0$.

Right $\sigma$-$(S, 1)$ ring can be defined in a similar way, i.e., for $a, b \in R$, $ab = 0$ implies $aRb = 0$ and $aR\sigma(b) = 0$. We say that $R$ is a $\sigma$-$(S, 1)$ ring if it is both left and right $\sigma$-$(S, 1)$ ring.

**Example 2.1.** Let $R = \left\{ \begin{pmatrix} a & 0 \\ 0 & b \end{pmatrix}; a, b \in \mathbb{H} \right\}$. Define $\sigma : R \to R$ by

$$\sigma \begin{pmatrix} a & 0 \\ 0 & b \end{pmatrix} = \begin{pmatrix} 0 & 0 \\ 0 & b \end{pmatrix}.$$ Then $\sigma$ is an endomorphism of $R$.

Now the only matrices $A \in R$ and $B \in R$ satisfying $AB = O$ are of the type $A = \begin{pmatrix} a & 0 \\ 0 & 0 \end{pmatrix}$ and $B = \begin{pmatrix} 0 & 0 \\ 0 & b \end{pmatrix}$, where $a, b \in \mathbb{H}$.

Now for all $J = \begin{pmatrix} c & 0 \\ 0 & d \end{pmatrix} \in R$,

$$AJB = \begin{pmatrix} a & 0 \\ 0 & 0 \end{pmatrix} \begin{pmatrix} c & 0 \\ 0 & d \end{pmatrix} \begin{pmatrix} 0 & 0 \\ 0 & b \end{pmatrix} = \begin{pmatrix} 0 & 0 \\ 0 & 0 \end{pmatrix}.$$

Also

$$\sigma(A)JB = \begin{pmatrix} 0 & 0 \\ 0 & 0 \end{pmatrix} \begin{pmatrix} c & 0 \\ 0 & d \end{pmatrix} \begin{pmatrix} 0 & 0 \\ 0 & b \end{pmatrix} = \begin{pmatrix} 0 & 0 \\ 0 & 0 \end{pmatrix}.$$

Therefore, $R$ is a left $\sigma$-$(S, 1)$ ring. It can be seen that $R$ is also a right $\sigma$-$(S, 1)$ ring, and hence a $\sigma$-$(S, 1)$ ring.

We note that a left or right $\sigma$-$(S, 1)$ ring is an $(S, 1)$-ring, but the converse is not true.

**Example 2.2.** Let $S$ be a ring without zero divisor and $R = S \times S$. The only elements $p, q \in R$ such that $pq = 0$ are of the form $p = (a, 0)$ and

$q = (0, b)$, for all $a, b \in S$. Now for all $t = (u, v) \in R$, $ptq = (0, 0)$. So $R$ is an $(S, 1)$-ring. Define $\sigma : R \to R$ by $\sigma(a, b) = (b, a)$. Then $\sigma$ is an endomorphism of $R$. Now $\sigma(p)tq = (0, a)(u, v)(0, b) = (0, avb)$. Therefore, $R$ is not a left $\sigma$-$(S, 1)$ ring. It can be seen that $R$ is not a right $\sigma$-$(S, 1)$ ring.

**Proposition 2.1.** *Let $R$ be a ring, $\sigma$ an automorphism of $R$ such that $R$ is a left or right $\sigma$-$(S, 1)$ ring. Then $R$ is 2-primal.*

**Proof.** $R$ is a left $\sigma$-$(S, 1)$ ring. Therefore $R$ is 2-primal (Theorem 1.5 of Ref. 3), which implies that $P(R)$ is completely semiprime. We give a sketch of proof.

We consider left case (right case is similar). Let $a \in N(R)$, say $a^n = 0$. If $a \notin P$ for some prime ideal $P$, then $ax_1a \notin P$ for some element $x_1 \in R$. Continuing the process we can find elements $x_i \in R$ such that $P$ does not contain $b = ax_1a...x_{n-1}a$. But, $R$ is an $(S, 1)$-ring, so $a^n = 0$ implies $b = 0$, hence $b \in P$, a contradiction. $\square$

**Definition 2.2.** (Kwak[1]). Let $R$ be a ring and $\sigma$ an endomorphism of $R$. Then $R$ is said to be a $\sigma(*)$-ring if $a\sigma(a) \in P(R)$ implies $a \in P(R)$ for $a \in R$.

**Example 2.3.** Let $R = \left\{ \begin{pmatrix} a & b \\ 0 & c \end{pmatrix}; a, b, c \in F, \text{a field} \right\}$. Now $P(R) = \left\{ \begin{pmatrix} a & 0 \\ 0 & c \end{pmatrix}; a, c \in F \right\}$. Define $\sigma : R \to R$ by $\sigma\begin{pmatrix} a & b \\ 0 & c \end{pmatrix} = \begin{pmatrix} a & 0 \\ 0 & c \end{pmatrix}$. Then $\sigma$ is an endomorphism of $R$, and it can be easily seen that $R$ is a $\sigma(*)$-ring.

**Proposition 2.2.** *Let $R$ be a ring, $\sigma$ an automorphism of $R$ such that $R$ is a left or right $\sigma$-$(S, 1)$ ring. Then $R$ is a $\sigma(*)$-ring.*

**Proof.** $R$ is 2-primal and $P(R)$ is completely semiprime by Proposition 2.1. We will show that $R$ is a weak $\sigma$-rigid ring. Let $a \in R$ be such that $a\sigma(a) \in N(R)$. Now $a\sigma(a)\sigma^{-1}(a\sigma(a)) \in N(R)$ implies that $a^2 \in N(R)$, and so $a \in N(R)$. Therefore, $R$ is a weak $\sigma$-rigid ring, and is also a $\sigma(*)$-ring. $\square$

**Remark 2.1.** Converse of Proposition 2.1 and Proposition 2.2 is not true. For example, the ring in Example 2.3 is a $\sigma(*)$-ring. Therefore it is also a 2-primal ring, but it is not a $\sigma$-$(S, 1)$ ring (even not an $(S, 1)$-ring).

## 3. One sided $\sigma - (S,1)$ rings

We now find the possible examples of a ring with an endomorphism $\sigma$ such that $R$ is a left $\sigma$-$(S,1)$ ring, but not a right $\sigma$-$(S,1)$ ring, and vice versa. Our attention goes towards the ring of matrices, and have the following:

Let $R = \left\{ \begin{bmatrix} l & e & f \\ 0 & m & g \\ 0 & 0 & n \end{bmatrix} \mid e, f, g, l, m, n \in F \right\}$, where $F$ is a field.

Define $\sigma : R \to R$ by $\sigma \left( \begin{bmatrix} l & e & f \\ 0 & m & g \\ 0 & 0 & n \end{bmatrix} \right) = \begin{bmatrix} \alpha_1 & 0 & 0 \\ 0 & \alpha_2 & 0 \\ 0 & 0 & \alpha_3 \end{bmatrix}$, $\alpha_i \in \{0, l, m, n\}$;

$i = 1, 2, 3$. Then $\sigma$ is an endomorphism on $R$.

Let $A = \begin{bmatrix} x & u & v \\ 0 & y & w \\ 0 & 0 & z \end{bmatrix}$, $B = \begin{bmatrix} x' & u' & v' \\ 0 & y' & w' \\ 0 & 0 & z' \end{bmatrix} \in R$ be such that $AB = ARB = O$

and either, but not both, $AR\sigma(B) = O$ or $\sigma(A) RB = O$.

Clearly, $A$ and $B$ are nonzero elements of $R$ and both cannot be of trivial trace otherwise, $\sigma(A) RB = O = AR\sigma(B)$. As a necessity for

$$
AB = \begin{bmatrix} x & u & v \\ 0 & y & w \\ 0 & 0 & z \end{bmatrix} \begin{bmatrix} x' & u' & v' \\ 0 & y' & w' \\ 0 & 0 & z' \end{bmatrix} \tag{1}
$$
$$
= \begin{bmatrix} xx' & uy' + xu' & vz' + uw' + xv' \\ 0 & yy' & wz' + yw' \\ 0 & 0 & zz' \end{bmatrix}
$$

to be $O$, $A$ or $B$ should have some zero diagonal elements. So, we have the following cases:

**Case (1)** $x = y = z = 0$.
**Case (2)** $x \neq 0$ and $y = z = 0$.
**Case (3)** $y \neq 0$ and $x = z = 0$.
**Case (4)** $z \neq 0$ and $x = y = 0$.

Now we will discuss the above cases:

**Case (1):** Let $x = y = z = 0$.

Then $A = \begin{bmatrix} 0 & u & v \\ 0 & 0 & w \\ 0 & 0 & 0 \end{bmatrix}$ and $B = \begin{bmatrix} x' & u' & v' \\ 0 & y' & w' \\ 0 & 0 & z' \end{bmatrix} \in R^* = R - \{O\}$, then for

$$AB = \begin{bmatrix} 0 & u & v \\ 0 & 0 & w \\ 0 & 0 & 0 \end{bmatrix} \begin{bmatrix} x' & u' & v' \\ 0 & y' & w' \\ 0 & 0 & z' \end{bmatrix} \tag{2}$$

$$= \begin{bmatrix} 0 & uy' & vz' + uw' \\ 0 & 0 & wz' \\ 0 & 0 & 0 \end{bmatrix} = O.$$

We have the following subcases:

**Subcase (1(i)):** If $u = 0$, $z' = 0$ and surely not both $x'$ and $y'$ are zeros (i.e., $x'^2 + y'^2 \neq 0$), then define $\sigma$ such that,

$$\sigma \left( \begin{bmatrix} l & e & f \\ 0 & m & g \\ 0 & 0 & n \end{bmatrix} \right) = lI_3, \text{ when } x' \neq 0$$

and

$$\sigma \left( \begin{bmatrix} l & e & f \\ 0 & m & g \\ 0 & 0 & n \end{bmatrix} \right) = mI_3 \text{ when } x' = 0.$$

Obviously, $\sigma(A) = O$ and

$$AR = \begin{bmatrix} 0 & 0 & v \\ 0 & 0 & w \\ 0 & 0 & 0 \end{bmatrix} \begin{bmatrix} l & e & f \\ 0 & m & g \\ 0 & 0 & n \end{bmatrix}$$

$$= \begin{bmatrix} 0 & 0 & nv \\ 0 & 0 & nw \\ 0 & 0 & 0 \end{bmatrix} \neq O \text{ as } A \neq O,$$

while

$$ARB = \begin{bmatrix} 0 & 0 & v \\ 0 & 0 & w \\ 0 & 0 & 0 \end{bmatrix} \begin{bmatrix} l & e & f \\ 0 & m & g \\ 0 & 0 & n \end{bmatrix} \begin{bmatrix} x' & u' & v' \\ 0 & y' & w' \\ 0 & 0 & 0 \end{bmatrix}$$

$$= O.$$

Therefore, $AB = O$, $ARB = O$, $\sigma(A) RB = O$, but $AR\sigma(B) \neq O$.

**Subcase (1(ii))**: If $u = 0$ and $z' \neq 0$, then from equation (1) $w = 0$ and $v = 0$, which leads to $A = O$.

**Subcase (1(iii))**: If $u \neq 0$, then from equation (1), $y' = 0$. Thus, for

$$AB = \begin{bmatrix} 0 & u & v \\ 0 & 0 & w \\ 0 & 0 & 0 \end{bmatrix} \begin{bmatrix} x' & u' & v' \\ 0 & 0 & w' \\ 0 & 0 & z' \end{bmatrix} = \begin{bmatrix} 0 & 0 & vz' + uw' \\ 0 & 0 & wz' \\ 0 & 0 & 0 \end{bmatrix} = O,$$

we have the following subcases:

**Subcase(1(iii-a))**: If $z' = 0$, then $w' = 0$. Thus, for

$$A = \begin{bmatrix} 0 & u & v \\ 0 & 0 & w \\ 0 & 0 & 0 \end{bmatrix} \text{ and } B = \begin{bmatrix} x' & u' & v' \\ 0 & 0 & 0 \\ 0 & 0 & 0 \end{bmatrix}, \tag{3}$$

where $u \neq 0$ and for $\sigma$ defined as $\sigma \left( \begin{bmatrix} l & e & f \\ 0 & m & g \\ 0 & 0 & n \end{bmatrix} \right) = l I_3$,

we have $AB = O$, $ARB = O$, $\sigma(A) = O$ and

$$AR = \begin{bmatrix} 0 & u & v \\ 0 & 0 & w \\ 0 & 0 & 0 \end{bmatrix} \begin{bmatrix} l & e & f \\ 0 & m & g \\ 0 & 0 & n \end{bmatrix}$$

$$= \begin{bmatrix} 0 & mu & gu + nv \\ 0 & 0 & nw \\ 0 & 0 & 0 \end{bmatrix} \neq O.$$

Therefore, $AB = O$, $ARB = O$, $\sigma(A)RB = O$, but $AR\sigma(B) \neq O$.

**Subcase(1(iii-b))**: Let $u \neq 0, y' = 0$ and $z' \neq 0$, then $w = 0$. Thus,

$$AB = \begin{bmatrix} 0 & u & v \\ 0 & 0 & 0 \\ 0 & 0 & 0 \end{bmatrix} \begin{bmatrix} x' & u' & v' \\ 0 & 0 & w' \\ 0 & 0 & z' \end{bmatrix}$$

$$= \begin{bmatrix} 0 & 0 & vz' + uw' \\ 0 & 0 & 0 \\ 0 & 0 & 0 \end{bmatrix} = O$$

implies that $vz' + uw' = 0$ and

$$ARB = \begin{bmatrix} 0 & u & v \\ 0 & 0 & 0 \\ 0 & 0 & 0 \end{bmatrix} \begin{bmatrix} l & e & f \\ 0 & m & g \\ 0 & 0 & n \end{bmatrix} \begin{bmatrix} x' & u' & v' \\ 0 & 0 & w' \\ 0 & 0 & z' \end{bmatrix}$$

$$= \begin{bmatrix} 0 & 0 & z'(gu+nv)+muw' \\ 0 & 0 & 0 \\ 0 & 0 & 0 \end{bmatrix}.$$

First, let $u \neq 0$, $y' = 0$, $z' \neq 0$, $w = 0$, with $vz' + uw' = 0$ and suppose $v = 0$, then $w' = 0$. Thus,

$$A = \begin{bmatrix} 0 & u & 0 \\ 0 & 0 & 0 \\ 0 & 0 & 0 \end{bmatrix} \text{ and } B = \begin{bmatrix} x' & u' & v' \\ 0 & 0 & 0 \\ 0 & 0 & z' \end{bmatrix},$$

then $AB = O$, but

$$ARB = \begin{bmatrix} 0 & u & 0 \\ 0 & 0 & 0 \\ 0 & 0 & 0 \end{bmatrix} \begin{bmatrix} l & e & f \\ 0 & m & g \\ 0 & 0 & n \end{bmatrix} \begin{bmatrix} x' & u' & v' \\ 0 & 0 & 0 \\ 0 & 0 & z' \end{bmatrix}$$

$$= \begin{bmatrix} 0 & 0 & guz' \\ 0 & 0 & 0 \\ 0 & 0 & 0 \end{bmatrix} \neq O,$$

which is not our case.

Thus, with $u \neq 0$, $y' = 0$, $z' \neq 0$, $w = 0$, with $vz' + uw' = 0$, we must have $v \neq 0$. Now, let $u \neq 0$, $y' = 0$, $z' \neq 0$, $w = 0$, with $vz' + uw' = 0$, $v \neq 0$. If $w' = 0$, then

$$AB = \begin{bmatrix} 0 & u & v \\ 0 & 0 & 0 \\ 0 & 0 & 0 \end{bmatrix} \begin{bmatrix} x' & u' & v' \\ 0 & 0 & 0 \\ 0 & 0 & z' \end{bmatrix} = \begin{bmatrix} 0 & 0 & vz' \\ 0 & 0 & 0 \\ 0 & 0 & 0 \end{bmatrix} \neq O,$$

which is not our case.

Therefore, for $u \neq 0$, $y' = 0$, $z' \neq 0$, $w = 0$, with $vz' + uw' = 0$, $v \neq 0$, we only have, $w' \neq 0$.

Thus, $A = \begin{bmatrix} 0 & u & v \\ 0 & 0 & 0 \\ 0 & 0 & 0 \end{bmatrix}$ and $B = \begin{bmatrix} x' & u' & v' \\ 0 & 0 & w' \\ 0 & 0 & z' \end{bmatrix}$, where $u$, $v$, $z'$, $w' \in R^*$, with $vz' + uw' = 0$ and $guz' + nvz' + muw' = 0$, which cannot be true for all $R$. As $x + y = 0$, $ax + by = c$ has unique solution $x = vz' = \dfrac{c}{a-b}$ and

$$y = uw' = -\frac{c}{a-b}, \text{ since } \begin{bmatrix} 1 & 1 & 0 \\ a & b & c \end{bmatrix}, \text{ has row echelon form } \begin{bmatrix} 1 & 0 & \dfrac{c}{a-b} \\ 0 & 1 & -\dfrac{c}{a-b} \end{bmatrix}.$$

**Case (2):** Let $x \neq 0$ and $y = z = 0$. From equation (1) $x' = 0$. Then for

$$AB = \begin{bmatrix} x & u & v \\ 0 & 0 & w \\ 0 & 0 & 0 \end{bmatrix} \begin{bmatrix} 0 & u' & v' \\ 0 & y' & w' \\ 0 & 0 & z' \end{bmatrix}$$

$$= \begin{bmatrix} 0 & uy' + xu' & vz' + uw' + xv' \\ 0 & 0 & wz' \\ 0 & 0 & 0 \end{bmatrix},$$

where $x \neq 0$.
We have the following subcases:

**Subcase((2(i)):** If $w \neq 0$, then $z' = 0$. Thus, we have

$$A = \begin{bmatrix} x & u & v \\ 0 & 0 & w \\ 0 & 0 & 0 \end{bmatrix} \text{ and } B = \begin{bmatrix} 0 & u' & v' \\ 0 & y' & w' \\ 0 & 0 & 0 \end{bmatrix}$$

where $x,\ w \neq 0$, and

$$AB = \begin{bmatrix} x & u & v \\ 0 & 0 & w \\ 0 & 0 & 0 \end{bmatrix} \begin{bmatrix} 0 & u' & v' \\ 0 & y' & w' \\ 0 & 0 & 0 \end{bmatrix}$$

$$= \begin{bmatrix} 0 & uy' + xu' & uw' + xv' \\ 0 & 0 & 0 \\ 0 & 0 & 0 \end{bmatrix}.$$

Now consider the following subcases:

**Subcase (2(i-a)):** If $u = 0$, then

$$AB = \begin{bmatrix} x & 0 & v \\ 0 & 0 & w \\ 0 & 0 & 0 \end{bmatrix} \begin{bmatrix} 0 & u' & v' \\ 0 & y' & w' \\ 0 & 0 & 0 \end{bmatrix}$$

$$= \begin{bmatrix} 0 & xu' & xv' \\ 0 & 0 & 0 \\ 0 & 0 & 0 \end{bmatrix} = O,$$

which implies that $u' = v' = 0$, but

$$ARB = \begin{bmatrix} x & 0 & v \\ 0 & 0 & w \\ 0 & 0 & 0 \end{bmatrix} \begin{bmatrix} l & e & f \\ 0 & m & g \\ 0 & 0 & n \end{bmatrix} \begin{bmatrix} 0 & 0 & 0 \\ 0 & y' & w' \\ 0 & 0 & 0 \end{bmatrix}$$

$$= \begin{bmatrix} 0 & xy'e & xw'e \\ 0 & 0 & 0 \\ 0 & 0 & 0 \end{bmatrix} \neq O,$$

unless $y' = w' = 0$ (i.e., $B = O$).

**Subcase (2(i-b)):** Let $u \neq 0$. Now since $x \neq 0$, $uy' + xu' = 0$ and $uw' + xv' = 0$. So, we have

$$ARB = \begin{bmatrix} x & u & v \\ 0 & 0 & w \\ 0 & 0 & 0 \end{bmatrix} \begin{bmatrix} l & e & f \\ 0 & m & g \\ 0 & 0 & n \end{bmatrix} \begin{bmatrix} 0 & u' & v' \\ 0 & y' & w' \\ 0 & 0 & 0 \end{bmatrix}$$

$$= \begin{bmatrix} 0 & y'(xe+mu) + lxu' & w'(xe+mu) + lxv' \\ 0 & 0 & 0 \\ 0 & 0 & 0 \end{bmatrix} \neq O,$$

for every element in $R$. For, let $X = uy'$, $Y = xu'$, $Z = uw'$, $W = xv'$, $p = -xye'$, $q = -xw'e$. The equations $uy' + xu' = 0$, $uw' + xv' = 0$, $muy' + lxu' = -xy'e$, and $muw' + lxv' = -xw'e$, becomes $X + Y = 0$, $Z + W = 0$, $mX + lY = p$, $mZ + lY = q$. Its coefficient matrix

$$\begin{bmatrix} 1 & 1 & 0 & 0 & 0 \\ 0 & 0 & 1 & 1 & 0 \\ m & l & 0 & 0 & p \\ 0 & 0 & m & l & q \end{bmatrix}, \text{ has the reduced row echelon form: } \begin{bmatrix} 1 & 0 & 0 & 0 & -\dfrac{p}{l-m} \\ 0 & 1 & 0 & 0 & \dfrac{p}{l-m} \\ 0 & 0 & 1 & 0 & -\dfrac{q}{l-m} \\ 0 & 0 & 0 & 1 & \dfrac{q}{l-m} \end{bmatrix}.$$

**Subcase (2(ii)):** Let $w = 0$, then

$$AB = \begin{bmatrix} x & u & v \\ 0 & 0 & 0 \\ 0 & 0 & 0 \end{bmatrix} \begin{bmatrix} 0 & u' & v' \\ 0 & y' & w' \\ 0 & 0 & z' \end{bmatrix}$$

$$= \begin{bmatrix} 0 & uy' + xu' & vz' + uw' + xv' \\ 0 & 0 & 0 \\ 0 & 0 & 0 \end{bmatrix}$$

**Subcase (2(ii-a)):** If $u = 0$, then $u' = 0$ as $x \neq 0$. Now for

$$AB = \begin{bmatrix} x & 0 & v \\ 0 & 0 & 0 \\ 0 & 0 & 0 \end{bmatrix} \begin{bmatrix} 0 & 0 & v' \\ 0 & y' & w' \\ 0 & 0 & z' \end{bmatrix}$$

$$= \begin{bmatrix} 0 & 0 & vz' + xv' \\ 0 & 0 & 0 \\ 0 & 0 & 0 \end{bmatrix} = O$$

and

$$ARB = \begin{bmatrix} x & 0 & v \\ 0 & 0 & 0 \\ 0 & 0 & 0 \end{bmatrix} \begin{bmatrix} l & e & f \\ 0 & m & g \\ 0 & 0 & n \end{bmatrix} \begin{bmatrix} 0 & 0 & v' \\ 0 & y' & w' \\ 0 & 0 & z' \end{bmatrix}$$

$$= \begin{bmatrix} 0 & xy'e & z'(fx + nv) + xw'e + lxv' \\ 0 & 0 & 0 \\ 0 & 0 & 0 \end{bmatrix} = O,$$

we should have $y' = 0$. Now

$$AB = \begin{bmatrix} x & 0 & v \\ 0 & 0 & 0 \\ 0 & 0 & 0 \end{bmatrix} \begin{bmatrix} 0 & 0 & v' \\ 0 & 0 & w' \\ 0 & 0 & z' \end{bmatrix} = \begin{bmatrix} 0 & 0 & vz' + xv' \\ 0 & 0 & 0 \\ 0 & 0 & 0 \end{bmatrix}$$

and

$$ARB = \begin{bmatrix} x & 0 & v \\ 0 & 0 & 0 \\ 0 & 0 & 0 \end{bmatrix} \begin{bmatrix} l & e & f \\ 0 & m & g \\ 0 & 0 & n \end{bmatrix} \begin{bmatrix} 0 & 0 & v' \\ 0 & 0 & w' \\ 0 & 0 & z' \end{bmatrix}$$

$$= \begin{bmatrix} 0 & 0 & z'(fx + nv) + xw'e + lxv' \\ 0 & 0 & 0 \\ 0 & 0 & 0 \end{bmatrix}$$

**Subcase (2(ii-a1)):** Let $x \neq 0$, $w = 0$, $u = 0$, $u' = 0$, $y' = 0$ and $v = 0$. Then, from (1.1), $v' = 0$. Thus, $AB = O$ and

$$ARB = \begin{bmatrix} x & 0 & 0 \\ 0 & 0 & 0 \\ 0 & 0 & 0 \end{bmatrix} \begin{bmatrix} l & e & f \\ 0 & m & g \\ 0 & 0 & n \end{bmatrix} \begin{bmatrix} 0 & 0 & 0 \\ 0 & 0 & w' \\ 0 & 0 & z' \end{bmatrix}$$

$$= \begin{bmatrix} 0 & 0 & xw'e + fxz' \\ 0 & 0 & 0 \\ 0 & 0 & 0 \end{bmatrix} \neq O$$

for every element of $R$, as $w'e + fz' \neq 0$ for $B \neq O$ and every $e, f \in F$.

**Subcase (2(ii-a2)):** Let $x \neq 0$, $w = 0$, $u = 0, u' = 0$, $y' = 0$ and $v \neq 0$. Then

$$
AB = \begin{bmatrix} x & 0 & v \\ 0 & 0 & 0 \\ 0 & 0 & 0 \end{bmatrix} \begin{bmatrix} 0 & 0 & v' \\ 0 & 0 & w' \\ 0 & 0 & z' \end{bmatrix} = \begin{bmatrix} 0 & 0 & vz' + xv' \\ 0 & 0 & 0 \\ 0 & 0 & 0 \end{bmatrix}
$$

and

$$
ARB = \begin{bmatrix} x & 0 & v \\ 0 & 0 & 0 \\ 0 & 0 & 0 \end{bmatrix} \begin{bmatrix} l & e & f \\ 0 & m & g \\ 0 & 0 & n \end{bmatrix} \begin{bmatrix} 0 & 0 & v' \\ 0 & 0 & w' \\ 0 & 0 & z' \end{bmatrix}
$$
$$
= \begin{bmatrix} 0 & 0 & z'\,(fx + nv) + xw'e + lxv' \\ 0 & 0 & 0 \\ 0 & 0 & 0 \end{bmatrix} \neq O,
$$

since the equations

$$ vz' + xv' = 0,\ n\,(vz') + (l)\,(xv') = -x\,(w'e + fz') $$

lead to the linear system $X + Y = 0$, $nX + lY = -x\,(w'e + fz')$, $n, l, e, f \in F$.

**Subcase(2(ii-b)):** Let $x \neq 0$, $w = 0$, $u \neq 0$, then

$$
AB = \begin{bmatrix} x & u & v \\ 0 & 0 & 0 \\ 0 & 0 & 0 \end{bmatrix} \begin{bmatrix} 0 & u' & v' \\ 0 & y' & w' \\ 0 & 0 & z' \end{bmatrix} = \begin{bmatrix} 0 & uy' + xu' & vz' + uw' + xv' \\ 0 & 0 & 0 \\ 0 & 0 & 0 \end{bmatrix}
$$

and

$$
ARB = \begin{bmatrix} x & u & v \\ 0 & 0 & 0 \\ 0 & 0 & 0 \end{bmatrix} \begin{bmatrix} l & e & f \\ 0 & m & g \\ 0 & 0 & n \end{bmatrix} \begin{bmatrix} 0 & u' & v' \\ 0 & y' & w' \\ 0 & 0 & z' \end{bmatrix}
$$
$$
= \begin{bmatrix} 0 & y'\,(xe + mu) + lxu' & w'\,(xe + mu) + z'\,(gu + fx + nv) + lxv' \\ 0 & 0 & 0 \\ 0 & 0 & 0 \end{bmatrix},
$$

where $(uy') + (xu') = 0$, $(uw') + (xv') + (vz') = 0$, $m\,(uy') + l\,(xu') + e\,(xy') = 0$, $m\,(uw') + l\,(xv') + n\,(vz') + e\,(xw') = -\,(gu + fx)\,z'$.
The equations $(uy') + (xu') = 0$ and $m\,(uy') + l\,(xu') + e\,(xy') = 0$, lead to the linear system $X + Y = 0$ and $mX + lY = -e\,(xy')$, which does not have a trivial solution for each $m, n, l \in F$, and hence $AB = O$ and $ARB = O$ is

not satisfied for each element in $R$.

**Case (3):** Let $y \neq 0$ and $x = z = 0$, then by equation (1), $y' = 0$. Now for

$$
AB = \begin{bmatrix} 0 & u & v \\ 0 & y & w \\ 0 & 0 & 0 \end{bmatrix} \begin{bmatrix} x' & u' & v' \\ 0 & 0 & w' \\ 0 & 0 & z' \end{bmatrix}
$$

$$
= \begin{bmatrix} 0 & 0 & vz' + uw' \\ 0 & 0 & wz' + yw' \\ 0 & 0 & 0 \end{bmatrix} = O,
$$

where $y \neq 0$, and

$$
ARB = \begin{bmatrix} 0 & u & v \\ 0 & y & w \\ 0 & 0 & 0 \end{bmatrix} \begin{bmatrix} l & e & f \\ 0 & m & g \\ 0 & 0 & n \end{bmatrix} \begin{bmatrix} x' & u' & v' \\ 0 & 0 & w' \\ 0 & 0 & z' \end{bmatrix}
$$

$$
= \begin{bmatrix} 0 & 0 & z'(gu + nv) + muw' \\ 0 & 0 & z'(gy + nw) + myw' \\ 0 & 0 & 0 \end{bmatrix} = O,
$$

we have $vz' + uw' = 0$, $wz' + yw = 0$, $guz' + n(vz') + m(uw') = 0$, $gyz' + nwz' + myw' = 0$.
The equations $vz' + uw' = 0$ and $n(vz') + m(uw') = -guz'$, lead to the linear system $X + Y = 0$ and $nX + mY = -guz'$, which does not have a trivial solution for each $m, n, g \in F$, and hence $AB = O$ and $ARB = O$ is not satisfied for each element in $R$.

**Case (4):** Let $z \neq 0$ and $x = y = 0$. From equation (1), $z' = 0$. Then for

$$
AB = \begin{bmatrix} 0 & u & v \\ 0 & 0 & w \\ 0 & 0 & z \end{bmatrix} \begin{bmatrix} x' & u' & v' \\ 0 & y' & w' \\ 0 & 0 & 0 \end{bmatrix} = \begin{bmatrix} 0 & uy' & uw' \\ 0 & 0 & 0 \\ 0 & 0 & 0 \end{bmatrix},
$$

where $z \neq 0$, we have the following subcases:

**Subcase (4(i)):** Let $u = 0$, so that

$$
A = \begin{bmatrix} 0 & 0 & v \\ 0 & 0 & w \\ 0 & 0 & z \end{bmatrix} \text{ and } B = \begin{bmatrix} x' & u' & v' \\ 0 & y' & w' \\ 0 & 0 & 0 \end{bmatrix}.
$$

Define $\sigma : R \to R$ by

$$\sigma\left(\begin{bmatrix} l & e & f \\ 0 & m & g \\ 0 & 0 & n \end{bmatrix}\right) = \begin{bmatrix} n & 0 & 0 \\ 0 & n & 0 \\ 0 & 0 & n \end{bmatrix}.$$

Then

$$AB = \begin{bmatrix} 0 & 0 & v \\ 0 & 0 & w \\ 0 & 0 & z \end{bmatrix} \begin{bmatrix} x' & u' & v' \\ 0 & y' & w' \\ 0 & 0 & 0 \end{bmatrix} = O,$$

$$ARB = \begin{bmatrix} 0 & 0 & v \\ 0 & 0 & w \\ 0 & 0 & z \end{bmatrix} \begin{bmatrix} l & e & f \\ 0 & m & g \\ 0 & 0 & n \end{bmatrix} \begin{bmatrix} x' & u' & v' \\ 0 & y' & w' \\ 0 & 0 & 0 \end{bmatrix} = O,$$

and $AR\sigma\left(B\right) = O$ but $\sigma\left(A\right)RB \neq O$.

**Subcase (4(ii)):** Let $u \neq 0$, then $y' = w' = z' = 0$. So,

$$A = \begin{bmatrix} 0 & u & v \\ 0 & 0 & w \\ 0 & 0 & z \end{bmatrix} \text{ and } B = \begin{bmatrix} x' & u' & v' \\ 0 & 0 & 0 \\ 0 & 0 & 0 \end{bmatrix}$$

Define $\sigma : R \to R$ by

$$\sigma\left(\begin{bmatrix} l & e & f \\ 0 & m & g \\ 0 & 0 & n \end{bmatrix}\right) = \begin{bmatrix} n & 0 & 0 \\ 0 & n & 0 \\ 0 & 0 & n \end{bmatrix},$$

we have $AB = O$, $ARB = O$, $\sigma\left(A\right)RB \neq O$ but $AR\sigma\left(B\right) = O$.
From above discussion we conclude that a left $\sigma\text{-}(S, 1)$ ring need not be a right $\sigma\text{-}(S, 1)$ ring, and vice versa. We put the above briefly as follows:

**Example 3.1.** Let $R = \left\{ \begin{bmatrix} l & e & f \\ 0 & m & g \\ 0 & 0 & n \end{bmatrix} : e, f, g, l, m, n \in F \right\}$, where $F$ is a

field. Define $\sigma : R \to R$ by $\sigma\left(\begin{bmatrix} l & e & f \\ 0 & m & g \\ 0 & 0 & n \end{bmatrix}\right) = \begin{bmatrix} \alpha_1 & 0 & 0 \\ 0 & \alpha_2 & 0 \\ 0 & 0 & \alpha_3 \end{bmatrix}$, $\alpha_i \in$

$\{0, l, m, n\}$; $i = 1, 2, 3$. Then, $\sigma$ is an endomorphism on $R$.

Let $A = \begin{bmatrix} x & u & v \\ 0 & y & w \\ 0 & 0 & z \end{bmatrix}$, $B = \begin{bmatrix} x' & u' & v' \\ 0 & y' & w' \\ 0 & 0 & z' \end{bmatrix} \in R$ satisfying $AB = ARB = O$ and either, but not both, $AR\sigma (B) = O$ or $\sigma (A) RB = O$.

Clearly, $A$ and $B$ are nonzero elements of $R$ and cannot both be of trivial trace otherwise, $\sigma (A) RB = O = AR\sigma (B)$. We have the following cases:

(1)

$$A = \begin{bmatrix} 0 & 0 & v \\ 0 & 0 & w \\ 0 & 0 & 0 \end{bmatrix} \text{ and } B = \begin{bmatrix} x' & u' & v' \\ 0 & y' & w' \\ 0 & 0 & 0 \end{bmatrix},$$

where $x'^2 + y'^2 \neq 0$ and $\sigma$ is defined as

$$\sigma \left( \begin{bmatrix} l & e & f \\ 0 & m & g \\ 0 & 0 & n \end{bmatrix} \right) = lI_3, \text{ when } x' \neq 0$$

and

$$\sigma \left( \begin{bmatrix} l & e & f \\ 0 & m & g \\ 0 & 0 & n \end{bmatrix} \right) = mI_3, \text{ when } x' = 0.$$

Then $\sigma(A)RB = 0$, but $AR\sigma(B) \neq 0$.

(2)

$$A = \begin{bmatrix} 0 & u & v \\ 0 & 0 & w \\ 0 & 0 & 0 \end{bmatrix} \text{ and } B = \begin{bmatrix} x' & u' & v' \\ 0 & 0 & 0 \\ 0 & 0 & 0 \end{bmatrix},$$

where $u \neq 0$, and $\sigma$ is defined as

$$\sigma \left( \begin{bmatrix} l & e & f \\ 0 & m & g \\ 0 & 0 & n \end{bmatrix} \right) = lI_3.$$

Then $\sigma(A)RB = 0$, but $AR\sigma(B) \neq 0$.

(3)

$$A = \begin{bmatrix} 0 & 0 & v \\ 0 & 0 & w \\ 0 & 0 & z \end{bmatrix} \text{ and } B = \begin{bmatrix} x' & u' & v' \\ 0 & y' & w' \\ 0 & 0 & 0 \end{bmatrix}$$

and $\sigma$ is defined as

$$\sigma\left(\begin{bmatrix} l & e & f \\ 0 & m & g \\ 0 & 0 & n \end{bmatrix}\right) = \begin{bmatrix} n & 0 & 0 \\ 0 & n & 0 \\ 0 & 0 & n \end{bmatrix}.$$

Then $\sigma(A)RB \neq O$, but $AR\sigma(B) = O$.

(4)

$$A = \begin{bmatrix} 0 & u & v \\ 0 & 0 & w \\ 0 & 0 & z \end{bmatrix} \text{ and } B = \begin{bmatrix} x' & u' & v' \\ 0 & 0 & 0 \\ 0 & 0 & 0 \end{bmatrix},$$

where $u \neq 0$ and $\sigma$ is defined as

$$\sigma\left(\begin{bmatrix} l & e & f \\ 0 & m & g \\ 0 & 0 & n \end{bmatrix}\right) = \begin{bmatrix} n & 0 & 0 \\ 0 & n & 0 \\ 0 & 0 & n \end{bmatrix}.$$

Then $\sigma(A)RB \neq O$, but $AR\sigma(B) = O$.

From above discussion, we conclude that there exist a ring with an endomorphism $\sigma$ which is left $\sigma - (S, 1)$ ring, but not a right $\sigma - (S, 1)$ ring, and vice versa.

The class of rings introduced, namely left/right $\sigma$-$(S, 1)$ rings needs further investigation towards invariance of ideals (particularly prime ideals) of these rings under endomorphism $\sigma$. That would lead to study of extension rings over these rings.

## References

1. Kwak, T. K. *Prime radicals of skew-polynomial rings*, Int. J. Math. Sci., Vol. 2(2) (2003), 219-227.
2. Kim, N. K., and Lee, Y., *On right quasi duo-rings which are II-regular*, Bull. Korean Math. Soc., Vol. 37(2) (2000), 217-227.
3. Shin, G. Y. *Prime ideals and sheaf representation of a pseudo symmetric ring*, Trans. Amer. Math. Soc., Vol. 184 (1973), 43-60.

# Embedding of post-Lie algebras into postassociative algebras

Vsevolod Gubarev

*University of Vienna*
*Oskar-Morgenstern-Platz 1, 1090 Vienna, Austria*
*Sobolev Institute of mathematics*
*Acad. Koptyug ave. 4, 630090 Novosibirsk, Russia*

*vsevolod.gubarev@univie.ac.at*

In honour of 80th anniversary of Professor Leonid Arkad'evich Bokut'

Applying Gröbner—Shirshov technique, we prove that any post-Lie algebra injectively embeds into its universal enveloping postassociative algebra.

*2010 Mathematics subject classification:* 16W99, 17B99.

*Keywords:* Rota—Baxter algebra, postassociative algebra, post-Lie algebra, Gröbner—Shirshov bases.

## 1. Introduction

Linear operator $R$ defined on an algebra $A$ over a field $\Bbbk$ is called a Rota—Baxter operator (RB-operator, for short) of a weight $\lambda \in \Bbbk$ if it satisfies the relation

$$R(x)R(y) = R(R(x)y + xR(y) + \lambda xy), \quad x, y \in A. \tag{1}$$

In this case, an algebra $A$ is called Rota—Baxter algebra (RB-algebra).

G. Baxter defined the notion of what is now called Rota—Baxter operator on a (commutative) algebra in 1960[4], solving an analytic problem. The relation (1) with $\lambda = 0$ appeared as a generalization of integration by parts formula. G.-C. Rota[34], P. Cartier[7] and others studied different combinatorial properties of RB-operators and RB-algebras. In 1980s, the deep connection between Lie RB-algebras and Yang—Baxter equation was found[5,35].

There are different constructions of the free commutative RB-algebra[7,24,34] generated by a given set. In 2008, K. Ebrahimi-Fard and L. Guo constructed the free associative RB-algebra[12]. In 2010, L.A. Bokut,

Yu. Chen and X. Deng[6] found a linear basis of the free associative RB-algebra by means of Gröbner—Shirshov technique. Linear bases of the free Lie RB-algebra were found in Ref. 16,22,33.

Pre-Lie algebras were introduced independently by E.B. Vinberg[37] in 1960 and M. Gerstenhaber[13] in 1963. Pre-Lie algebras also known as left-symmetric algebras satisfy the identity $(x_1x_2)x_3 - x_1(x_2x_3) = (x_2x_1)x_3 - x_2(x_1x_3)$.

In 2001, J.-L. Loday[28] defined the notion of a dendriform dialgebra (pre-associative algebra) as a vector space endowed with two bilinear operations $\succ, \prec$ satisfying

$$(x_1 \succ x_2 + x_1 \prec x_2) \succ x_3 = x_1 \succ (x_2 \succ x_3),$$
$$(x_1 \succ x_2) \prec x_3 = x_1 \succ (x_2 \prec x_3),$$
$$x_1 \prec (x_2 \succ x_3 + x_2 \prec x_3) = (x_1 \prec x_2) \prec x_3.$$

In 1995, J.-L. Loday[27] also defined Zinbiel algebra (precommutative algebra), on which the identity $(x_1 \succ x_2 + x_2 \succ x_1) \succ x_3 = x_1 \succ (x_2 \succ x_3)$ holds. A preassociative algebra with the identity $x \succ y = y \prec x$ is a precommutative algebra and under the product $x \cdot y = x \succ y - y \prec x$ is a pre-Lie algebra.

In 2004, dendriform trialgebra (postassociative algebra) was introduced[31], i.e., an algebra with bilinear operations $\prec, \succ, \cdot$ satisfying seven certain axioms. A space $A$ with two bilinear operations $[,]$ and $\cdot$ is called a post-Lie algebra (B. Vallette, 2007[36]) if $[,]$ is a Lie bracket and the next identities hold

$$(x\cdot y)\cdot z - x\cdot(y\cdot z) - (y\cdot x)\cdot z + y\cdot(x\cdot z) = [y,x]\cdot z, \quad x\cdot[y,z] = [x\cdot y, z] + [y, x\cdot z].$$

Let us explain the choice of terminology. Given a binary quadratic operad $\mathcal{P}$, the defining identities for pre- and post-$\mathcal{P}$-algebras were found in Ref. 2. One can define the operad of pre- and post-$\mathcal{P}$-algebras as $\mathcal{P} \bullet \text{PreLie}$ and $\mathcal{P} \bullet \text{PostLie}$ respectively. Here PreLie and PostLie denote the operads of pre-Lie algebras and post-Lie algebras respectively, $\bullet$ denotes the black Manin product of operads[14]. By pre- or postalgebra we will mean pre- or post-$\mathcal{P}$-algebra for some operad $\mathcal{P}$.

In 2000, M. Aguiar[1] stated that an associative algebra with a given Rota—Baxter operator $R$ of weight zero under the operations $a \succ b = R(a)b$, $a \prec b = aR(b)$ is a preassociative algebra. In 2002, K. Ebrahimi-Fard[11] showed that an associative RB-algebra of nonzero weight $\lambda$ under the same two products $\succ, \prec$ and the third operation $a \cdot b = \lambda ab$ is a postassociative algebra. The analogue of the Aguiar construction for the pair of pre-Lie algebras and Lie RB-algebras of weight zero was stated in 2000 by M. Aguiar[1] and by I.Z. Golubchik, V.V. Sokolov[15]. In 2010[3], this construction for the pair of post-Lie algebras and Lie RB-algebras of nonzero weight was extended.

In 2013[2], the construction of M. Aguiar and K. Ebrahimi-Fard was generalized for the case of arbitrary variety.

In 2008, the notion of universal enveloping RB-algebras of pre- and postassociative algebras was introduced[12]. In Ref. 12, it was also proved that the universal enveloping of a free pre- or postassociative algebra is free.

In 2010, with the help of Gröbner—Shirshov bases Ref. 6, Yu. Chen and Q. Mo proved that every preassociative algebra over a field of characteristic zero injectively embeds into its universal enveloping RB-algebra[8].

In 2013[20], given a variety Var, it was proved that every pre-Var-algebra (post-Var-algebra) injectively embeds into its universal enveloping Var-RB-algebra of weight $\lambda = 0$ ($\lambda \neq 0$). Based on the last result, we formulate

**Problem 1.1.** Construct the universal enveloping RB-algebra of a pre- or postalgebra.

In Ref. 23, L. Guo actually stated the following problem for Var = As:

**Problem 1.2.** Clarify if the pairs of varieties (RBVar, preVar) and (RB$_\lambda$Var, postVar) for $\lambda \neq 0$ are Poincaré—Birkhoff—Witt (PBW-) pairs in the sense of Ref. 32.

Here RBVar (RB$_\lambda$Var) denotes the variety of Var-algebras endowed with an RB-operator of (non)zero weight $\lambda$.

**Problem 1.3** (Guo et al[25]). Given a variety Var of algebras, whether the variety of RB-Var-algebras is Schreier, i.e., whether every subalgebra of the free algebra is free itself?

**Problem 1.4.** a) Prove that every pre-Lie (post-Lie) algebra injectively embeds into its universal enveloping preassociative (postassociative) algebra.

b) Construct the universal enveloping preassociative (postassociative) algebra for given pre-Lie (post-Lie) algebra.

The answer on Problem 1.4b will cover Problem 1.4a. For pre-Lie algebras, Problem 1.4b and Problem 1.2 were stated in Ref. 26. The discussion of Problem 1.4 in the case of restricted pre-Lie algebras can be found in Ref. 9. The analogues of Problem 1.4 for Koszul-dual objects, di- and trialgebras, were solved in Ref. 21,30.

Problems 1.1–1.3 were solved by author in commutative[17], associative[18], and Lie[19] cases. So, the question of L. Guo[23] is completely answered.

The main result of the current work is the affirmative answer on Problem 1.4a in postalgebra case. In April 2018, the proof that the pair of varieties of pre-Lie and preassociative algebras is a PBW-pair was announced [10].

## 2. Postalgebras

A *postassociative algebra* is a linear space with three bilinear operations $\cdot$, $\succ$, $\prec$ satisfying seven identities:

$$(x \prec y) \prec z = x \prec (y \succ z + y \prec z + y \cdot z), \quad (x \succ y) \prec z = x \succ (y \prec z),$$
$$(x \succ y + y \succ x + x \cdot y) \succ z = x \succ (y \succ z),$$
$$x \succ (y \cdot z) = (x \succ y) \cdot z, \quad (x \prec y) \cdot z = x \cdot (y \succ z),$$
$$(x \cdot y) \prec z = x \cdot (y \prec z), \quad (x \cdot y) \cdot z = x \cdot (y \cdot z). \tag{2}$$

A *post-Lie algebra* is a vector space endowed with two bilinear products $[,]$ and $\cdot$, the bracket $[,]$ is Lie, and the following identities are fulfilled:

$$(x \cdot y) \cdot z - x \cdot (y \cdot z) - (y \cdot x) \cdot z + y \cdot (x \cdot z) = [y, x] \cdot z, \quad x \cdot [y, z] = [x \cdot y, z] + [y, x \cdot z].$$

## 3. Embedding of pre- and postalgebras into RB-algebras

**Theorem 3.1**[1–3,11,15,29]. Let $A$ be an RB-algebra of a variety Var and weight $\lambda = 0$ ($\lambda \neq 0$). With respect to the operations

$$x \succ y = R(x)y, \quad x \prec y = xR(y) \quad (x \cdot y = \lambda xy) \tag{3}$$

$A$ is a pre-Var-algebra (post-Var-algebra).

Denote the pre- and post-Var-algebra obtained in Theorem 3.1 as $A_\lambda^{(R)}$.

Given a pre-Var-algebra $\langle C, \succ, \prec \rangle$, universal enveloping RB-Var-algebra $U$ of $C$ is the universal algebra in the class of all RB-Var-algebras of weight zero such that there exists homomorphism from $C$ to $U_0^{(R)}$. Analogously universal enveloping RB-Var-algebra of a post-Var-algebra is defined.

**Theorem 3.2**[20]. Every pre-Var-algebra (post-Var-algebra) could be embedded into its universal enveloping RB-algebra of the variety Var and weight $\lambda = 0$ ($\lambda \neq 0$).

Let us consider the idea of the proof of Theorem 3.2 in the postalgebra case. Suppose $\langle A, \succ, \prec, \cdot \rangle$ is a post-Var-algebra. Then the direct sum of two isomorphic copies of $A$, the space $\hat{A} = A \oplus A'$, endowed with a binary operation

$$a * b = a \succ b + a \prec b + a \cdot b, \quad a * b' = (a \succ b)',$$
$$a' * b = (a \prec b)', \quad a' * b' = (\lambda a \cdot b)' \tag{4}$$

for $a, b \in A$, is proved to be an algebra of the variety Var. Moreover, the map $R(a') = \lambda a$, $R(a) = -\lambda a$ is an RB-operator of weight $\lambda$ on $\hat{A}$. The injective embedding of $A$ into $\hat{A}$ is given by $a \mapsto a'$, $a \in A$. However, $\hat{A}$ is not a universal enveloping RB-algebra of $A$.

## 4. Embedding of post-Lie algebras into postassociative algebras

Let $RAs\langle X \rangle$ denote the free associative algebra generated by a set $X$ with a linear map $R$ in the signature. One can construct a linear basis of $RAs\langle X \rangle$ (see, e.g., Ref. 25) by induction. First, all elements from $S(X)$ lie in the basis. Next, if we have basic elements $a_1, a_2, \ldots, a_k$, $k \geq 1$, then the word $w_1 R(a_1) w_2 \ldots w_k R(a_k) w_{k+1}$ lies in the basis of $RAs\langle X \rangle$. Here $w_1, \ldots, w_{k+1} \in S(X) \cup \emptyset$. Let us denote the basis obtained as $RS(X)$. Given a word $u$ from $RS(X)$, the number of appeareances of the symbol $R$ in $u$ is denoted by $\deg_R(u)$, $R$-degree of $u$. We call an element from $RS(X)$ of the form $R(w)$ as $R$-letter. By $X_\infty$ we denote the union of $X$ and the set of all $R$-letters. Given $u \in RS(X)$, define $\deg u$ (degree of $u$) as the length of $u$ in the alphabet $X_\infty$.

Suppose that $X$ is a well-ordered set with respect to $<$. Let us introduce by induction the deg-lex order on $S(X)$. Firstly, we compare two words $u$ and $v$ by the length: $u < v$ if $|u| < |v|$. Secondly, when $|u| = |v|$, $u = x_i u'$, $v = x_j v'$, $x_i, x_j \in X$, we have $u < v$ if either $x_i < x_j$ or $x_i = x_j$, $u' < v'$. We compare two words $u$ and $v$ from $RS(X)$ by $R$-degree: $u < v$ if $\deg_R(u) < \deg_R(v)$. If $\deg_R(u) = \deg_R(v)$, we compare $u$ and $v$ in deg-lex order as words in the alphabet $X_\infty$. Here we define each $x$ from $X$ to be less than all $R$-letters and $R(a) < R(b)$ if and only if $a < b$.

Let $*$ be a symbol not containing in $X$. By a $*$-bracketed word on $X$, we mean a basic word from $RAs\langle X \cup \{*\} \rangle$ with exactly one occurrence of $*$. The set of all $*$-bracketed words on $X$ is denoted by $RS^*(X)$. For $q \in RS^*(X)$ and $u \in RAs\langle X \rangle$, we define $q|_u$ as $q|_{* \to u}$ to be the bracketed word obtained by replacing the letter $*$ in $q$ by $u$.

The order defined above is monomial, i.e., from $u < v$ it follows that $q|_u < q|_v$ for all $u, v \in RS(X)$ and $q \in RS^*(X)$.

Given $f \in RAs\langle X \rangle$, by $\bar{f}$ we mean the leading word in $f$.

**Definition 4.1** [25]. Let $f, g \in RAs\langle X \rangle$. If there exist $\mu, \nu, w \in RS(X)$ such that $w = \bar{f}\mu = \nu\bar{g}$ with $\deg w < \deg(\bar{f}) + \deg(\bar{g})$, then we define $(f, g)_w$ as $f\mu - \nu g$ and call it the *composition of intersection* of $f, g$ with respect to $(\mu, \nu)$. If there exist $q \in RS^*(X)$ and $w \in RS(X)$ such that

$w = \bar{f} = q|_{\bar{g}}$, then we define $(f, g)_w^q$ as $f - q|_g$ and call it the *composition of inclusion* of $f, g$ with respect to $q$.

**Definition 4.2** [25]. Let $S$ be a subset of monic elements from $RAs\langle X \rangle$ and $w \in RS(X)$.

(1) For $u, v \in RAs\langle X \rangle$, we call $u$ and $v$ congruent modulo $(S, w)$ and denote this by $u \equiv v \mod (S, w)$ if $u - v = \sum c_i q_i|_{s_i}$ with $c_i \in \Bbbk$, $q_i \in RS^*(X)$, $s_i \in S$ and $q_i|_{\overline{s_i}} < w$.

(2) For $f, g \in RAs\langle X \rangle$ and suitable $w, \mu, \nu$ or $q$ that give a composition of intersection $(f, g)_w$ or a composition of inclusion $(f, g)_w^q$, the composition is called trivial modulo $(S, w)$ if $(f, g)_w$ or $(f, g)_w^q \equiv 0 \mod (S, w)$.

(3) The set $S \subset RAs\langle X \rangle$ is called a *Gröbner—Shirshov basis* if, for all $f, g \in S$, all compositions of intersection $(f, g)_w$ and all compositions of inclusion $(f, g)_w^q$ are trivial modulo $(S, w)$.

**Theorem 4.3** [25]. Let $S$ be a set of monic elements in $RAs\langle X \rangle$, let $<$ be a monomial ordering on $RS(X)$ and let $Id(S)$ be the $R$-ideal of $RAs\langle X \rangle$ generated by $S$. If $S$ is a Gröbner—Shirshov basis in $RAs\langle X \rangle$, then $RAs\langle X \rangle = \Bbbk Irr(S) \oplus Id(S)$ where $Irr(S) = RS(X) \setminus \{q|_{\bar{s}} \mid q \in RS^*(X), s \in S\}$ and $Irr(S)$ is a linear basis of $RAs\langle X \rangle / Id(S)$.

Let $A$ be an associative algebra with an RB-operator $R$. Then the algebra $A^{(-)}$ is a Lie RB-algebra under the product $[x, y] = xy - yx$ and the same action of $R$. So, we state the analogue of Problem 1.4. How does the left adjoint functor from the category of Lie RB-algebras to the category of associative RB-algebras (of the same weight) look like? Is it the embedding always injective?

Let $\hat{L} = L \oplus L'$ be exactly the Lie algebra with the RB-operator $R$ of weight $-1$ constructed in the sketch of the proof of Theorem 3.2. Then $R(y_\alpha) = y_\alpha$, $R(x_\alpha) = 0$, and $X_\Lambda \cup Y_\Lambda = \{x_\alpha, \alpha \in \Lambda\} \cup \{y_\alpha, \alpha \in \Lambda\}$ is a linear basis of $\hat{L}$. Note that $\mathrm{Span}\{X\}$ and $\mathrm{Span}\{Y\}$ are Lie subalgebras of $\hat{L}$. Suppose that $\Lambda$ is well-ordered set. Extend the order to $X_\Lambda \cup Y_\Lambda$ as follows: $t_\alpha < t_\beta$, $t \in \{x, y\}$, if and only if $\alpha < \beta$ and $y_\alpha < x_\beta$ for all $\alpha, \beta \in \Lambda$.

Consider the set $S$ of the following elements in $RAs\langle X_\Lambda \cup Y_\Lambda \rangle$:

$$x_\alpha x_\beta - x_\beta x_\alpha - [x_\alpha, x_\beta], \quad y_\alpha y_\beta - y_\beta y_\alpha - [y_\alpha, y_\beta], \quad \beta < \alpha, \tag{5}$$

$$x_\alpha y_\beta - y_\beta x_\alpha - [x_\alpha, y_\beta],$$

$$R(\vec{y}_\alpha) - \vec{y}_\alpha, \tag{6}$$

$$R(\vec{x}_\alpha), \tag{7}$$

$$R(u) - u, \quad u = R(a_1)\vec{y}_{\alpha_1} R(a_2) \dots R(a_s)\vec{y}_{\alpha_s} R(a_{s+1}), \tag{8}$$

$$R(R(a)\vec{x}_\alpha R(b)) - R(R(a)\vec{x}_\alpha b + a\vec{x}_\alpha R(b) - a\vec{x}_\alpha b), \tag{9}$$

$$R(R(a)\vec{x}_\alpha) - R(a\vec{x}_\alpha), \quad R(\vec{x}_\alpha R(b)) - R(\vec{x}_\alpha b), \tag{10}$$

$$R(a)R(b) - R(R(a)b + aR(b) - ab). \tag{11}$$

Here $a, b, a_i$ are elements from $RS'(X_\Lambda \cup Y_\Lambda) = RS(X_\Lambda \cup Y_\Lambda) \setminus (S(X_\Lambda) \cup S(Y_\Lambda))$. By $\vec{y}_{\alpha_i}$ or $\vec{x}_\alpha$ we mean a monomial from $\Bbbk[Y_\Lambda]$ or $\Bbbk[X_\Lambda]$ respectively.

**Theorem 4.4.** The set $S$ is a a Gröbner—Shirshov basis in $RAs\langle X_\Lambda \cup Y_\Lambda \rangle$.

PROOF. It is known that all compositions between two elements from (5) are trivial, as it is the method to construct the classical universal enveloping algebra.

Compositions of intersection between two expressions from (11) are trivial by Ref. 25. Thus, all compositions of intersection which are not at the same time compositions of inclusion are trivial.

Let us consider a composition of inclusion between (7) and (11). Let $\vec{x}_\alpha \in C(X)$, $a \in RS'(X_\Lambda \cup Y_\Lambda)$, $w = R(a)R(\vec{x}_\alpha)$. By the following, we get that the composition of inclusion is trivial:

$$R(a)R(\vec{x}_\alpha) \overset{(7)}{\equiv} 0 \mod (S, w),$$

$$R(a)R(\vec{x}_\alpha) \overset{(11)}{\equiv} R(R(a)\vec{x}_\alpha + aR(\vec{x}_\alpha) - a\vec{x}_\alpha) \overset{(10)}{\equiv} R(aR(\vec{x}_\alpha)) \overset{(7)}{\equiv} 0 \mod (S, w).$$

A composition of inclusion between (6) and (11) is analogously trivial.

Compute a composition of inclusion between (8) and (11). For $\vec{y}_{\alpha_i} \in \Bbbk[Y]$, $w = R(b)R(R(a_1)\vec{y}_{\alpha_1} \ldots \vec{y}_{\alpha_s} R(a_{s+1}))$, $b, a_1, \ldots, a_{s+1} \in RS'(X_\Lambda \cup Y_\Lambda)$, we have

$$R(b)R(R(a_1)\vec{y}_{\alpha_1} R(a_2) \ldots R(a_s)\vec{y}_{\alpha_s} R(a_{s+1}))$$

$$\overset{(11)}{\equiv} R(R(b)R(a_1)\vec{y}_{\alpha_1} R(a_2) \ldots R(a_s)\vec{y}_{\alpha_s} R(a_{s+1}))$$
$$+ R(bR(R(a_1)\vec{y}_{\alpha_1} R(a_2) \ldots R(a_s)\vec{y}_{\alpha_s} R(a_{s+1})))$$
$$- R(bR(a_1)\vec{y}_{\alpha_1} R(a_2) \ldots R(a_s)\vec{y}_{\alpha_s} R(a_{s+1}))$$

$$\overset{(8)}{\equiv} R(R(b)R(a_1)\vec{y}_{\alpha_1} R(a_2) \ldots R(a_s)\vec{y}_{\alpha_s} R(a_{s+1}))$$

$$\overset{(11)}{\equiv} R(R(R(b)a_1 + bR(a_1) - ba_1)\vec{y}_{\alpha_1} R(a_2) \ldots R(a_s)\vec{y}_{\alpha_s} R(a_{s+1}))$$

$$\overset{(8)}{\equiv} R(R(b)a_1 + bR(a_1) - ba_1)\vec{y}_{\alpha_1} R(a_2) \ldots R(a_s)\vec{y}_{\alpha_s} R(a_{s+1}) \mod (S, w);$$

$$R(b)R(R(a_1)\vec{y}_{\alpha_1} R(a_2) \ldots R(a_s)\vec{y}_{\alpha_s} R(a_{s+1}))$$

$$\overset{(8)}{\equiv} R(b)R(a_1)\vec{y}_{\alpha_1} R(a_2) \ldots R(a_s)\vec{y}_{\alpha_s} R(a_{s+1})$$

$$\overset{(11)}{\equiv} R(R(b)a_1 + bR(a_1) - ba_1)\vec{y}_{\alpha_1} R(a_2) \dots R(a_s)\vec{y}_{\alpha_s} R(a_{s+1}) \quad \text{mod } (S, w).$$

Thus, the corresponding composition of inclusion is trivial modulo $(S, w)$.

Let us calculate a composition of inclusion between (9) and (11). Let $\vec{x}_\alpha \in C(X)$, $a, b, c \in RS'(X_\Lambda \cup Y_\Lambda)$, $w = R(a)R(R(b)\vec{x}_\alpha R(c))$. Modulo $(S, w)$ we have

$$R(a)R(R(b)\vec{x}_\alpha R(c)) \overset{(9)}{\equiv} R(a)R(R(b)\vec{x}_\alpha c + b\vec{x}_\alpha R(c) - b\vec{x}_\alpha c)$$
$$\overset{(11)}{\equiv} R(R(a)R(b)\vec{x}_\alpha c + R(a)b\vec{x}_\alpha R(c) - R(a)b\vec{x}_\alpha c)$$
$$+ R(aR(R(b)\vec{x}_\alpha c + b\vec{x}_\alpha R(c) - b\vec{x}_\alpha c))$$
$$- R(aR(b)\vec{x}_\alpha c + ab\vec{x}_\alpha R(c) - ab\vec{x}_\alpha c). \quad (12)$$

For the first summand of the RHS of (12), we obtain

$$R(R(a)R(b)\vec{x}_\alpha c) \overset{(11)}{\equiv} R(R(R(a)b + aR(b) - ab)\vec{x}_\alpha c) \quad \text{mod } (S, w). \quad (13)$$

Also, modulo $(S, w)$, we have

$$R(a)R(R(b)\vec{x}_\alpha R(c)) \overset{(11)}{\equiv} R(R(a)R(b)\vec{x}_\alpha R(c))$$
$$+ aR(R(b)\vec{x}_\alpha R(c)) - aR(b)\vec{x}_\alpha R(c)) \overset{(9)}{\equiv} R(R(a)R(b)\vec{x}_\alpha R(c))$$
$$+ R(aR(R(b)\vec{x}_\alpha c + b\vec{x}_\alpha R(c) - b\vec{x}_\alpha c)) - R(aR(b)\vec{x}_\alpha R(c)). \quad (14)$$

Further, the first summand of the RHS of (14) is congruent modulo $(S, w)$ to

$$R(R(a)R(b)\vec{x}_\alpha R(c)) \overset{(11)}{\equiv} R(R(R(a)b + aR(b) - ab)\vec{x}_\alpha R(c))$$
$$\overset{(9)}{\equiv} R(R(R(a)b\vec{x}_\alpha R(c) + aR(b)\vec{x}_\alpha R(c) - ab\vec{x}_\alpha R(c))$$
$$+ R(R(R(a)b + aR(b) - ab)\vec{x}_\alpha c)$$
$$- R(R(a)b\vec{x}_\alpha c + aR(b)\vec{x}_\alpha c - ab\vec{x}_\alpha c). \quad (15)$$

Note that substitution of (15) into (14) gives exactly the same as the substitution of (13) into (12). Compositions of inclusion between (10) and (11) are trivial by similar calculations. $\square$

**Corollary 4.5.** The quotient $A$ of $RAs\langle X_\Lambda \cup Y_\Lambda \rangle$ by $Id(S)$ is the universal enveloping associative RB-algebra of the Lie algebra $\hat{L}$ with the RB-operator $R$. Moreover, $\hat{L}$ injectively embeds into $A^{(-)}$.

PROOF. The RB-identity (1) when at least one of arguments lies in $S(X_\Lambda) \cup S(Y_\Lambda)$ follows from (5)–(7), (10) and the fact that the space $\Bbbk X_\Lambda$ is a subalgebra in $\hat{L}$. Within (11), we get an asssociative RB-algebra structure on $A$. By (5)–(7), we have that $A$ is enveloping of $\hat{L}$ for both: the Lie bracket $[,]$ and the action of $R$.

Let us prove that $A$ is the universal enveloping one. At first, $A$ is generated by $\hat{L}$. At second, all elements from $S$ are identities in the universal enveloping associative RB-algebra $U_{RB}(\hat{L})$. Indeed, (5) are enveloping conditions for the product, (11) is the RB-identity. Let us show by induction that (6) and (7) follow from the enveloping conditions for the action of $R$ on $L'$ and RB-identity. Suppose that we have proved that $R(\vec{y}_\alpha) = \vec{y}_\alpha$ and $R(\vec{y}_\beta) = \vec{y}_\beta$, then

$$\vec{y}_\alpha \vec{y}_\beta = R(\vec{y}_\alpha)R(\vec{y}_\beta) = R(R(\vec{y}_\alpha)\vec{y}_\beta + \vec{y}_\alpha R(\vec{y}_\beta) - \vec{y}_\alpha \vec{y}_\beta) = R(\vec{y}_\alpha \vec{y}_\beta). \quad (16)$$

Analogously we deduce $R(\vec{x}_\alpha) = 0$ for $\vec{x}_\alpha \in S(X)$.

The relations (10) follow from (7) and (1) immediately. Now we state (9):

$$0 = R(a)R(\vec{x}_\alpha)R(b) = R(R(a)\vec{x}_\alpha - a\vec{x}_\alpha)R(b)$$
$$= R(R(a)\vec{x}_\alpha R(b) - a\vec{x}_\alpha R(b) + R(R(a)\vec{x}_\alpha - a\vec{x}_\alpha)b - R(a)\vec{x}_\alpha b + a\vec{x}_\alpha b)$$
$$= R(R(a)\vec{x}_\alpha R(b) - a\vec{x}_\alpha R(b) - R(a)\vec{x}_\alpha b + a\vec{x}_\alpha b).$$

Finally, let us prove (8) by induction on $s$ analogously to (16):

$$R(a_1)\vec{y}_{\alpha_1} R(a_2) \ldots R(a_s)\vec{y}_{\alpha_s} R(a_{s+1})$$
$$= R(R(a_1)\vec{y}_{\alpha_1} \ldots R(a_s))R(\vec{y}_{\alpha_s} R(a_{s+1}))$$
$$= R(R(a_1)\vec{y}_{\alpha_1} R(a_2) \ldots R(a_s)\vec{y}_{\alpha_s} R(a_{s+1})).$$

Hence, the quotient of $RAs\langle X_\Lambda \cup Y_\Lambda \rangle$ by $Id(S)$ is the universal enveloping associative RB-algebra for $\hat{L}$. By Theorems 4.3 and 4.4 we get the injectivity of embedding $\hat{L}$ into $A^{(-)}$. $\quad\square$

**Remark 4.6**. Analogously to Corollary 4.5, one can find the linear basis of the universal enveloping associative RB-algebra for a Lie algebra endowed with an RB-operator $R$ of nonzero weight $\lambda$ satisfying $R^2 = -\lambda R$.

**Corollary 4.7**. A post-Lie algebra injectively embeds into its universal enveloping postassociative algebra.

PROOF. Let $L$ be a post-Lie algebra. By Theorem 3.2, $L$ can be injectively embedded into $\hat{L}^{(R)}$ with the RB-operator $R$ of weight $-1$. Then, by Corollary 4.5, we embed the Lie RB-algebra $\hat{L}$ into the associative algebra $A$ with the RB-operator $P$. Thus, the subalgebra (in postalgebra sense) $T$ in $A^{(P)}$ generated by the set $L' = \text{Span}\{x_\alpha - y_\alpha \mid \alpha \in \Lambda\}$ is an injective enveloping postassociative algebra of the initial post-Lie algebra $L$. $\quad\square$

The question whether the pair (postLie, postAs) is a PBW-pair is still open.

## Acknowledgements

The author expresses his gratitude to P. Kolesnikov for the useful comments and remarks.

This work was supported by the Austrian Science Foundation FWF grant P28079.

## References

1. M. Aguiar, Pre-Poisson algebras, *Lett. Math. Phys.* **54** (2000) 263–277.
2. C. Bai, O. Bellier, L. Guo, X. Ni, Splitting of operations, Manin products, and Rota—Baxter operators, *Int. Math. Res. Notices* **3** (2013) 485–524.
3. C. Bai, L. Guo, X. Ni, $\mathcal{O}$-operators on associative algebras, associative Yang—Baxter equations and dendriform algebras, Conference Proceedings *Quantized Algebra and Physics* (2012) 10–51.
4. G. Baxter, An analytic problem whose solution follows from a simple algebraic identity, *Pacific J. Math.* **10** (1960) 731–742.
5. A.A. Belavin, V.G. Drinfel'd, Solutions of the classical Yang-Baxter equation for simple Lie algebras, *Funct. Anal. Appl. 16* (3) (1982) 159–180.
6. L.A. Bokut, Yu. Chen, X. Deng, Gröbner-Shirshov bases for Rota-Baxter algebras, *Sib. Math. J.* **51** (6) (2010) 978–988.
7. P. Cartier, On the structure of free Baxter algebras, *Adv. Math.* **9** (1972) 253–265.
8. Yu. Chen, Q. Mo, Embedding dendriform algebra into its universal enveloping Rota—Baxter algebra, *Proc. Amer. Math. Soc.* **139** (12) (2011) 4207–4216.
9. I. Dokas, Pre-Lie algebras in positive characteristic, *J. Lie Theory* **23** (4) (2013) 937–952.
10. V. Dotsenko, P. Tamaroff, Endofunctors and Poincaré-Birkhoff-Witt theorems, arXiv:1804.06485 [math.CT].
11. K. Ebrahimi-Fard, Loday-type algebras and the Rota-Baxter relation, *Lett. Math. Phys.* **61** (2002) 139–147.
12. K. Ebrahimi-Fard, L. Guo, Rota—Baxter algebras and dendriform algebras, *J. Pure Appl. Algebra* **212** (2) (2008) 320–339.
13. M. Gerstenhaber, The cohomology structure of an associative ring, *Ann. of Math.* **78** (1963) 267–288.
14. V. Ginzburg, M. Kapranov, Koszul duality for operads, *Duke Math. J.* **76** (1) (1994) 203–272.
15. I.Z. Golubchik, V.V. Sokolov, Generalized operator Yang-Baxter equations, integrable ODEs and nonassociative algebras, *J. Nonlinear Math. Phys.* **7** (2) (2000) 184–197.

16. V.Yu. Gubarev, Free Lie Rota-Baxter algebras, *Sib. Math. J.* **57** (5) (2016) 809–818.
17. V. Gubarev, Universal enveloping commutative Rota—Baxter algebras of pre- and post-commutative algebras, *Axioms* **6** (4) (2017) 1–33.
18. V. Gubarev, Universal enveloping associative Rota—Baxter algebras of pre-associative and postassociative algebras, *J. Algebra* **516** (2018) 298–328.
19. V. Gubarev, Universal enveloping Lie Rota—Baxter algebras of pre-Lie and post-Lie algebras, *Algebra and Logic* **58** (1) (2019) 3–21.
20. V. Gubarev, P. Kolesnikov, Embedding of dendriform algebras into Rota—Baxter algebras, *Cent. Eur. J. Math.* **11** (2) (2013) 226–245.
21. V. Gubarev, P. Kolesnikov, Operads of decorated trees and their duals, *Comment. Math. Univ. Carolin.* **55** (4) (2014) 421–445.
22. V. Gubarev, P. Kolesnikov, Groebner—Shirshov basis of the universal enveloping Rota—Baxter algebra of a Lie algebra, *J. Lie Theory* **27** (3) (2017) 887–905.
23. L. Guo, *An Introduction to Rota—Baxter Algebra*, Intern. Press, Somerville, MA; Higher education press, Beijing, 2012.
24. L. Guo, W. Keigher, Baxter algebras and shuffle products, *Adv. Math.* **150** (2000) 117–149.
25. L. Guo, W. Sit, and R. Zhang, Differential type operators and Gröbner-Shirshov bases, *J. Symb. Comput.* **52** (2013) 97–123.
26. P. Kolesnikov, Gröbner—Shirshov bases for pre-associative algebras, *Commun. Algebra* **45** (12) (2017) 5283–5296.
27. J.-L. Loday, Cup-product for Leibniz cohomology and dual Leibniz algebras, *Math. Scand.* **77** (2) (1995) 189–196.
28. J.-L. Loday, *Dialgebras, Dialgebras and related operads, 1–61*, Springer-Verl., Berlin, 2001.
29. J.-L. Loday, On the algebra of quasi-shuffles, *Manuscripta Math.* **123** (2007) 79–93.
30. J.-L. Loday, T. Pirashvili, Universal enveloping algebras of Leibniz algebras and (co)homology, *Math. Ann.* **296** (1993) 139–158.
31. J.-L. Loday, M. Ronco, Trialgebras and families of polytopes, *Comtep. Math.* **346** (2004) 369–398.
32. A.A. Mikhalev, I.P. Shestakov, PBW-pairs of varieties of linear algebras, *Commun. Algebra* **42** (2) (2014) 667–687.
33. J. Qiu, Yu. Chen, Gröbner-Shirshov bases for Lie $\Omega$-algebras and free Rota-Baxter Lie algebras, *J. Algebra Appl.* **16** (2) (2017) 175–190.
34. G.-C. Rota, Baxter algebras and combinatorial identities I, *Bull. Amer. Math. Soc.* **75** (1969) 325–329.
35. M.A. Semenov-Tyan-Shanskii, What is a classical $r$-matrix? *Funct. Anal. Appl.* **17** (4) (1983) 259–272.
36. B. Vallette, Homology of generalized partition posets, *J. Pure Appl. Algebra* **208** (2) (2007) 699–725.
37. E.B. Vinberg, Homogeneous cones, *Sov. Math. Dokl.* **1** (1960) 787–790.

# Free Left GC-lpp Semigroups

Junying Guo

*College of Science and Technology, Jiangxi Normal University, Nanchang,Jiangxi 330022, P.R. China.*

*651945171@qq.com*

Xiaojiang Guo

*Department of Mathematics, Jiangxi Normal University, Nanchang,Jiangxi 330022, P.R. China.*

*xjguo@jxnu.edu.cn, xjguo1967@sohu.com*

Xiaoting He

*Department of Mathematics, Jiangxi Normal University, Nanchang, Jiangxi 330022, P.R. China.*

*446248332@qq.com*

K. P. Shum

*Institute of Mathematics, Yunnan University, Kunming, Yunnan, 650091, P.R. China.*

*kpshum@ynu.edu.cn*

The left GC-lpp semigroups are known as the common generalizations of left ample semigroups and right inverse semigroups. In this paper, we will prove that the class of left GC-lpp semigroups is a quasivarity. Furthermore, the structure of free left GC-lpp semigroups will be described and investigated. The main result in this paper is that any left GC-lpp monoid has a F-cover.

*Keywords*: Left GC-lpp semigroup; Quasivarity; Free unary semigroup.

*2010 Mathematics subject classification:* 20M05, 20M07, 20M10

## 1. Introduction

Recall that the relation $\mathcal{R}^*$ on a semigroup $S$ is defined by $a\mathcal{R}^*b$ if and only if the elements $a$ and $b$ of $S$ are related by the Green's relation $\mathcal{R}$ in some oversemigroup of $S$. As in Ref. 1, a semigroup $S$ is said to be an *left abundant* semigroup if each $\mathcal{R}^*$-class of $S$ contains at least one idempotent. Naturally, a semigroup $S$ is called *abundant* if it is both left and right abundant. It is easy to see that if $S$ is a left abundant semigroup whose set of idempotents forms a left regular sub-band, then each $\mathcal{R}^*$-class of $S$ contains precisely one idempotent. We use $a^\dagger$ to denote the typical idempotent in the $\mathcal{R}^*$-class containing $a$. Moreover, a left abundant semigroup $S$ is called *left GC-lpp* if (1) $E(S)$ (the set of idempotents of $S$) forms a left regular band; and (2) for all elements $a \in S$ and idempotent $f \in S$, we have $af = (af)^\dagger a$, where $(af)^\dagger$ is the idempotent in the $\mathcal{R}^*$-class of $S$ containing $af$. Obviously, a semigroup is left ample if and only if it is a left GC-lpp semigroup whose band of idempotents is a semilattice (see Ref. 2); and a semigroup is $\mathcal{R}$-unipotent if and only if it is a left GC-lpp semigroup which is regular (for example, see Ref. 4). Consequently, we may regard a left GC-lpp semigroup as an algebra with a multiplication and a unitary operation $\dagger$ defined by the rule that $\dagger : a \mapsto a^\dagger$. From this point of views, the class of left GC-lpp semigroup can be regarded as a quasi-variety (Proposition 2.1). Thus, by the general results of universal algebra, the free left GC-lpp semigroups exist. Hence, the following question naturally arises: what is the structure of the free left GC-lpp semigroups? Our aim of this paper is to provide an answer towards to the above question.

Let $E(S)$ be the set of all idempotents of the semigroup $S$. Then,we call a semigroup $S$ *E-unitary* if $E(S)$ is a unitary subset of $S$, that is, for all $x \in S$ and $e \in E(S)$, one of $xe$ and $ex$ in $E(S)$ implies $x \in E(S)$. If $S$ is a regular semigroup, then this concept is equivalent to requiring that $ea \in E(S)$ implies $a \in E(S)$ for all $a \in S$ and $e \in E(S)$. As motivated by the main result of Saito in Ref. 5, Around 1974, McAlister initiated the study of $E$-unitary inverse semigroups in Ref. 6 and described them, up to isomorphism, as the so-called $P$-semigroups. His results were later further extended to $E$-unitary $\mathcal{R}$-unipotent semigroups by Takizawa[7] and Szendrei[4], and generalized to the more general $E$-unitary regular semigroups by Szendrei[8] and also to the weakly $E$-unitary locally inverse semigroups by Billhart and Szendrei[9]. The most remarkable result is that the free inverse semigroups are $E$-unitary inverse semigroups whose semilattice of idempotents forms a free semilattice and in which the homomorphic image of the smallest group congruence is a free group (see Ref. 10).

We call a left abundant semigroup *left adequate* if the set of idempotents forms a semilattice. Moreover, a left adequate semigroup $S$ is called *left ample* if for all $a \in S$ and $f \in E(S)$, $af = (af)^\dagger a$, where $(af)^\dagger$ is the idempotent in the $\mathcal{R}^*$-class containing $af$. The left ample semigroups are also known as the *left type A semigroup*. Dually, the *right ample semigroup* can be defined. If $S$ is both a left ample and a right ample semigroup, then we call this semigroup the *ample semigroup*. Formerly, we call this semigroup the *type A semigrpup*. An inverse semigroup is also called *proper* if $\sigma \cap \mathcal{R} = id_S$, where $\sigma$ is the minimum group congruence on $S$ and we use $id_S$ to denote the identical relation on $S$. For inverse semigroups, being proper is equivalent to being $E$-unitary. This means that the "proper" property is kind of the "$E$-unitary" property. The well-known structure theorem of $E$-unitary inverse semigroups was first established by McAlister, and hence, we sometimes referred this theorem as the McAlister's $P$-theorem. A parallel theory of McAlister on proper inverse semigroups was later developed by Fountain and Gomes in Refs. 11,12 from inverse semigroups to proper left ample semigroups and to proper ample semigroups, respectively. We call a left ample semigroup $S$ *proper* if (1) there exists a smallest right cancellative congruence $\sigma$; and (2) $\mathcal{R}^* \cap \sigma = id_S$. Fountain and Gomes[11] generalized the $P$-theorem of McAlister to proper left ample semigroups. We notice that Gomes and Szendrei[13] considered another kind of proper left ample semigroups, namely, the *almost factorizable weakly ample semigroups*. In Refs. 14,15, Fountain established the structure of free left ample semigroups and pointed out that any free left ample semigroup is a proper left ample semigroup whose semilattice of idempotents is a free semilattice and in which the homomorphic image of the smallest right cancellative congruence is a free semigroup.

As an analogy of the proper left ample semigroups, Guo[16] first named a left GC-lpp semigroup *proper* if (1) there exists a smallest right cancellative monoid congruence $\sigma$ on $S$; and (2) $\mathcal{R}^* \cap \sigma = id_S$. In the same reference, he further considered the structure of proper abundant left GC-lpp semigroups. On the other hand, Guo and Shum[17] studied the superabundant semigroups whose band of idempotents is a left regular band. In 1999, Guo and Tian[18] proved that any left GC-lpp semigroup has P-covers, they further described the structure of P-covers for left GC-lpp semigroups and they also called this kind of semigroup the *left regular type A semigroups*. In Ref. 19, Guo, Ni and Shum constructed a theorem for the left GC-lpp monoids which are the F-rpp semigroups.

For F-rpp semigroups, the reader is referred to the Refs. 20,21. Recently, Guo and Shum[22] have further established a $P$-theorem of proper left GC-lpp semigroups. As inspiring by the results of Fountain on free left ample semigroups, and also by the results of J. B. Fountain, Gomes and Gould on free ample semigroups, we naturally have the following problem: whether the free left GC-lpp semigroups are semidirect products of left regular bands and free semigroups? The answer to this problem is positive; see Theorem 3.1.

Now, we call an F-inverse (E-unitary inverse) semigroup $T$ an *F-cover* or call a (*E-unitary cover*) for an inverse semigroup $S$ if $S$ is an idempotent-separating homomorphic image of $T$. It is well known that any inverse semigroup has a F-cover (E-unitary cover). Along this direction, it can be found that any orthodox semigroup has E-unitary covers. In this aspect, Guo[23] established the structure of E-unitary covers for an orthodox semi-group. On the other hand, Guo and Xie[24], El Qallali and Fountain[25], and Fountain and Gomes[12] all concentrated on the structure of proper covers of left ample semigroups.

It was also stated by Fountain that any left ample semigroup has proper covers (such proper covers are analogues of E-unitary covers for a left ample semigroup). Recently, Cui and Guo[26] proved that any left ample monoid has F-covers and the structure of F-covers of a left ample monoid was discussed and considered by them. The following question naturally arises: whether any left GC-lpp monoid has F-covers? As an application in this paper, we answer affirmatively the above problem in Theorem 5.2.

## 2. Quasi-variety of left GC-lpp semigroups

Throughout this section, we use the terminology and notation given in Refs. 1,27,28. For those definitions and notions not mentioned, the readers are referred to M. Petrich[29].

We begin by some elementary facts about left GC-lpp semigroups. The alternative characterizations of the relation $\mathcal{R}^*$ were given by the following lemma in Refs. 30,31; and the dual for the relation $\mathcal{L}^*$. We first state a crucial lemma.

**Lemma 2.1.** *Let $S$ be a semigroup and $a, b \in S$. Then the following conditions are equivalent:*

(1) $a\mathcal{R}^*b$.

(2) *For all $x, y \in S^1$, $xa = ya$ if and only if $xb = yb$.*

The following lemma is an easy consequence of Lemma 2.1.

**Lemma 2.2.** [2] *Let $S$ be a semigroup and $a, e^2 = e \in S$. Then the following conditions are equivalent:*

(1) $a\mathcal{R}^*e$.

(2) $ea = a$ *and for all $x, y \in S^1$, $xa = ya$ implies that $xe = ye$.*

It is well known that $\mathcal{R}^*$ is a left congruence while $\mathcal{L}^*$ is a right congruence. In general, $\mathcal{L} \subseteq \mathcal{L}^*$ and $\mathcal{R} \subseteq \mathcal{R}^*$. But when $a$ and $b$ are both regular elements of $S$, $a\mathcal{R}$ ($\mathcal{L}$) $b$ if and only if $a\mathcal{R}^*$ ($\mathcal{L}^*$) $b$. In particular, when $S$ is regular , we have $\mathcal{L} = \mathcal{L}^*$ and $\mathcal{R} = \mathcal{R}^*$.

**Lemma 2.3.** [1] *Let $S$ be a left GC-lpp semigroup. Then the following relation*

$$\sigma = \{(x, y) \in S \times S : ex = ey \text{ for some } e \in E(S)\}$$

*is the smallest right cancellative monoid congruence on $S$.*

Any left GC-lpp semigroup $S$ can be regarded as an algebra of type $(2, 1)$, where the unary operation † is given by $a \mapsto a^\dagger$. If we consider a left GC-lpp semigroup in this way, we can call such an algebra a †-*semigroup*. As in universal algebra, we now adopt the terminologies †-*subsemigroup*, †-*homomorphism* and †-*congruence*. Now, we state the following statements of semigroups by using the above terminologies.

- a †-*semigroup* is a subsemigroup $T$ in which for all $t \in T$, $t^\dagger \in T$.
- a †-*homomorphism* of a left GC-lpp semigroup $S$ into another one $U$ is a homomorphism $\varphi$ of $S$ into $U$ satisfying that for all $s \in S$, $\varphi(s^\dagger) = (\varphi(s))^\dagger$.
- a †-*congruence* on $S$ is a congruence $\rho$ on $S$ in which for all $s, t \in S$, if $(s, t) \in \rho$, then $(s^\dagger, t^\dagger) \in \rho$.

By definition, if $S$ is a left GC-lpp semigroup, then so is any †-subsemigroup of $S$.

An *implication* is an ordered pair of families of identities is expressed in the form

$$\{u_\alpha = v_\alpha\}_{\alpha \in A} \Rightarrow \{w_\beta = z_\beta\}_{\beta \in B}. \tag{1}$$

An algebra $\mathfrak{A}$ *satisfies implication* (1) if for any substitution of variables by elements of $\mathfrak{A}$ in all words $u_\alpha, v_\alpha, w_\beta, z_\beta$, the resulting elements $\widehat{u_\alpha}, \widehat{v_\alpha}, \widehat{w_\beta}, \widehat{z_\beta}$ of $\mathfrak{A}$ have the following property:

$$\widehat{w_\beta} = \widehat{z_\beta} \text{ for all } \beta \in B$$

whenever

$$\widehat{u_\alpha} = \widehat{v_\alpha} \text{ for all } \alpha \in A.$$

Recall that a class $\mathfrak{C}$ of algebras of the same type is a *quasivariety* of algebras if there exists a family $\mathfrak{J}$ of implications such that $\mathfrak{C}$ consists of all algebras which satisfy each implication. In this case, we say that $\mathfrak{C}$ *is defined by the family $\mathfrak{J}$ of implications* (for more details, the reader is referred to Ref. 32(p.219)). It is easy to verify that any identity is an implication, and so the varieties are quasivarieties.

The following proposition is immediate.

**Proposition 2.1.** *The class of left GC-lpp semigroups is a quasi-variety whose quasi-identities are*

$$(xy)z = x(yz),\ (x^\dagger)^\dagger = x^\dagger,\ x^\dagger x = x,\ xy^\dagger = (xy^\dagger)^\dagger x,\ x^\dagger x^\dagger = x^\dagger,$$
$$(x^2 = x \wedge y^2 = y) \Rightarrow xy = xyx,$$
$$xy = zy \Rightarrow xy^\dagger = zy^\dagger.$$

**Proof.** We first observe the following facts:

- $E(S)$ is a left regular band if and only if $S$ satisfies the implication: $(x^2 = x \wedge y^2 = y) \Rightarrow xy = xyx$.
- $y\mathcal{R}^* y^\dagger$ if and only if $S$ satisfies the implication: $y^\dagger y^\dagger = y^\dagger \wedge y^\dagger y = y \wedge (xy = zy \Rightarrow xy^\dagger = zy^\dagger)$.

The proof is a routine check and we hence omit the details. $\qquad\square$

By a general result in universal algebra, one can easily see that a free left GC-lpp semigroup always exists.

Let $\rho$ be a †-congruence on a left GC-lpp semigroup $S$. We consider the smallest †-congruence having the same partition of $E(S)$ as $\rho$. It is to adopt the corresponding results for inverse semigroups and for the left ample semigroups. Define the relation $\rho_{min}$ on $S$ by

$$a\rho_{min}b \text{ if and only if } ea = eb \text{ for some } e \in E(S) \text{ with } e\rho a^\dagger \rho b^\dagger.$$

Now, we have the following proposition for a †-congruence on a left GC-lpp semigrpup $S$.

**Proposition 2.2.** *Let $\rho$ be a †-congruence on a left GC-lpp semigroup $S$. Then $\rho_{min}$ is a †-congruence on $S$ such that $\rho_{min}|_{E(S)} = \rho|_{E(S)}$ and $\rho_{min} \subseteq \tau$ for any †-congruence $\tau$ with $\tau|_{E(S)} = \rho_{E(S)}$. Furthermore, $S/\rho_{min}$ is left GC-lpp and $E(S/\rho_{min}) = \{e\rho_{min} : e \in E(S)\}$.*

**Proof.** Clearly, the relation $\rho_{min}$ is reflexive and symmetric. If $a\rho_{min}b,\ b\rho_{min}c$ then there are idempotents $e, f$ such that $e\rho a^\dagger \rho b^\dagger \rho c^\dagger \rho f$, $ea = eb$ and $fb = fc$. Hence, $e = e^2 \rho ef\rho f^2 = f$ so that $ef\rho a^\dagger \rho c^\dagger$ and $efa = efb = efc$, whence $a\rho_{min}c$.

Now let $a\rho_{min}b$ and $c \in S$. Then $ea = eb$ for some idempotent $e$ with $e\rho a^\dagger \rho b^\dagger$. Certainly, $(ac)^\dagger e \bullet ac = (ac)^\dagger e \bullet bc$. Since $a^\dagger ac = ac$, we have $a^\dagger (ac)^\dagger = (ac)^\dagger$, so that we arrive $(ac)^\dagger e \, \rho \, (ac)^\dagger a^\dagger = (ac)^\dagger a^\dagger (ac)^\dagger = (ac)^\dagger$; dually, $(bc)^\dagger e \rho (bc)^\dagger$. Compute

$$(ac)\rho = (ac)\rho = (eac)\rho = (ebc)\rho = (e\rho)(b\rho)(c\rho) = (b^\dagger \rho)(b\rho)(c\rho) = (bc)\rho.$$

By the definition of $\dagger$-congruence, we have $(ac)^\dagger \rho = (bc)^\dagger \rho$. Thus, by $(ac)^\dagger e \bullet ac = (ac)^\dagger e \bullet bc$, we can deduce that $ac\rho_{min}bc$. On the other hand, since $S$ is a left GC-lpp semigroup, we deduce that $(ce)^\dagger \bullet ca = cea = ceb = (ce)^\dagger \bullet bc$. By $e\rho a^\dagger$, we get $ce\rho ca^\dagger$ and because $\rho$ is a $\dagger$-congruence, we have $(ce)^\dagger \rho = (ca)^\dagger \rho$. Similarly, we have $(ce)^\dagger \rho (cb)^\dagger$. Therefore, we obtain $(ce)^\dagger \bullet ca = (ce)^\dagger \bullet cb$, this equality implies that $ca\rho cb$. Thus, $\rho_{min}$ is indeed a congruence on $S$. Observe that $ea^\dagger b^\dagger \cdot a^\dagger = ea^\dagger b^\dagger = ea^\dagger b^\dagger \cdot b^\dagger$ and by $e\rho a^\dagger \rho b^\dagger$, we have $ea^\dagger b^\dagger \rho a \rho b^\dagger$. Hence, the above result leads to $a^\dagger \rho_{min} b^\dagger$. Now, it follows that $\rho_{min}$ is a $\dagger$-congruence on $S$.

If $x, y \in S$ and $(xa)\rho = (ya)\rho$, then $fxa = fya$ for some $f \in E(S)$ with $f\rho(xa)^\dagger \rho(ya)^\dagger$, hence by $a\mathcal{R}^* a^\dagger$, we see that $f(xa^\dagger) = f(ya^\dagger)$. By applying $\mathcal{R}^*$ is a left congruence, we have $(xa)^\dagger \mathcal{R}^* xa \mathcal{R}^* xa^\dagger \mathcal{R}^* (xa^\dagger)^\dagger$, and $(xa)^\dagger = (xa^\dagger)^\dagger$ since each $\mathcal{R}^*$-class of a left GC-lpp semigroup contains exactly one idempotent. It follows that $f\rho(xa^\dagger)^\dagger$; similarly, we have $f\rho(ya^\dagger)^\dagger$. Thus, by using $f(xa^\dagger) = f(ya^\dagger)$, we can show that $(xa^\dagger)\rho_{min} = (ya^\dagger)\rho_{min}$ and together with $a^\dagger \rho \bullet a\rho = a\rho$, we hence deduces that $a^\dagger \rho_{min} \mathcal{R}^* a\rho_{min}$. Therefore, we have proved that $S/\rho$ is a left abundant semigroup.

If $e, f \in E(S)$ and $e\rho f$, then $ef\rho e\rho f$ and $ef \bullet e = ef = ef \bullet f$ since $E(S)$ is a left regular band, hence $e\rho_{min}f$. Conversely, if $e\rho_{min}f$, then it is immediate that $e\rho f$. Thus $\rho|_{E(S)} = \rho_{min}|_{E(S)}$.

Now, we may assume that $\tau$ is a $\dagger$-congruence on $S$ such that $\tau|_{E(S)} = \rho_{min}|_{E(S)}$. If $a, b \in S$ and $a\rho_{min}b$, then $ea = eb$ for some $e \in E(S)$ with $e\rho a^\dagger \rho b^\dagger$. Now, by using the above proof, we easily see that $e\rho_{min}a^\dagger \rho_{min}b^\dagger$ and by our hypothesis, we deduce that $e\tau a^\dagger \tau b^\dagger$. Thus, we obtain the following equalities:

$$a\tau = (a^\dagger a)\tau = (a^\dagger \tau)(a\tau) = (e\tau)(a\tau)$$
$$= (ea)\tau = (eb)\tau = (e\tau)(b\tau) = (b^\dagger \tau)(b\tau) = b\tau,$$

and whence, $\rho_{min} \subseteq \tau$.

Suppose that $a \in S$ and $a^2 \rho_{min} a$. Then by $a^\dagger \rho_{min} \mathcal{R}^* a\rho_{min}$, we get $a^\dagger \rho_{min} = (aa^\dagger)\rho_{min}$. It follows that $a^\dagger \rho_{min} = (aa^\dagger)^\dagger \rho_{min} a\rho_{min}$ since $S$ is a left GC-lpp semigroup. By the foregoing arguments, we have $a^\dagger \rho_{min} \mathcal{R}^* a\rho_{min}$ and so $a^\dagger \rho_{min} \mathcal{R} a\rho_{min}$. Thus , $a^\dagger \rho_{min} = a\rho_{min} a^\dagger \rho_{min} =$

$(aa^\dagger)\rho_{min}$, thereby we have $a^\dagger\rho_{min} = (aa^\dagger)\rho_{min}$ since $\rho_{min}$ is a $\dagger$-congruence. Now, we have $a^\dagger\rho_{min} = (aa^\dagger)^\dagger\rho_{min}a\rho_{min} = a^\dagger\rho_{min}a\rho_{min} = a\rho_{min}$. Consequently, we further deduce that $E(S/\rho_{min}) = \{e\rho_{min} : e \in E(S)\}$. Thus, we have finally proved that $E(S/\rho_{min})$ is a left regular band.

It remains to verify that $S/\rho_{min}$ is a left GC-lpp semigroup. We have already proved that $S/\rho_{min}$ is a left abundant semigroup whose band of idempotents is a left regular band. For $e \in E(S), a \in S$, we have $ae = (ae)^\dagger a$ and so $a\rho_{min} \cdot e\rho_{min} = (ae)^\dagger\rho_{min} \cdot a\rho_{min}$. By the above proof and arguments, we have $((ae)\rho)^\dagger = (ae)^\dagger\rho_{min}$. So that we deduce that $a\rho_{min} \cdot e\rho_{min} = ((ae)\dagger\rho_{min})^\dagger \cdot a\rho_{min}$. Thus, $S/\rho_{min}$ is a left GC-lpp semigroup. $\square$

A congruence $\rho$ on a semigroup $S$ is called a *right cancellative monoid congruence* on $S$ if $S/\rho$ is a right cancellative monoid. It is well known that the universal relation $\omega$ is a $\dagger$-congruence. By Proposition 2.2, the following corollary follows immediate.

**Corollary 2.1.** *Let $S$ be a left GC-lpp semigroup. Then the smallest right cancellative congruence $\sigma$ on $S$ is $\omega_{min}$.*

By Lemma 2.3, the following proposition follows straightforwardly.

**Proposition 2.3.** *The class of proper left GC-lpp semigroups is a quasi-variety whose quasi-identities are*

$$(xy)z = x(yz), \ (x^\dagger)^\dagger = x^\dagger, \ x^\dagger x = x, \ xy^\dagger = (xy^\dagger)^\dagger x, \ x^\dagger x^\dagger = x^\dagger,$$
$$(x^\dagger y = x^\dagger z \wedge y^\dagger = z^\dagger) \Rightarrow y = z,$$
$$(x^2 = x \wedge y^2 = y) \Rightarrow xy = xyx,$$
$$xy = zy \Rightarrow xy^\dagger = zy^\dagger.$$

**Proof.** It is not difficult to see that for a left GC-lpp semigroup $S$, $E(S) = \{x^\dagger : x \in S\}$. Hence, for any $y, z \in S$, we have the following facts:

- $(y, z) \in \sigma$ if and only if there exists $x \in S$ such that $x^\dagger y = x^\dagger z$.
- $y\mathcal{R}^* z$ if and only if $y^\dagger = z^\dagger$.

By using the above facts, we can derive that $(y, z) \in \sigma \sqcap \mathcal{R}^*$ if and only if $(x^\dagger y = x^\dagger z \wedge y^\dagger = z^\dagger)$. Therefore, $S$ satisfies $\sigma \sqcap \mathcal{R}^* = id_S$ if and only if it satisfies the implication: $(x^\dagger y = x^\dagger z \wedge y^\dagger = z^\dagger) \Rightarrow y = z$. The rest part of the proof follows from Proposition 2.1. $\square$

Let $S$ be a left GC-lpp semigroup. By a *proper $\dagger$-congruence* on $S$, we mean a $\dagger$-congruence $\rho$ on $S$ such that $S/\rho$ is proper.

We now have the following proposition.

**Proposition 2.4.** *Let $S$ be a proper left GC-lpp semigroup. If $\rho$ is a †-congruence on $S$, then $\rho_{min}$ is a proper †-congruence on $S$.*

**Proof.** Denote by $\sigma_1$ the smallest right cancellative congruence on $S/\rho_{min}$ and by $\mathcal{R}_1^*$ the relations $\mathcal{R}^*$ on $S/\rho_{min}$. It suffices to verify that $\sigma_1 \cap \mathcal{R}_1^* = id_{S/\rho_{min}}$. Now, we simply let $a, b \in S$ such that $(a\rho_{min}, b\rho_{min}) \in \sigma_1 \cap \mathcal{R}_1^*$. By the proof of Proposition 2.2, $a^\dagger \rho_{min} \mathcal{R}_1^* a\rho_{min}, b^\dagger \rho_{min} \mathcal{R}_1^* b\rho_{min}$ and $a^\dagger \rho_{min} \mathcal{R}_1^* b^\dagger \rho_{min}$. It follows that $a^\dagger \rho_{min} = b^\dagger \rho_{min}$ since $S/\rho_{min}$ is a left GC-lpp semigroup. Hence, we have $fa^\dagger = fb^\dagger$ for some $f \in E(S)$ with $f\rho a^\dagger \rho b^\dagger$. Thus, $(fa)^\dagger = fa^\dagger = fb^\dagger = (fb)^\dagger$. But $(a\rho_{min}, b\rho_{min}) \in \sigma_1$, so we have $(ea)\rho_{min}(eb)$ for some $e \in E(S)$. It is easy to see that $a\sigma b$ in $S$, and whence $fa\sigma fb$. As a consequence, we have $(fa, fb) \in \mathcal{R}^* \cap \sigma$, and $fa = fb$ since $S$ is proper, we obtain $a\rho_{min} = b\rho_{min}$. This shows that $S/\rho_{min}$ is proper.  $\square$

Let $S$ be a left abundant semigroup. Then, we define a relation $\leq_r$ on $S$ by

$$x \leq_r y \text{ if and only if } R^*(x) \subset R^*(y) \text{ and there exists an element}$$
$$f \in E(S) \cap R^*(x) \text{ such that } x = fy,$$

where $R^*(x)$ is the right *-ideal of $S$ generated by $x$ (for the right *-ideal, see Ref. 27). Dually, we can define the $\leq_l$ on right abundant semigroups. If $S$ is abundant, we define $\leq \, = \, \leq_l \cap \leq_r$. Indeed, $\leq_r$ ($\leq_l$; $\leq$) are all partial orders extending the natural order on the set of idempotents. We call these orders the *Lawson orders* (for more information on Lawson orders, the reader is referred to Ref. 33). We are now able to provide the following lemma.

**Lemma 2.4.** *Let $S$ be a left abundant semigroup and $x, y \in S$. Then $x \leq_r y$ if and only if for some [for all] $y^\dagger$, there exists $f \in E(S) \cap R_x^*$ such that $f\omega y^\dagger$ (that is, $f = fy^\dagger = y^\dagger f$) and $x = fy$.*

Recall Ref. 20 that a left abundant semigroup $S$ is *F-lpp* if there exists a right cancellative monoid congruence $\rho$ on $S$ each class of which contains a greatest element under the Lawson order $\leq_r$. In this case, $\rho$ is indeed the smallest right cancellative monoid congruence of the F-lpp semigroup. It is not difficult to see that a semigroup is an F-regular semigroup if and only if it is a F-lpp semigroup being regular. Dually, we can define *F-rpp semigroup*. And, an abundant semigroup $S$ is called an *F-abundant semigroup* if there exists a cancellative congruence $\rho$ on $S$ such that each $\rho$-class of $S$ contains a greatest element under Lawson order $\leq$. In fact, a semigroup is F-abundant if and only if it is both F-lpp and F-rpp. In what follows, we call a left GC-lpp semigroup being F-lpp an *FLGC-lpp semigroup*.

Now $S$ is considered as a FLGC-lpp semigroup. For $s \in S$, we denote by $s^\circ$ the greatest element of the $\sigma$-class of $S$ containing $s$. With the above notion, we may regard a FLGC-lpp semigroup as an algebra of type $(2,1,1)$ with the multiplication, the unary operation $\dagger$ and the unary operation $\diamond$ defined by the rule that $\diamond : \ s \mapsto s^\circ$. Regarded a FLGC-lpp semigroup in the way, we will have such an algebra a $(\dagger, \diamond)$-*semigroup*. If $T$ is a $\dagger$-subsemigroup of $S$, then by Lemma 2.3, $\sigma^T = \sigma^S \cap T \times T$ and by Lemma 2.4, for any $x, y \in T$, $x \leq_r y$ in $T$ if and only if $x \leq_r y$. This shows that we can view an FLGC-lpp semigroup as an algebra of type $(2,1,1)$. Therefore we have the following proposition.

**Proposition 2.5.** *If the FLGC-lpp semigroups are regarded as the algebras of type $(2,1,1)$, then the class of FLGC-lpp semigroups is a quasi-variety whose quasi-identities are*

$$(xy)z = x(yz), \ (x^\dagger)^\dagger = x^\dagger, \ x^\dagger x = x, \ xy^\dagger = (xy^\dagger)^\dagger x, \ x^\dagger x^\dagger = x^\dagger,$$
$$(x^2 = x \wedge y^2 = y) \Rightarrow xy = xyx,$$
$$xy = zy \Rightarrow xy^\dagger = zy^\dagger,$$
$$(x^2 = x \wedge xy = xz) \Rightarrow (y^\circ = z^\circ \wedge y = y^\dagger y^\circ).$$

**Proof.** By the proof of Proposition 2.3, wwe know $(y, z) \in \sigma$ if and only if $xy = xz$ for some $x \in E(S)$. If $S$ is an FLGC-lpp semigroup, then as $y^\circ, z^\circ$ are both the greatest in the same $\sigma$-class, we get $y^\circ = z^\circ$ and by Lemma 2.4, $y = y^\dagger y^\circ$. Conversely, if the implication: $(x^2 = x \wedge xy = xz) \Rightarrow (y^\circ = z^\circ \wedge y = y^\dagger y^\circ)$ holds, then for any $a, b \in S$, by $(a, b) \in \sigma$, we can easily derive that $a^\circ$ is the greatest element in $\sigma_a$. Hence, $S$ is FLGC-lpp. The rest part for the proofs follow from Proposition 2.1. $\quad\square$

It is well known that a semigroup is a left regular band if and only if it is a left GC-lpp semigroup being a band. On the other hand, we have $x = x^\dagger$ for a left regular band $L$ and $x \in L$. Associating with Proposition 2.1, we have the following proposition.

**Proposition 2.6.** *The class of left regular bands is a quasi-variety whose quasi-identities are*

$$(xy)z = x(yz), \ (x^\dagger)^\dagger = x^\dagger, \ x^\dagger x = x, \ xy^\dagger = (xy^\dagger)^\dagger x, \ x = x^\dagger, \ x^\dagger x^\dagger = x^\dagger,$$

$$(x^2 = x \wedge y^2 = y) \Rightarrow xy = xyx,$$
$$xy = zy \Rightarrow xy^\dagger = zy^\dagger.$$

**Proof.** We first observe that for a left GC-lpp semigroup $S$, $S$ is a band if and only if $S$ satisfies the implications: $x^\dagger x^\dagger = x^\dagger, x = x^\dagger$. Hence, $S$ is a left GC-lpp semigroup being a band, that is, $S$ is a left regular band, if and only if it satisfies the implications in the theorem. The above results lead to the following theorem. □

## 3. Free left GC-lpp semigroups

If $X$ is a non-empty set, then, we use $F(X)$ to denote the set of all non-empty finite words $x_1 x_2 ... x_n$ in the 'alphabet' $X$. We call the word without alphabets the *empty word* on $X$, says 1. Put $F^*(X) = F(X) \cup \{1\}$. A binary operation is defined on $F^*(X)$ by juxtaposition:

$$(x_1 x_2 ... x_m)(y_1 y_2 ... y_n) = x_1 ... x_m y_1 ... y_n.$$

With respect to this operation, the set $F(X)$ clearly forms a semigroup, called the *free semigroup* on the set $X$, and $F^*(X)$ is a monoid, called the *free monoid* on $X$.

Let $X^\dagger = \{(x)^\dagger : x \in X\}$. Then, we use $U(X)$ to denote the least subset of $F(X \sqcup X^\dagger)$ satisfying the following conditions:

(U1)　$X \subseteq U(X)$;
(U2)　$u, v \in U(X) \Rightarrow uv \in U(X)$;
(U3)　$u \in U(X) \Rightarrow (u)^\dagger \in U(X)$;

and by $\iota_X$ the identity mapping on $X$. The result [34](Theorem I.1.5, p51) leads to $(U(X), \iota_X)$ forms a free unary semigroup on $X$, whose unary operation $\dagger : u \mapsto (u)^\dagger$. For the sake of simplicity, we simply write $(u)^\dagger$ as $u^\dagger$.

Now let $\gamma$ be the $\dagger$-congruence on $U(X)$ generated by the following relations:

$$\{(u^\dagger, u^\dagger u^\dagger) : u \in U(X)\} \tag{2}$$
$$\{(u^\dagger v^\dagger, u^\dagger v^\dagger u^\dagger) : u, v \in U(X)\} \tag{3}$$
$$\{(uv^\dagger, (uv^\dagger)^\dagger u) : u, v \in U(X)\} \tag{4}$$
$$\{(u^\dagger u, u) : u \in u \in U(X)\} \tag{5}$$
$$\{(u^\dagger, (u^\dagger)^\dagger) : u \in U(X)\} \tag{6}$$
$$\{((u^\dagger v)^\dagger, u^\dagger v^\dagger) : u, v \in U(X)\} \tag{7}$$
$$\{((uv^\dagger)^\dagger, (uv)^\dagger) : u, v \in U(X)\} \tag{8}$$

In what follows, we use $FLGC(X)$ to denote the quotient semigroup $U(X)/\gamma$ of $U(X)$ over $\gamma$. For the sake of convenience, the image of $u \in FLGC(X)$ under $\gamma$ is simply denoted by $u$. And, for $u, v \in FLGC(X)$, if $(u, v) \in \gamma$, we shall write $u \equiv v$.

We state below an useful lemma.

**Lemma 3.1.** *The idempotents of $FLGC(X)$ are of the form:* $u_1^\dagger u_2^\dagger \cdots u_n^\dagger$, *where $u_i \in F(X)$ and $n$ is a positive integer.*

**Proof.** We first show that

$$\left( c_1 p_{11}^\dagger p_{12}^\dagger \cdots p_{1m_1}^\dagger c_2 p_{21}^\dagger p_{22}^\dagger \cdots p_{2m_2}^\dagger \cdots c_n p_{n1}^\dagger p_{n2}^\dagger \cdots p_{nm_n}^\dagger \right)^\dagger$$
$$\equiv \left( \prod_{i=1}^{n} \prod_{j=1}^{m_i} (c_1 c_2 \cdots c_i p_{ij})^\dagger \right) \left( \prod_{k=1}^{n} c_k \right)^\dagger . \quad (9)$$

By induction, we only need verify the case $n = 2$. Indeed, the proof follows immediately from computation:

$$\left( c_1 p_{11}^\dagger p_{12}^\dagger \cdots p_{1m_1}^\dagger c_2 p_{21}^\dagger p_{22}^\dagger \cdots p_{2m_2}^\dagger \right)^\dagger$$
$$\overset{\text{by (4)}}{\equiv} \left( (c_1 p_{11})^\dagger c_1 p_{12}^\dagger \cdots p_{1m_1}^\dagger c_2 p_{21}^\dagger p_{22}^\dagger \cdots p_{2m_2}^\dagger \right)^\dagger$$

$$\cdots$$

$$\overset{\text{by (4)}}{\equiv} \left( (c_1 p_{11})^\dagger (c_1 p_{12})^\dagger \cdots (c_1 p_{1m_1})^\dagger c_1 c_2 p_{21}^\dagger p_{22}^\dagger \cdots p_{2m_2}^\dagger \right)^\dagger$$
$$\overset{\text{by (4)}}{\equiv} \left( (c_1 p_{11})^\dagger (c_1 p_{12})^\dagger \cdots (c_1 p_{1m_1})^\dagger (c_1 c_2 p_{21})^\dagger c_1 c_2 p_{22}^\dagger \cdots p_{2m_2}^\dagger \right)^\dagger$$

$$\cdots$$

$$\equiv \left( (c_1 p_{11})^\dagger (c_1 p_{12})^\dagger \cdots (c_1 p_{1m_1})^\dagger (c_1 c_2 p_{21})^\dagger (c_1 c_2 p_{22})^\dagger \cdots c_1 c_2 p_{2m_2}^\dagger \right)^\dagger$$
$$\overset{\text{by (4)}}{\equiv} \left( (c_1 p_{11})^\dagger (c_1 p_{12})^\dagger \cdots (c_1 p_{1m_1})^\dagger (c_1 c_2 p_{21})^\dagger (c_1 c_2 p_{22})^\dagger \cdots (c_1 c_2 p_{2m_2})^\dagger c_1 c_2 \right)^\dagger$$
$$\overset{\text{by (7)}}{\equiv} (c_1 p_{11})^\dagger (c_1 p_{12})^\dagger \cdots (c_1 p_{1m_1})^\dagger (c_1 c_2 p_{21})^\dagger (c_1 c_2 p_{22})^\dagger \cdots (c_1 c_2 p_{2m_2})^\dagger (c_1 c_2)^\dagger$$

On the other hand, by the definition of $\gamma$, the idempotents of $FLGC(X)$ are of the form: $(u)^\dagger$ with $u \in U(X)$, which is indeed the iterations of elements of the forms:

$$(c_1 p_{11}^\dagger p_{12}^\dagger \cdots p_{1m_1}^\dagger c_2 p_{21}^\dagger p_{22}^\dagger \cdots p_{2m_2}^\dagger \cdots c_n p_{n1}^\dagger p_{n2}^\dagger \cdots p_{nm_n}^\dagger)^\dagger;$$

or $(u^\dagger v)^\dagger$ with $u, v \in U(X)$; or $(uv^\dagger)^\dagger$ with $u, v \in U(X)$. By the relations (7) and (8), $(u^\dagger v)^\dagger \equiv u^\dagger v^\dagger$ and $(uv^\dagger)^\dagger = (uv)^\dagger$, we can easily see that any idempotent of $FLGC(X)$ is indeed the iteration of elements of the form

$$(c_1 p_{11}^\dagger p_{12}^\dagger \cdots p_{1m_1}^\dagger c_2 p_{21}^\dagger p_{22}^\dagger \cdots p_{2m_2}^\dagger \cdots c_n p_{n1}^\dagger p_{n2}^\dagger \cdots p_{nm_n}^\dagger)^\dagger.$$

By (9), it is not difficult to see that the iteration of elements of the left-side form in (9) is still of this form. Hence the lemma holds. $\square$

**Lemma 3.2.** *The elements of $FLGC(X)$ are of the forms: $u_1^\dagger u_2^\dagger \cdots u_n^\dagger v$, where $u_i \in F(X)$, $v \in F^*(X)$ and $n$ is a positive integer.*

**Proof.** By the relation (4), any element of $FLGC(X)$ can be expressed by an element of the form $u^\dagger v$ with $u \in U(X)$ and $v \in F^*(X)$, and by Lemma 3.1, the lemma follows immediately. $\square$

We now call a word $w$ in $FLGC(X)$ is *reducible* if $w \equiv uv$ for some $u, v \in FLGC(X)$ and if otherwise *irreducible*. A word $p$ in $F(X)$ is a *subword* of $q \in U(X)$ if one the following three cases holds:

**(1)** $q \in U(X)z^\dagger p$ for some $z \in U(X)$;
**(2)** $q \in pz^\dagger U(X)$ for some $z \in U(X)$;
**(3)** $q \in U(X)z^\dagger p y^\dagger U(X)$ for some $z, y \in U(X)$.

**Lemma 3.3.** *Let $u_1^\dagger u_2^\dagger \cdots u_m^\dagger v, w_1^\dagger w_2^\dagger \cdots w_n^\dagger z \in FLGC(X)$ with $u_i, w_j \in F(X), v, z \in F^*(X)$. Then $u_1^\dagger u_2^\dagger \cdots u_m^\dagger v \equiv w_1^\dagger w_2^\dagger \cdots w_n^\dagger z$ if and only if $v = z$ and $u_1^\dagger u_2^\dagger \cdots u_m^\dagger v^\dagger \equiv w_1^\dagger w_2^\dagger \cdots w_n^\dagger z^\dagger$.*

**Proof.** Because we have

$$u_1^\dagger u_2^\dagger \cdots u_m^\dagger v \equiv u_1^\dagger u_2^\dagger \cdots u_m^\dagger v^\dagger v \equiv w_1^\dagger w_2^\dagger \cdots w_n^\dagger z^\dagger z \equiv w_1^\dagger w_2^\dagger \cdots w_n^\dagger z,$$

it remains to verify the necessity part. For this purpose, we assume that $u_1^\dagger u_2^\dagger \cdots u_m^\dagger v \equiv w_1^\dagger w_2^\dagger \cdots w_n^\dagger z$ Then $v\sigma = (u_1^\dagger u_2^\dagger \cdots u_m^\dagger v)\sigma = (w_1^\dagger w_2^\dagger \cdots w_n^\dagger z)\sigma = z\sigma$ and it is trivial that $v = z$. On the other hand, since $\gamma$ is a $\dagger$-congruence, we get

$$u_1^\dagger u_2^\dagger \cdots u_m^\dagger v^\dagger \equiv \left(u_1^\dagger u_2^\dagger \cdots u_m^\dagger v\right)^\dagger \equiv \left(w_1^\dagger w_2^\dagger \cdots w_n^\dagger z\right)^\dagger \equiv w_1^\dagger w_2^\dagger \cdots w_n^\dagger z^\dagger.$$

$\square$

**Proposition 3.1.** *In the semigroup $FLGC(X)$, the following statements hold:*

(1) $u_1^\dagger u_2^\dagger \cdots u_m^\dagger w$ *is an idempotent if and only if $w = 1$. Moreover, $E(FLGC(X)) = \{u_1^\dagger u_2^\dagger \cdots u_m^\dagger : u_i \in F(X)\}$ is a left regular band.*

(2) $u_1^\dagger u_2^\dagger \cdots u_m^\dagger w \mathcal{R}^* u_1^\dagger u_2^\dagger \cdots u_m^\dagger w^\dagger$.

(3) $FLGC(X)$ *is a left GC-lpp semigroup.*

(4) *For any* $u_1^\dagger u_2^\dagger \cdots u_m^\dagger w_1, v_1^\dagger v_2^\dagger \cdots v_n^\dagger w_2$ *with* $u_i, v_j \in F(X), w_1, w_2 \in F^*(X),$ $u_1^\dagger u_2^\dagger \cdots u_m^\dagger w_1 \sigma v_1^\dagger v_2^\dagger \cdots v_n^\dagger w_2$ *if and only if* $w_1 = w_2$.

(5) $\sigma \cap \mathcal{R}^* = id_{FLGC(X)}$. *Moreover,* $FLGC(X)$ *is a proper left GC-lpp semigroup.*

(6) $u_1^\dagger u_2^\dagger \cdots u_m^\dagger w_1 \quad \leq_r \quad v_1^\dagger v_2^\dagger \cdots v_n^\dagger w_2 \quad$ *if and only if* $u_1^\dagger u_2^\dagger \cdots u_m^\dagger w_1^\dagger \omega v_1^\dagger v_2^\dagger \cdots v_n^\dagger w_2^\dagger$ *and* $w_1 = w_2$.

**Proof.** (1) If $u_1^\dagger u_2^\dagger \cdots u_m^\dagger w$ is an idempotent, then

$$u_1^\dagger u_2^\dagger \cdots u_m^\dagger w \equiv u_1^\dagger u_2^\dagger \cdots u_m^\dagger w u_1^\dagger u_2^\dagger \cdots u_m^\dagger w$$

$$\equiv u_1^\dagger u_2^\dagger \cdots u_m^\dagger (wu_1)^\dagger (wu_2)^\dagger \cdots (wu_m)^\dagger w^2,$$

hence by Lemma 3.3, we have $w = w^2$ and thereby , $w = 1$. Conversely, if $w = 1$, then $u_1^\dagger u_2^\dagger \cdots u_m^\dagger w \equiv u_1^\dagger u_2^\dagger \cdots u_m^\dagger$ is an idempotent. Obviously, $E(FLGC(X)) = \{u_1^\dagger u_2^\dagger \cdots u_m^\dagger : u_i \in FSU(X)\}$, and whence $E(FLGC(X))$ is a left regular band.

(2) If $x_1^\dagger x_2^\dagger \cdots x_r^\dagger w_1, y_1^\dagger y_2^\dagger \cdots y_t^\dagger w_2 \in FLGC(X)$ and

$$x_1^\dagger x_2^\dagger \cdots x_r^\dagger w_1 \circ u_1^\dagger u_2^\dagger \cdots u_m^\dagger w \equiv y_1^\dagger y_2^\dagger \cdots y_t^\dagger w_2 \circ u_1^\dagger u_2^\dagger \cdots u_m^\dagger w,$$

then

$$x_1^\dagger x_2^\dagger \cdots x_r^\dagger (w_1 u_1)^\dagger (w_1 u_2)^\dagger \cdots (w_1 u_m)^\dagger w_1 w$$

$$\equiv y_1^\dagger y_2^\dagger \cdots y_t^\dagger (w_2 u_1)^\dagger (w_2 u_2)^\dagger \cdots (w_2 u_m)^\dagger w_2 w,$$

and by Lemma 3.3,

$$x_1^\dagger x_2^\dagger \cdots x_r^\dagger (w_1 u_1)^\dagger (w_1 u_2)^\dagger \cdots (w_1 u_m)^\dagger (w_1 w)^\dagger$$

$$\equiv y_1^\dagger y_2^\dagger \cdots y_t^\dagger (w_2 u_1)^\dagger (w_2 u_2)^\dagger \cdots (w_2 u_m)^\dagger (w_2 w)^\dagger,$$

so that we have

$$x_1^\dagger x_2^\dagger \cdots x_r^\dagger (w_1 u_1)^\dagger (w_1 u_2)^\dagger \cdots (w_1 u_m)^\dagger (w_1 w^\dagger)^\dagger$$

$$\equiv y_1^\dagger y_2^\dagger \cdots y_t^\dagger (w_2 u_1)^\dagger (w_2 u_2)^\dagger \cdots (w_2 u_m)^\dagger (w_2 w^\dagger)^\dagger;$$

and $w_1 w = w_2 w$, so that $w_1 = w_2$. Thus, we deduce that

$$x_1^\dagger x_2^\dagger \cdots x_r^\dagger w_1 \circ u_1^\dagger u_2^\dagger \cdots u_m^\dagger w^\dagger$$

$$\equiv x_1^\dagger x_2^\dagger \cdots x_r^\dagger (w_1 u_1)^\dagger (w_1 u_2)^\dagger \cdots (w_1 u_m)^\dagger w_1 w^\dagger$$

$$\equiv x_1^\dagger x_2^\dagger \cdots x_r^\dagger (w_1 u_1)^\dagger (w_1 u_2)^\dagger \cdots (w_1 u_m)^\dagger (w_1 w^\dagger)^\dagger (w_1 w^\dagger)$$

$$\equiv y_1^\dagger y_2^\dagger \cdots y_t^\dagger (w_2 u_1)^\dagger (w_2 u_2)^\dagger \cdots (w_2 u_m)^\dagger (w_2 w^\dagger)^\dagger (w_2 w^\dagger)$$

$$\equiv y_1^\dagger y_2^\dagger \cdots y_t^\dagger (w_2 u_1)^\dagger (w_2 u_2)^\dagger \cdots (w_2 u_m)^\dagger (w_2 w^\dagger)$$
$$\equiv y_1^\dagger y_2^\dagger \cdots y_t^\dagger w_2 \circ u_1^\dagger u_2^\dagger \cdots u_m^\dagger w^\dagger.$$

This, together with

$$u_1^\dagger u_2^\dagger \cdots u_m^\dagger w^\dagger \circ u_1^\dagger u_2^\dagger \cdots u_m^\dagger w \equiv u_1^\dagger u_2^\dagger \cdots u_m^\dagger w^\dagger w \equiv u_1^\dagger u_2^\dagger \cdots u_m^\dagger w$$

implies that $u_1^\dagger u_2^\dagger \cdots u_m^\dagger w \mathcal{R}^* u_1^\dagger u_2^\dagger \cdots u_m^\dagger w^\dagger$ by Lemma 2.2.

(3) Compute

$$u_1^\dagger u_2^\dagger \cdots u_m^\dagger w \circ x_1^\dagger x_2^\dagger \cdots x_r^\dagger \equiv u_1^\dagger u_2^\dagger \cdots u_m^\dagger (wx_1)^\dagger (wx_2)^\dagger \cdots (wx_r)^\dagger w$$
$$\equiv u_1^\dagger u_2^\dagger \cdots u_m^\dagger (wx_1)^\dagger (wx_2)^\dagger \cdots (wx_r)^\dagger w^\dagger w$$
$$\equiv u_1^\dagger u_2^\dagger \cdots u_m^\dagger (wx_1)^\dagger (wx_2)^\dagger \cdots (wx_r)^\dagger w^\dagger \circ u_1^\dagger u_2^\dagger \cdots u_m^\dagger w$$
$$\equiv \left( u_1^\dagger u_2^\dagger \cdots u_m^\dagger w \circ x_1^\dagger x_2^\dagger \cdots x_r^\dagger \right)^\dagger \left( u_1^\dagger u_2^\dagger \cdots u_m^\dagger w \right).$$

Again by (1), $E(FLGC(X))$ is a left regular band. Thus $FLGC(X)$ is a left GC-lpp semigroup.

(4) If $u_1^\dagger u_2^\dagger \cdots u_m^\dagger w_1 \sigma v_1^\dagger v_2^\dagger \cdots v_n^\dagger w_2$ then by Lemmas 2.3 and 3.3,

$$x_1^\dagger x_2^\dagger \cdots x_r^\dagger \circ u_1^\dagger u_2^\dagger \cdots u_m^\dagger w_1 \equiv x_1^\dagger x_2^\dagger \cdots x_r^\dagger \circ v_1^\dagger v_2^\dagger \cdots v_n^\dagger w_2$$

for some $x_1^\dagger x_2^\dagger \cdots x_r^\dagger$, so that $w_1 = w_2$. Conversely, if $w_1 = w_2$, then we have

$$u_1^\dagger u_2^\dagger \cdots u_m^\dagger v_1^\dagger v_2^\dagger \cdots v_n^\dagger \circ u_1^\dagger u_2^\dagger \cdots u_m^\dagger w_1 \equiv u_1^\dagger u_2^\dagger \cdots u_m^\dagger v_1^\dagger v_2^\dagger \cdots v_n^\dagger w_1$$
$$\equiv u_1^\dagger u_2^\dagger \cdots u_m^\dagger v_1^\dagger v_2^\dagger \cdots v_n^\dagger \circ v_1^\dagger v_2^\dagger \cdots v_n^\dagger w_2,$$

and so $u_1^\dagger u_2^\dagger \cdots u_m^\dagger w_1 \sigma v_1^\dagger v_2^\dagger \cdots v_n^\dagger w_2$.

(5) If $\left( u_1^\dagger u_2^\dagger \cdots u_m^\dagger w_1, v_1^\dagger v_2^\dagger \cdots v_n^\dagger w_2 \right) \in \sigma \cap \mathcal{R}^*$, then by (2) and (4), $u_1^\dagger u_2^\dagger \cdots u_m^\dagger w_1^\dagger \equiv v_1^\dagger v_2^\dagger \cdots v_n^\dagger w_2^\dagger$ and $w_1 = w_2$, hence

$$u_1^\dagger u_2^\dagger \cdots u_m^\dagger w_1 \equiv u_1^\dagger u_2^\dagger \cdots u_m^\dagger w_1^\dagger w_1 \equiv v_1^\dagger v_2^\dagger \cdots v_n^\dagger w_2^\dagger w_2 \equiv v_1^\dagger v_2^\dagger \cdots v_n^\dagger w_2,$$

whence we have $\sigma \cap \mathcal{R}^* = id_{FLGC(X)}$.

(6) If $u_1^\dagger u_2^\dagger \cdots u_m^\dagger w_1 \leq_r v_1^\dagger v_2^\dagger \cdots v_n^\dagger w_2$ then by Lemma 2.4 and since each $\mathcal{R}^*$-class of a left GC-lpp semigroup contains exactly one idempotent, $u_1^\dagger u_2^\dagger \cdots u_m^\dagger w_1^\dagger \omega v_1^\dagger v_2^\dagger \cdots v_n^\dagger w_2^\dagger$ and

$$u_1^\dagger u_2^\dagger \cdots u_m^\dagger w_1 \equiv u_1^\dagger u_2^\dagger \cdots u_m^\dagger w_1^\dagger \circ v_1^\dagger v_2^\dagger \cdots v_n^\dagger w_2,$$

hence $w_1 = w_2$. The converse statement follows immediately by routine checking. $\qquad\square$

We now proceed to prove the main theorem of this paper.

**Theorem 3.1.** *Let $X$ be a non-empty set. Then $FLGC(X)$ is a free left GC-lpp semigroup on $X$.*

**Proof.** Let $T$ be a left GC-lpp semigroup and $\varphi$ a function of $X$ into $T$. Define $\theta : X \cup X^{\dagger} \to T$ by

$$\theta(x) = \varphi(x) \text{ and } \theta(x^{\dagger}) = (\varphi(x))^{\dagger}.$$

Because $U(X)$ is a free unary semigroup, the function $\theta$ can be extended to a homomorphism $\theta : U(X) \to T$ in the way: $\theta(x_1 x_2 ... x_n) = \theta(x_1)\theta(x_2)...\theta(x_n)$ for any $x_1, x_2, ..., x_n \in X$. It is straightforward to check that $\theta$ preserves the unary operation $\dagger$. Hence the kernel $Ker\theta$ of $\theta$ is a $\dagger$-congruence on $U(X)$. Thus by Proposition 2.3, the relations $(2)-(8)$ are contained in $Ker\theta$, and $\gamma \subseteq Ker\theta$, thereby there is a $\dagger$-homomorphism $\phi$ of $FLGC(X)$ onto $T$ such that $\theta = \phi\gamma^{\#}$, where $\gamma^{\#}$ is the $\dagger$-homomorphism of $U(X)$ onto $FLGC(X)$ induced by $\gamma$. Therefore, we have shown that $FLGC(X)$ is a free left GC-lpp semigroup. Our proof is completed. $\square$

## 4. Semidirect product and free left GC-lpp semigroups

Let $T$ be a monoid acting on the left of a left regular band $L$ by morphisms. Then, there is a map from $T \times L$ into $L$, given by $(t, x) \mapsto t.x$ such that for all $s, t \in T$ and for all $x, y \in L$,

$$1.x = x, \ (st).x = s.(t.x), \text{ and } s.(xy) = (s.x)(s.y).$$

The *semidirect product* $L * T$ is the semigroup whose underline set is $L \times T$ with a binary operation given by

$$(x, s)(y, t) = (x(s.y), st).$$

**Proposition 4.1.** *Let $M$ be a right cancellative monoid and $L$ a left regular band. Then the semidirect product $L * M$ is a left GC-lpp semigroup whose band of idempotents is isomorphic to $L$ and such that $L * M/\sigma \cong M$.*

**Proof.** By computation, we have $E(L * M) = \{(x, 1) : x \in L\}$ and further, we observe that it is a left regular band.

Let $(x, a) \in L*M$. If $(u, b), (w, c) \in L*M$ and $(u, b)(x, a) = (w, c)(x, a)$, then $(u(b.x), ba) = (w(c.x), ca)$ and by comparing their components, we have $u(b.x) = w(c.x)$ and $ba = ca$. The second equality shows that $b = c$ since $M$ is right cancellative. Thus, we have

$$(u, b)(x, 1) = (u(b.x), b) = (w(c.x), c) = (w, c)(x, 1).$$

Together with $(x,1)(x,a) = (x(1.x),a) = (x,a)$, we can easily deduce that $(x,1)\mathcal{R}^*(x,a)$. Therefore, $L * M$ is a left abundant semigroup.

Finally, if $(y,1) \in E(L * M)$, then

$$(x,a)(y,1) = (x(a.y),a) = (x(a.y)x, a = (x(a.y),1)(x,a).$$

On the other hand, by the foregoing proof, we have $(x(a.y),1)\mathcal{R}^*(x(a.y),a)$. Thus,

$$(x,a)(y,1) = ((x,a)(y,1))^\dagger(x,a).$$

Consequently, $L * M$ is a left GC-lpp semigroup. □

**Theorem 4.1.** *Let $X$ be a non-empty set. Then the free left GC-lpp semigroup $FLGC(X)$ can be embedded into a semidirect product of the left regular band $E(FLGC(X))$ and the free monoid $F^*(X)$ on $X$.*

**Proof.** Define an action of $F^*(X)$ on $FLGB(X)$ as follows: for all $w \in E(FLGC(X))$ and $a \in F^*(X)$,

$$a.w = (aw)^\dagger.$$

Notice that $a(uv) = (au)^\dagger av$ for $u,v \in E(FLGC(X))$. We observe that $(a(uv))^\dagger = (au)^\dagger(av)^\dagger$, thereby, wqe have $a.(uv) = (a.u)(a.v)$. This shows that the above action is a homomorphism. Consider the mapping

$$\theta : FLGC(X) \to E(FLGC(X)) * F^*(X); \quad u_1^\dagger u_2^\dagger \cdots u_m^\dagger a \mapsto (u_1^\dagger u_2^\dagger \cdots u_m^\dagger a^\dagger, a),$$

where $ua = red(ua)$.

By the multiplication of $FLGC(X)$ and by routine checking one can easily shows that $\theta$ is a homomorphism and by Lemma 3.1, $\theta$ is injective. The proof is completed. □

## 5. A proper cover theorem for left GC-lpp semigroups

Similar as in Ref. 15, we call the homomorphism $\varphi : S \to T$ of semigroups an $\mathcal{R}^*$-*homomorphism* if $a\varphi = b\varphi$ implies that $a\mathcal{R}^*b$. When $S$ and $T$ are left GC-lpp semigroups, it is easy to check that $\varphi$ is an $\mathcal{R}^*$-homomorphism if $\varphi$ is an idempotent-separating †-homomorphism. Now let $S$ be a left GC-lpp semigroup. If there exist a proper left GC-lpp semigroup (F-lpp semigroup) $U$ and a surjective $\mathcal{R}^*$then we shall say that $U$ is a *proper cover* (an *F-cover*) of $S$.

The following theorem is the main result in Ref. 18. We provide here a simple proof.

**Theorem 5.1.** *Any left GC-lpp semigroup has proper covers.*

**Proof.** Every left GC-lpp semigroup is a †-homomorphic image of some free left GC-lpp semigroup. Thus $S \cong FLGC(X)/\rho$ for some $X$ and some †-congruence $\rho$ on $FLGC(X)$. Now $\rho_{min} \subseteq \rho$; so $FLGC(X)/\rho \cong (FLGC(X)/\rho_{min})/(\rho/\rho_{min})$. Since $FLGC(X)$ is a proper left GC-lpp semigroup, it follows from Proposition 2.4 that $FLGC(X)/\rho_{min}$ is a proper left GC-lpp semigroup. By Proposition 2.2, $\rho|_{E(FLGC(X))} = \rho_{min}|_{E(FLGC(X))}$, whence $\rho/\rho_{min}$ is an idempotent-separating †-congruence, thus the homomorphism induced by $\rho/\rho_{min}$ is an $\mathcal{R}^*$-homomorphism. Consequently, $FLGC(X)/\rho_{min}$ is a proper cover of $S$. $\square$

**Lemma 5.1.** $FLGC(X)^1$ *is an F-lpp semigroup.*

**Proof.** By Proposition 3.1 (4), the $\sigma$-class of $FLGC(X)$ containing $u_1^\dagger u_2^\dagger \cdots u_m^\dagger v$ is

$$\{x_1^\dagger x_2^\dagger \cdots x_n^\dagger v : x_i \in F(X), n \text{ is a positive integer}\},$$

and by Proposition 4.1 (6), $v$ is its greatest element. On the other hand, the set of all idempotents of $FLGC(X)$ is one $\sigma$-class in which 1 is the greatest element. Thus $FLGC(X)^1$ is an F-lpp semigroup. $\square$

We now return to prove Theorem 5.1. If $S$ is a left GC-lpp semigroup, then there exists a set $X$ and a surjective †-homomorphism $\phi$ of $FLGC(X)$ onto $S$. Define

$$\varphi : FLGC(X)^1 \to S; \ x \mapsto \varphi(x) = \begin{cases} \phi(x) \text{ if } x \in FLGC(X); \\ \iota \quad \text{ if } x = 1, \end{cases}$$

where $\iota$ is the identity of $S$. It is not difficult to see that $\varphi$ is a surjective †-homomorphism of $FLGC(X)^1$ onto $S$. Now by using the same arguments as in the proof of Theorem 5.1, we are able to state the main theorem of this paper.

**Theorem 5.2.** *Any left GC-lpp monoid has a F-cover.*

### Acknowledgement

This research is jointly supported by the National Natural Science Foundation of China (grant: 11761034; 11361027; 11661042); the Natural Science Foundation of Jiangxi Province (grant: 20161BAB201018) and the Science Foundation of the Education Department of Jiangxi Province, China (grant: GJJ14251).

## References

1. X.J. Guo, Y.Q. Guo and K. P. Shum, Left abundant semigroups, *Comm. Algebra* 32(2004), 2061-2085.
2. J.B. Fountain, Adequate semigroups, *Proc. Edinburgh Math. Soc.* 22(1979), 113-125.
3. M. Petrich, *Inverse semigroups*, Wiley and Sons, New York, 1985.
4. M.B. Szendrei, $E$-unitary $\mathcal{R}$-unipotent semigroups, *Semigroup Forum* 32(1985), 87-96.
5. T. Saito, Proper ordered inverse semigroups, *Pacific J. Math.* 15(1965), 649-666.
6. D.B. McAlister, Groups, semilattices and inverse semigroups, $I$; $II$, *Trans. Amer. Math. Soc.* 192(1974), 227-244; ibid. 196(1974), 351-370.
7. K. Takizawa, $E$-unitary $\mathcal{R}$-unipotent semigroups, *Bull. Tokyo Gakugei University* 30(1978), 21-33.
8. M.B. Szendrei, $E$-unitary regular semigroups, *Proc. Royal Soc. Edingburgh* 106A(1987), 89-102.
9. B. Billhart and M.B. Szendrei, Weakly $E$-unitary locally inverse semigrops, *J. Algebra* 267(2003), 559-576.
10. H.E. Scheiblich, Free inverse semigroups, *Proc. Amer. Math. Soc.* 38(1973), 1-7.
11. J.B. Fountain and G.M.S. Gomes, Proper left type-A monoids revisited, *Glasgow Math. J.* 35(1993), 293-306.
12. J.B. Fountain and G.M.S. Gomes, Proper covers of ample monoids, *Proc. Edinb. Math. Soc. (ser. 2)* 49(2006), 277-289.
13. G.M.S. Gomes and M.B. Szendrei, Almost factorizable weakly ample semigroups, *Comm. Algebra* 35(2007), 3503-3523.
14. J.B. Fountain, Free right h-adequate semigroups, In: *Semigroups, theory and applications*, pp. 97-120, Lecture Notes in Mathematics 1320, Springer, 1988.
15. J.B. Fountain, Free right type A semigroups, *Glasgow math. J.* 33(1991), 135-148.
16. X.J. Guo, Abundant left C-lpp proper semigroups, *Southeast Asian Bull. Math.* 23(2000), 41-50.
17. X.J. Guo and K. P. Shum, On left cyber groups, *Inter. Math. J.* 5(2004), 705-717.
18. X.J. Guo and Z.J. Tian, Left regular bands, right cancellative monoids and left regular type-A monoids, *J. Math. Res. Exposition* 19(1999), 563-568.
19. X.J. Guo, X.F. Ni and K. P. Shum, Generalizations of E-unitary inverse semigroups by using McAlister's approach, *Asian-European J. Math.* 1(2008), 535-553.
20. X.J. Guo, X.P. Li and K. P. Shum, F-rpp semigroups, *Inter. Math. Forum* 1(2006), 1571-1585.
21. H. Huang, X.J. Guo and H. Chen, The structure of F-rpp semigroups, *Inter. Math. Forum* 2(2007), 771-779.
22. X.J. Guo and K. P. Shum, Proper left GC-lpp semigroups and a generalized version of McAlister's P-theorem, *Algebra Colloquium* 18(2011), 475-486.

23. X.J. Guo, On the structure of E-unitary covers for an orthodox semigroup, *Adv. Math. (China)*, 26(1997), 181-182.

24. X.J. Guo and X.Y. Xie, A note on left type-A semigroups, *Semigroup Forum* 58(1999), 313-316.

25. A. El Qallali and J.B. Fountain, Proper covers for left ample semigroups, *Semigroup Forum* 71(2005), 411-427.

26. R.R. Cui and X.J. Guo, F-covers for right type-A semigroups, *J. Math. Res. Exposition* 30(2010), 791-798.

27. J.B. Fountain, Abundant semigroups, *Proc. London Math. Soc.* (3)44(1982), 103-129.

28. J.M. Howie, *An introduction to semigroup theory*, Academic Press, London, 1976.

29. M. Petrich, *Lectures in semigroups*, Ackademic Verlag, Berlin, 1977.

30. D.B. McAlister, One-to-one partial right translations of a right cancellative semigroup, *J. Algebra* 43(1976), 231-251.

31. F. Pastijn, A representation of a semigroup of matrices over a group, *J. Algebra* 43(1974), 351-370.

32. S. Burries and H.P. Sankappanavar, *A course in Universsal Algebra*, World Scientific Publishing Co. Pte. Ltd., Singapore, 1981.

33. X.J. Guo and F.Y. Luo, The natural partial orders on abundant semigroups, *Adv. Math. (China)* 34(2005), 297-308.

34. M. Petrich, N.R. Reilly. *Completely regular semigroups*. John Wiley & Sons, INC., 1999.

# On the cover-avoiding properties in finite groups*

Xiuyun Guo

*Department of Mathematics, Shanghai University, Shanghai 200444, P. R. China*

*xyguo@staff.shu.edu.cn*

Lingling Han

*Department of Mathematics, Shanghai University, Shanghai 200444, P. R. China*

*hanll@shu.edu.cn*

K. P. Shum

*Institute of Mathematics, Yunnan University, Kunming 650091, P. R. China*

*kpshum@ynu.edu.cn*

This article is dedicated to celebrate the 80th birthday of Professor Leoind Bokut

In this paper, we give a survey concerning the influence and the related progress of the cover-avoiding subgroups on the structure of finite groups.

*Keywords*: cover-avoiding property, semi cover-avoiding subgroup

## 1. Introduction

Throughout this paper, all groups considered are finite groups.

Recall that a subgroup $H$ of a group $G$ covers a chief factor $A/B$ of $G$ if $HA = HB$ and $H$ is said to be avoiding a chief factor $A/B$ of $G$ if $H \cap A = H \cap B$. We say that a subgroup $H$ of a group $G$ has the cover-avoiding property in $G$ if $H$ either covers or avoids every chief factor of $G$, in this case we may also say that $H$ is a cover-avoiding subgroup of $G$ (in short, a CAP-subgroup). Hall in 1937[25] defined and investigated *system normalizers* of groups, and he proved that in a soluble group the system normalizers are nilpotent and any two are conjugate. Furthermore,

*Supported by the National Natural Science Foundation of China(11771271).

he also proved that a system normalizer of a soluble group $G$ covers the central chief factors and avoids the noncentral chief factors of $G$. Carter found Carter subgroups, and he noticed that any two Carter subgroups of a soluble group are conjugate and a Carter subgroup of a soluble group must contain a system normalizer[10,11]. W. Gaschütz [G][15] introduced a certain conjugacy class of subgroups of a solvable group which he called pre-Frattini subgroups, and he proved that these subgroups have the property that they not only avoid the complemented chief factors of the solvable group $G$ but also cover the rest of its chief factors. Thereafter, many authors studied this property, for example, Gillam[16] and Tomkinson[42]. In these papers, the main aim was to find some kind of subgroups of a soluble group having the cover and avoidance properties. However, the question arises whether we can obtain structural insight into a group when some of its subgroups have the cover-avoiding properties. In fact, in recently years, many interesting results have been given and proved under the assumption that some kind of families of subgroups have the cover-avoiding properties.

In this survey article, we discuss the influence and mention the related progress of the cover-avoiding-property subgroups on the structure of groups. The notations and terminologies that we use in this paper are standard.

## 2. Cover-avoiding properties and the structure of groups

In this section we mainly concentrate on those results which are closely related to the cover-avoiding property of subgroups of a given group.

In 1993, Ezquerro[13] first gave some characterizations for a group $G$ to be a $p$-supersolvable and a supersolvable group based on the assumption that all maximal subgroups of some Sylow subgroup of $G$ have the cover-avoiding properties.

In 2003, Guo and Shum[21] pushed further this approach and obtained some characterizations for a solvable group based on the assumption that some of its maximal subgroups or 2-maximal subgroups have the cover-avoiding properties. They also investigated the $p$-solvability and $p$-nilpotency of a group provided some of its subgroups have the cover-avoiding properties. After the above results have been published, it attracts the attention of many authors and they have later joined in the research in cover-avoiding properties of groups.

We now state some of the main results in detail.

**Definition 2.1.** Let $H$ be a subgroup of a group $G$ and $N/K$ a normal factor of $G$. We say that:

(1) $H$ covers $N/K$ if $HN = HK$.

(2) $H$ avoids $N/K$ if $H \cap N = H \cap K$.

We say that $H$ has the cover-avoiding properties in $G$ if $H$ either covers or avoids every chief factor of $G$. In short, we simply say that $H$ is a CAP-subgroup.

Now we list some families of subgroups:

Let $G$ be a group and $p$ a prime number. Let

$\mathcal{F} = \{M \mid M \text{ is a maximal subgroup in } G \}$

$\mathcal{F}_n = \{M \mid M \in \mathcal{F} \text{ and } M \text{ is non nilpotent } \}$

$\mathcal{F}_c = \{M \mid M \in \mathcal{F} \text{ whose index } |G : M| \text{ is composite } \}$

$\mathcal{F}^p = \{M \mid M \in \mathcal{F} \text{ and } N_G(P) \leq M \text{ for a Sylow } p\text{-subgroup } P \text{ of } G\}$

$\mathcal{F}^{op} = \bigcup\limits_{p \in \pi(G) - \{2\}} \mathcal{F}^p$

$\mathcal{F}^{pcn} = \mathcal{F}^p \bigcap \mathcal{F}_c \bigcap \mathcal{F}_n$

$\mathcal{F}^{ocn} = \mathcal{F}^{op} \bigcap \mathcal{F}_c \bigcap \mathcal{F}_n$

Then we define the following:

$$S^{pcn}(G) = \bigcap\{M \mid M \in \mathcal{F}^{pcn}\}$$

if $\mathcal{F}^{pcn}$ is non-empty; otherwise $S^{pcn}(G) = G$.

$$S^{ocn}(G) = \bigcap\{M \mid M \in \mathcal{F}^{ocn}\}$$

if $\mathcal{F}^{ocn}$ is non-empty; otherwise $S^{ocn}(G) = G$.

It is clear that $S^{pcn}(G)$ and $S^{ocn}(G)$ are characteristic subgroups of $G$, and

$$\Phi(G) \leq S^{ocn}(G) \leq S^{pcn}(G).$$

**Theorem 2.1.** [21] *(Corollary 2.5)* *Let $p$ be the largest prime number dividing the order of the group $G$. If every maximal subgroup $M$ of $G$ in $\mathcal{F}^p \cap \mathcal{F}_c$ is nilpotent, then $G$ is $p$-solvable.*

**Theorem 2.2.** [21] *(Lemma 2.6)* *For any group $G$, $S^{ocn}(G)$ is solvable.*

**Theorem 2.3.** [21] *(Corollary 2.7)* *If every maximal subgroup $M$ of a group $G$ in $\mathcal{F}^{op} \cap \mathcal{F}_c$ is nilpotent, then $G$ is solvable.*

**Theorem 2.4.** [21] *(Lemma 2.8)* *Let $N$ be a minimal normal subgroup and $M$ a maximal subgroup of a group $G$. If $M$ is solvable and $M \cap N = 1$, then $G$ is solvable.*

**Theorem 2.5.** [21 *(Corollary 2.9)*] *A group $G$ is solvable if and only if there exists a maximal subgroup $M$ of $G$ such that $M$ is a solvable $CAP$-subgroup of $G$.*

**Theorem 2.6.** [21 *(Theorem 3.1)*] *A group $G$ is solvable if and only if every maximal subgroup $M$ of $G$ in $\mathcal{F}^{ocn}$ is a $CAP$-subgroup of $G$.*

**Theorem 2.7.** [21 *(Theorem 3.2)*] *Let $H_1$ and $H_2$ be two Hall subgroups of a group $G$ such that $G = H_1 H_2$. Then $G$ is a solvable group if and only if $H_1$ and $H_2$ are both solvable $CAP$-subgroups of $G$.*

**Theorem 2.8.** [21 *(Theorem 3.4)*] *If every 2-maximal subgroup of a group $G$ is a $CAP$-subgroup of $G$, then $G$ is solvable.*

**Example 2.1.** [21 (Example 3.5)] *Let $G = A_4$, the alternating group of degree 4. Also let $H$ be a Sylow 2-subgroup of $G$. Then $|H| = 2^2$ and $H$ is a minimal normal subgroup of $G$. It is clear that every minimal subgroup of $H$ is a 2-maximal subgroup of $G$ but it is not a $CAP$-subgroup of $G$.*

**Theorem 2.9.** [21 *(Theorem 3.7)*] *A group $G$ is solvable if and only if there exists a solvable 2-maximal subgroup $L$ of $G$ such that $L$ is a $CAP$-subgroup of $G$.*

**Theorem 2.10.** [21 *(Theorem 3.8)*] *Let $G$ be a group and $p$ the largest prime number dividing the order of $G$. If every maximal subgroup $M$ of $G$ in $\mathcal{F}^{pcn}$ is a $CAP$-subgroup of $G$, then $G$ is $p$-solvable.*

**Theorem 2.11.** [21 *(Theorem 3.9)*] *Let $p$ be a prime dividing the order of the group $G$ and $P$ a Sylow $p$-subgroup of $G$. Then $G$ is $p$-solvable if and only if $P$ is a $CAP$-subgroup of $G$.*

**Theorem 2.12.** [21 *(Theorem 3.11)*] *Let $H$ be a normal subgroup of a group $G$ and $p$ the smallest prime number dividing the order of $H$. If all 2-maximal subgroups of every Sylow $p$-subgroup of $H$ are $CAP$-subgroups of $G$ and $G$ is $A_4$-free, then $H$ is $p$-nilpotent.*

**Theorem 2.13.** [21 *(Theorem 3.14)*] *Let $H$ be a normal subgroup of a group $G$ and $p$ the smallest prime number dividing the order of $H$. If all maximal subgroups of every Sylow $p$-subgroup of $H$ are $CAP$-subgroups of $G$, then $H$ is $p$-nilpotent.*

**Theorem 2.14.** [21 *(Theorem 3.15)*] *Let $\mathcal{F}$ be the class of groups with Sylow tower of supersolvable type and $H$ a normal subgroup of a group $G$ such* that $G/H \in \mathscr{F}$. If $G$ is $A_4$-free and all 2-maximal subgroups of every Sylow subgroup of $H$ are $CAP$-subgroups of $G$, then $G$ is in $\mathscr{F}$.

There are also some other investigations. We now list in below some important results.

**Theorem 2.15.** [29(Theorem 1)] Let $H/K$ be a chief factor of $G$. Then the following statements are equivalent in pairs.
(1) $H/K$ is soluble.
(2) Every maximal subgroup of $G$ covers or avoids $H/K$.
(3) Every maximal subgroup of $G$ in $\mathcal{F}^{ocn}$ covers or avoids $H/K$.
(4) Every Hall subgroup of $G$ covers or avoids $H/K$.
(5) There exists a prime $p \in \pi(H/K)$ and $P \in Syl_p(G)$ such that $P$ covers or avoids $H/K$.

Let $p$ be a prime dividing $|G|$ and let $P$ be a Sylow $p$-subgroup of $G$. Consider $\mathcal{M}(P)$ to be the set of all maximal subgroups of $P$.

**Definition 2.2.** [35(Definition 1.1)] Let $d$ be the minimum number of generators of $P$. Then there exist maximal subgroups $P_1, \cdots, P_d$ of $P$ such that $\bigcap_{i=1}^{d} P_i = \Phi(P)$, the Frattini subgroup of $P$. By $\mathcal{M}_d(P)$ denote the family of subgroups $P_1, \cdots, P_d$ with the above property.

The following theorems are core theorems of CAP-subgroups as they constitute the theory of CAP subgroups of a group.

**Theorem 2.16.** [35(Theorem 3.1)] Let $G$ be a $p$-solvable group and let $P$ be a Sylow $p$-subgroup of $G$, where $p$ is a fixed prime. Then the following statements are equivalent:
($i$) $G$ is $p$-supersolvable.
($ii$) Every member in $\mathcal{M}(P)$ is a CAP-subgroup of $G$.
($iii$) Every member in a fixed $\mathcal{M}_d(P)$ is a CAP-subgroup of $G$.

**Theorem 2.17.** [35(Theorem 3.2)] Let $G$ be a group. Then the following statements are equivalent:
($i$) $G$ is supersolvable.
($ii$) For each Sylow subgroup $P$ of $G$, every member in $\mathcal{M}(P)$ is a CAP-subgroup of $G$.
($iii$) For each Sylow subgroup $P$ of $G$, every member in a fixed $\mathcal{M}_d(P)$ is a CAP-subgroup of $G$.

## 3. Semi cover-avoiding properties and the structure of groups

In the above section, we have already seen that the cover-avoiding properties of subgroups can greatly influence the structure of groups, however, there exists a subgroup of a finite group which is not a CAP-subgroup. We give the following interesting example:

**Example 3.1.** [14 (Example 4.2)] Let $A_4$ denote the alternative group of degree 4, and $C_2 = \langle c \rangle$ be a cyclic group of order 2 generated by $c$; let $G = C_2 \times A_4$. Then $A_4 = K_4 \cdot \langle t \rangle$ with $K_4 = \langle a, b \rangle$ the Klein Four Group with generators elements $a$ and $b$ of order 2. Take $H = \langle ac \rangle$ be the subgroup of $G$ generated by $ac$. Then it is easy to see that $H$ covers or avoids every chief factor of the following chief series

$$1 < K_4 < A_4 < G.$$

However,

$$HC_2 = \langle a, c \rangle \neq H \cdot (K_4 \times C_2)$$

and

$$H \cap (K_4 \times C_2) = H \neq 1 = H \cap C_2 .$$

Thus $H$ is not a CAP-subgroup in $G$ since $(K_4 \times C_2)/C_2$ is a chief factor of $G$.

In the above example, we see immediately that the following definition is necessary.

**Definition 3.1.** [14 (Definition 2.1)] Let $H$ be a subgroup of a group $G$. Then $H$ is said to be *semi cover-avoiding* in $G$ if there is a chief series

$$1 = G_0 < G_1 < \cdots < G_\ell = G$$

of $G$ such that for every $i = 1, \cdots, \ell$, either $H$ covers $G_j/G_{j-1}$ or $H$ avoids $G_j/G_{j-1}$. In short, $H$ is said to be a semi CAP-subgroup of $G$.

Naturally, we have the local cases.

**Definition 3.2.** [14 (Definition 2.2)] Let $G$ be a group and $H$ a subgroup of $G$. Then, the subgroup $H$ is said to be semi $p$-cover-avoiding in $G$ if there is a chief series $1 = G_0 < G_1 < \ldots < G_t = G$ of $G$ such that for every $i = 1, \cdots, t$, if $G_i/G_{i-1}$ is $p$-singular, then either $H$ covers $G_i/G_{i-1}$ or $H$ avoids $G_i/G_{i-1}$. In short, we call $H$ a semi $p$-CAP-subgroup.

**Definition 3.3.** [46](Definition 2.1) Let $L$ be a subgroup of a group $G$. If there is a chief series of $G$ such that $L$ covers or avoids every $\pi$-chief factor of this series, then we call $L$ a semi $\pi$-cover-avoiding subgroup of $G$, or say that $L$ has the semi $\pi$-cover-avoidance properties in $G$.

Recently, many authors have investigated the structure of groups by using the semi CAP-subgroups, semi $p$-CAP-subgroups or semi $\pi$-CAP-subgroups (One should notice that some authors call semi CAP-subgroups and semi $p$-CAP-subgroups as the partially CAP-subgroups and partially $p$-CAP-subgroups respectively), for example, the reader can refer to [1, 2, 5-7, 12, 14, 15, 17-19, 22-27, 36-39, 41, 44-46, 49, 50], etc.

We now list below some frequently cited theorems which are related to the semi cover-avoiding subgroups in a given group.

**Theorem 3.1.** [14](Theorem 2.2) *Let $G$ be a group. Then, the following three statements are equivalent:*
  *(1) $G$ is a solvable group.*
  *(2) Every Sylow subgroup of $G$ is semi cover-avoiding in $G$.*
  *(3) Every maximal subgroup of $G$ is semi cover-avoiding in $G$.*

**Theorem 3.2.** [24](Corollary 2.11) *A group $G$ is solvable if and only if there exists a maximal subgroup $M$ of $G$ such that $M$ is a solvable semi cover-avoiding subgroup of $G$.*

**Theorem 3.3.** [24](Theorem 3.1) *Let $G$ be a group and $p$ the largest prime number dividing the order of $G$. Then $G$ is $p$-solvable if and only if every maximal subgroup $M$ of $G$ in $\mathcal{F}^{pcn}$ is a semi $p$-cover-avoiding subgroup of $G$.*

**Theorem 3.4.** [24](Theorem 3.2) *Let $G$ be a group and $p$ be an odd prime dividing the order of $G$. Then $G$ is $p$-solvable if and only if $M$ is a semi $p$-cover-avoiding subgroup of $G$ for every maximal subgroup $M$ of $G$ in $\mathcal{F}^p \bigcap \mathcal{F}_n$.*

**Theorem 3.5.** [24](Theorem 3.3) *Let $G$ be a group. Then $G$ is solvable if and only if $M$ is a semi 2-cover-avoiding subgroup of $G$ for every maximal subgroup $M$ of $G$ in $\mathcal{F}^2$.*

**Theorem 3.6.** [19](Theorem 3.2) *Let $p$ be the smallest prime dividing the order of the group $G$ and let $P$ be a $p$-Sylow subgroup of $G$. If $P$ is cyclic or every maximal subgroup of $P$ has the semi cover-avoiding property in $G$, then $G$ is $p$-nilpotent.*

**Theorem 3.7.** [19 *(Theorem 3.5)*] *Let $p$ be an odd prime dividing the order of the group $G$ and $P$ a $p$-Sylow subgroup of $G$. If $N_G(P)$ is $p$-nilpotent and every maximal subgroup of $P$ has the semi cover-avoiding property in $G$, then $G$ is $p$-nilpotent.*

**Theorem 3.8.** [19 *(Theorem 3.7)*] *Let $p$ be a prime dividing the order of the $p$-solvable group $G$ and let $P$ be a $p$-Sylow subgroup of $G$. If $P$ is a cyclic group or every maximal subgroup of $P$ has the semi cover-avoiding property in $G$, then $G$ is $p$-supersolvable.*

**Theorem 3.9.** [19 *(Theorem 3.12)*] *Let $\mathfrak{F}$ be a saturated formation containing the class of supersolvable groups $\mathfrak{U}$ and let $N$ be a normal subgroup of the group $G$ such that $G/N$ belongs to $\mathfrak{F}$. Suppose that, for every prime $p$ dividing the order of $N$, there is a $p$-Sylow subgroup $P$ of $N$ such that every minimal subgroup of $P \cap O_p(G)$ has the semi cover-avoiding property in $G$ and that, when $p = 2$, either every cyclic group of order $4$ of $P \cap O_p(G)$ also has the semi cover-avoiding property in $G$ or that $P \cap O_p(G)$ is quaternion-free. Then $G$ belongs to $\mathfrak{F}$.*

**Theorem 3.10.** [45 *(Theorem 3.1)*] *Let $G$ be a group and $p$ a prime divisor of the order of $G$ such that $(|G|, p^2 - 1) = 1$. Then, $N$ is a normal subgroup of $G$ such that $G/N$ is $p$-nilpotent, if every cyclic subgroup of $N$ with order $\leq p^{\mu_p}$ $(\mu_p = 1 + \frac{1+(-1)^p}{2})$ is semi $p$-cover-avoiding in $G$, then $G$ is $p$-nilpotent.*

We now consider the so called $ss$-quasi normal subgroups of a given group $G$. With the aid of the formation theory of groups, generalized semi CAP-subgroups can be further studied.

The followings are some typical results.

**Definition 3.4.** [27] We call a subgroup $H$ of $G$ an $ss$-quasi normal subgroup of $G$ if there is a subgroup $B$ of $G$ such that $G = HB$ and $H$ permutes with every Sylow subgroup of $B$.

**Theorem 3.11.** [27 *(Theorem 3.1)*] *Let $G$ be a group and $p$ a prime divisor of $|G|$ with $(|G|, p - 1) = 1$. Let $P$ be a Sylow $p$-subgroup of $G$. If all maximal subgroups of $P$ are either semi $p$-cover-avoiding or $ss$-quasi normal subgroups in $G$, then $G$ is $p$-nilpotent.*

**Theorem 3.12.** [27 *(Theorem 3.6)*] *Let $\mathfrak{F}$ be a saturated formation containing $\mathfrak{U}$, the class of all supersoluble groups. Suppose that $G$ is a group with a solvable normal subgroup $H$ such that $G/H \in \mathfrak{F}$. If all maximal subgroups* *of all Sylow subgroups of $F(H)$ are either semi p-cover-avoiding or ss-quasinormal in $G$, then $G \in \mathfrak{F}$.*

**Theorem 3.13.** [27(Theorem 3.7)] *Let $\mathfrak{F}$ be a saturated formation containing $\mathfrak{U}$, the class of all supersoluble groups. Suppose that $G$ is a group with a normal subgroup $H$ such that $G/H \in \mathfrak{F}$. If all maximal subgroups of all Sylow subgroups of $F^*(H)$ (the generalized Fitting subgroup of $H$) are either semi p-cover-avoiding or ss-quasinormal in $G$, then $G \in \mathfrak{F}$.*

Now we list below another families of subgroups:

Let $G$ be a group, $H$ a normal subgroup of $G$ and $p$ a prime number. Let

$$\mathcal{F}^\pi = \bigcup_{p \in \pi} \mathcal{F}^p$$
$$\mathcal{F}_p = \{M \mid M \in \mathcal{F} \text{ and } |G : M|_p = 1\}$$
$$\mathcal{F}_{pc} = \mathcal{F}_p \cap \mathcal{F}_c$$
$$\mathcal{F}_{pcn} = \mathcal{F}_p \cap \mathcal{F}_c \cap \mathcal{F}_n$$
$$\mathcal{F}_H = \{M \mid M \in \mathcal{F} \text{ and } H \not\leq M\}$$

The following theorems constitute the fundamental theory of semi $\pi$-covering-avoiding subgroups of a given group $G$.

**Theorem 3.14.** [46(Theorem 3.1)] *A group $G$ is $\pi$-solvable if and only if $M$ is a semi $\pi$-cover-avoiding subgroup of $G$ for any $M \in \mathcal{F}^\pi(G)$.*

**Theorem 3.15.** [46(Theorem 3.5)] *A group $G$ is $\pi$-solvable if and only if $M$ is a semi $\pi$-cover-avoiding subgroup of $G$ for any $M \in \mathcal{F}_{pcn}(G)$, where $p$ is the largest prime dividing the order of $G$.*

**Theorem 3.16.** [46(Theorem 3.7)] *A group $G$ is $\pi$-separable if and only if there exists a Hall $\pi$-subgroup $H$ of $G$ such that $H$ is semi $\pi$-cover-avoiding in $G$.*

**Theorem 3.17.** [46(Theorem 3.8)] *A group $G$ is $\pi$-solvable if there exists a solvable Hall $\pi$-subgroup $H$ of $G$ such that $H$ is semi $\pi$-cover-avoiding in $G$.*

**Theorem 3.18.** [46(Theorem 3.9)] *A group $G$ is $\pi$-solvable if and only if every Sylow subgroup of $G$ is semi $\pi$-cover-avoiding in $G$.*

**Theorem 3.19.** [46(Theorem 3.11)] *Let $G$ be a group. If every 2-maximal subgroup of $G$ is semi $\pi$-cover-avoiding in $G$, then $G$ is $\pi$-solvable.*

**Theorem 3.20.** [46(Theorem 3.14)] *Suppose that $\mathfrak{F}$ is a saturated formation containing the class of supersolvable groups. Let $H$ be a normal subgroup* of a group $G$ such that $G/H \in \mathfrak{F}$. If for any Sylow subgroup $S$ of $H$, either $S$ is cyclic or every maximal subgroup of $S$ is semi-cover-avoiding in $G$, then $G \in \mathfrak{F}$.

**Theorem 3.21.** [1 (Main theorem)] Let $p$ be a prime number, let $G$ be a group, and let $G^+ = G/O_{p'}(G)$. Then every subgroup of $G$ of order $p^2$ is a semi CAP-subgroup of $G$ if and only if one of the following statements holds:

(1) the order of the Sylow $p$-subgroups of $G$ is at most $p$;

(2) $G$ is a $p$-supersoluble group;

(3) $\Phi(G^+) = 1$ and, if $P$ is a Sylow $p$-subgroup of $G$, $P^+ = Soc(G^+) = V_1 \times \cdots \times V_r$, where $V_1, \cdots, V_r$ are minimal normal subgroups of $G^+$ which are $G^+$-isomorphic to a 2-dimensional irreducible $G^+$-module $V$ over the Galois field $GF(p)$. Furthermore, $V$ is not an absolutely irreducible $G^+$-module when $r > 1$.

**Theorem 3.22.** [5 (Theorem 3.2))] Let $p$ be a prime dividing the order of a group $G$. Suppose that all maximal subgroups of every Sylow $p$-subgroup of $G$ are semi CAP-subgroups of $G$. Then, either $G$ is a group whose Sylow $p$-subgroups are cyclic groups of order $p$, or $G$ is a $p$-supersoluble group.

**Definition 3.5.** [6 (Definition 3.8)] Given a group $G$, a subgroup $K$ of $G$ is called a second maximal subgroup, or 2-maximal subgroup, if there exists a maximal subgroup $M$ of $G$ such that $K \leq M$ and $K$ is maximal in $M$.

**Theorem 3.23.** [7 (Theorem 15)] Let $G$ be a $p$-soluble group. The following statements are equivalent.

(1) Every 2-maximal subgroup of each Sylow $p$-subgroup of $G$ is a semi CAP-subgroup of $G$.

(2) One of the following statements holds.

(a) $G$ is $p$-supersoluble.

(b) If $P$ is a Sylow $p$-subgroup of $G$ and $Q$ is a 2-maximal subgroup of $P$, then

$$\Phi(G/O_{p'}(G)) \leq QO_{p'}(G)/O_{p'}(G);$$

i. if $\Phi(G/O_{p'}(G)) = QO_{p'}(G)/O_{p'}(G)$, then every chief series of the group $G$ has exactly one complemented $p$-chief factor; moreover, this $p$-chief factor has order $p^2$;

ii. if $\Phi(G/O_{p'}(G)) < QO_{p'}(G)/O_{p'}(G)$, then all complemented $p$-chief factors of $G$ are $G$-isomorphic to a 2-dimensional irreducible $G$-module $V$ which is not an absolutely irreducible $G$-module.

**Theorem 3.24.** [36] *(Theorem 3.1)* Let $G$ be a group, $p$ a prime dividing the order of $G$, and $P$ a Sylow $p$-subgroup of $G$.

(1) If $G$ is $p$-nilpotent, then every maximal subgroup of $P$ is a semi CAP-subgroup of $G$.

(2) If $(|G|, p-1) = 1$ and every maximal subgroup of $P$ is a semi CAP-subgroup of $G$, then $G$ is $p$-nilpotent.

## 4. Further investigations

In this section we will introduce some further investigations:

**Example 4.1.** [20] *(Example 1.1)* Let $N = \langle a \rangle \times \langle b \rangle$ be an elementary abelian 5-group of order $5^2$ and $c \in Aut(N)$ such that $a^c = b^2$, $b^c = a$. Then the semidirect product $G = N \rtimes \langle c \rangle$ is of order $5^2 \times 2^3$. It is clear that $K = \langle b \rangle$ is not a semi $CAP$-subgroup of $G$, but $K$ is a semi $CAP$-subgroup of $H = N \rtimes \langle c^2 \rangle$.

We pose the following open problem:

**Question 4.1.** [20] Describe the structure of groups whenever some kinds of subgroups are semi CAP-subgroups in some subgroup?

We discuss some recent results of the topic.

**Theorem 4.1.** [20] *(Theorem 3.1)* Let $p$ be the smallest prime dividing the order of a group $G$ and let $P$ be a Sylow $p$-subgroup of $G$. Then $G$ is $p$-nilpotent if and only if every maximal subgroup of $P$ is a semi CAP-subgroup in $N_G(P)$ and $P'$ is a semi CAP-subgroup in $G$.

**Remark 4.1.** [20] *(Remark 3.1)* The hypothesis that $P'$ is a semi CAP-subgroup in above Theorem is essential. In fact, Let $G = PSL(2,7)$ and $P$ be a Sylow 2-subgroup of $G$. Since $P = N_G(P)$, every maximal subgroup of $P$ is a semi CAP-subgroup in $N_G(P)$, but $G$ is not 2-nilpotent.

**Remark 4.2.** [20] *(Remark 3.1)* Even if $G$ is a solvable group and $p$ is an odd prime, the hypothesis that $P'$ is a semi CAP-subgroup in the above theorem is also essential. For example, let $H = C_3 \times C_3 \times C_3$ be an elementary abelian group of order $3^3$. Then there is a subgroup $C_{13} \rtimes C_3$ in the automorphism group of $H$, where $C_{13} \rtimes C_3$ is a semidirect product. Let $G = (C_3 \times C_3 \times C_3) \rtimes (C_{13} \rtimes C_3)$ be the corresponding semidirect product and $P \in Syl_3(G)$. Clearly, $P = N_G(P)$. It follows that every maximal subgroup of $P$ is a semi CAP-subgroup in $N_G(P)$, but $G$ is not 3-nilpotent.

**Remark 4.3.** [20](Remark 3.1) Furthermore we can not remove the hypothesis that $p$ is the smallest prime dividing the order of a group $G$ in the above Theorem. In fact, let $P$ be a Sylow 3-subgroup of $A_5$, the alternating group of degree 5. Then every maximal subgroup of $P$ is a semi CAP-subgroup in $N_G(P)$ and $P' = 1$ is a semi CAP-subgroup in $G$, but $A_5$ is not 3-nilpotent.

We list another important results.

**Theorem 4.2.** [20](Theorem 3.3) *Let $\mathfrak{F}$ be a saturated formation containing the class of all supersolvable groups $\mathfrak{U}$ and let $H$ be a normal subgroup of a group $G$ such that $G/H \in \mathfrak{F}$. Suppose that, for all primes $p$ dividing the order of $H$ and for all $P \in Syl_p(H)$, every maximal subgroup of $P$ is a semi CAP-subgroup in $N_G(P)$ and $P'$ is a semi CAP-subgroup in $G$. Then $G \in \mathfrak{F}$.*

We indicate an another possible approach to investigate the extended CAP-subgroups. We first consider the following example.

**Example 4.2.** [32](Example 1.1) Let $P = \langle a, b | a^4 = b^4 = [a, b] = 1 \rangle$ be a direct product of two cyclic groups of order 4 and $c \in Aut(P)$ such that $a^c = a^2b^3$, $b^c = a^3b$. Then the semidirect product: $K = P \rtimes \langle c \rangle$ is of order $2^4 \times 3$. We set $G = K \times C_2$, a direct product of $K$ and a cyclic group $C_2 = \langle d \rangle$ of order 2.

An easy proof shows that $\Phi(G) = \langle a^2, b^2 \rangle$ is a minimal normal subgroup of $G$. We consider the subgroup $H = \langle a^2 \rangle$ of order 2 of $G$. We can see that $H$ covers or avoids every non-Frattini chief factors of $G$. However, $H \cap \Phi(G) = H \neq 1 = H \cap 1$ and $H\Phi(G) = \Phi(G) \neq H$.

Furthermore, If a subgroup $H$ of a group $G$ covers or avoids every non-Frattini chief factor of $G$ and $H \leq M \leq G$, then it does not necessarily follow that $H$ covers or avoids every non-Frattini chief factor of $M$. In fact, we also have the following example:

**Example 4.3.** [32](Example 2.1) Let $G = A_4 \times A_4$, a direct product of two Alternating groups of four letters. We write $V_4 \times V_4$ as $\langle x, y \rangle \times \langle a, b \rangle$ with generators $x, y, a$ and $b$ of order 2. Let $H = \{(1, 1), (x, a), (y, b), (xy, ab)\}$. Then $H$ covers or avoids every non-Frattini chief factor of $G$. Put $M = A_4 \times V_4$. If $K = V_4 \times \langle a \rangle$ and $L = 1 \times \langle a \rangle$, then $K/L$ is a non-Frattini chief factor of $M$. We can deduce that $K/L$ is neither covered nor avoided by $H$.

The above examples show that we may investigate the structure of groups by using some kinds of subgroups which cover or avoid the non-

Frattini chief factors of these groups. In fact, there are plenty of results of this topic, the reader can refer to (Ref. 30–34,43). For the sake of convenience of the readers, we list some of these results.

**Definition 4.1.**[32](Definition 1.1) A subgroup $H$ of a group $G$ is said to be a $CAP^*$-subgroup of $G$ if, for any non-Frattini chief factor $K/L$ of $G$, we have $H$ either covers $K/L$ or avoids $K/L$.

**Theorem 4.3.**[32](Theorem 3.1) *Let $G$ be a group. Then the following statements are equivalent:*
    (1) *$G$ is solvable;*
    (2) *Every Hall subgroup of $G$ is a $CAP^*$-subgroup of $G$;*
    (3) *Every Sylow subgroup of $G$ is a $CAP^*$-subgroup of $G$;*
    (4) *Every maximal subgroup of $G$ is a $CAP^*$-subgroup of $G$.*

**Theorem 4.4.**[33](Theorem 3.1) *Let $H$ be a normal subgroup of $G$. If every maximal subgroup of any Sylow subgroup of $H$ is a $CAP^*$-subgroup of $G$, then $H \leq Z_{\mathfrak{U}_\phi}(G)$ ($Z_{\mathfrak{U}_\phi}(G)$ is the product of all normal subgroups $K$ of $G$ such that all non-Frattini $G$-chief factors of $K$ have prime order).*

**Definition 4.2.**[31](Definition 2.1) A subgroup $H$ of a group $G$ is said to be a semi $CAP^*$-subgroup of $G$ if there is a chief series $1 = G_0 < G_1 < \cdots < G_m = G$ of $G$ such that for every non-Frattini chief factor $G_i/G_{i-1}$, where $i = 1, 2, \cdots, m$, $H$ either covers $G_i/G_{i-1}$ or avoids $G_i/G_{i-1}$.

**Theorem 4.5.**[31](Theorem 3.1) *Let $H$ be a normal subgroup of a group $G$ and let $p$ be the largest prime dividing the order of $G$. If every maximal subgroup $M$ of $G$ in $\mathcal{F}_{pc} \cap \mathcal{F}_H$ is a semi $CAP^*$-subgroup of $G$, then $H$ is solvable.*

We also list some other results here.

**Theorem 4.6.**[3](Theorem 1.3) *Let $\mathfrak{F}$ be a saturated formation containing $\mathfrak{U}$, the class of all supersoluble groups, and let $G$ be a group with a normal subgroup $H$ such that $G/H \in \mathfrak{F}$. Then $G \in \mathfrak{F}$ if every cyclic subgroup of the generalized Fitting subgroup $F^*(E)$ of prime order or order 4 is a semi $CAP$-subgroup of $G$.*

**Theorem 4.7.**[3](Theorem 1.4) *Let $p$ be a prime number and let $G$ be a group. If every cyclic subgroup of $G$ of order $p$ or order 4 is a semi $CAP$-subgroup of $G$, then $G$ is $p$-supersoluble.*

**Definition 4.3.** [4 (Definition 1)] Let $A$ be a subgroup of a group $G$. We say that $A$ is a strong CAP-subgroup of $G$ if $A$ is a CAP-subgroup of any subgroup of $G$ containing $A$.

**Theorem 4.8.** [4 (Theorem A)] *Let $\mathfrak{F}$ be a saturated formation containing all supersoluble groups and $G$ a group with a normal subgroup $E$ such that $G/E \in \mathfrak{F}$. Suppose that every noncyclic Sylow subgroup $P$ of the generalized Fitting subgroup $F^*(E)$ of $E$ has a subgroup $D$ such that $1 < |D| < |P|$ and all subgroups $H$ of $P$ with order $|H| = |D|$ and with order $2|D|$ (if $P$ is a nonabelian 2-group) are strong CAP-subgroups of $G$. Then $G \in \mathfrak{F}$.*

**Theorem 4.9.** [28 (Main Theorem)] *Let $\mathfrak{F}$ be a saturated formation containing $\mathfrak{U}$, the class of all supersoluble groups, and $E$ a normal subgroup of $G$ such that $G/E \in \mathfrak{F}$. Suppose that, for every non-cyclic Sylow subgroup $P$ of $F^*(E)$ (the generalized Fitting subgroup of $E$), $P$ has a subgroup $D$ such that $1 < |D| < |P|$ and all subgroups $H$ of $P$ with order $|H| = |D|$ and with order $|H| = 2|D|$ (if $P$ is a non-abelian 2-group and $|P : D| > 2$) satisfy the cover-avoidance property in $G$. Then $G \in \mathfrak{F}$.*

**Theorem 4.10.** [9 (Theorem A)] *Let $G$ be a group and $P$ a Sylow $p$-subgroup of $G$ and assume that $1 < d < |P|$. Suppose that every subgroup of order $d$ is a semi CAP-subgroup of $G$ and every cyclic subgroup of order 4 of $G$ is a semi CAP-subgroup of $G$ when $d = 2$ and $P$ is non-abelian. Then $G$ has $p$-length at most 1.*

**Theorem 4.11.** [40 (Theorem 9)] *Let $U \leq G_1 \times G_2$. Then $U$ is a CAP-subgroup of $G_1 \times G_2$ if and only if $UM \cap G_1$ is a CAP-subgroup of $G_1$ for every $M \trianglelefteq G_2$, and $UN \cap G_2$ is a CAP-subgroup of $G_2$ for every $N \trianglelefteq G_1$.*

**Definition 4.4.** [47 (Definition 2.2)] Let $H$ be a subgroup of a group $G$. $H$ is called an $\mathscr{H}$-subgroup of $G$ if the following condition is satisfied:

$$N_G(H) \cap H^g \leq H; \text{ for all } g \in G.$$

The set of all $\mathscr{H}$-subgroups of a group $G$ will be denoted by $\mathscr{H}(G)$.

**Theorem 4.12.** [47 (Theorem 3.1)] *Let $G$ be a $p$-solvable group and $p$ a prime divisor of $|G|$ such that $(|G|, p - 1) = 1$. If every maximal subgroup of Sylow $p$-subgroup $P$ of $G$ is semi $p$-cover-avoiding in $G$ or belongs to $\mathscr{H}(G)$, then $G$ is $p$-nilpotent.*

**Theorem 4.13.** [47 (Theorem 4.1)] *Let $G$ be a $p$-solvable group and $P \in Syl_p(F_p(G))$, where $p$ is a prime divisor of $|G|$. If every maximal sub-*

group of $P$ is semi $p$-cover-avoiding in $G$ or belongs to $\mathscr{H}(G)$, then $G$ is $p$-supersolvable.

**Theorem 4.14.** [47(Theorem 4.5)] *Let $\mathfrak{F}$ be a saturated formation containing the class of all supersolvable groups $\mathfrak{U}$ and let $H$ be a solvable normal subgroup of a group $G$ such that $G/H \in \mathfrak{F}$. If, for every prime $p$ dividing the order of $F(H)$, every maximal subgroup of Sylow $p$-subgroups of $F(H)$ is semi $p$-cover-avoiding in $G$ or belongs to $\mathscr{H}(G)$, then $G \in \mathfrak{F}$.*

**Theorem 4.15.** [47(Theorem 4.7)] *Let $p$ be the smallest prime divisor of the order of a $p$-solvable group $G$. If every maximal subgroup of Sylow $p$-subgroups of $F_p(G)$ is semi $p$-cover-avoiding in $G$ or belongs to $\mathscr{H}(G)$, then $G$ is $p$-nilpotent.*

In closing this article, we must point out that there are still some ongoing researches in the topics which are related to CAP-subgroups of a given group. Many of these interesting results and research progresses have not yet mentioned and given in this article, for example, see Ref. 8,48,51, etc. The readers are always referred to the bibliography of this article.

## References

1. A. Ballester-Bolinches, R. Esteban-Romero and Y. M. Li, On second minimal subgroups of Sylow subgroups of finite groups. *J. Algebra*, **342:1**(2011), 134-146.
2. A. Ballester-Bolinches, R. Esteban-Romero and Y. M. Li, Cover and avoidance properties and the structure of finite groups. *Proceedings of the International Conference on Algebra*, 2010, 26-42, World Sci. Publ., Hackensack, NJ, 2012.
3. A. Ballester-Bolinches, R. Esteban-Romero and Y. M. Li, A question on partial CAP-subgroups of finite groups. *Sci. China Math.*, **55:5**(2012), 961-966.
4. A. Ballester-Bolinches, L. M. Ezquerro and A. N. Skiba, Subgroups of finite groups with a strong cover-avoidance property. *Bull. Aust. Math. Soc.*, **79:3**(2009), 499-506.
5. A. Ballester-Bolinches, L. M. Ezquerro and A. N. Skiba, Local embeddings of some families of subgroups of finite groups. *Acta Math. Sin. (Engl. Ser.)*, **25:6**(2009), 869-882.
6. A. Ballester-Bolinches, L. M. Ezquerro and A. N. Skiba, On subgroups which cover or avoid chief factors of a finite group. *Algebra Discrete Math.*, **no.4**(2009), 18-28.
7. A. Ballester-Bolinches, L. M. Ezquerro and A. N. Skiba, On second maximal subgroups of Sylow subgroups of finite groups. *J. Pure Appl. Algebra*, **215:4**(2011), 705-714.

8. A. Ballester-Bolinches, S. F. Kamornikov, O. L. Shemetkova and X. Yi, On subgroups of finite groups with a cover and avoidance property. *Sib. lektron. Mat. Izv.*, **13**(2016), 950-954.

9. A. Ballester-Bolinches, M. Luis, Y. M. Li and N. Su, On partial CAP-subgroups of finite groups. *J. Algebra*, **431:5**(2015), 196-208.

10. R. W. Carter, Nilpotent self-normalizing subgroups of soluble groups. *Math. Z.*, **75:1**(1960/1961), 136-139.

11. R. W. Carter, Nilpotent self-normalizing subgroups and system normalizers. *Proc. London Math. Soc.*, **12:3**(1962), 535-563.

12. Y. Q. Cui and X. Y. Guo, Influence of the semi cover-avoiding property on the structure of finite groups. *Int. J. Algebra*, **2:5-8**(2008), 395-402.

13. L. M. Ezquerro, A contribution to the theory of finite supersolvable groups. *Rend. Sem. Mat. Univ. Padova*, **89**(1993), 161-170.

14. Y. Fan, X. Y. Guo and K. P. Shum, Remarks on two generalizations of normality of subgroups. *Chinese Ann. Math. Ser. A*, **27:2**(2006), 169-176.

15. W. Gaschütz, Praefrattinigruppen. *Arch. Math.*, **13:1**(1962), 418-426.

16. J. D. Gillam, Cover-avoid subgroups in finite solvable groups. *J. Algebra*, **29:2**(1974), 324-329.

17. X. Y. Guo and Y. Q. Cui, Semi cover-avoiding property and structure of finite groups. *(Chinese) J. Shanghai Univ. Nat. Sci.*, **14:3**(2008), 244-247.

18. P. F. Guo and X. Y. Guo, *CAP*-embedded subgroups of *p*-fitting subgroups. (Chinese) *Math. Pract. Theory*, **40:24**(2010), 207-212.

19. X. Y. Guo, P. F. Guo and K. P. Shum, On semi cover-avoiding subgroups of finite groups. *J. Pure Appl. Algebra*, **209:1**(2007), 151-158.

20. X. Y. Guo, J. J. Liu and Q. L. Li, On local semi *CAP*-subgroups of finite groups. *Publ. Math. Debrecen*, **79:1-2**(2011), 119-131.

21. X. Y. Guo and K. P. Shum, Cover-avoidance properties and the structure of finite groups. *J. Pure Appl. Algebra*, **181:2-3**(2003), 297-308.

22. W. B. Guo and A. N. Skiba, On finite quasi-$\mathcal{F}$-groups. *Comm. Algebra*, **37:2**(2009), 470-481.

23. X. Y. Guo and L. L. Wang, On finite groups with some semi cover-avoiding subgroups. *Acta Math. Sin. (Engl. Ser.)*, **23:9**(2007), 1689-1696.

24. X. Y. Guo, J. X. Wang and K. P. Shum, On semi-cover-avoiding maximal subgroups and solvability of finite groups. *Comm. Algebra*, **34:9**(2006), 3235-3244.

25. P. Hall, On The system normalizers of a soluble group, *Proc. London Math. Soc.*, **43:2**(1937), 507-528.

26. X. L. He and Y. M. Wang, On *p*-cover-avoid and $\mathcal{F}$-quasinormally embedded subgroups in finite groups. *J. Math. Res. Exposition*, **30:4**(2010), 743-750.

27. Q. J. Kong and X. Y. Guo, Finite groups with some semi-p-cover-avoiding or ss-quasinormal subgroups. *Bull. Korean Math. Soc.*, **51:4**(2014), 943-948.

28. Y. M. Li, On cover-avoiding subgroups of Sylow subgroups of finite groups. *Rend. Semin. Mat. Univ. Padova*, **123**(2010), 249-258.

29. X. L. Liu and N. Q. Ding, On chief factors of finite groups. (English summary) *J. Pure Appl. Algebra*, **210:3**(2007), 789-796.

30. J. J. Liu and X. Y. Guo, A note on cover-avoiding properties of finite groups.

*East-West J. Math.*, **12:2**(2010), 109-115.

31. J. J. Liu, X. Y. Guo and Q. L. Li, On non-Frattini chief factors and solvability of finite groups. *Proc. Indian Acad. Sci. Math. Sci.*, **122:2**(2012), 163-173.

32. J. J. Liu, X. Y. Guo and S. R. Li, The influence of $CAP^*$-subgroups on the solvability of finite groups. *Bull. Malays. Math. Sci. Soc.*, **(2)35:1**(2012), 227-237.

33. Y. M. Li and Y. J. Huang, Finite groups with some generalized CAP-subgroups. *Bull. Malays. Math. Sci. Soc.*, **39:4**(2016), 1413-1420.

34. S. R. Li and J. J. Liu, A generalization of cover-avoiding properties in finite groups. *Comm. Algebra*, **39:4**(2011), 1455-1464.

35. J. J. Liu, S. R. Li, Z. C. Shen and X. C. Liu, Finite groups with some CAP-subgroups. *Indian J. Pure Appl. Math.*, **42:3**(2011), 145-156.

36. Y. M. Li, L. Miao and Y. M. Wang, On semi cover-avoiding maximal subgroups of Sylow subgroups of finite groups. *Comm. Algebra*, **37:4**(2009), 1160-1169.

37. Y. M. Li and G. J. Wang, A note on maximal CAP-subgroups of finite groups. *Algebras Groups Geom.*, **23:3**(2006), 285-290.

38. X. H. Li and Y. W. Yang, Semi $CAP$-subgroups and the structure of finite groups. (Chinese) *Acta Math. Sinica (Chin. Ser.)*, **51:6**(2008), 1181-1186.

39. C. W. Li and X. L. Yi, On the semi cover-avoiding property and $\mathcal{F}$-supplementation. *Int. J. Group Theory*, **1:3**(2012), 21-31.

40. J. Petrillo, CAP-subgroups in a direct product of finite groups. *J. Algebra*, **306:2**(2006), 432-438.

41. J. Petrillo, The embedding of CAP-subgroups in finite groups. *Ric. Mat.*, **59:1**(2010), 97-107.

42. M. J. Tomkinson, Cover-avoidance properties in finite soluble groups. *Canad. Math. Bull.*, **19:2**(1976), 213-216.

43. X. Z. Tang and W. B. Guo, On partial $CAP^*$-subgroups of finite groups. *J. Algebra Appl.*, **16:1**(2017), 1750009, 12 pp.

44. J. X. Wang, Semi-2-cover-avoiding properties correlative to solvability of finite groups. (Chinese) *J. Jilin Univ. Sci.*, **48:3**(2010), 384-388.

45. L. L. Wang and G. Y. Chen, Some properties of finite groups with some (semi-p-) cover-avoiding subgroups. *J. Pure Appl. Algebra*, **213:5**(2009), 686-689.

46. J. X. Wang and X. Y. Guo, Semi-cover-avoiding properties and the structure of finite groups. *J. Math. Res. Exposition*, **30:3**(2010), 527-535.

47. L. L. Wang, A. F. Wang and G. Y. Chen, On semi p-cover-avoiding subgroups and H-subgroups of finite groups. *Ital. J. Pure Appl. Math.*, **no.36**(2016), 219-230.

48. Y. Xu and G. Y. Chen, On nearly CAP-embedded subgroups of finite groups. *Ital. J. Pure Appl. Math.*, **no.37**(2017), 59-68.

49. Y. Xu and X. H. Li, Maximal subgroups of a Sylow $p$-subgroup and $p$-nilpotency of finite groups. *J. Algebra Appl.*, **9:3**(2010), 383-391.

50. T. Zhao and X. H. Li, Semi cover-avoiding properties of finite groups. *Front. Math. China*, **5:4**(2010), 793-800.

51. G. Zhong and S. X. Lin, On $c^{\#}$-normal subgroups of finite groups. *J. Algebra Appl.*, **16:8**(2017), 1750160, 11 pp.

# The length of the group algebra of the group $Q_8$*

A. E. Guterman

*Lomonosov Moscow State University, Moscow, 119991, Russia*
*Moscow Institute of Physics and Technology, Dolgoprudny, 141701, Russia*

*alexander.guterman@gmail.com*

O. V. Markova

*Lomonosov Moscow State University, Moscow, 119991, Russia*
*Moscow Institute of Physics and Technology, Dolgoprudny, 141701, Russia*

*ov_markova@mail.ru*

We compute the length function of the group algebra of quaternions over an arbitrary field.

*2010 Mathematics subject classification:* 13E10, 16S50, 20C05, 20K01.

*Keywords*: Finite-dimensional algebras, Lengths of sets and algebras, Group algebras, Non-Abelian groups

## 1. Introduction

All algebras under consideration are associative, unitary and finite-dimensional.

Let $\mathbb{F}$ be a field, and let $\mathcal{A}$ be an $\mathbb{F}$-algebra. First we recall the notion of the *length* of the algebra $\mathcal{A}$.

For a nonempty finite set $B = \{b_1, \ldots, b_m\}$ (i.e. a finite alphabet) we refer to finite sequences of letters from $B$ as *words*. Let $B^*$ be the set of all words over $B$, and let $F_B$ be the free semigroup over $B$, that is, the set $B^*$ equipped with the concatenation operation.

**Definition 1.1.** The *length* of the word $b_{i_1} \ldots b_{i_t}$, where $b_{i_j} \in B$, is equal to $t$. We adopt the convention that the empty word $1$ is a word of the length $0$ over $B$.

---

*The work of the first author is supported by the grant RFBR 17-01-00895, the work of the second author is supported by the grant RFBR 16-01-00113

**Notation 1.1.** Let $B^i$ denote the set of all words of length not greater than $i$ over $B$, $i \geq 0$.

Let $S$ be a finite subset of the algebra $\mathcal{A}$.

The products of elements of the set $S$ can be viewed as images of elements of the free semigroup $F_S$ under the natural homomorphism. We can refer them as words in the elements of the set $S$ and use the natural notation $S^i$ defined above.

**Notation 1.2.** Put $\mathcal{L}_i(S) = \langle S^i \rangle$, where $\langle S \rangle$ denotes the linear span of a set $S$ in some vector space over the field $\mathbb{F}$. Note that $\mathcal{L}_0(S) = \langle 1 \rangle = \mathbb{F}$. Let also $\mathcal{L}(S) = \bigcup_{i=0}^{\infty} \mathcal{L}_i(S)$, that is, $\mathcal{L}(S)$ coincides with the algebra generated by $S$.

**Remark 1.1.** The set $S$ is a generating system for $\mathcal{A}$ if and only if $\mathcal{A} = \mathcal{L}(S)$. It follows from the definition of $S^i$ for $i, j > 0$ that

$$S^{i+j} = S^i S^j,$$

$$\mathcal{L}_{i+j}(S) = \langle \mathcal{L}_i(S)\mathcal{L}_j(S) \rangle.$$

**Definition 1.2.** The *length of the generating set* $S$ of the algebra $\mathcal{A}$ is $l(S) = \min\{k \in \mathbb{Z}_+ : \mathcal{L}_k(S) = \mathcal{A}\}$.

**Definition 1.3.** The *length of the algebra* $\mathcal{A}$ is $l(\mathcal{A}) = \max\{l(S) : \mathcal{L}(S) = \mathcal{A}\}$.

A trivial upper bound for the length of an arbitrary associative algebra $\mathcal{A}$ is $\dim \mathcal{A} - 1$. Moreover, $l(\mathcal{A}) = \dim \mathcal{A} - 1$ if and only if the algebra $\mathcal{A}$ is a single-generated algebra, i.e. if $\mathcal{A}$ contains an element $a \in \mathcal{A}$ such that $\mathcal{A} = \mathcal{L}(\{a\})$ (see Ref. 16(Lemma 5.3)). The problem of improving length estimates by considering a greater number of numerical characteristics of algebras arises in a natural way. So, in Ref. 19(Theorem 3.1), Pappacena obtained an upper bound for the length of an arbitrary algebra $\mathcal{A}$ in the following form. Let $d$ be the dimension of $\mathcal{A}$ over $\mathbb{F}$, and $m$ be the maximal degree of a minimal polynomial among the elements of $\mathcal{A}$.

**Theorem 1.1.** [19(Theorem 3.1)] *Let $\mathbb{F}$ be an arbitrary field and let*

$$f(d, m) = m\sqrt{\frac{2d}{m-1} + \frac{1}{4}} + \frac{m}{2} - 2.$$

*Then $l(\mathcal{A}) < f(\dim \mathcal{A}, m(\mathcal{A}))$.*

For example, for the matrix algebra the theorem above provides a bound with asymptotic behavior $O(n^{3/2})$:

**Theorem 1.2.** [19](*Corollary 3.2*) *Let* $\mathbb{F}$ *be an arbitrary field. Then*

$$l(M_n(\mathbb{F})) < n\sqrt{\frac{2n^2}{n-1} + \frac{1}{4}} + \frac{n}{2} - 2.$$

The last bound is much better than the trivial bound which provides the value $n^2 - 1$, however, it is conjectured by Paz in Ref. 20 that actually the length of the matrix algebra is a linear function in $n$.

For the case of commutative algebras, a smaller upper bound for the length as a function of the same two invariants was obtained in Ref. 16(Theorem 3.11). This bound is given by the function

$$g(d,m) = \begin{cases} (m-1)[\log_m d] + [m^{\{\log_m d\}}] - 1 \text{ for } m \geq 2; \\ 0 \text{ for } m = 1. \end{cases}$$

and it is sharper than the estimate in Ref. 19(Theorem 3.1). In Ref. 18(Theorem 1, Corollary 1) the length of algebras are estimated by the nilpotency index of its ideal and the length of the quotient algebra by this ideal, in particular, with the nilpotency index of its Jacobson radical.

However, the lengths of the matrix algebras are the most well-investigated as of yet. In particular, commutative matrix subalgebras, quasi-commutative matrix subalgebras and many other classes of matrix subalgebras were considered, see Ref. 5,9–11,13,14,16,19.

The question of length evaluation for algebras is closely related to other combinatorial problems in theory of rings, semigroups, and representation theory. In particular in Ref. 4 for the study of finite dimensional representations of semigroups, namely the connection between the character of a finite dimensional representation of a semigroup and its composition factors, the length function is used along with Shirshov's Theorem, see Ref. 1,4 and references therein. It turns out that the composition factors are determined by the traces of elements of length at most $2l(M_n(\mathbb{F}))+1$. Thus any upper bound for the length is actual for this problem. We remark, that the applications of the length function are not restricted by the problems mentioned above.

In the papers [7,8] we started the systematic study of the length function of group algebras of finite groups. In these papers we obtain general estimates and several concrete computations of the length for Abelian groups in the cases of semisimple group algebras[7] and in modular case[8]. In this paper we firstly address a non-Abelian group and obtain exact values of the length for arbitrary fields.

**Notation 1.3.** Let $\mathbb{F}$ be an arbitrary field and let $G$ be a finite group. By $\mathbb{F}G$ (or sometimes by $\mathbb{F}[G]$ if the notation of the group is rather complex) we denote the *group algebra* of the group $G$ over the field $\mathbb{F}$.

By the definition $\dim \mathbb{F}G = |G|$ and for the length of the group algebra of a finite Abelian group $G$ over an arbitrary field we always have the trivial upper bound $l(\mathbb{F}G) \leq \dim \mathbb{F}G - 1 = |G| - 1$, and for non-Abelian group $l(\mathbb{F}G) \leq \dim \mathbb{F}G - 2 = |G| - 2$ (for the details see Ref. 17(Lemma 4.34, Corollary 4.36)).

In particular, for the length of the algebra $\mathbb{F}\mathbf{Q}_8$ we obtain the upper-bound $l(\mathbb{F}\mathbf{Q}_8) \leq 6$. It turns out that it is possible to improve this bound and to compute the length of $\mathbb{F}\mathbf{Q}_8$ exactly. Namely we will show that depending on the field of coefficients $\mathbb{F}$ the length of the group algebra $\mathbb{F}\mathbf{Q}_8$ can attain only the values 3 and 4. More precisely, in our paper we prove the following main result:

**Theorem 1.3.** *Let $\mathbb{F}$ be an arbitrary field. Then*

$$l(\mathbb{F}\mathbf{Q}_8) = \begin{cases} 4, & \text{if } \operatorname{char} \mathbb{F} \neq 2 \text{ and } \exists\, \alpha, \beta \in \mathbb{F}\colon \alpha^2 + \beta^2 = -1; \\ 3, & \text{otherwise.} \end{cases}$$

Denote the elements of the group $\mathbf{Q}_8$ in the following way: the neutral element is denoted by $e$, its square root is denoted by $m$, the other elements are $i, j, k, mi, mj, mk$. The reason is to distinguish $\pm 1$ in the field of coefficients with the neutral element and its square root (which are usually denoted by 1 and $-1$) in of the group $\mathbf{Q}_8$.

Our paper is organized as follows. In Section 2 we collect some known results useful for our investigations. In Section 3 we obtain an auxiliary result concerning the length of the algebra $M_2(\mathbb{F}) \oplus D_n(\mathbb{F})$. Section 4 describes special generating systems of the algebra of diagonal matrices. Section 5 introduces generalized quaternion algebras and characterizes the group algebra $\mathbb{F}\mathbf{Q}_8$ via these algebras. In Section 6 we present some technical results for the later use. In Section 7 we estimate the length of the algebra $\mathcal{D} \oplus D_4(\mathbb{F})$, where $\mathcal{D}$ is a generalized quaternion algebra with the parameters $(-1, -1)$ over $\mathbb{F}$. Section 8 contains the lower bound for $l(\mathbb{F}\mathbf{Q}_8)$ and the proof of the main result if the characteristic of the ground field is different from 2. In Section 9 we consider fields of characteristic 2 and conclude the proof.

## 2. Some known facts

Further we will several times use the description of length behavior under changes of generators, see Ref. 6(Propositions 2.1 and 2.2). For the convenience of the reader we provide here the corresponding results with their proofs.

**Proposition 2.1.** [6(*Proposition 2.1*)] *Let* $\mathbb{F}$ *be an arbitrary field and let* $\mathcal{A}$ *be a finite-dimensional algebra over* $\mathbb{F}$. *If* $\mathcal{S} = \{a_1, \ldots, a_k\}$ *is a generating system of this algebra and* $C = \{c_{i,j}\} \in M_k(\mathbb{F})$ *is a nonsingular matrix, then the set of the coordinates of the vector*

$$
C \begin{pmatrix} a_1 \\ \vdots \\ a_k \end{pmatrix} = \begin{pmatrix} c_{1,1}a_1 + c_{1,2}a_2 + \ldots + c_{1,k}a_k \\ \vdots \\ c_{k,1}a_1 + c_{k,2}a_2 + \ldots + c_{k,k}a_k \end{pmatrix}, \tag{1}
$$

*i.e. the set*

$$
\mathcal{S}_c = \{c_{1,1}a_1 + c_{1,2}a_2 + \ldots + c_{1,k}a_k, \ldots, c_{k,1}a_1 + c_{k,2}a_2 + \ldots + c_{k,k}a_k\},
$$

*is a system of generators for the algebra* $\mathcal{A}$ *and* $l(\mathcal{S}_c) = l(\mathcal{S})$.

**Proof.** Let us prove the equality $\mathcal{L}_n(\mathcal{S}) = \mathcal{L}_n(\mathcal{S}_c)$ by induction on $n$.

*The base:* For $n = 0$ the statement holds by definition. Let $n = 1$. By the definition we have that for any linear combination $\gamma_1 a_1 + \ldots + \gamma_k a_k \in \mathcal{L}_1(\mathcal{S})$, $\gamma_1, \ldots, \gamma_k \in \mathbb{F}$. Therefore, $\mathcal{L}_1(\mathcal{S}_c) \subseteq \mathcal{L}_1(\mathcal{S})$. Since the matrix $C$ is nonsingular, there exists the matrix $C^{-1} = (d_{ij})$. Then by equation (1) and by the definition of the linear span

$$
a_i = (d_{i1}, d_{i2}, \ldots, d_{ik}) \begin{pmatrix} c_{11}a_1 + c_{12}a_2 + \ldots + c_{1k}a_k \\ \vdots \\ c_{k1}a_1 + c_{k2}a_2 + \ldots + c_{kk}a_k \end{pmatrix} \in \mathcal{L}_1(\mathcal{S}_c),
$$

$$
\text{for all } i = 1, \ldots, \ k.
$$

Consequently, $\mathcal{L}_1(\mathcal{S}) \subseteq \mathcal{L}_1(\mathcal{S}_c)$. That is, $\mathcal{L}_1(\mathcal{S}_c) = \mathcal{L}_1(\mathcal{S})$.

*The step.* Let us assume that $n > 1$ and the statement of the proposition holds for all $m < n$. Then

$$
\mathcal{L}_n(\mathcal{S}) = \langle \mathcal{L}_1(\mathcal{S})\mathcal{L}_{n-1}(\mathcal{S}) + \mathcal{L}_1(\mathcal{S})\rangle = \langle \mathcal{L}_1(\mathcal{S}_c)\mathcal{L}_{n-1}(\mathcal{S}_c) + \mathcal{L}_1(\mathcal{S}_c)\rangle = \mathcal{L}_n(\mathcal{S}_c).
$$

$\square$

**Proposition 2.2.** [6(*Proposition 2.2*)] *Let* $\mathbb{F}$ *be an arbitrary field and let* $\mathcal{A}$ *be a finite-dimensional unitary algebra over* $\mathbb{F}$. *Assume that* $\mathcal{S} = \{a_1, \ldots, a_k\}$ *is* *a system of generators for this algebra such that* $1_A \notin \langle a_1, \ldots, a_k \rangle$. *Then for any* $\gamma_1, \ldots, \gamma_k \in \mathbb{F}$ *the set*

$$S_1 = \{a_1 + \gamma_1 1_A, \ldots, a_k + \gamma_k 1_A\}$$

*is a generating system of the algebra* $A$ *and* $l(S_1) = l(S)$.

**Proof.** As in the previous statement, we prove the equality $\mathcal{L}_n(S) = \mathcal{L}_n(S_1)$ for all $n$ by the induction on $n$.

*The base.* Since $1_A \in \mathcal{L}_0(S) = \mathcal{L}_0(S_1)$, we have $\mathcal{L}_1(S_1) = \mathcal{L}_1(S)$.

*The step.* Let us assume that $n > 1$ and the statement of the proposition holds for all $m < n$. Then

$$\mathcal{L}_n(S) = \langle \mathcal{L}_1(S)\mathcal{L}_{n-1}(S) \rangle = \langle \mathcal{L}_1(S_1)\mathcal{L}_{n-1}(S_1) \rangle = \mathcal{L}_n(S_1).$$

$\square$

**Proposition 2.3.** [17(Proposition 3.19)] *Let* $A$ *be a finite-dimensional unitary algebra over a field* $\mathbb{F}$. *Then the inequality* $l(A_\mathbb{F}) \leq l(A_\mathbb{K})$ *holds for an arbitrary extension* $\mathbb{K}$ *of the field* $\mathbb{F}$.

**Proof.** Let us write $l = l(A_\mathbb{F})$. Then it follows from the definition of the length of an algebra that there exists a generating system $S = \{a_1, \ldots, a_k\}$ of the length $l$ in the algebra $A_\mathbb{F}$. It is clear that the set $S' = \{a_1 \otimes 1, \ldots, a_k \otimes 1\}$ is a systems of generators in the algebra $A_\mathbb{K}$. We claim that $l(S') = l$.

Since $l(S) = l$, there exists a word $v = v(a_1, \ldots, a_k)$ of the length $l$, which is irreducible in $\mathcal{L}_l(S)$. Suppose $v \otimes 1$ is reducible in $\mathcal{L}_l(S')$. Then there exist words $v_1', \ldots, v_m' \in (S')^{l-1}$ and coefficients $\alpha_1, \ldots, \alpha_m \in \mathbb{K}$ such that $v \otimes 1 = \sum_{i=1}^{m} \alpha_i v_i'$. However, by the choice of the system $S'$, for each word $v_i'$ we have

$$v_i' = (a_{i_1} \otimes 1) \cdots (a_{i_j} \otimes 1) = (a_{i_1} \cdots a_{i_j}) \otimes 1,$$

i.e. each word $v_i'$ has the form $v_i' = v_i \otimes 1$ for some $v_i \in S^{l-1}$. Then $v \otimes 1 = \sum_{i=1}^{m} v_i \otimes \alpha_i$. From here we obtain $\alpha_i \in \mathbb{F}$ and $v = \sum_{i=1}^{m} \alpha_i v_i$. This is a contradiction with the irreducibility of the word $v$, which shows that $v'$ is irreducible in $\mathcal{L}_l(S')$, i.e. $l = l(S) \leq l(S')$.

The opposite inequality follows from the fact that any element of the algebra $A_\mathbb{K}$ is a linear combination of elements of the form $a \otimes 1$, where $a \in A_\mathbb{F}$, whether any such element is a linear combination of words from $(S')^l$.

Consequently, $l(A_\mathbb{K}) \geq l(S') = l = l(A_\mathbb{F})$. $\square$

**Theorem 2.1.** [15 (Theorem 2)] *Let $\mathcal{A}$ and $\mathcal{B}$ be finite-dimensional associative algebras over a field $\mathbb{F}$ with the lengths $l_{\mathcal{A}}$ and $l_{\mathcal{B}}$, correspondingly. Then the following inequalities are true:*

$$\max\{l_{\mathcal{A}}, l_{\mathcal{B}}\} \leq l(\mathcal{A} \oplus \mathcal{B}) \leq l_{\mathcal{A}} + l_{\mathcal{B}} + 1. \tag{2}$$

**Proof.** Let us denote $p = l_{\mathcal{A}}$, $q = l_{\mathcal{B}}$.

To prove the lower bound we consider two generating systems $\{a_1, \ldots, a_k\}$ and $\{b_1, \ldots, b_m\}$ of the lengths $p$ and $q$ for algebras $\mathcal{A}$ and $\mathcal{B}$, correspondingly. Then the set $\{(a_1, 0), \ldots, (a_k, 0), (0, b_1), \ldots, (0, b_m)\}$ is a required generating system in $\mathcal{A} \oplus \mathcal{B}$ of the length $\max\{p, q\}$.

Now we consider an arbitrary generating system $S = \{(c_1, d_1), \ldots, (c_n, d_n)\}$ for the algebra $\mathcal{A} \oplus \mathcal{B}$ and claim that any word in elements from $S$ of the length $p + q + 2$ is reducible. We note that the set $c_1, \ldots, c_n$ is a generating system for the algebra $\mathcal{A}$, while the set $d_1, \ldots, d_n$ is a generating system for the algebra $\mathcal{B}$. We write $N = p + q + 2$. Assume that $v = (c_{i_1}, d_{i_1}) \ldots (c_{i_N}, d_{i_N}) = (c_{i_1} \ldots c_{i_N}, d_{i_1} \ldots d_{i_N})$, $i_j \in \{1, \ldots, n\}$, $j = 1, \ldots, N$. Since the length of $\mathcal{A}$ is equal to $p$, the word $c_{i_1} \ldots c_{i_{p+1}}$ is reducible, i.e. $c_{i_1} \ldots c_{i_{p+1}} = \alpha_1 c_{i_1} \ldots c_{i_p} + \ldots + \alpha_{M-1} c_n + \alpha_M 1_{\mathcal{A}}$. Since the length of $\mathcal{B}$ is equal to $q$, then the word $d_{i_{p+2}} \ldots d_{i_N}$ is reducible, i.e. $d_{i_{p+2}} \ldots d_{i_N} = \beta_1 d_{i_{p+2}} \ldots d_{i_{N-1}} + \ldots + \beta_{K-1} d_n + \beta_K 1_{\mathcal{B}}$. After substituting the representations for $c_{i_1} \ldots c_{i_{p+1}}$ and $d_{i_{p+2}} \ldots d_{i_N}$ into $v$, we obtain the equation

$$\{(c_{i_1} \ldots c_{i_{p+1}}, \, d_{i_1} \ldots d_{i_{p+1}}) - \alpha_1 (c_{i_1} \ldots c_{i_p}, \, d_{i_1} \ldots d_{i_p}) - \ldots -$$

$$\alpha_{M-1}(c_n, d_n) - \alpha_M(1_A, 1_B)\}\{(c_{i_{p+2}} \ldots c_{i_N}, \, d_{i_{p+2}} \ldots d_{i_N}) - \beta_K(1_A, 1_B) -$$

$$\beta_{K-1}(c_n, d_n) - \ldots - \beta_1(c_{i_{p+2}} \ldots c_{i_{N-1}}, \, d_{i_{p+2}} \ldots d_{i_{N-1}})\} = (0, x)(y, 0) = 0.$$

Therefore the word $v$ can be represented as a linear combination of the words of smaller length. Since $v$ is chosen arbitrary, we get $l(\mathcal{A} \oplus \mathcal{B}) \leq p + q + 1$.

$\square$

Let $M_n(\mathbb{F})$ denotes the matrix algebra of size $n$ over the field $\mathbb{F}$, $D_n(\mathbb{F})$ denotes the algebra of diagonal matrices of size $n$ over the field $\mathbb{F}$; $I$, $I_n$ denote the identity matrix in $M_n(\mathbb{F})$, $O$, $O_n$ denote the zero matrix in $M_n(\mathbb{F})$; $E_{i,j}$ is $(i, j)$-th matrix unit, i.e. the matrix with 1 on $(i, j)$-th position and 0 on the other positions.

To prove the upper bound we need some facts about the lengths of diagonal matrices and the algebra $M_2(\mathbb{F}) \oplus D_n(\mathbb{F})$. For $n = 1, 2$ the length of these algebras is known already:

**Lemma 2.1.** [17(Lemmas 3.17–3.18)] *Let $\mathbb{F}$ be an arbitrary field. Then*
*1. $l(M_2(\mathbb{F}) \oplus \mathbb{F}) = 2$;*
*2. $l(M_2(\mathbb{F}) \oplus D_2(\mathbb{F})) = 3$.*

**Theorem 2.2.** [16(Theorem 5.4)] *Let $\mathbb{F}$ be an arbitrary field.*
*1. If $\mathbb{F}$ is infinite, then $l(D_n(\mathbb{F})) = n - 1$.*

$$2.\ l(D_n(\mathbb{F}_q)) = \begin{cases} n - 1, & \text{for } q \geq n; \\ (q - 1)[\log_q n] + [q^{\{\log_q n\}}] - 1, & \text{for } q < n. \end{cases}$$

## 3. Length of the algebra $M_2(\mathbb{F}) \oplus D_n(\mathbb{F})$

**Lemma 3.1.** *Let $\mathbb{F}$ be an arbitrary field. Then $l(M_2(\mathbb{F}) \oplus D_n(\mathbb{F})) \leq \max\{4, l(D_n(\mathbb{F}))\}$.*

**Proof.** Let us consider an arbitrary system of generators $\mathcal{S}$ for the given algebra and prove that $l(\mathcal{S}) \leq \max\{4, l(D_n(\mathbb{F}))\}$. By definition

$$\mathcal{S} = \{C_i = A_i \oplus D_i | A_i \in M_2(\mathbb{F}), D_i \in D_n(\mathbb{F}), \ i = 1, \ldots, k\},$$

in this case the sets $\mathcal{S}_A = \{A_i, i = 1, \ldots, k\}$ and $\mathcal{S}_D = \{D_i, i = 1, \ldots, k\}$ are systems of generators for algebras $M_2(\mathbb{F})$ and $D_n(\mathbb{F})$, correspondingly. In particular $\dim \mathcal{L}_1(\mathcal{S}_A) \in \{3, 4\}$ since the full matrix algebra is noncommutative.

I. Let $\dim \mathcal{L}_1(\mathcal{S}_A) = 4$, this implies that $\mathcal{L}_1(\mathcal{S}_A) = M_2(\mathbb{F})$. Applying Proposition 2.1 [6(Proposition 2.1)], without loss of generality, we can assume that $C_1 = E_{12} \oplus D_1$, $C_2 = E_{21} \oplus D_2$, $C_3 = E_{22} \oplus D_3$. Consider the commutator of matrices $C_1$ and $C_2$:

$$[C_1, C_2] = [E_{12}, E_{21}] \oplus O = (E_{11} - E_{22}) \oplus O.$$

We have

$$[C_1, C_2]C_1 = E_{12} \oplus O \in \mathcal{L}_3(\mathcal{S}),$$

$$C_2[C_1, C_2] = E_{21} \oplus O \in \mathcal{L}_3(\mathcal{S}),$$

$$-[C_1, C_2]C_3 = E_{22} \oplus O \in \mathcal{L}_3(\mathcal{S}),$$

$$[C_1, C_2](I - C_3) = E_{11} \oplus O \in \mathcal{L}_3(\mathcal{S}).$$

So, all four matrix units lie in $\mathcal{L}_3(\mathcal{S})$. Hence, $M_2(\mathbb{F}) \oplus O \subseteq \mathcal{L}_3(\mathcal{S})$.

Moreover $\mathcal{L}_{l(D_n(\mathbb{F}))}(\mathcal{S}_D) = D_n(\mathbb{F})$, thus for an arbitrary $D \in D_n(\mathbb{F})$ there exist a certain matrix $A \in M_2(\mathbb{F})$ such that $A \oplus D \in \mathcal{L}_{l(D_n(\mathbb{F}))}(\mathcal{S})$. As it was shown previously, $A \oplus O \in \mathcal{L}_3(\mathcal{S})$, therefore, $O \oplus D \in \mathcal{L}_{\max\{3,l(D_n(\mathbb{F}))\}}(\mathcal{S})$. It follows that $\mathcal{L}_{\max\{3,l(D_n(\mathbb{F}))\}}(\mathcal{S}) = M_2(\mathbb{F}) \oplus D_n(\mathbb{F})$ and $l(\mathcal{S}) \leq \max\{3, l(D_n(\mathbb{F}))\} \leq \max\{4, l(D_n(\mathbb{F}))\}$.

II. Let $\dim \mathcal{L}_1(\mathcal{S}_A) = 3$. For an element $A \in M_2(\mathbb{F}) \oplus D_n(\mathbb{F})$ we use the notation $A = (a_{i,j})$ identifying $A$ with a matrix from $M_{2+n}(\mathbb{F})$.

By Proposition 2.2[6 (Proposition 2.2)] the change of the generating system to $\mathcal{S}_1 = \{S - s_{1,1}I \mid S = (s_{i,j}) \in \mathcal{S}\}$ preserves the length.

i. Assume that there are two matrices $X = (x_{i,j})$, $Y = (y_{i,j}) \in \mathcal{S}_1$, such that vectors

$$c_1 = (x_{1,2}, \ x_{2,1}),$$

$$c_2 = (y_{1,2}, \ y_{2,1})$$

are linearly independent. In this case there exists an invertible matrix $F = (f_{k,m}) \in M_2(\mathbb{F})$ such that $(1,0) = f_{1,1}c_1 + f_{1,2}c_2$, $(0,1) = f_{2,1}c_1 + f_{2,2}c_2$. Set $H_r = f_{r,1}X + f_{r,2}Y$, $r = 1,2$ and by Proposition 2.1[6 (Proposition 2.1)] we can change the generating system to $\mathcal{S}_2 = \{H_1, H_2, S \mid S \in \mathcal{S}_1, S \neq X, S \neq Y\}$. After that we change this system to

$$\mathcal{S}_3 = \{H_1, H_2, S - s_{1,2}H_1 - s_{2,1}H_2 \mid S = (s_{i,j}) \in \mathcal{S}_2, S \neq H_1, S \neq H_2\}.$$

Then the length is preserved on each step. For the simplicity of notations we assume that $\mathcal{S} = \mathcal{S}_3$ and $H_1 = A_1 \oplus D_1$, $H_2 = A_2 \oplus D_2$, where $A_1 = E_{12} + aE_{22}$, $A_2 = E_{21} + bE_{22} \in M_2(\mathbb{F})$, $a, b \in \mathbb{F}$, $D_1, D_2 \in D_n(\mathbb{F})$. Then

$$H_1 H_2 = (E_{11} + bE_{12} + aE_{21} + abE_{22}) \oplus D_1 D_2, \quad H_2 H_1 = ((1+ab)E_{22}) \oplus D_1 D_2.$$

If $ab = -1$, then $A_2 A_1 = O$. Therefore, $A_1 A_2 = (A_1 + A_2)^2 - A_1^2 - A_2^2 \in \mathcal{L}_1(\{A_1, A_2\})$ ( by Cayley-Hamilton theorem), this implies that $A_1, A_2$ do not generate the full matrix algebra.

Thus further we can assume that $1 + ab = \alpha \neq 0$. Then

$$[H_1, H_2]H_1 = \alpha E_{12} \oplus O \in \mathcal{L}_3(\mathcal{S}),$$

$$H_2[H_1, H_2] = \alpha E_{21} \oplus O \in \mathcal{L}_3(\mathcal{S}),$$

$$H_2[H_1, H_2]H_1 = \alpha E_{22} \oplus O \in \mathcal{L}_4(\mathcal{S}),$$

$$(H_2 - bI)[H_1, H_2](H_1 - aI) + [H_1, H_2] = \alpha E_{11} \oplus O \in \mathcal{L}_4(\mathcal{S}).$$

Analogously to Item I, we can prove that $l(\mathcal{S}) \leq \max\{4, l(D_n(\mathbb{F}))\}$.

ii. If the condition from Item i is not true then there exists a matrix $C = (c_{i,j})$ in $\mathcal{S}_1$ such that the vector $(c_{1,2}, c_{2,1})$ has two non-zero coordinates and all other vectors $(s_{1,2}, s_{2,1})$, $S = (s_{i,j}) \in \mathcal{S}_1$ can be expressed as a linear combinations of this vector $(c_{1,2}, c_{2,1})$. Without loss of generality (up to the division by non-zero constant) we can assume that $c_{1,2} = 1$, $c_{2,1} = c$, $c \neq 0$. By Proposition 2.1 [6 (Proposition 2.1)] we can change the generating system to $\mathcal{S}_2 = \{C, S - s_{1,2}C \mid S \in \mathcal{S}_1, S \neq C\}$ and the length is preserved. Assume $H_1 = C = A_1 \oplus D_1$. There is a matrix $H_2 = A_2 \oplus D_2'$ in $\mathcal{S}_2$, such that the matrices $A_1, A_2$ are linearly independent. In this case we have $A_2 = dE_{22}$, by the construction and therefore, $d \neq 0$. It is possible to assume that $A_2 = E_{22}$ and $H_2 = A_2 \oplus D_2$. We denote $H_1 = (h_{i,j})$. Then again by Proposition 2.1 [6 (Proposition 2.1)] we pass to the generating system $\mathcal{S}_3 = \{H_1 - h_{2,2}H_2, S \mid S \in \mathcal{S}_2, S \neq H_1\}$ without change of the length. For simplicity we again assume that $\mathcal{S} = \mathcal{S}_3$. Then

$$H_2[H_1, H_2] = (-cE_{21}) \oplus O \in \mathcal{L}_3(\mathcal{S}),$$

$$[H_1, H_2]H_2 = E_{12} \oplus O \in \mathcal{L}_3(\mathcal{S}),$$

$$[H_1, H_2]H_1H_2 = (-cE_{22}) \oplus O \in \mathcal{L}_4(\mathcal{S}),$$

$$[H_1, H_2](H_1H_2 - H_1) = (-cE_{11}) \oplus O \in \mathcal{L}_4(\mathcal{S}).$$

Therefore as in Item II.i we conclude that $l(\mathcal{S}) \leq \max\{4, l(D_n(\mathbb{F}))\}$. $\square$

**Lemma 3.2.** *Let $\mathbb{F}$ be a field, $\operatorname{char}\mathbb{F} \neq 2$ and $n \geq 4$. Then $l(M_2(\mathbb{F}) \oplus D_n(\mathbb{F})) = \max\{4, l(D_n(\mathbb{F}))\}$.*

**Proof.** Let us prove at first the statement in the case $n = 4$ and show that $l(M_2(\mathbb{F}) \oplus D_4(\mathbb{F})) = 4$. By Theorem 2.2 [16 (Theorem 5.4)] we have $l(D_4(\mathbb{F})) \leq 3$, thus $\max\{4, l(D_4(\mathbb{F}))\} = 4$. By Lemma 3.1 it is sufficient to find a generating system for this algebra of the length 4. We consider $M_2(\mathbb{F}) \oplus D_4(\mathbb{F})$ as a subalgebra of $M_6(\mathbb{F})$ generated by $E_{12}, E_{21}, E_{11}, \ldots, E_{66}$. Then to construct a generating system one can take the matrices $A = E_{12} + E_{21} + E_{33} + E_{44} - E_{55} - E_{66}$, $B = E_{22} + E_{44} + E_{66}$. It can be shown directly that the pair $A, B$ is generating the algebra $M_2(\mathbb{F}) \oplus D_4(\mathbb{F})$. We have $A^2 = I$, $B^2 = B$ and therefore all the words in $A, B$, containing the squares of the letters can be changed by the words of smaller length. Then $\dim \mathcal{L}_3(\{A, B\}) = \dim\langle I, A, B, AB, BA, ABA, BAB\rangle \leq 7$. Thus $l(\{A, B\}) > 3$.

Assume now $n \geq 5$. Then we can decompose $M_2(\mathbb{F}) \oplus D_n(\mathbb{F}) = (M_2(\mathbb{F}) \oplus D_4(\mathbb{F})) \oplus D_{n-4}(\mathbb{F})$. Apply now two times Theorem 2.1 [15(Theorem 2)] about the length of direct sum of algebras. On the one hand,

$$l(M_2(\mathbb{F}) \oplus D_n(\mathbb{F})) = l((M_2(\mathbb{F}) \oplus D_4(\mathbb{F})) \oplus D_{n-4}(\mathbb{F}))$$
$$\geq \max\{l(M_2(\mathbb{F}) \oplus D_4(\mathbb{F})), l(D_{n-4}(\mathbb{F}))\}$$

hence,

$$l(M_2(\mathbb{F}) \oplus D_n(\mathbb{F})) \geq l(M_2(\mathbb{F}) \oplus D_4(\mathbb{F})) = 4.$$

On the other hand,

$$l(M_2(\mathbb{F}) \oplus D_n(\mathbb{F})) \geq \max\{l(M_2(\mathbb{F})), l(D_n(\mathbb{F}))\} \geq l(D_n(\mathbb{F})).$$

It follows that $l(M_2(\mathbb{F}) \oplus D_n(\mathbb{F})) \geq \max\{4, l(D_n(\mathbb{F}))\}$. By Lemma 3.1 the result follows. $\qquad\square$

## 4. On generators of diagonal matrices

**Lemma 4.1.** *Let $\mathbb{F}$ be an arbitrary field, $n \in \mathbb{N}$; $n \geq 2$ and $\mathcal{S} = \{D_1, \ldots, D_k\}$ be the generating system for the algebra $D_n(\mathbb{F})$. Here we assume that the identity is considered as a word of the length 0 in generators. Consider the subalgebra $\mathcal{A} \subseteq D_n(\mathbb{F})$, generated as a vector space by the words in $\mathcal{S}$ of the length not smaller than 1. Then $\mathcal{A} \neq D_n(\mathbb{F})$ if and only if there exists an index $i \in \{1, \ldots, n\}$, for which $(D_r)_{i,i} = 0$ simultaneously holds for all $r = 1, \ldots, k$.*

**Proof.** Suppose that all matrices from $\mathcal{S}$ contain zero element on the $i$-th position. Then this condition holds for any linear combination of the words from $\mathcal{S}$ of the positive length, therefore $\mathcal{A} \neq D_n(\mathbb{F})$.

Let us prove the converse implication. By conditions of our lemma we have that $\dim \mathcal{A} = \dim D_n(\mathbb{F}) - 1 = n - 1$. Therefore $\mathcal{A}$ as a vector space over $\mathbb{F}$ can be defined by a single linear equation. Let us write this equation: $a_1 x_1 + \ldots + a_n x_n = 0$. By conditions $\mathcal{A}$ is not only a vector space but also is an algebra. This implies that any Hadamard product of the solutions is also a solution of this equation. Let $a_j$ be a first non-zero coefficient of this equation. Let us prove that $i = j$.

If $j = n$ then the equation has the form $a_n x_n = 0$ and since $a_n \neq 0$ we have $x_n = 0$. In this case we can indeed set $i = n$.

Let $j < n$. Consider normal fundamental system of solutions for this equation $f_1, \ldots, f_{n-1}$, that is the standard basis of the solution space obtained by substituting coordinates of the standard basis vectors from $\mathbb{F}^{n-1}$

as free variables $x_1, \ldots, x_{j-1}, x_{j+1}, \ldots, x_n$. Then $f_1 = e_1, \ldots, f_{j-1} = e_{j-1}$, where $e_m$ is the $m$-th vector of the standard basis in $\mathbb{F}^n$. For all $s \geq j$ we have $f_s = \alpha_s e_j + e_{s+1}$.

Suppose for some indices $p, q$, $p < q$ we have $\alpha_p \alpha_q \neq 0$, then a nonzero solution vector $f_p \circ f_q = \alpha_p \alpha_q e_j$ corresponds to the zero set of free variables, a contradiction. Therefore no more than one vector from the fundamental system of solutions has 2 nonzero coordinates. Since $\alpha_s = -a_i^{-1} a_s$, then no more than 2 coefficients of this equation are different from zero. If this is the only coefficient $a_j$, then the equation has the form $x_j = 0$ and we can assume $i = j$.

Let us show that the situation when there are exactly 2 nonzero coefficients is not realizable. Indeed assume the opposite: let $a_j x_j + a_t x_t = 0$, $a_j a_t \neq 0$. Equivalently, $x_j = \alpha_t x_t$, $\alpha_t \neq 0$. Then we have the solutions of the form $f_t = \alpha_t e_j + e_{t+1}$ and $v_t = f_t \circ f_t = \alpha_t^2 e_j + e_{t+1}$. Moreover $\langle f_1, \ldots, f_{n-1} \rangle$ is the solution space thus $v_t \in \langle f_1, \ldots, f_{n-1} \rangle$, that is $v_t = \beta_1 f_1 + \cdots + \beta_{n-1} f_{n-1}$ for some coefficients $\beta_1, \ldots, \beta_{n-1} \in \mathbb{F}$. Comparing the coordinates of the vectors in different sides of this equation we obtain $\beta_l = 0$ when $l \neq t$ and $\beta_t = 1$, therefore $v_t = f_t$ and $\alpha_t^2 = \alpha_t$. This implies that $\alpha_t = 1$. Then the algebra $\mathcal{A}$ is defined by the equation $x_j = x_t$. Matrix $I$ is also a solution of the equation $x_j = x_t$, therefore the system $\mathcal{S} \cup \{I\}$ generates the proper subalgebra $\mathcal{A}$ in $D_n(\mathbb{F})$, a contradiction.    $\square$

## 5. Generalized quaternion algebras

We show that the length problem for the quaternion group algebra depends on the structure of the field, namely such invariant of a field as Stufe (or level) (for details see, for example, Ref. 12(Chapter XI), Ref. 22(Chapter 2) and Ref. 3 for matrix interpretation).

**Definition 5.1.**[22(Chapter 2, Definition 2.1)] *Stufe* (or sometimes called *level*) $S(\mathbb{F})$ of a field $\mathbb{F}$ is the least positive integer $s$, such that the equation

$$a_1^2 + a_2^2 + \ldots + a_s^2 = -1, \ (a_i \in \mathbb{F})$$

has a solution. If this equation does not have a solution for any $s > 0$, then it is said that $S(\mathbb{F}) = \infty$.

**Definition 5.2.**[22(Chapter 2, Definition 2.1)] A field $\mathbb{F}$ satisfying $S(\mathbb{F}) = \infty$ is called *formally real*.

**Theorem 5.1.** *(Pfister Theorem*[22(Chapter 2, Theorem 2.2)]*) Let $\mathbb{F}$ be a field, $S(\mathbb{F}) < \infty$. Then $S(\mathbb{F}) = 2^k$ for some $k \in \mathbb{Z}_+$. Conversely, for any integer* *of the form* $2^k$ *there is a field* $\mathbb{F}$ *satisfying the condition* $S(\mathbb{F}) = 2^k$.

**Definition 5.3.** [21(§1.6, p. 14)]Let $\mathbb{F}$ be a field of characteristic different from 2 and let $a$ and $b$ be nonzero elements from $\mathbb{F}$. A *generalized quaternion algebra* $\mathcal{A} = \left( \frac{a,b}{\mathbb{F}} \right)$ over $\mathbb{F}$ is a 4-dimensional $\mathbb{F}$-vector space with the basis $\{1, \mathbf{i}, \mathbf{j}, \mathbf{k}\}$ and the following multiplication rules:

$$\mathbf{i}^2 = a, \ \mathbf{j}^2 = b, \ \mathbf{ij} = -\mathbf{ji} = \mathbf{k}.$$

**Definition 5.4.** In the case $a = b = -1$ we denote $\mathcal{D} = \left( \frac{-1,-1}{\mathbb{F}} \right)$.

**Proposition 5.1.** [21(§1.6, p. 15)] *Let* $\mathcal{A} = \left( \frac{a,b}{\mathbb{F}} \right)$ *be a generalized quaternion algebra. Then the following conditions are equivalent:*
*(i)* $\mathcal{A}$ *is a division algebra;*
*(iii) if the triple* $(c_0, c_1, c_2) \in \mathbb{F}^3$ *satisfies the condition* $c_0^2 = ac_1^2 + bc_2^2$, *then* $c_0 = c_1 = c_2 = 0$.

**Corollary 5.1.** *Let* $\mathbb{F}$ *be a field of characteristic different from* 2 *and satisfying* $S(\mathbb{F}) > 2$. *Then the algebra* $\mathcal{D}$ *is a division algebra.*

**Notation 5.1.** Let $\mathbb{F}$ be a field of characteristic different from 2 and satisfying $S(\mathbb{F}) > 2$. Then we would refer to $\mathcal{D}$ as the *quaternion (division) algebra* and use the notation $\mathbf{1}, \mathbf{i}, \mathbf{j}, \mathbf{k}$ for its standard basis elements.

**Notation 5.2.** Let $Z(\mathcal{A})$ denote the *center* of an $\mathbb{F}$-algebra $\mathcal{A}$.

We provide here for completeness the known decomposition of the group algebra $\mathbb{F}\mathbf{Q}_8$ into direct sum of matrix algebras.

**Lemma 5.1.** *Let* $\mathbb{F}$ *be a field,* char $\mathbb{F} \neq 2$. *Then*
$\mathbb{F}\mathbf{Q}_8 \cong M_2(\mathbb{F}) \oplus D_4(\mathbb{F})$ *if* $S(\mathbb{F}) \leq 2$, *or otherwise* $\mathbb{F}\mathbf{Q}_8 \cong \mathcal{D} \oplus D_4(\mathbb{F})$.

**Proof.** Applying Mashke Theorem [21(§3.6, p. 51)] we conclude that $\mathbb{F}\mathbf{Q}_8$ is semisimple. Therefore Artin-Wedderburn Theorem [21(§3.5, p. 49)] is applicable to $\mathbb{F}\mathbf{Q}_8$. We compute the decomposition of the group algebra $\mathbb{F}\mathbf{Q}_8$ into direct sum of matrix algebras directly.

Taking orthogonal central idempotents $e_1 = (e + m)/2$ and $e_2 = (e - m)/2$, we have $e_1 + e_2 = 1$ and $\mathbb{F}\mathbf{Q}_8 = (\mathbb{F}\mathbf{Q}_8)e_1 \oplus (\mathbb{F}\mathbf{Q}_8)e_2$.

Since the cyclic subgroup $\langle m \rangle$ is a normal subgroup in $Q_8$ (being its center), then $(\mathbb{F}\mathbf{Q}_8)e_1 \cong \mathbb{F}[Q_8/\langle m \rangle]$ (see Ref. 21(§3.6, p. 53, Exercise 3)). The group $\mathbf{Q}_8/\langle m \rangle$ is a non-cyclic group of order 4, hence it is the Klein group $\mathbb{Z}_2 \oplus \mathbb{Z}_2$. Since the field $\mathbb{F}$ contains $-1$, we have $\mathbb{F}\mathbb{Z}_2 = \mathbb{F} \oplus \mathbb{F}$ and $\mathbb{F}[\mathbb{Z}_2 \oplus \mathbb{Z}_2] = (\mathbb{F}\mathbb{Z}_2)[\mathbb{Z}_2] = (\mathbb{F} \oplus \mathbb{F})[\mathbb{Z}_2] = \mathbb{F}\mathbb{Z}_2 \oplus \mathbb{F}\mathbb{Z}_2 = \mathbb{F} \oplus \mathbb{F} \oplus \mathbb{F} \oplus \mathbb{F} \cong D_4(\mathbb{F})$.

Notice that $(\mathbb{F}\mathbf{Q}_8)e_2 \cong \mathbb{F}\mathbf{Q}_8/I \cong \mathcal{D}$ where $I$ is the ideal of $\mathbb{F}\mathbf{Q}_8$, generated by $e_1$ (we have a natural homomorphism of the quaternion group algebra to the quaternion algebra $\mathcal{D}$ by identifying $m$ with $-1$).

If $S(\mathbb{F}) > 2$ then $\mathcal{D}$ is a division algebra by Corollary 5.1 and we have $\mathbb{F}\mathbf{Q}_8 \cong \mathcal{D} \oplus D_4(\mathbb{F})$.

Suppose $S(\mathbb{F}) \leq 2$, which implies that there exist elements $\alpha, \beta \in \mathbb{F}$, such that $\alpha^2 + \beta^2 = -1$. Consider the mapping $\varphi : \mathcal{D} \to M_2(\mathbb{F})$ defined on the basis of $\mathcal{D}$ by $\varphi(\mathbf{1}) = I$, $\varphi(\mathbf{i}) = \begin{pmatrix} \alpha & \beta \\ \beta & -\alpha \end{pmatrix}$, $\varphi(\mathbf{j}) = \begin{pmatrix} 0 & -1 \\ 1 & 0 \end{pmatrix}$, $\varphi(\mathbf{k}) = \begin{pmatrix} \beta & -\alpha \\ -\alpha & -\beta \end{pmatrix}$ and continued by linearity. Direct calculations show that $\varphi(\mathbf{i})^2 = \varphi(\mathbf{j})^2 = \varphi(\mathbf{k})^2 = -I = -\varphi(\mathbf{1})$, $\varphi(\mathbf{i})\varphi(\mathbf{j}) = -\varphi(\mathbf{j})\varphi(\mathbf{i}) = \varphi(\mathbf{k})$, thus $\varphi$ is a homomorphism of algebras. Moreover, we have

$$E_{11} = (I - \alpha\varphi(\mathbf{i}) - \beta\varphi(\mathbf{k}))/2, \ E_{22} = (\alpha\varphi(\mathbf{i}) + \beta\varphi(\mathbf{k}) + I)/2,$$

$$E_{12} = (\alpha\varphi(\mathbf{k}) - \beta\varphi(\mathbf{i}) - \varphi(\mathbf{j}))/2, \ E_{21} = (\alpha\varphi(\mathbf{k}) - \beta\varphi(\mathbf{i}) + \varphi(\mathbf{j}))/2,$$

that is $\varphi(\mathcal{D}) = M_2(\mathbb{F})$. Consequently, we have $\mathbb{F}\mathbf{Q}_8 \cong M_2(\mathbb{F}) \oplus D_4(\mathbb{F})$. $\qquad\square$

**Lemma 5.2.** *Let $\mathbb{F}$ be a field,* char $\mathbb{F} \neq 2$. *Then $l(\mathbb{F}\mathbf{Q}_8) \leq 4$. Moreover, if $S(\mathbb{F}) \leq 2$, then $l(\mathbb{F}\mathbf{Q}_8) = 4$.*

**Proof.** I. Let $S(\mathbb{F}) \leq 2$, then in the field $\mathbb{F}$ there are elements $\alpha, \beta$, such that $\alpha^2 + \beta^2 = -1$. Then we have $\mathbb{F}\mathbf{Q}_8 \cong M_2(\mathbb{F}) \oplus D_4(\mathbb{F})$ by Lemma 5.1.

Hence by Lemma 3.2 we obtain

$$l(\mathbb{F}\mathbf{Q}_8) = l(M_2(\mathbb{F}) \oplus D_4(\mathbb{F})) = 4.$$

In particular, $S(\mathbb{F}) = 1$ and $l(\mathbb{F}\mathbf{Q}_8) = 4$ if $\mathbb{F}$ is algebraically closed.

II. Let $\mathbb{F}$ be a field which is not algebraically closed and $\bar{\mathbb{F}}$ be its algebraic closure. From Proposition 2.3 and by the argument from item I we obtain

$$l(\mathbb{F}\mathbf{Q}_8) \leq l(\bar{\mathbb{F}}\mathbf{Q}_8) = 4.$$

$\qquad\square$

## 6. Some technical results

**Proposition 6.1.** *Let $\mathbb{F}$ be a field of characteristic different from 2 and $S(\mathbb{F}) > 2$. Consider the arbitrary generating system $S$ for the algebra $\mathcal{A} = \mathcal{D} \oplus D_4(\mathbb{F})$. Then there exist a generating system $\tilde{S} = \{C_i = A_i \oplus D_i | C_i \in$*

$\mathcal{D}, D_i \in D_4(\mathbb{F})$, $i = 1, \ldots, k\}$ *for algebra* $\mathcal{A}$ *such that*

1. $l(\tilde{\mathcal{S}}) = l(\mathcal{S})$;
2. $A_1 = \mathbf{i} + \alpha\mathbf{k}$, $A_2 = \mathbf{j} + \beta\mathbf{k}$, $A_3 = \gamma\mathbf{k}$, $\alpha, \beta, \gamma \in \mathbb{F}$, $A_s = 0$ *for all* $s \geq 3$.

**Proof.** By definition

$$\mathcal{S} = \{M_i = B_i \oplus F_i | B_i \in \mathcal{D}, F_i \in D_4(\mathbb{F}), \ i = 1, \ldots, r\},$$

in this case $\mathcal{S}_B = \{B_i, i = 1, \ldots, r\}$ and $\mathcal{S}_F = \{F_i, i = 1, \ldots, r\}$ are generating systems for algebras $\mathcal{D}$ and $D_n(\mathbb{F})$, respectively.

Applying Proposition 2.2[6(Proposition 2.2)] we change the generating set to $\mathcal{S}_1 = \{S - (S)_1 1_{\mathcal{A}} | \ S \in \mathcal{S}\}$, where $(S)_1$ is the coefficient at the basis vector $\mathbf{1}$ in the decomposition of the first component of $S$ under the chosen basis of $\mathcal{D}$, leaving the length invariant and thus we can always assume that $\mathcal{S} = \mathcal{S}_1$.

Since the division ring $\mathcal{D}$ is not one-generated as $\mathbb{F}$-algebra we have $3 \leq \dim \mathcal{L}_1(\mathcal{S}_B) \leq 4$.

I. Let $\dim \mathcal{L}_1(\mathcal{S}_B) = 4$, this condition implies that $\mathcal{L}_1(\mathcal{S}_B) = \mathcal{D}$. Then choose the elements $B_{i_1}, B_{i_2}, B_{i_3} \in \mathcal{S}_B$, which together with $\mathbf{1}$ form a basis of $\mathcal{D}$. By construction $B_{i_1}, B_{i_2}, B_{i_3} \in \langle \mathbf{i}, \mathbf{j}, \mathbf{k} \rangle$, therefore these elements form a basis of a given linear span.

Then using Proposition 2.1[6(Proposition 2.1)] we can change the generating system to $\mathcal{S}_2 = \{C_1 = \mathbf{i} \oplus G_1, C_2 = \mathbf{j} \oplus G_2, C_3 = \mathbf{k} \oplus G_3, S | S \in \mathcal{S} \backslash \{M_{i_1}, M_{i_2}, M_{i_3}\}\}$ leaving the length invariant. Applying the same Proposition 2.1[6(Proposition 2.1)] we choose

$$\tilde{\mathcal{S}} = \Big\{ C_1, C_2, C_3, S - (\omega_{s,1}C_1 + \omega_{s,2}C_2 + \omega_{s,3}C_3) | S \in \mathcal{S} \setminus \{M_{i_1}, M_{i_2}, M_{i_3}\} \Big\},$$

where $S = B_i \oplus F_i$ and $B_i = \omega_{s,1}\mathbf{i} + \omega_{s,2}\mathbf{j} + \omega_{s,3}\mathbf{k}$. Thus the item 2 is true for the values $\alpha = 0, \beta = 0, \gamma = 1$.

II. Let $\dim \mathcal{L}_1(\mathcal{S}_A) = 3$.

By conditions there are two elements $M_i$, $M_j \in \mathcal{S}$ (we can choose them as $M_1$ and $M_2$) such that the system $\mathbf{1}, B_1, B_2$ is a basis of $\mathcal{L}_1(\mathcal{S}_B)$. If $B_1 = \alpha_1\mathbf{i} + \alpha_2\mathbf{j} + \alpha_3\mathbf{k}$, $B_2 = \beta_1\mathbf{i} + \beta_2\mathbf{j} + \beta_3\mathbf{k}$, then the matrix $\begin{pmatrix} \alpha_1 & \alpha_2 & \alpha_3 \\ \beta_1 & \beta_2 & \beta_3 \end{pmatrix}$ has a nonzero minor of order 2. Since it is possible to choose an arbitrary order of the imaginary units in the basis of $\mathcal{D}$ we can assume that $\begin{vmatrix} \alpha_1 & \alpha_2 \\ \beta_1 & \beta_2 \end{vmatrix} \neq 0$.

In this case there is a non-singular matrix $F = (f_{k,m}) \in M_2(\mathbb{F})$ such that $(1,0) = f_{1,1}(\alpha_1, \alpha_2) + f_{1,2}(\beta_1, \beta_2)$, $(0,1) = f_{2,1}(\alpha_1, \alpha_2) + f_{2,2}(\beta_1, \beta_2)$. Assume that $C_r = f_{r,1}M_1 + f_{r,2}M_2$, $r = 1, 2$ and by Proposition 2.1[6(Proposition 2.1)] we can pass to the generating system $\mathcal{S}_2 =$

$\{C_1, C_2, S| \ S \in \mathcal{S}, \ S \neq M_1, M_2\}$ leaving the length invariant. Then by construction $C_1 = \mathbf{i} + \alpha\mathbf{k} \oplus D_1$, $C_2 = \mathbf{j} + \beta\mathbf{k} \oplus D_2$.

After that by Proposition 2.1[6](Proposition 2.1) leaving the length invariant we can pass to the generating system $\mathcal{S}_3 = \{C_1, C_2, S - (S)_\mathbf{i}C_1 - (S)_\mathbf{j}C_2| \ S \in \mathcal{S}_2, \ S \neq C_1, C_2\}$, where again $(S)_\mathbf{i}$ and $(S)_\mathbf{j}$ denote the coefficients at the basis vectors $\mathbf{i}$ and $\mathbf{j}$ in the decomposition of the first component of $S$ under the chosen basis $\mathcal{D}$, respectively. Setting $\mathcal{S}_3 = \{C_1, C_2, \tilde{C}_3, \dots, \tilde{C}_k\}$ we have that the first component of any $\tilde{C}_m$ is $\gamma_m\mathbf{k}$, $m = 3, \dots, k$. If at least one $\gamma_m \neq 0$, then we take $C_3 = \tilde{C}_m$ and $C_r = \tilde{C}_r - \gamma_m^{-1}\gamma_r\tilde{C}_3$ and $\tilde{\mathcal{S}} = \{C_1, \dots, C_k\}$. Otherwise, $\tilde{\mathcal{S}} = \mathcal{S}_3$ and $\gamma = 0$. $\qquad\square$

In order to prove the next result on the length of the algebra $\mathcal{D} \oplus D_4(\mathbb{F})$ we need the following technical lemma. As usual for a matrix $A \in M_{n,m}(\mathbb{F})$ we denote by $A[i_1, \dots, i_k|j_1, \dots, j_l]$ the $k \times l$ submatrix of $A$ which is located in the intersection of rows with the indices $i_1, \dots, i_k$ and columns with the indices $j_1, \dots, j_l$. Also $A(i_1, \dots, i_k|j_1, \dots, j_l)$ denotes the $(n - k) \times (m - l)$ matrix obtained from $A$ by deleting the rows with the indices $i_1, \dots, i_k$ and the columns with the indices $j_1, \dots, j_l$.

**Lemma 6.1.** *Let $\mathbb{F}$ be a field of characteristic different from $2$ and $S(\mathbb{F}) > 2$, $\omega_1, \omega_2, \omega_3, \nu_1, \nu_2, \nu_3, \beta_1, \beta_2, \gamma_1, \gamma_2 \in \mathbb{F}$ be such that $(\omega_1, \omega_2, \omega_3) \neq (0, 0, 0)$, $(\nu_1, \nu_2, \nu_3) \neq (0, 0, 0)$ and the matrix $\Gamma = \begin{pmatrix} \gamma_1 & \beta_1 \\ \gamma_2 & \beta_2 \end{pmatrix}$ has neither a zero row nor a zero column and $\beta_1 \neq \beta_2$. We consider*

$$M = \begin{pmatrix} 1 & & 1 & 1 & 1 & 1 \\ 0 & & 1 & 0 & \beta_1 & \beta_2 \\ 0 & & 0 & 1 & \gamma_1 & \gamma_2 \\ -\omega_1^2 - \omega_2^2 - \omega_3^2 & & 1 & 0 & \beta_1^2 & \beta_2^2 \\ -\nu_1^2 - \nu_2^2 - \nu_3^2 & & 0 & 1 & \gamma_1^2 & \gamma_2^2 \\ -\omega_1\nu_1 - \omega_2\nu_2 - \omega_3\nu_3 & & 0 & 0 & \gamma_1\beta_1 & \gamma_2\beta_2 \\ 0 & & 0 & 0 & \gamma_1^2\beta_1 & \gamma_2^2\beta_2 \\ 0 & & 0 & 0 & \gamma_1\beta_1^2 & \gamma_2\beta_2^2 \\ 0 & & 0 & 1 & \gamma_1^3 & \gamma_2^3 \\ 0 & & 1 & 0 & \beta_1^3 & \beta_2^3 \end{pmatrix} \in M_{10,5}(\mathbb{F}) \ .$$

*Then the row $(1, 0, 0, 0, 0)$ lies in the linear span of the rows of $M$.*

**Proof.** We consider the submatrix

$$M_1 = M[2,3,7,8,9,10|2,3,4,5] = \begin{pmatrix} 1 & 0 & \beta_1 & \beta_2 \\ 0 & 1 & \gamma_1 & \gamma_2 \\ 0 & 0 & \gamma_1^2\beta_1 & \gamma_2^2\beta_2 \\ 0 & 0 & \gamma_1\beta_1^2 & \gamma_2\beta_2^2 \\ 0 & 1 & \gamma_1^3 & \gamma_2^3 \\ 1 & 0 & \beta_1^3 & \beta_2^3 \end{pmatrix}.$$

If $\mathrm{rk}(M_1) = 4$, then we can consider the submatrix of $M$ obtained by union of $M_1$ with the first row and the column $(1,0,0,0,0,0,0)^t$, i.e., the matrix $M_2 = M[1,2,3,7,8,9,10|1,2,3,4,5]$. Then $\mathrm{rk}(M_2) = \mathrm{rk}(M_1) + 1 = 5$, and hence $\mathrm{rk}(M) = 5$. Therefore, any vector lies in the linear span of the rows of $M$, so, the vector $(1,0,0,0,0)$ lies in the linear span of the rows of $M$. Thus, it remains to consider the case $\mathrm{rk}(M_1) < 4$.

Let us consider different $4 \times 4$ minors of $M_1$. It is straightforward to see that

$$\det(M_1[1,2,3,4|]) = -\det(M_1[3,4,5,6|])$$
$$= \det(M_1[1,3,4,5|]) \tag{3}$$
$$= \beta_1\beta_2\gamma_1\gamma_2(\gamma_1\beta_2 - \gamma_2\beta_1)$$
$$\det(M_1[1,2,4,5|]) = -\gamma_1\gamma_2(-\gamma_2^2\beta_1^2 + \gamma_1^2\beta_2^2 - \beta_2^2 + \beta_1^2) \tag{4}$$
$$\det(M_1[1,2,5,6|]) = \gamma_1^3\beta_2^3 - \gamma_2^3\beta_1^3 - \gamma_1\beta_2^3 + \gamma_2\beta_1^3 + \beta_1\gamma_2^3$$
$$- \beta_2\gamma_1^3 - \beta_1\gamma_2 + \gamma_1\beta_2. \tag{5}$$

All of them are 0 since $\mathrm{rk}(M_1) < 4$. We examine the condition $\det(M_1[1,2,3,4|]) = 0$. The general case splits into the following subcases.

1. Let $\beta_1 = 0$. Substituting this to the expressions (4) and (5) we obtain that

$$-\gamma_1\gamma_2(\gamma_1^2\beta_2^2 - \beta_2^2) = -\gamma_1\gamma_2\beta_2^2(\gamma_1^2 - 1) = 0$$

and

$$\gamma_1^3\beta_2^3 - \gamma_1\beta_2^3 - \beta_2\gamma_1^3 + \gamma_1\beta_2 = \gamma_1\beta_2(\gamma_1^2\beta_2^2 - \beta_2^2 - \gamma_1^2 + 1)$$
$$= \gamma_1\beta_2(\gamma_1^2 - 1)(\beta_2^2 - 1) = 0.$$

Since $\Gamma$ has neither zero rows nor zero columns, it follows that $\gamma_1\beta_2 \neq 0$. Therefore, the following cases appear:

* $\beta_2 = 1$. Then $M[2,4|] = \begin{pmatrix} 0 & 1\,0\,0\,1 \\ -\omega_1^2 - \omega_2^2 - \omega_3^2 & 1\,0\,0\,1 \end{pmatrix}$. By the condition $S(\mathbb{F}) > 2$ we have $-\omega_1^2 - \omega_2^2 - \omega_3^2 \neq 0$. Then $\frac{-1}{\omega_1^2 + \omega_2^2 + \omega_3^2}(M[4|] - M[2|])$ is the required vector.

* $\gamma_1 = 1$. Then $M[3,5,8|] = \begin{pmatrix} 0 & 0\,1\,1 & \gamma_2 \\ -\nu_1^2 - \nu_2^2 - \nu_3^2 & 0\,1\,1 & \gamma_2^2 \\ 0 & 0\,0\,0 & \gamma_2\beta_2^2 \end{pmatrix}$, and the re-

qured vector is a scalar multiple of $M[5|] - M[3|] - \beta_2^{-2}(\gamma_2 - 1)M[8|]$ since $-\nu_1^2 - \nu_2^2 - \nu_3^2 \neq 0$ by the condition $S(\mathbb{F}) > 2$.

* $\beta_2 = -1$. Then $M[2,4,8|] = \begin{pmatrix} 0 & 1\,0\,0\,-1 \\ -\omega_1^2 - \omega_2^2 - \omega_3^2 & 1\,0\,0\,1 \\ 0 & 0\,0\,0\,\gamma_2 \end{pmatrix}$. If $\gamma_2 \neq 0$,

then the required vector is the nonzero scalar multiple of $M[4|] - M[2|] - 2\gamma_2^{-1}M[8|]$.

Assume now that $\gamma_2 = 0$. Since the case $\gamma_1 = 1$ is already considered we can WLOG assume that $\gamma_1 \neq 1$. Then

$$M[2,3,4,5,6,9|] = \begin{pmatrix} 0 & 1\,0\,0\,-1 \\ 0 & 0\,1\,\gamma_1\,0 \\ -\omega_1^2 - \omega_2^2 - \omega_3^2 & 1\,0\,0\,1 \\ -\nu_1^2 - \nu_2^2 - \nu_3^2 & 0\,1\,\gamma_1^2\,0 \\ -\omega_1\nu_1 - \omega_2\nu_2 - \omega_3\nu_3 & 0\,0\,0\,0 \\ 0 & 0\,1\,\gamma_1^3\,0 \end{pmatrix}.$$

In this case if $\gamma_1 \neq -1$, then the rows $M[3|]$ and $M[9|]$ are linearly independent, and then the required vector is $M[5|] - (\gamma_1 + 1)^{-1}M[3|] - (\gamma_1 + 1)^{-1}M[9|]$.

If $\gamma_1 = -1$, then we consider

$$M[1,4,5,6|] = \begin{pmatrix} 1 & 1\,1\,1\,1 \\ -\omega_1^2 - \omega_2^2 - \omega_3^2 & 1\,0\,0\,1 \\ -\nu_1^2 - \nu_2^2 - \nu_3^2 & 0\,1\,1\,0 \\ -\omega_1\nu_1 - \omega_2\nu_2 - \omega_3\nu_3 & 0\,0\,0\,0 \end{pmatrix}$$

and the row $M[4|] + [5|] + 2M[6|] - M[1|] = (-(\omega_1 + \nu_1)^2 - (\omega_2 + \nu_2)^2 - (\omega_3 + \nu_3)^2 - 1, 0, 0, 0, 0)$ is required by the condition on the field.

* $\gamma_1 = -1$. If in this case $\beta_2 = -1$ then we are done by the previous case. So we may assume that $\beta_2 \neq -1$. Hence the rows $M[2|]$ and $M[10|]$ are linearly independent, and thus the required vector can be obtained from $M[4|] - (\beta_2 + 1)^{-1}M[2|] - (\beta_2 + 1)^{-1}M[10|]$.

Thus all the possibilities for $\beta_1 = 0$ are considered, and in any case we found that $(1, 0, 0, 0, 0)$ is a linear combination of rows of $M$.

2. The case $\beta_2 = 0$ can be reduced to the case 1 by interchange of 4th and 5th columns.

3. The case $\gamma_1 = 0$ can be reduced to the case 1 by interchange of 2nd and 3rd columns.

4. The case $\gamma_2 = 0$ can be reduced to the case 1 by pairwise interchange of 2nd and 3rd columns and 4th and 5th columns.

5. Now it remains to consider $\beta_1 \beta_2 \gamma_1 \gamma_2 \neq 0$. Then by (3) it holds that $\gamma_1 \beta_2 - \gamma_2 \beta_1 = 0 = \det(\Gamma)$. Since there are no zero rows in $\Gamma$, it follows that there exists a coefficient $\kappa \in \mathbb{F}$, $\kappa \neq 0$, such that $\gamma_1 = \kappa \beta_1, \gamma_2 = \kappa \beta_2$. We can substitute these values to $M_1$, which produces:

$$
M_1 = \begin{pmatrix}
1 & 0 & \beta_1 & \beta_2 \\
0 & 1 & \kappa\beta_1 & \kappa\beta_2 \\
0 & 0 & \kappa^2\beta_1^3 & \kappa^2\beta_2^3 \\
0 & 0 & \kappa\beta_1^3 & \kappa\beta_2^3 \\
0 & 1 & \kappa^3\beta_1^3 & \kappa^3\beta_2^3 \\
1 & 0 & \beta_1^3 & \beta_2^3
\end{pmatrix}.
$$

By elementary row transformations of $M_1$ we obtain:

$$
M_1 \underset{\sim}{\overset{\kappa^{-2}(3)}{\kappa^{-1}(4)}} \begin{pmatrix}
1 & 0 & \beta_1 & \beta_2 \\
0 & 1 & \kappa\beta_1 & \kappa\beta_2 \\
0 & 0 & \beta_1^3 & \beta_2^3 \\
0 & 0 & \beta_1^3 & \beta_2^3 \\
0 & 1 & \kappa^3\beta_1^3 & \kappa^3\beta_2^3 \\
1 & 0 & \beta_1^3 & \beta_2^3
\end{pmatrix} \underset{\sim}{\overset{(4)-(3)}{\underset{(5)-\kappa^3(3)}{(6)-(3)}}} \begin{pmatrix}
1 & 0 & \beta_1 & \beta_2 \\
0 & 1 & \kappa\beta_1 & \kappa\beta_2 \\
0 & 0 & \beta_1^3 & \beta_2^3 \\
0 & 0 & 0 & 0 \\
0 & 1 & 0 & 0 \\
1 & 0 & 0 & 0
\end{pmatrix} \underset{\sim}{\overset{(1)-(6)}{(2)-(5)}}
$$

$$
\begin{pmatrix}
0 & 0 & \beta_1 & \beta_2 \\
0 & 0 & \kappa\beta_1 & \kappa\beta_2 \\
0 & 0 & \beta_1^3 & \beta_2^3 \\
0 & 0 & 0 & 0 \\
0 & 1 & 0 & 0 \\
1 & 0 & 0 & 0
\end{pmatrix} \underset{\sim}{\overset{(2)-\kappa(1)}{}} \begin{pmatrix}
0 & 0 & \beta_1 & \beta_2 \\
0 & 0 & 0 & 0 \\
0 & 0 & \beta_1^3 & \beta_2^3 \\
0 & 0 & 0 & 0 \\
0 & 1 & 0 & 0 \\
1 & 0 & 0 & 0
\end{pmatrix} = \tilde{M}_1. \quad (6)
$$

Since $\operatorname{rk}(\tilde{M}_1) = \operatorname{rk}(M_1) < 4$ we obtain that

$$
0 = \det(\tilde{M}_1[1, 3, 5, 6|]) = \begin{vmatrix} \beta_1 & \beta_2 \\ \beta_1^3 & \beta_2^3 \end{vmatrix} = \beta_2\beta_1^3 - \beta_1\beta_2^3 = \beta_1\beta_2(\beta_1^2 - \beta_2^2),
$$

hence, by the conditions of our case $\beta_1^2 - \beta_2^2 = 0$. By the condition $\beta_1 \neq \beta_2$. Then $\beta_1 = -\beta_2$. By substituting the last equality to $M$ and doing the elementary row transformations from (6) with $M$ we obtain

$$\tilde{M}[1,4,9,10|] = \begin{pmatrix} 1 & 1 & 1 & 1 & 1 \\ -\omega_1^2 - \omega_2^2 - \omega_3^2 & 1 & 0 & \beta_1^2 & \beta_1^2 \\ 0 & 0 & 1 & 0 & 0 \\ 0 & 1 & 0 & 0 & 0 \end{pmatrix},$$

which implies $\tilde{M}[4|] + (\beta_1^2 - 1)\tilde{M}[10|] + \beta_1^2\tilde{M}[3|] - \beta_1^2\tilde{M}[1|] = (-\omega_1^2 - \omega_2^2 - \omega_3^2 - \beta_1^2, 0, 0, 0, 0)$, note that $-\omega_1^2 - \omega_2^2 - \omega_3^2 - \beta_1^2 \neq 0$ by the condition on the field. Therefore the required row is obtained in this last case as well, which concludes the proof. ∎

## 7. Length of $\mathcal{D} \oplus D_4(\mathbb{F})$

Recall that $\mathcal{D} = \left(\frac{-1,-1}{\mathbb{F}}\right)$ is a division algebra.

**Proposition 7.1.** *Let $\mathbb{F}$ be a field of characteristic different from 2 and $S(\mathbb{F}) > 2$. Consider the arbitrary generating system $S$ for the algebra $\mathcal{A} = \mathcal{D} \oplus D_4(\mathbb{F})$. Then $\mathbf{i} \oplus O, \mathbf{j} \oplus O, \mathbf{k} \oplus O \in \mathcal{L}_3(S)$.*

**Proof.** The given generating system $S$ for the algebra $\mathcal{A} = \mathcal{D} \oplus D_4(\mathbb{F})$ has the following form:

$$S = \{C_i = A_i \oplus D_i | A_i \in \mathcal{D}, D_i \in D_4(\mathbb{F}), \ i = 1, \ldots, k\}.$$

By Proposition 6.1 we can change the set $S$ to $\tilde{S}$ if necessary. Notice that by Propositions 2.1–2.2 and Ref. 6(Propositions 2.1, 2.2) we have $\mathcal{L}_3(S) = \mathcal{L}_3(\tilde{S})$. Thus it is sufficient to prove the statement for $\tilde{S}$. In order to simplify the notations we denote this new set again by $S$. Then we assume that for matrices $C_i$ the statement (2) from Proposition 6.1 is true.

Direct computations show that

$$[C_1, C_2] = [\mathbf{i} + \alpha\mathbf{k}, \mathbf{j} + \beta\mathbf{k}] \oplus O = (-2\alpha\mathbf{i} - 2\beta\mathbf{j} + 2\mathbf{k}) \oplus O \in \mathcal{L}_2(S),$$

$$C_1[C_1, C_2] = (\mathbf{i} + \alpha\mathbf{k})(-2\alpha\mathbf{i} - 2\beta\mathbf{j} + 2\mathbf{k}) \oplus O$$

$$= (2\alpha\beta\mathbf{i} - 2(1 + \alpha^2)\mathbf{j} - 2\beta\mathbf{k}) \oplus O \in \mathcal{L}_3(S),$$

$$C_2[C_1, C_2] = (\mathbf{j} + \beta\mathbf{k})(-2\alpha\mathbf{i} - 2\beta\mathbf{j} + 2\mathbf{k}) \oplus O$$

$$= (2(1 + \beta^2)\mathbf{i} - 2\alpha\beta\mathbf{j} + 2\alpha\mathbf{k}) \oplus O \in \mathcal{L}_3(S)$$

and

$$\begin{vmatrix} -2\alpha & -2\beta & 2 \\ 2\alpha\beta & -2(1+\alpha^2) & -2\beta \\ 2(1+\beta^2) & -2\alpha\beta & 2\alpha \end{vmatrix} = 2^3(\beta^2 + 1 + \alpha^2)^2 \neq 0$$

by conditions on the field $\mathbb{F}$. Therefore

$$\dim_{\mathbb{F}}\langle [C_1, C_2], C_1[C_1, C_2], C_2[C_1, C_2]\rangle = 3$$

and

$$\langle [C_1, C_2], C_1[C_1, C_2], C_2[C_1, C_2]\rangle = \langle \mathbf{i} \oplus O, \mathbf{j} \oplus O, \mathbf{k} \oplus O\rangle.$$

The last equality implies that inclusions $\mathbf{i} \oplus O, \mathbf{j} \oplus O, \mathbf{k} \oplus O \in \mathcal{L}_3(\mathcal{S})$ are proved.

$\square$

**Lemma 7.1.** *Let* $\mathbb{F}$ *be a field of characteristic different from* 2 *and assume that* $S(\mathbb{F}) > 2$. *Then* $l(\mathcal{D} \oplus D_4(\mathbb{F})) \leq 3$.

**Proof.** To prove the lemma we consider an arbitrary generating system

$$\mathcal{S} = \{C_i = A_i \oplus D_i | A_i \in \mathcal{D}, D_i \in D_4(\mathbb{F}), \ i = 1, \ldots, k\}, \tag{7}$$

for the algebra $\mathcal{A} = \mathcal{D} \oplus D_4(\mathbb{F})$ and show that $l(\mathcal{S}) \leq 3$.

By Proposition 6.1 we can change the set $\mathcal{S}$ to $\tilde{\mathcal{S}}$ if necessary. In order to simplify the notations we denote this new set $\mathcal{S}$ again. Then without loss of generality we assume that for matrices $C_i$ the statement (2) from Proposition 6.1 is true.

By definition the sets $\mathcal{S}_A = \{A_i, i = 1, \ldots, k\}$ and $\mathcal{S}_D = \{D_i, i = 1, \ldots, k\}$ are generating sets for the algebras $\mathcal{D}$ and $D_n(\mathbb{F})$, respectively.

By Proposition 7.1 we already have $\mathbf{i} \oplus O, \mathbf{j} \oplus O, \mathbf{k} \oplus O \in \mathcal{L}_3(\mathcal{S})$. It remains to show that a certain basis of the algebra $0 \oplus D_4(\mathbb{F})$ is contained in $\mathcal{L}_3(\mathcal{S})$. If this is done, we can consider its basis together with $1_{\mathcal{A}}$ and elements $\mathbf{i} \oplus O, \mathbf{j} \oplus O, \mathbf{k} \oplus O$. These elements constitute a basis of $\mathcal{A}$ which is contained in $\mathcal{L}_3(\mathcal{S})$, and thus prove that $\mathcal{A} \subseteq \mathcal{L}_3(\mathcal{S})$.

Observe that $l(D_4(\mathbb{F})) \leq 3$ by Theorem 2.2. Since $\mathcal{S}$ is a generating set, it follows that there are the elements

$$Z_i = U_i \oplus V_i \in \mathcal{S}^3, \ i = 1, 2, 3, \tag{8}$$

such that $I, V_1, V_2, V_3$ is a basis of the algebra $D_4(\mathbb{F})$. Hence it is sufficient to show that $1 \oplus O \in \mathcal{L}_3(\mathcal{S})$. Then the result will follow by Proposition 7.1.

Our aim is to consider all possible cases for the dimension $\dim \mathcal{L}_1(\mathcal{S}_D) \in \{2, 3, 4\}$. We remind that $D_1, D_2$ are the diagonal matrices from (7).

1. Assume that $\dim\langle I, D_1, D_2\rangle \leq 2$. Let us show that $\mathbf{1} \oplus O \in \mathcal{L}_2(\mathcal{S})$ in this case.

1a. Assume firstly that $\dim\langle I, D_1, D_2\rangle = 1$. Then $D_1 = \lambda I$. In this case $C_1 = (\mathbf{i}+\alpha\mathbf{k})\oplus\lambda I$, $C_1^2 = (-1-\alpha^2)\mathbf{1}\oplus\lambda^2 I$, $C_1^2 - \lambda^2 \mathbf{1}_\mathcal{A} = (-1-\alpha^2-\lambda^2)\mathbf{1}\oplus O \in \mathcal{L}_2(\mathcal{S})$, and by the conditions on $\mathbb{F}$, the scalar $-1 - \alpha^2 - \lambda^2 \neq 0$. Therefore $\mathbf{1} \oplus O \in \mathcal{L}_2(\mathcal{S})$, as required.

1b. Now, let $\dim\langle I, D_1, D_2\rangle = 2$. Then up to the interchange of $D_1$ and $D_2$ without loss of generality we may assert that $D_2 = \mu D_1 + \lambda I$. We can assume that $\mu \neq 0$, since otherwise the arguments from Item 1a work for the element $C_2$ instead of $C_1$. Hence,

$$C_2 - \mu C_1 = (\mathbf{j} - \mu\mathbf{i} + (\beta - \mu\alpha)\mathbf{k}) \oplus \lambda I.$$

By Proposition 7.1 it holds that

$$(\mathbf{j} - \mu\mathbf{i} + (\beta - \mu\alpha)\mathbf{k}) \oplus O \in \mathcal{L}_3(\mathcal{S}).$$

Therefore, $0 \oplus \lambda I \in \mathcal{L}_3(\mathcal{S})$.

If $\lambda \neq 0$, then

$$\mathbf{1} \oplus O = \mathbf{1}_\mathcal{A} - \lambda^{-1}(0 \oplus \lambda I) \in \mathcal{L}_3(\mathcal{S})$$

and the result follows. If $\lambda = 0$, then

$$(C_2 - \mu C_1)^2 = (-1 - \mu^2 - (\beta - \mu\alpha)^2)\mathbf{1} \oplus O.$$

By the conditions on $\mathbb{F}$, the scalar multiple $-1 - \mu^2 - (\beta - \mu\alpha)^2 \neq 0$. Hence,

$$\mathbf{1} \oplus O \in \mathcal{L}_2(\mathcal{S}) \subseteq \mathcal{L}_3(\mathcal{S}).$$

Let us consider the general case now.

2. Let $\dim \mathcal{L}_1(\mathcal{S}_D) = 2$. Since $\langle I, D_1, D_2\rangle \subseteq \mathcal{L}_1(\mathcal{S}_D)$ by the definition, we get that conditions of Item 1 are satisfied, and the lemma is proved.

3. Now we consider the case $\dim \mathcal{L}_1(\mathcal{S}_D) = 3$. If $\dim\langle I, D_1, D_2\rangle \leq 2$ then we are done by Item 1. Hence, we may further assume that $\dim\langle I, D_1, D_2\rangle = 3$. In this case we obtain that $\mathcal{L}_1(\mathcal{S}_D) = \langle I, D_1, D_2\rangle$, and thus the matrices $D_1$ and $D_2$ generate the algebra $D_4(\mathbb{F})$. Hence we choose in (8) the elements $Z_1 = C_1$, $Z_2 = C_2$, $Z_3$ is a word of the length 2 in $Z_1$ and $Z_2$.

3a. Let $D_1, D_2$ (without $I$) does not generate $D_4(\mathbb{F})$.

In this case by Lemma 4.1 the matrices $D_1$ and $D_2$ have a common zero coordinate. Hence up to the order of coordinates the vectors of standard basis of the space $\langle D_1, D_2\rangle$ have the form $(0, 1, 0, \alpha_1)$, $(0, 0, 1, \alpha_2)$, and $\alpha_i \neq 0$ for at least one $i = 1, 2$. In this notations we may assume that $\langle \mathcal{S}\rangle$ contains the elements $H_1 = W_1 \oplus \mathrm{diag}\{0,1,0,\alpha_1\}$, $H_2 = W_2 \oplus \mathrm{diag}\{0,0,1,\alpha_2\}$, $W_1, W_2 \in \langle \mathbf{i}, \mathbf{j}, \mathbf{k}\rangle$. By construction we have $\dim\langle A_1, A_2\rangle = 2$, hence, $\dim\langle W_1, W_2\rangle = 2$, i.e., $W_1 \neq 0$ and $W_2 \neq 0$.

If $\alpha_s = 0$ for some $s \in \{1,2\}$, then we consider $W_s$ and $H_s$ to prove that $\mathbf{1} \oplus O \in \mathcal{L}_3(\mathcal{S})$. We denote $W_s = (\omega_1 \mathbf{i} + \omega_2 \mathbf{j} + \omega_3 \mathbf{k})$. By Proposition 7.1 we know that $W_s \oplus O \in \mathcal{L}_3(\mathcal{S})$. Hence,

$$H_s - (W_s \oplus O) = 0 \oplus E_{s+1,s+1} \in \mathcal{L}_3(\mathcal{S}).$$

Since by direct computations $H_s^2 = W_s^2 \oplus E_{s+1,s+1} = (-\omega_1^2 - \omega_2^2 - \omega_3^2)\mathbf{1} \oplus E_{s+1,s+1} \in \mathcal{L}_2(\mathcal{S})$, it follows that $H_s^2 - 0 \oplus E_{s+1,s+1} = (-\omega_1^2 - \omega_2^2 - \omega_3^2)\mathbf{1} \oplus O \in \mathcal{L}_3(\mathcal{S})$. By the condition $S(\mathbb{F}) > 2$ it follows that $-\omega_1^2 - \omega_2^2 - \omega_3^2 \neq 0$. Hence $\mathbf{1} \oplus O \in \mathcal{L}_3(\mathcal{S})$, as required.

Assume now that $\alpha_1 \alpha_2 \neq 0$. We denote $W_1 = \omega_1 \mathbf{i} + \omega_2 \mathbf{j} + \omega_3 \mathbf{k}$. By Proposition 7.1 $W_i \oplus O \in \mathcal{L}_3(\mathcal{S})$ if $i = 1, 2$. Hence, $H_1 - (W_1 \oplus O) = 0 \oplus \mathrm{diag}\{0,1,0,\alpha_1\} \in \mathcal{L}_3(\mathcal{S})$. Let us consider the word

$$\begin{aligned} H_1^2 H_2 &= W_1^2 W_2 \oplus \mathrm{diag}\{0,0,0,\alpha_1\alpha_2\} \\ &= (-\omega_1^2 - \omega_2^2 - \omega_3^2)W_2 \oplus \mathrm{diag}\{0,0,0,\alpha_1\alpha_2\} \in \mathcal{L}_3(\mathcal{S}). \end{aligned}$$

By Proposition 7.1

$$H_1^2 H_2 + (\omega_1^2 + \omega_2^2 + \omega_3^2)(W_2 \oplus O) = 0 \oplus \mathrm{diag}\{0,0,0,\alpha_1\alpha_2\} \in \mathcal{L}_3(\mathcal{S}).$$

Hence,

$$0 \oplus E_{2,2} \text{ and } 0 \oplus E_{4,4} \in \mathcal{L}_3(\mathcal{S}). \tag{9}$$

By the direct computation we have

$$H_1^2 = W_1^2 \oplus \mathrm{diag}\{0,1,0,\alpha_1^2\} = (-\omega_1^2 - \omega_2^2 - \omega_3^2)\mathbf{1} \oplus \mathrm{diag}\{0,1,0,\alpha_1^2\} \in \mathcal{L}_2(\mathcal{S}).$$

Hence by (9)

$$H_1^2 - 0 \oplus \mathrm{diag}\{0,1,0,\alpha_1^2\} = (-\omega_1^2 - \omega_2^2 - \omega_3^2)\mathbf{1} \oplus O \in \mathcal{L}_3(\mathcal{S}).$$

Since $S(\mathbb{F}) > 2$ by conditions, it follows that $-\omega_1^2 - \omega_2^2 - \omega_3^2 \neq 0$. Thus $\mathbf{1} \oplus O \in \mathcal{L}_3(\mathcal{S})$ is again true.

3b. Let $D_1, D_2$ do not have a common zero coordinate. Then up to the order of coordinates the standard basis vectors of the space $\langle D_1, D_2 \rangle$ have the form $(1, 0, \beta_1, \beta_2)$, $(0, 1, \gamma_1, \gamma_2)$, where the matrix $\begin{pmatrix} \beta_1 & \beta_2 \\ \gamma_1 & \gamma_2 \end{pmatrix}$ does not have zero columns. Then $\langle \mathcal{S} \rangle$ contains the elements

$$\begin{aligned} H_1 &= W_1 \oplus \mathrm{diag}\{1,0,\beta_1,\beta_2\} = W_1 \oplus F_1, \\ H_2 &= W_2 \oplus \mathrm{diag}\{0,1,\gamma_1,\gamma_2\} = W_2 \oplus F_2, \end{aligned} \tag{10}$$

where $W_1, W_2 \in \langle \mathbf{i}, \mathbf{j}, \mathbf{k} \rangle$, and $W_1 W_2 \neq 0$. We denote $W_1 = \omega_1 \mathbf{i} + \omega_2 \mathbf{j} + \omega_3 \mathbf{k}$ and $W_2 = \nu_1 \mathbf{i} + \nu_2 \mathbf{j} + \nu_3 \mathbf{k}$.

If the matrix $\Gamma = \begin{pmatrix} \beta_1 & \beta_2 \\ \gamma_1 & \gamma_2 \end{pmatrix}$ has $i$-th zero row, then $D_i = E_{i,i}, i \in \{1,2\}$. In this case $\mathbf{1} \oplus O \in \mathcal{L}_3(\mathcal{S})$ similarly to the beginning of Item 3a, where $H_s$ is $H_i$. Therefore, further we may assume that there are no zero rows in $\Gamma$.

Let us compute words and combinations of words in $H_1$, $H_2$ of the length 2:

$$(H_1 H_2 + H_2 H_1)/2 = (-\omega_1 \nu_1 - \omega_2 \nu_2 - \omega_3 \nu_3) \mathbf{1} \oplus \mathrm{diag}\{0, 0, \beta_1 \gamma_1, \beta_2 \gamma_2\},$$

$$H_1^2 = (-\omega_1^2 - \omega_2^2 - \omega_3^2) \mathbf{1} \oplus \mathrm{diag}\{1, 0, \beta_1^2, \beta_2^2\},$$

$$H_2^2 = (-\nu_1^2 - \nu_2^2 - \nu_3^2) \mathbf{1} \oplus \mathrm{diag}\{0, 1, \gamma_1^2, \gamma_2^2\}.$$

By Proposition 7.1 it follows that $W_1 \oplus O, W_2 \oplus O \in \mathcal{L}_3(\mathcal{S})$. Hence,

$$0 \oplus F_1, \ 0 \oplus F_2, 0 \oplus F_1^3, \ 0 \oplus F_2^3, 0 \oplus F_1 F_2^2, \ 0 \oplus F_1^2 F_2 \in \mathcal{L}_3(\mathcal{S}),$$

where $F_1, F_2$ are defined in (10).

Let us consider $10 \times 5$ matrix $M$ whose rows are the coordinates of different words in $H_1, H_2$ of the lengths $0, 1, 2, 3$ in the basis $\mathbf{1}, E_{i,i}, i = 1, 2, 3, 4$. For the convenience of the readers we write down the corresponding words in $H_1, H_2$ on the right hand side of the matrix:

$$M = \begin{pmatrix} 1 & 1 & 1 & 1 & 1 \\ 0 & 1 & 0 & \beta_1 & \beta_2 \\ 0 & 0 & 1 & \gamma_1 & \gamma_2 \\ -\omega_1^2 - \omega_2^2 - \omega_3^2 & 1 & 0 & \beta_1^2 & \beta_2^2 \\ -\nu_1^2 - \nu_2^2 - \nu_3^2 & 0 & 1 & \gamma_1^2 & \gamma_2^2 \\ -\omega_1 \nu_1 - \omega_2 \nu_2 - \omega_3 \nu_3 & 0 & 0 & \gamma_1 \beta_1 & \gamma_2 \beta_2 \\ 0 & 0 & 0 & \gamma_1^2 \beta_1 & \gamma_2^2 \beta_2 \\ 0 & 0 & 0 & \gamma_1 \beta_1^2 & \gamma_2 \beta_2^2 \\ 0 & 0 & 1 & \gamma_1^3 & \gamma_2^3 \\ 0 & 1 & 0 & \beta_1^3 & \beta_2^3 \end{pmatrix} \quad \begin{matrix} \mathbf{1} \oplus I \\ H_1 \\ H_2 \\ H_1^2 \\ H_2^2 \\ (H_1 H_2 + H_2 H_1)/2 \\ 0 \oplus F_1 F_2^2 \\ 0 \oplus F_2 F_1^2 \\ 0 \oplus F_2^3 \\ 0 \oplus F_1^3 \end{matrix}$$

Note that $\beta_1 \neq \beta_2$ since otherwise $D_1, D_2$ do not generate $D_4(\mathbb{F})$. Hence, Lemma 6.1 is applicable. It follows that $(1, 0, 0, 0, 0)$ lies in the linear span of the rows of $M$, hence $\mathbf{1} \oplus O \in \mathcal{L}_3(\mathcal{S})$ as required.

4. It remains to consider the case $\dim \mathcal{L}_1(\mathcal{S}_D) = 4$, i.e. $\mathcal{L}_1(\mathcal{S}_D) = D_4(\mathbb{F})$. As in Item 2 we note that if $\dim \langle I, D_1, D_2 \rangle \leq 2$, then by Item 1 the result is proved. Hence we may further assume that $\dim \langle I, D_1, D_2 \rangle = 3$ and $Z_1 = C_1, Z_2 = C_2$. Since $\dim \mathcal{L}_1(\mathcal{S}_D) = 4$ we can choose an element from

$\mathcal{S}$ as $Z_3$ also. By Proposition 7.1 we have $U_i \oplus O \in \mathcal{L}_3(\mathcal{S})$. It follows that $0 \oplus V_i \in \mathcal{L}_3(\mathcal{S})$.

4a. At first we assume that the matrices $D_1$, $D_2$, $V_3$ do not generate $D_4(\mathbb{F})$. Then by Lemma 4.1 they have a common zero coordinate. Hence, $D_1^2 \in \langle V_1, V_2, V_3 \rangle$. Also direct computations show that $C_1^2 = (-1 - \alpha^2)\mathbf{1} \oplus D_1^2 \in \mathcal{S}^2$. Therefore,

$$\mathbf{1} \oplus O(1 + \alpha^2)^{-1}(-C_1^2 + D_1^2) \in \mathcal{L}_2(\mathcal{S}).$$

Therefore, $\mathcal{L}_3(\mathcal{S})$ contains a basis of $\mathcal{A}$.

4b. Now we assume that $D_1$, $D_2$, $V_3$ do not have a common zero coordinate. Then up to the order of coordinates the standard basis of the linear space $\langle D_1, D_2, V_3 \rangle$ have the form $(1, 0, 0, \alpha_1)$, $(0, 1, 0, \alpha_2)$, $(0, 0, 1, \alpha_3)$, and for at least one $i \in \{1, 2, 3\}$ we have $\alpha_i \neq 0$. In this notations $\langle \mathcal{S} \rangle$ contains the elements $H_1 = W_1 \oplus \operatorname{diag}\{1, 0, 0, \alpha_1\}$, $H_2 = W_2 \oplus \operatorname{diag}\{0, 1, 0, \alpha_2\}$, $H_3 = W_3 \oplus \operatorname{diag}\{0, 0, 1, \alpha_3\}$, where $W_1, W_2, W_3 \in \langle \mathbf{i}, \mathbf{j}, \mathbf{k} \rangle$.

If there exist $s \neq t \in \{1, 2, 3\}$ such that $\alpha_s \alpha_t \neq 0$, then

$$(H_s H_t + H_t H_s)H_s = \mu H_s \oplus \operatorname{diag}\{0, 0, 0, \alpha_s^2 \alpha_t\} \in \mathcal{L}_3(\mathcal{S}).$$

From Proposition 7.1 we obtain that $0 \oplus \operatorname{diag}\{0, 0, 0, 1\} \in \mathcal{L}_3(\mathcal{S})$ and together with $0 \oplus V_i$, $i = 1, 2, 3$ it constitutes a basis for $0 \oplus D_4(\mathbb{F})$. Hence, $\mathcal{L}_3(\mathcal{S})$ contains the basis of $\mathcal{A}$.

Therefore it remains to consider the case $\alpha_s \neq 0$, $\alpha_t = \alpha_q = 0$ for some $s, t, q$, $\{s, t, q\} = \{1, 2, 3\}$. By construction for the matrices $A_1, A_2$ introduced in the formula (7) we have $\dim \langle A_1, A_2 \rangle = 2$. Then $\dim \langle W_1, W_2, W_3 \rangle \geq 2$. Hence it is impossible that $W_t = 0$ and $W_q = 0$ hold simultaneously. Let $W_t = (\omega_1 \mathbf{i} + \omega_2 \mathbf{j} + \omega_3 \mathbf{k}) \neq 0$. Then repeating the arguments from Item 2.b for the matrix $H_t$ we obtain that $\mathbf{1} \oplus O \in \mathcal{L}_3(\mathcal{S})$ which concludes the proof. $\qquad\square$

## 8. Length of the quaternion group algebra

**Lemma 8.1.** *Let* $\mathbb{F}$ *be an arbitrary field. Then* $l(\mathbb{F}\mathbf{Q}_8) \geq 3$.

**Proof.** To prove the lemma it is sufficient to find a system of generators for the algebra $\mathbb{F}\mathbf{Q}_8$ with the length 3.

The generating set $\mathcal{S} = \{i, j\}$ for the group $\mathbf{Q}_8$ will be also a generating system for its group algebra. Consider the words in generators $\{i, j\}$ with the length starting from 0:

0. $e$;

1. $i, j$;
2. $i^2 = m, j^2 = m, ij = k, ji = mk$;
3. $i^3 = mi, j^3 = mj$.

Since both generators are of the order 4, further considerations are not useful. Also note that $i^3$ can not be expressed as a word of the length 2 in $i$ and $j$. It follows that $\mathcal{L}_2(\mathcal{S}) \subset \mathcal{L}_3(\mathcal{S}) = \mathbb{F}\mathbf{Q}_8$ therefore $l(\mathcal{S}) = 3$. □

**Theorem 8.1.** *Let $\mathbb{F}$ be a field, $\operatorname{char}\mathbb{F} \neq 2$ and $S(\mathbb{F}) > 2$. Then $l(\mathbb{F}\mathbf{Q}_8) = 3$.*

**Proof.** The lower bound $l(\mathbb{F}\mathbf{Q}_8) \geq 3$ follows from Lemma 8.1.

By Lemma 5.1 we have $\mathbb{F}\mathbf{Q}_8 \cong \mathcal{D} \oplus D_4(\mathbb{F})$. Hence, from Lemma 7.1 we have

$$l(\mathbb{F}\mathbf{Q}_8) = l(\mathcal{D} \oplus D_4(\mathbb{F})) \leq 3.$$

□

## 9. Fields of characteristics 2

**Lemma 9.1.** *Let $\mathbb{F}$ be a field, $\operatorname{char}\mathbb{F} = 2$. Then $l(\mathbb{F}\mathbf{Q}_8) = 3$.*

**Proof.** The lower bound $l(\mathbb{F}\mathbf{Q}_8) \geq 3$ follows from Lemma 8.1.

Let us prove the upper bound $l(\mathbb{F}\mathbf{Q}_8) \leq 3$.

I. Let $\mathbb{F}$ be algebraically closed.

In this case by Ref. 2(§1) we can consider the generating system $x, y$ for the Jacobson radical $J = J(\mathbb{F}\mathbf{Q}_8)$ such that

$$x^2 = yxy, \ y^2 = xyx, \ xy^2 = y^2x = x^2y = yx^2 = x^4 = y^4 = 0 \qquad (11)$$

and the set $E_1 = e$, $E_2 = x$, $E_3 = y$, $E_4 = xy$, $E_5 = yx$, $E_6 = x^2$, $E_7 = y^2$, $E_8 = x^3 = xyxy$ is a basis of $\mathbb{F}\mathbf{Q}_8$. Namely, we consider $x = ij + \omega i + \omega^2 j$, $y = ij + \omega^2 i + \omega j$, where $\omega \in \mathbb{F}$ is a root of $x^2 + x + 1$.

It follows from (11) that the nilpotency index of $J$ is 5, and $J^4 = \langle E_8 \rangle$.

Let $\mathcal{S} = \{z_1, \ldots, z_k\}$, $k \geq 2$ be a generating system of $\mathbb{F}\mathbf{Q}_8$. Up to the linear change of variables of the form $z_i = \alpha_i e + z_i'$, $z_i' \in J(\mathbb{F}\mathbf{Q}_8)$ we may assume that $z_i \in J(\mathbb{F}\mathbf{Q}_8)$ for all $i = 1, \ldots, k$ since by Proposition 2.2[6(Proposition 2.2)] $l(\mathcal{S}) = l(\{z_1', \ldots, z_k'\})$.

We are going to prove that all basis elements $E_i$ lie in $\mathcal{L}_3(\mathcal{S})$.

By definition, $E_1 \in \mathcal{L}_0(\mathcal{S}) \subset \mathcal{L}_3(\mathcal{S})$.

Since $\langle \mathcal{S}^2 \setminus \mathcal{S} \rangle \subseteq J^2$ there are elements $z, w \in \langle \mathcal{S} \rangle$, such that $z = x + \rho$, $w = y + \tau$, $\rho, \tau \in J^2$. Then $z^3 = x^3$, since $\rho^3 = x\rho^2 = \rho^2 x = \rho x \rho = 0$, and

$xE_ix = x^2E_i = E_ix^2 = 0$ for all $i \geq 4$ by (11). Therefore, $x\rho x = x^2\rho = \rho x^2 = 0$. Thus,

$$E_8 \in \mathcal{L}_3(\mathcal{S}).$$

Note that

$$zwz = (x + \rho)(y + \tau)(x + \rho) = xyx + \rho yx + x\tau x + xy\rho,$$

$$wzw = (y + \tau)(x + \rho)(y + \tau) = yxy + \tau xy + y\rho y + yx\tau,$$

and $\rho yx + x\tau x + xy\rho, \tau xy + y\rho y + yx\tau \in J^4$. Hence,

$$E_7 = y^2 = xyx = zwz - aE_8, \ E_6 = x^2 = yxy = wzw - bE_8 \in \mathcal{L}_3(\mathcal{S}).$$

In particular we proved that $J^3 \subseteq \mathcal{L}_3(\mathcal{S})$.

Now we consider the words in $z, w$ of the length 2: $zw = xy + x\tau + \rho y + \rho\tau = xy + u$, $u \in J^3 \subseteq \mathcal{L}_3(\mathcal{S})$, hence,

$$E_4 = xy = zw - u \in \mathcal{L}_3(\mathcal{S}).$$

Similarly,

$$E_5 = yx \in \mathcal{L}_3(\mathcal{S}).$$

Therefore, $J^2 \subseteq \mathcal{L}_3(\mathcal{S})$. Hence $\rho, \tau \in \mathcal{L}_3(\mathcal{S})$. It follows that

$$E_2 = x = z - \rho \in \mathcal{L}_3(\mathcal{S})$$

and

$$E_3 = y = w - \tau \in \mathcal{L}_3(\mathcal{S}).$$

II. Let $\mathbb{F}$ be not algebraically closed. We denote by $\bar{\mathbb{F}}$ its algebraic closure. From Proposition 2.3 and Item I we obtain the bound

$$l(\mathbb{F}\mathbf{Q}_8) \leq l(\bar{\mathbb{F}}\mathbf{Q}_8) \leq 3.$$

$\square$

**Proof of Theorem 1.3.** The result follows by Lemmas 5.2, 9.1 and Theorem 8.1.

$\square$

## References

1. L.A. Bokut, Y. Chen, *Gröbner-Shirshov bases and PBW theorems*, J. Sib. Fed. Univ. Math. Phys. **6**:4 (2013), 417–427.
2. E.C. Dade, *Une extension de la théorie de Hall et Higman*, J. Algebra **20**(1972), 570–609.
3. S. Feigelstock, *N-real fields*, Algebra Discrete Math. **3**(2003), 1–6.
4. A. Freedman, R.N. Gupta, R.M. Guralnick, *Shirshov's theorem and representations of semigroups*, Pacific J. Math. **181**:3(1997), 159–176.
5. A. Guterman, T. Laffey, O. Markova, H. Šmigoch, *A resolution of Paz's conjecture in the presence of a nonderogatory matrix*, Linear Algebra Appl. **543** (2018) 234–250.
6. A. Guterman, O. Markova, *Commutative matrix subalgebras and length function*, Linear Algebra Appl. **430** (2009) 1790–1805.
7. A. Guterman, O. Markova, M. Khrystik, *On the lengths of group algebras of finite Abelian groups in the semi-simple case*, Preprint, 2018.
8. A. Guterman, O. Markova, M. Khrystik, *On the lengths of group algebras of finite Abelian groups in the modular case*, Preprint, 2018.
9. A. Guterman, O. Markova, *The realizability problem for values of the length function for quasi-commuting matrix pairs*, Zap. Nauchn. Sem. POMI. **439** (2015) 59–73 [In Russian]; English transl. in J. Math. Sci. (N. Y.), **216**:6 (2016), 761–769.
10. A. Guterman, O. Markova, V. Mehrmann, *Lengths of quasi-commutative pairs of matrices*, Linear Algebra Appl. **498** (2016) 450–470.
11. T. Laffey, Simultaneous reduction of sets of matrices under similarity, Linear Algebra Appl., **84**(1986), 123–138.
12. T.Y. Lam, Introduction to Quadratic Forms over Fields. Graduate Studies in Mathematics. 67. American Mathematical Society, 2005.
13. W.E. Longstaff, *Burnside's theorem: irreducible pairs of transformations*, Linear Algebra Appl. **382**(2004) 247–269.
14. W.E. Longstaff, P. Rosenthal, *On the lengths of irreducible pairs of complex matrices*, Proc. Am. Math. Soc., **139**:11(2011), 3769–3777.
15. O. Markova, *On the length of the algebra of upper-triangular matrices*, Uspekhi Mat. Nauk, **60**:5(365) (2005), 177–178; English transl. in Russian Math. Surveys, **60**:5 (2005), 984–985.
16. O. Markova, *Upper bound for the length of commutative algebras*, Mat. Sb., **200**:12 (2009), 41–62; English transl. in Sb. Math., **200**:12 (2009), 1767–1787.
17. O. Markova, *The length function and matrix algebras*, Fundam. Prikl. Mat. **17**:6(2012), 65-173; English transl. in J. Math. Sci. (N. Y.) **193**:5(2013) 687–768.
18. O. Markova, *On the relationship between the length of an algebra and the index of nilpotency of its Jacobson radical*, Mat. Zametki, **94**:5 (2013), 682–688; English transl. in Math. Notes, **94**:5 (2013), 636–641.
19. C. J. Pappacena, *An upper bound for the length of a finite-dimensional algebra*, J. Algebra, **197**(1997), 535–545.
20. A. Paz, *An application of the Cayley–Hamilton theorem to matrix polynomials*

*in several variables*, Linear Multilinear Algebra, **15**(1984), 161–170.

21. R. Pierce, Associative algebras, Springer-Verlag, Berlin, 1982.
22. A.R. Rajwade, Squares. London Math. Soc. Lecture Notes 171. Cambridge Univ. Press, 1993.

# Results on intersecting families of subsets, a survey

Gyula O. H. Katona*

*MTA Rényi Institute, Budapest, Hungary*

*katona.gyula.oh@renyi.mta.hu*

Dedicated to the 80th birthday of Professor L. Bokut.

## 1. Introduction

The underlying set will be $\{1, 2, \ldots, n\}$. The family of all $k$-element subsets of $[n]$ is denoted by $\binom{[n]}{k}$. Its subfamilies are called *uniform*. A family $\mathcal{F}$ of some subsets of $[n]$ is called *intersecting* if $F \cap G \neq \emptyset$ holds for every pair $F, G \in \mathcal{F}$. The whole story has started with the seminal paper of Erdős, Ko and Rado[7]. Their first observation was that an intersecting family in $2^{[n]}$ can contain at most one of the complementing pairs, therefore the size of an intersecting family cannot exceed the half of the number of all subsets of $[n]$.

**Observation 1.1.** (Erdős, Ko, Rado[7]) If $\mathcal{F} \subset 2^{[n]}$ is intersecting then $|\mathcal{F}| \leq 2^{n-1} = 2^n/2$.

The following trivial construction shows that the bound is sharp.

Construction 1.1. Take all subsets of $[n]$ containing the element 1.

However there are many other construction giving equality in Observation 1.1. The following one will be interesting for our further investigations.

Construction 1.2. If $n$ is odd take all sets of size at least $\frac{n-1}{2}$. If $n$ is even then choose all the sets of size at least $\frac{n}{2} - 1$ and the sets of size $\frac{n}{2}$ not containing the element $n$.

*This research was supported by the National Research, Development and Innovation Office – NKFIH Fund No's SSN117879, NK104183 and K116769.

The analogous problem when the intersecting subsets have size exactly $k$, that is the case of uniform families, is not so trivial when $k \leq \frac{n}{2}$. (Otherwise all $k$-element subsets can be chosen.)

**Theorem 1.1.** *(Erdős, Ko, Rado*[7]*) If $\mathcal{F} \subset \binom{[n]}{k}$ is intersecting where $k \leq \frac{n}{2}$ then*

$$|\mathcal{F}| \leq \binom{n-1}{k-1}.$$

The original proof uses the so called *shifting method*. There is a shorter proof based on the *cycle method* in Ref. 20. It can also be found in the books[2,5]. We will give an algebraic (using eigenvalues) proof in Section 4. In the case of Theorem 1.1 there is only one extremal construction, mimicking Construction 1.1.

Construction 1.3. Take all subsets of $[n]$ having size $k$ and containing the element 1.

It is worth mentioning that the results contained in Ref. 7 were obtained in the late 1930's when all three authors worked in England, but they did not publish them because they did not think that the mathematical community would find them interesting. They sent the paper for publication only in 1960 when they realized that the "mathematical climate" has changed: Combinatorics became a science. It was a very good idea, this paper is the second most cited paper of Erdős according to MathSciNet, although the competition is tough.

We say that a family $\mathcal{F}$ is *trivially intersecting* if there is an element $a \in [n]$ such that all members of $\mathcal{F}$ contain $a$. Constructions 1.1 and 1.3 are trivially intersecting. Construction 1.2 shows that in the non-uniform case non-trivially intersecting families can be as large as the trivially intersecting one. But this seemed not to be true for the uniform families. Ref. 7 posed the problem of finding the largest $k$-uniform non-trivially intersecting family. It was found by Hilton and Milner.

**Theorem 1.2.** [15] *If $\mathcal{F}$ is an intersecting but not a trivially intersecting family, $\mathcal{F} \subset \binom{[n]}{k}(2k \leq n)$ then*

$$|\mathcal{F}| \leq 1 + \binom{n-1}{k-1} - \binom{n-k-1}{k-1}.$$

The construction giving equality is the following.

Construction 1.4. Let $K = \{2, 3, \ldots, k+1\}$. The extremal family will consist of all $k$-element sets containing 1 and intersecting $K$.

The goal of the present paper is to survey only some directions of this theory. A comprehensive survey would be a book. (Let us call the reader's attention to the forthcoming book of Gerbner and Patkós[14].) The author, of course, selected the directions according his own interest, covering his own results. There is an overlapping with the paper[21].

## 2. *t*-intersecting families

Already Erdős, Ko and Rado[7] considered a more general problem. A family $\mathcal{F} \subset 2^{[n]}$ is *t-intersecting* if $|F \cap G| \geq t$ holds for every pair $F, G \in \mathcal{F}$. They posed a conjecture for the maximal size of non-uniform a $t$-intersecting family. This conjecture was justified in the following theorem.

**Theorem 2.1.** *(Katona[18]) If $\mathcal{F} \subset 2^{[n]}$ is t-intersecting then*

$$|\mathcal{F}| \leq \begin{cases} \sum_{i=\frac{n+t}{2}}^{n} \binom{n}{i} & \text{if } n+t \text{ is even} \\ \sum_{i=\frac{n+t+1}{2}}^{n} \binom{n}{i} + \binom{n-1}{\frac{n+t-1}{2}} & \text{if } n+t \text{ is odd} . \end{cases}$$

Here the generalization of Construction 1.1 gives only $2^{n-t}$, less than the upper bound in Theorem 2.1 (if $t > 1$). In order to obtain a sharp construction we have to mimic Construction 1.2.

Construction 2.1. If $n+t$ is even, take all sets of size at least $\frac{n+t}{2}$. If $n+t$ is odd then choose all the sets of size at least $\frac{n+t+1}{2}$ and the sets of size $\frac{n+t-1}{2}$ not containing the element $n$.

The uniform case is harder, again. A natural trial to obtain the best construction is the obvious generalization of Construction 1.3.

Construction 2.2. Take all subsets of $[n]$ having size $k$ and containing $[t]$ as a subset.

This is really $t$-intersecting, but Ref. 7 contains an example when this construction is not the best. Let $n = 8, k = 4, t = 2$. Divide the underlying set into two parts $[8] = X_1 \cup X_2$ where $X_1 = [4], X_2 = \{5, 6, 7, 8\}$. Let $\mathcal{F}$ consist of all 4-element subsets $F$ satisfying $|X_1 \cap F| \geq 3$. This family is 2-intersecting and has 17 members, while Construction 2.2 has only $\binom{6}{2} = 15$ in this case.

But Erdős, Ko and Rado[7] were able to prove that Construction 2.2 gives the largest family when $n$ is large with respect to $k$. The dependence of the threshold on $t$ is not interesting here since $1 \leq t < k$ can be supposed.

**Theorem 2.2.** [7] *If* $\mathcal{F} \subset \binom{[n]}{k}$ *is t-intersecting and* $n \geq n(k)$ *then*

$$|\mathcal{F}| \leq \binom{n-t}{k-t}. \tag{1}$$

The next step towards the better understanding of the situation was when Frankl[9] and Wilson[29] determined the exact value of the threshold $n(k)$ in Theorem 2.1.

Let us now consider the following generalization of the counter-example above.

**Construction 2.3.** Choose a non-negative integer parameter $i$ and define the family

$$\mathcal{A}(n,k,t,i) = \{A: \ |A| = k, |A \cap [t+2i]| \geq t+i\}. \tag{2}$$

It is easy to see that $\mathcal{A}(n,k,t,i)$ is $t$-intersecting for each $i$.

Introduce the following notation:

$$\max_{0 \leq i} |\mathcal{A}(n,k,t,i)| = \mathrm{AK}(n,k,t).$$

This is the size of the best of the constructions (2). Frankl[9] conjectured that this construction gives the largest $k$-uniform $t$-intersecting family. Frankl and Füredi[11] proved the conjecture for a very large class of parameters but the full conjecture remained open until 1996 when it became a theorem.

**Theorem 2.3.** *(Ahlswede and Khachatrian[1]) Let* $\mathcal{F} \subset \binom{[n]}{k}$ *be a t-intersecting family. Then*

$$|\mathcal{F}| \leq AK(n,k,t)$$

*holds.*

Of course Theorem 2.3 has many consequences. We will exhibit only one result of us, in Section 3, where this theorem is used and plays a role even in the formulation of the statement.

## 3. Largest union-intersecting families and related problems

The following problem was asked by János Körner.

Let $\mathcal{F} \subset 2^{[n]}$ and suppose that if $F_1, F_2, G_1, G_2 \in \mathcal{F}, F_1 \neq F_2, G_1 \neq G_2$ holds then

$$(F_1 \cup F_2) \cap (G_1 \cup G_2) \neq \emptyset.$$

What is the maximum size of such a family?

He conjectured that the following construction gives the largest one.

**Construction 3.1.** If $n$ is odd then take all sets of size at least $\frac{n-1}{2}$. If $n$ is even then choose all the sets of size at least $\frac{n}{2}$ and the sets of size $\frac{n}{2} - 1$ containing the element 1.

We solved the problem in a more general setting. A family $\mathcal{F} \subset 2^{[n]}$ is called a *union-t-intersecting* if

$$|(F_1 \cup F_2) \cap (G_1 \cup G_2)| \geq t$$

holds for any four members such that $F_1 \neq F_2, G_1 \neq G_2$.

**Theorem 3.1.** *(Katona-D.T. Nagy[24]) If $\mathcal{F} \subset 2^{[n]}$ is a union-t-intersecting family then*

$$|\mathcal{F}| \leq \begin{cases} \sum_{i=\frac{n+t}{2}-1}^{n} \binom{n}{i} & \text{if } n+t \text{ is even} \\ \sum_{i=\frac{n+t-1}{2}}^{n} \binom{n}{i} + AK\left(n, \frac{n+t-3}{2}, t\right) & \text{if } n+t \text{ is odd .} \end{cases}$$

The following construction shows that the estimate is sharp.

**Construction 3.2.** If $n+t$ is even, take all the sets with size at least $\frac{n+t}{2} - 1$. Otherwise choose all the sets of size at least $\frac{n+t-1}{2}$ and the sets of size $\frac{n+t-3}{2}$ following Construction 2.3 where $k = \frac{n+t-3}{2}$ and $i$ chosen to maximize (2).

Since the result contains the AK-function, it is obvious that Theorem 2.3 must be used in the proof of this theorem.

As before, the uniform case is more difficult. Yet, we will treat it in an even more general form. A family $\mathcal{F} \subset 2^{[n]}$ is called a *(u,v)-union-intersecting* if for different members $F_1, \ldots, F_u, G_1, \ldots, G_v$ the following holds:

$$\left(\cup_{i=1}^{u} F_i\right) \cap \left(\cup_{j=1}^{v} G_j\right) \neq \emptyset.$$

**Theorem 3.2.** *(Katona-D.T. Nagy[24]) Let $1 \leq u \leq v$ and suppose that the family $\mathcal{F} \subset \binom{[n]}{k}$ is a (u,v)-union-intersecting family then*

$$|\mathcal{F}| \leq \binom{n-1}{k-1} + u - 1$$

*holds if $n > n(k, v)$.*

The following construction shows that the estimate is sharp.

Construction 3.3. Take all $k$-element subsets containing the element 1, and choose $u - 1$ distinct sets non containing 1.

The theorem does not give a solution for small values.

Open Problem 3.1.[21] Is there an Ahlswede-Khachatrian type theorem here, too?

A new result of Alishahi and Taherkhani[3] gave Theorem 3.2 a wider perspective. The *Kneser graph* $K(n, k)(1 \leq k \leq \frac{n}{2})$ is the graph whose vertices are all $k$-element subsets of an $n$-element set, where two vertices are adjacent iff the corresponding sets are disjoint. Using this terminology Erdős-Ko-Rado theorem claims that the largest independent set of vertices in this graph has size $\binom{n-1}{k-1}$. In other words, if a set $S$ of vertices of the Kneser graph $K(n, k)$ induces the empty graph then $|S| \leq \binom{n-1}{k-1}$.

What is now the maximum of the size of $S$ if it does not induce a star $S_r$ (a graph with $r + 1$ vertices, in which one vertex (the center) is adjacent to all other ones)? This was answered by Gerbner, Lemons, Palmer, Patkós, and Szécsi[13] in the following theorem. (Formulated by intersecting subsets, again.)

**Theorem 3.3.**[13] *Let $\mathcal{F} \subset \binom{[n]}{k}$ be a family in which no member is disjoint to $r$ other members. If $n \geq n(k, r)$ then*

$$|\mathcal{F}| \leq \binom{n-1}{k-1}.$$

Suppose now that $S$ is such a set of vertices of the Kneser graph $K(n, k)$ that it does not induce a complete bipartite graph $K_{u,v}(u \leq v)$. The maximum size of $S$ under this condition is determined by Theorem 3.2 for large enough $n$.

But Ref. 3 solves the problem in full generality, for an arbitrary graph $G$ instead of a complete bipartite graph. Let $\chi(G)$ be the chromatic number of $G$, furthermore let $\eta(G)$ be the size of the smallest color class for all proper colorings with $\chi(G)$ colors.

**Theorem 3.4.**[3] *Let $S$ be a set of vertices of the Kneser graph $K(n, k)$ not inducing $G$ as a subgraph. If $n$ is large enough ($n \geq n(k, G)$) then*

$$|S| \leq \binom{n}{k} - \binom{n - \chi(G) + 1}{k} + \eta(G) - 1.$$

## 4. Two- or more-part intersecting families

Before starting the real subject of the present section, we will give an algebraic proof of Theorem 1.1. This proof is spectral, based on the approach in Ref. 28 and in Ref. 4. Before really starting the proof we have to remind the reader some known definitions and facts from the literature.

Let $G$ be a simple graph on $N$ vertices with adjacency matrix $A$. The number $\lambda$ is called an *eigenvalue* of a matrix $A$ if there is a non-zero vector $x$ such that $Ax = \lambda x$. The vector $x$ is the *associated eigenvector*. If $I$ is the identity matrix of the same size then $\det(A - \lambda I)$ is a polynomial of $\lambda$. All the roots of the equation $\det(A - \lambda I) = 0$ are real. These roots are the eigenvalues and their number with multiplicities is $N$. Index them according to their natural ordering: $\lambda_1 \geq \lambda_2 \geq \ldots \geq \lambda_N$. It is known that if $G$ is a regular graph then $\lambda_1$ is the common degree. (See e.g. Ref. 27.) If $\alpha(G)$ is the maximum number of independent vertices then (see Ref. 16,28)

$$\alpha(G) \leq -N \frac{\lambda_N}{\lambda_1 - \lambda_N} \tag{3}$$

holds.

**Proof of Theorem 1.1.** Let $n \geq 2k$ be positive integers. The *Kneser graph* $K(n, k)$ is the graph whose vertices are all $k$-element subsets of an $n$-element set, where two vertices are adjacent iff the corresponding sets are disjoint. (It was define in the previous section, we repeated here for the case, the reader did not read the whole paper.) The number of vertices of this graph is thus $N = \binom{n}{k}$, and it is known that its eigenvalues are all numbers of the form $(-1)^j \binom{n-k-j}{k-j}$, for $j \in \{0, 1, \ldots, k\}$ (See, e.g., Ref. 28 for a proof. They have different multiplicities.) In particular, the largest eigenvalue is the degree of regularity $d = \lambda_1 = \binom{n-k}{k}$ and the smallest (most negative) one is $\lambda_N = -\binom{n-k-1}{k-1} = -\frac{k}{n-k}d$. Substituting these values into (3)

$$-\binom{n}{k} \frac{-\frac{k}{n-k}d}{d + \frac{k}{n-k}d} = \binom{n-1}{k-1}$$

is obtained, finishing the elegant algebraic proof of Theorem 1.1. (See Ref. 28.) $\qquad\qquad\square$

The goal of this section is to consider the problem when the underlying set is partitioned into two (or more) parts $X_1, X_2$ and the sets $F \in \mathcal{F}$ have fixed sizes in both parts. For some motivation see Ref. 22 (Section 4). More precisely let $X_1$ and $X_2$ be disjoint sets of $n_1$, respectively $n_2$ elements.

Ref. 10 considered such subsets of $X = X_1 \cup X_2$ which had $k$ elements in $X_1$ and $\ell$ elements in $X_2$. The family of all such sets is denoted by

$$\binom{X_1, X_2}{k, \ell} = \binom{X_1}{k} \uplus \binom{X_2}{\ell} = \{F \subset X_1 \cup X_2 : |F \cap X_1| = k, |F \cap X_2| = \ell\}.$$

The construction above, taking all possible sets containing a fixed element also works here. If the fixed element is in $X_1$ then the number of these sets is

$$\binom{n_1 - 1}{k - 1}\binom{n_2}{\ell},$$

otherwise it is

$$\binom{n_1}{k}\binom{n_2 - 1}{\ell - 1}.$$

The following theorem of Frankl[10] claims that the larger one of these is the best.

**Theorem 4.1.** *Let $X_1, X_2$ be two disjoint sets of $n_1$ and $n_2$ elements, respectively. The positive integers $k, \ell$ satisfy the inequalities $2k \leq n_1, 2\ell \leq n_2$. If $\mathcal{F}$ is an intersecting subfamily of $\binom{X_1, X_2}{k, \ell}$ then*

$$|\mathcal{F}| \leq \max\left\{\binom{n_1 - 1}{k - 1}\binom{n_2}{\ell}, \binom{n_1}{k}\binom{n_2 - 1}{\ell - 1}\right\}.$$

Actually his theorem is formulated for an arbitrary number of parts.

**Theorem 4.2.** [10] *Let $p \geq 2$ and suppose $n_1 \geq 2k_1, n_2 \geq 2k_2, \ldots, n_p \geq 2k_p$. Let $X_1, X_2, \ldots, X_p$ be $p$ pairwise disjoint sets, where $|X_i| = n_i$ and let $X = \cup_{i=1}^k X_i$ be their union. Let $\mathcal{F}$ be an intersecting family of subsets of $X$, where each $F \in \mathcal{F}$ has exactly $k_i$ elements in $X_i$ for all $1 \leq i \leq p$. Then*

$$|\mathcal{F}| \leq \max_{1 \leq i \leq p} \frac{k_i}{n_i} \prod_{i=1}^{p} \binom{n_i}{k_i}.$$

**Proof of Theorem 4.2**[10]. The (categorical) product $G_1 \times G_2$ of two graphs $G_1$ and $G_2$ is the graph whose vertex set consists of the pairs $(v_1, v_2)$ where $v_i$ is a vertex of $G_i$ and two vertices $(u_1, u_2)$ and $(v_1, v_2)$ are adjacent iff $\{u_1, v_1\}$ is an edge in $G_1$ and $\{u_2, v_2\}$ is an edge in $G_2$.

Let $A = (a_{ij})$ and $B$ be $p \times r$ and $s \times t$ matrices, respectively. The *tensor product* $A \otimes B$ is an $ps \times rt$ matrix obtained by blocks of copies of $B$ multiplied by the entries $a_{ij}$. It is easy to see that the adjacency matrix of $G_1 \times G_2$ is the tensor product of the respective adjacency matrices.

Suppose that $\lambda$ and $\mu$ are eigenvalues of $A$ and $B$, respectively. That is $Ax = \lambda x$ and $By = \mu y$ hold for some non-zero vectors $x$ and $y$. Then we have $(A \otimes B)(x \otimes y) = Ax \otimes By = \lambda x \otimes \mu y = \lambda \mu x \otimes y$ showing that $\lambda \mu$ is an eigenvalue associated with the eigenvector $x \otimes y$. One can see that all eigenvalues of $A \otimes B$ can be obtained in this way, see e.g. Ref. 17.

Using the (obvious extension of) our previous notation, the members of the family in the theorem are elements of

$$\binom{X_1, \ldots, X_p}{k_1, \ldots, k_p} = \biguplus_{i=1}^{p} \binom{X_i}{k_i}. \tag{4}$$

It is easy to see that the vertices of the product graph

$$\mathrm{K}(n_1, k_1) \times \mathrm{K}(n_2, k_2) \times \ldots \times \mathrm{K}(n_p, k_p) \tag{5}$$

are exactly the elements of (4). The number of vertices is $N^* = \prod_{i=1}^{p} \binom{n_i}{k_i}$. Two vertices are adjacent iff they are adjacent in every factor that is the corresponding subsets, elements of $\mathcal{F}$ are disjoint. Therefore the aim of the theorem is to determine the independence number of the graph (5).

The upper estimate (3) will be used. We made some easy remarks above, on the products of two graphs, their adjacency matrices and eigenvalues. These statements can be extended to the product of more graphs by induction. Hence the eigenvalues of the graph (5) will be products of eigenvalues of the Kneser graphs, one eigenvalue from each $\mathrm{K}(n_i, k_i)$. Therefore the largest eigenvalue of the product will be the product of the largest eigenvalues of $\mathrm{K}(n_i, k_i)$'s:

$$\lambda_1^* = d^* = \prod_{i=1}^{k} \binom{n_i - k_i}{k_i}. \tag{6}$$

The smallest eigenvalue must be negative, therefore the number of odd indices $j_i$ in the corresponding product

$$\prod_{i=1}^{p} (-1)^{j_i} \binom{n_i - k_i - j_i}{k_i - j_i} \tag{7}$$

must be odd. If $j_i$ is even then $\binom{n_i - k_i - j_i}{k_i - j_i}$ can be replaced by $\binom{n_i - k_i}{k_i}$ decreasing (making more negative) the product (7). If $j_u$ and $j_v$ are both odd then

$$(-1)^2 \binom{n_u - k_u - j_u}{k_u - j_u} \binom{n_v - k_v - j_v}{k_v - j_v}$$

can be replaced by

$$\binom{n_u - k_u}{k_u}\binom{n_v - k_v}{k_v},$$

decreasing the product, again. If all these changes are carried out, we have all $j_i$'s 0 with one exception where it is 1. The smallest of these is the smallest eigenvalue:

$$\lambda^*_{N^*} = -\max_{1 \le i \le p}\left\{\frac{k_i}{n_i - k_i}\right\}\prod_{i=1}^{p}\binom{n_i - k_i}{k_i} = -\max_{1 \le i \le p}\left\{\frac{k_i}{n_i - k_i}\right\}d^*. \quad (8)$$

Substituting (6) and (8) into (3) gives:

$$\prod_{i=1}^{p}\binom{n_i}{k_i}\frac{\max_{1 \le i \le p}\left\{\frac{k_i}{n_i - k_i}\right\}}{1 + \max_{1 \le i \le p}\left\{\frac{k_i}{n_i - k_i}\right\}}.$$

The last factor is equal to

$$\max_{1 \le i \le p}\left\{\frac{k_i}{n_i}\right\},$$

completing the proof of the theorem. $\qquad\square$

Theorems 4.1 and 4.2 could be formulated in such a way that the largest subfamily of (4) is one of the trivially intersecting families. It is natural to ask what is the largest non-trivially intersecting subfamily. For sake of simplicity let us first consider the case of two parts.

Take a Hilton-Milner family (Construction 1.4) in $X_1$, denote it by $HM(X_1, k)$. Extend its members in all possible ways by $\ell$-element subsets chosen from $X_2$:

$$HM_1(X_1, k; X_2, \ell) = \{F \cup G : F \in HM(X_1, k), G \subset X_2, |G| = \ell\}.$$

Define, similarly,

$$HM_2(X_1, k; X_2, \ell) = \{F \cup G : F \subset X_1, |F| = k, G \in HM(X_2, \ell)\}.$$

It was conjectured in Ref. 22 that either $HM_1(X_1, k; X_2, \ell)$ or $HM_2(X_1, k; X_2, \ell)$ is the largest nontrivially intersecting subfamily of $\binom{X_1, X_2}{k, \ell}$. Kwan, Sudakov and Vieira[26] showed that this is not true: there are other, "mixed" Hilton-Milner families which are better in some cases.

Fix an element $a \in X_1$, a set $A \subset X_1$ such that $a \notin A, |A| = k$ and a set $B \subset X_2$ such that $|B| = \ell$ and define

$$HM_1^{\mathrm{mix}}(X_1, k; X_2, \ell) = \{F : |F \cap X_1| = k, |F \cap X_2| = \ell, a \in F, F \cap (A \cup B) \ne \emptyset\}.$$

$\mathrm{HM}_2^{\mathrm{mix}}(X_1, k; X_2, \ell)$ is the symmetric construction.

**Theorem 4.3.** *(Kwan, Sudakov, Vieira*[26]*) If both* $|X_1|$ *and* $|X_2|$ *are large enough then the largest non-trivially intersecting subfamily of* $\binom{X_1, X_2}{k, \ell}$ *is one of* $\mathrm{HM}_1(X_1, k; X_2, \ell), \mathrm{HM}_2(X_1, k; X_2, \ell), \mathrm{HM}_1^{\mathrm{mix}}(X_1, k; X_2, \ell)$ *and* $\mathrm{HM}_2^{\mathrm{mix}}(X_1, k; X_2, \ell)$.

Their result actually claims the analogous statement for more parts. The proof uses the shifting method.

Let us consider now the case when other sizes are also allowed that is the family consists of sets satisfying $|F \cap X_1| = k_i, |F \cap X_2| = \ell_i$ for certain pairs $(k_i, \ell_i)$ of positive integers. Using the notation above, we will consider subfamilies of

$$\bigcup_{i=1}^{m} \binom{X_1, X_2}{k_i, \ell_i}.$$

The generalization is however a little weaker at one point. In Theorem 4.1 the thresholds $2k \le n_1, 2\ell \le n_2$ for validity are natural. If either $n_1$ or $n_2$ is smaller then the problem becomes trivial, all such sets can be selected in $\mathcal{F}$. In the generalization below there is no such natural threshold. There will be another difference in the formulation. We give the construction of the extremal family rather than the maximum number of sets.

**Theorem 4.4.** [22] *Let* $X_1, X_2$ *be two disjoint sets of* $n_1$ *and* $n_2$ *elements, respectively. Some positive integers* $k_i, \ell_i (1 \le i \le m)$ *are given. Define* $b = \max_i\{k_i, \ell_i\}$. *Suppose that* $9b^2 \le n_1, n_2$. *If* $\mathcal{F}$ *is an intersecting subfamily of*

$$\bigcup_{i=1}^{m} \binom{X_1, X_2}{k_i, \ell_i} \tag{9}$$

*then* $|\mathcal{F}|$ *cannot exceed the size of the largest trivially intersecting family satisfying the conditions.*

The family (9), in general, cannot be given in a product form. This is why the eigenvalues cannot be as easily determined as in the case above. The proof of Theorem 4.4 is based on the cycle method, more precisely on lemmas on direct products of cycles.

Theorems 1.1, 4.1, 4.2 and 4.4 state that the largest intersecting subfamily of a certain uniform family is a trivially intersecting one.

Open Problem 4.1. Find a general sufficient condition for uniform families $\mathcal{F}$ which ensures that the largest intersecting subfamily of $\mathcal{F}$ is trivially intersecting.

We do not even have a conjecture of this type, unlike in the case of non-uniform families. A family $\mathcal{F}$ is called *hereditry* (or *downset*) if $G \subset F \in \mathcal{F}$ implies $G \in \mathcal{F}$.

**Conjecture 4.1.** *(Chvátal[6]) If $\mathcal{F} \subset 2^{[n]}$ is a hereditary family then its largest intersecting subfamily is a trivially intersecting one.*

Many special cases are settled, but the conjecture is still open in its full generality.

## 5. Minimum shadows of $t$-intersecting families

Suppose that $\mathcal{F}$ is a $k$-uniform family: $\mathcal{F} \subset \binom{[n]}{k}$. Its shadow is a $k-1$-uniform family obtained by removing single elements from the members of $\mathcal{F}$.

$$\sigma(\mathcal{F}) = \{G : |G| = k - 1, G \subset F \text{ for some } F \in \mathcal{F}\}.$$

The shadow problem asks for the minimum of $|\sigma(\mathcal{F})|$, given $n, k$ and $|\mathcal{F}|$. The shadow theorem[19,25] determines the exact minimum for all cases, but here we give only a special case.

**Theorem 5.1.** *(Special case of the Shadow Theorem[19,25]) Let $n, k$ be integers and suppose that $\mathcal{F} \in \binom{[n]}{k}$ has the size $\binom{a}{k}$ for some integer $a$. Then*

$$\min |\sigma(\mathcal{F})| = \binom{a}{k-1}.$$

The construction giving equality is simply $\mathcal{F} = \binom{[a]}{k}$ that is the family of all $k$-element subsets of an $a$-element part of $[n]$. (It does not depend on $n$.)

It is natural to ask what is the minimum of $|\sigma(\mathcal{F})|$ under the condition that $\mathcal{F}$ is $t$-intersecting. The old construction does not work here if $a > 2k$. The answer is somewhat disappointing: we do not know this minimum value. But we can answer a more modest question, we are able to determine the minimum of the ratio

$$\frac{|\sigma(\mathcal{F})|}{|\mathcal{F}|}.$$

**Theorem 5.2.** [18] *If* $\mathcal{F} \subset \binom{[n]}{k}$ *is a t-intersecting family, then*

$$\frac{|\sigma(\mathcal{F})|}{|\mathcal{F}|} \geq \frac{\binom{2k-t}{k-1}}{\binom{2k-t}{k}} = \frac{k}{k-t+1}.$$

The problem is not just for itself. The proof of Theorem 2.1 is based on (a more general form of) Theorem 5.1. This estimate is sharp. If $\mathcal{F}$ consists of all $k$-element subsets of a $2k - t$-element set then the size of the shadow is $\binom{2k-t}{k-1}$, the ratio is exactly the above one. In this construction however the size $|\mathcal{F}|$ of the family is "small", does not depend on $n$. What happens if we suppose that $|\mathcal{F}|$ is large? We have a slight improvement in this case.

**Theorem 5.3.** [23] *If* $\mathcal{F} \subset \binom{[n]}{k}$ *is a t-intersecting family,* $1 \leq t$ *then*

$$|\sigma(\mathcal{F})| \geq |\mathcal{F}|\frac{k-1}{k-t} - c(k,t)$$

*where* $c(k,t)$ *does not depend on* $n$ *and* $|\mathcal{F}|$.

This is an improvement only when $t > 1$. A better multiplicative constant cannot be expected as the following example shows.

Divide $[n]$ into two parts, $X_1, X_2$ where $|X_1| = 2k - t - 2, |X_2| = n - 2k + t + 2$ and define $\mathcal{F}$ as the family of all $k$-element sets $F$ such that $|F \cap X_1| = k - 1, |F \cap X_2| = 1$. Here $|\mathcal{F}| = \binom{2k-t-2}{k-1}(n-2k+t+2), |\sigma(\mathcal{F})| = \binom{2k-t-2}{k-2}(n-2k+t+2) + \binom{2k-t-2}{k-1}$. Their ratio tends to $\frac{k-1}{k-t}$.

## References

1. Ahlswede, R., Khachatrian, L.H.: The Complete Intersection Theorem for Systems of Finite Sets, Europ. J. Combinatorics 18(1997) 125-136.
2. Aigner, Martin and Ziegler M. Günter: *Proofs form THE BOOK* Springer-Verlag, Berlin-Heidelberg, 1998.
3. Alishahi, Meysam and Taherkhani, Ali: Extremal $G$-free induced subgraphs of Kneser graphs, arXiv:1801.03972v1.
4. Alon, N., Dinur, I, Friedgut, E. and Sudakov, B.: Graph products, Fourier Analysis and Spectral techniques, Geometric and Functional Analysis 14 (2004), 913-940.
5. Alon, Noga and Spencer, Joel H.: *The probabilistic method*, Wiley - Interscience Series in Discrete Mathematics and Optimization, John Wiley & Sons, Inc. New York, 1992.
6. Chvátal, V.: Intersecting families of edges in hypergraphs having the hereditary property, in *Hypergraph Seminar*, Lecture Notes in Math., Vol. 41 I, pp. 61-66, Springer-Verlag, Berlin, 1974.

7. Erdős, P., Ko, Chao, Rado, R.: Intersection theorems for systems of finite sets, The Quarterly Journal of Mathematics, Oxford. Second Series 12(1961) 313-320.

8. Erdős, Peter L., Frankl, P. and Katona G.O. H.: Extremal hypergraph problems and convex hulls, Combinatorica 5(1985) 011-026.

9. Frankl, P.: The Erdős-Ko-Rado theorem is true for n=ckt. Combinatorics (Proc. Fifth Hungarian Colloq., Keszthely, 1976), Vol. I, 365-375. Colloq. Math. Soc. Jnos Bolyai, 18 , North-Holland, Amsterdam-New York, (1978).

10. Frankl, P.: An Erdős Ko Rado Theorem for Direct Products, Europ. J. Combinatorics (1996) 17 , 727 730.

11. Frankl, P., Füredi, Z.: Beyond the Erdős-Ko-Rado theorem. J. Combin. Theory Ser. A 56(1991) 182-194.

12. Gerbner, Dániel: Profile polytopes of some classes of families, Combinatorica, 33(2013) 199-216.

13. Gerbner, D., Lemons, N., Palmer, C., Patkós, B. and Szécsi, V.: Almost intersecting families of sets, *SIAM J. Discrete Math.* **26**(4)(2012) 1657-1669.

14. Gerbner, Dániel and Patkós, Balázs: *Extremal Finite Set Theory*, CRC Press, November, 2018.

15. Hilton A.J.W. and Milner, E.C.: Some intersection theorems for systems of finite sets, *Quarterly J. of Math. (Oxford)* **18**(1967) 369-384.

16. Hoffman, A.J.: On eigenvalues and colorings of graphs, B. Harris Ed., Graph Theory and its Applications, Academic, New York and London, 1970, 79-91.

17. Horn, R.A., and Johnson, C.R. *Topics in Matrix Analysis*, Cambridge University Press, New York, 1991.

18. Katona, G.: Intersection theorems for systems of finite sets, Acta Math. Acad. Sci. Hungar. 15(1964) 329-337.

19. Katona, G.: A theorem on finte sets, Theory of Graphs, Proc. Coll. held at Tihany, 1966, Akadémiai Kiadó, pp. 187-207.

20. Katona, G.O. H.: A simple proof of the Erdős-Chao Ko-Rado theorem, J. Combin. Theory Ser. B 13(1972) 183-184.

21. Katona, Gyula O. H.: Around the Complete Intersection Theorem, *Discrete Applied Mathematics*, **216**(3)(2017) 618-621. (Special Volume: Levon Khachatrian's legacy in extremal combinatorics, edited by Zoltán Füredi and Gyula O. H. Katona)

22. Katona, Gyula O. H.: A general 2-part Erdős-Ko-Rado theorem, *Opuscula Mathematica* **37**(4)(2017) 577-588.

23. Katona, Gyula O. H.: Results on the shadow of intersecting families, in preparation.

24. Katona, Gyula O. H. and Nagy, Daniel T.: Union-intersecting set systems, *Graphs and Combinatorics* **31**(2015) 1507-1516.

25. Kruskal, J.B.: The number of simplices in a complex, *Mathematical Optimization Techniques*, (University of California, 1963) 251-278.

26. Kwan, Metthew, Sudakov, Benny and Vieira, Pedro: Non-trivially intersecting multi-part families, *J. Combin. Theory Ser A* **156**(2018) 44-60.

27. Lovász, László, *Combinatorial Problems and Exercises* Akadémiai Kiadó, 1979. Problem 11.14.

28. Lovász, László, On the Shannon capacity of a graph, IEEE Transactions on Information Theory IT-25, (1979), 1-7.
29. Wilson, R.M.: The exact bound in the Erdős-Ko-Rado theorem, Combinatorica 4(1984) 247-257.

# Powers of Monomial Ideals and Combinatorics

Le Tuan Hoa

*Institute of Mathematics, VAST,*
*18 Hoang Quoc Viet, 10307 Hanoi, Vietnam*

*lthoa@math.ac.vn*

Dedicated to the 80th birthday of Professor L. Bokut.

This is an exposition of some new results on associated primes and the depth of different kinds of powers of monomial ideals in order to show a deep connection between commutative algebra and some objects in combinatorics such as simplicial complexes, integral points in polyhedrons and graphs.

*2010 Mathematics subject classification:* 13D45, 05C90.

*Keywords*: Associated prime, depth, monomial ideal, integral closure, simplicial complex, integer linear programming

## 1. Introduction

The interaction between commutative algebra and combinatorics has a long history. It goes back at least to Macaulay's article[37]. Stanley's solution of the so-called Upper Bound Conjecture for spheres gives a new impulse for the study in this direction. Since then, many books devoted to various topics of this interaction are published, see, e.g., Ref. 8,22,41,56,65. People even talk about the birth of a new area of mathematics called "Combinatorial Commutative Algebra".

This exposition is based on my talk at the "Third International Congress in Algebras and Combinatorics (ICAC 2017)" held in Hong Kong. The aim of the workshop is clear from its title: to know better about the interaction between various areas of mathematics, which include associative algebra, commutative algebra and combinatorics. So, the purpose of this paper is to provide some further interaction from current research interest. Two basic notions in commutative algebra are concerned here: the associated primes and the depth of a (graded or local) ring. Associated prime ideals of a ring play a role like prime divisors of a natural number in the Number Theory, while the depth measures how far the ring from being Cohen-Macaulay.

For simplicity, we work with homogeneous ideals $I$ in a polynomial ring $R = K[X_1, ..., X_r]$. On the way to give a counter-example to Conjecture 2.1 in Ref. 47, Brodmann[5] proves that $\text{Ass}(R/I^n)$ becomes stable for $n \gg 0$. This stable set is denoted by $\text{Ass}^\infty(R/I)$. Since associated primes are closely related to the depth, almost at the same time Brodmann[6] proves that $\text{depth}(R/I^n)$ becomes constant for all $n \gg 0$. This constant is denoted by $\lim_{n\to\infty} \text{depth}(R/I^n)$. It is however not known, when the sequences $\{\text{Ass}(R/I^n)\}$ and $\{\text{depth}(R/I^n)\}$ become stable, and it is little known about $\text{Ass}^\infty(R/I)$ and $\lim_{n\to\infty} \text{depth} R/I^n$. Therefore, it is of great interest to bound the least place $\text{astab}(I)$ (resp. $\text{dstab}(I)$) when the stability of $\text{Ass}(R/I^n)$ (resp. $\text{depth}(R/I^n)$) occurs (see Definition 2.2 and Definition 2.1), as well as to determine $\text{Ass}^\infty(R/I)$ and $\lim_{n\to\infty} \text{depth}(R/I^n)$. For an arbitrary ideal, these problems are very difficult, because it is not known how to compute all associated primes of a ring and there is no effective way to compute the depth. Only few restricted results were obtained in the general case, see Ref. 40,53 and Ref. 44(Theorem 2.2).

Luckily, in the case of monomial ideals (i.e. ideals generated by monomials) one can use combinatorics to compute associated primes as well as the depth. Even it is not a trivial task, it opens up a way to use combinatorics to deal with these problems. This paper is mainly devoted to bounding $\text{astab}(I)$ and $\text{dstab}(I)$ for monomial ideals. This topic attracts many researchers during the last decade. From this study some times one can get surprising relationships between seemingly unrelated notions of commutative algebra and combinatorics. For an example, Ref. 58(Theorem 1.2) states that the ring $R/I_\Delta^n$ is Cohen-Macaulay for some fixed $n \geq 3$ if and only if the simplicial complex $\Delta$ is a complete intersection (see Theorem 4.10). Together with stating some main results we also give some hints for their proofs. Techniques from combinatorics used to obtain results presented in this paper are so broad, that in the most cases we cannot go to the details. We only explain in more details how the existence of integer solutions of systems of linear constrains related to bounding $\text{astab}(I)$ and $\text{dstab}(I)$. Besides these two problems, we also list some results on properties of the sequences $\{\text{Ass}(R/I^n)\}$ and $\{\text{depth}(R/I^n)\}$, because they are useful in determining $\text{Ass}^\infty(R/I)$ and $\lim_{n\to\infty} \text{depth} R/^n$. Similar problems for integral closures of powers as well as symbolic powers are also considered in this paper.

We would like to mention that recently there is an intensive research on the so-called Castelnuovo-Mumford regularity of monomial ideals, which also involves a lot of combinatorics. The interested readers can consult the survey paper[2].

The paper is organized as follows. In Section 2 we recall some basic notions and facts from commutative algebra and formulate two main problems considered in this paper. In particular, the above mentioned Brodmann's results are stated here. Section 3 is devoted to bounding $\mathrm{astab}(I)$. This section is divided to three subsections: bounds on $\mathrm{astab}(I)$ and $\overline{\mathrm{astab}}(I)$ are presented in the first two subsections. These bounds are huge ones. Good bounds on these invariants for some classes of monomial ideals are given in the last subsection. The stability of the depth function is presented in Section 4. The first three subsections are devoted to three kind of powers. The last subsection is concerning with Cohen-Macaulay property of square-free monomials ideals.

## 2. Preliminaries

Let $R$ be either a Noetherian local ring with maximal ideal $\mathfrak{m}$ and $K = R/\mathfrak{m}$, or a standard graded finitely generated $K$-algebra with graded maximal ideal $\mathfrak{m}$, where $K$ is an infinite field (standard grading means $R = \oplus_{i \geq 0} R_i$ such that $R_0 = K$, $R_i R_j \subseteq R_{i+j}$ for all $i, j \geq 0$ and $R$ is generated by $R_1$ over $K$). A non-zero divisor $x \in R$ is called an $R$-regular element. A sequence of elements $x_1, ..., x_s$ of $R$ is called $R$-regular sequence if $x_i$ is an $R/(x_1, ..., x_{i-1})$-regular element for $i = 1, ..., s$, and $R \neq (x_1, ..., x_s)$. Then all maximal $R$-regular sequences in $\mathfrak{m}$ have the same length and this length is called the *depth* of $R$, denoted by $\mathrm{depth}(R)$. Moreover, in the graded case, one can choose a maximal $R$-regular sequence consisting of homogeneous elements. Rees[49] showes that

$$\mathrm{depth}(R) = \min\{i|\ \mathrm{Ext}_R^i(K, R) \neq 0\}.$$

One can also define $\mathrm{depth}(R)$ by using local cohomology:

$$\mathrm{depth}(R) = \min\{i|\ H_\mathfrak{m}^i(R) \neq 0\}.$$

The reader can consult the book[7] for the definition and a detailed algebraic introduction to Grothendieck's local cohomology theory.

The Krull dimension $\dim(R)$ of $R$ and $\mathrm{depth}(R)$ are two basic invariants of $R$. One has $\mathrm{depth}\,R \leq \dim R$. When the equality holds, $R$ is called a Cohen-Macaulay ring. "The notion of (local) Cohen-Macaulay ring is a workhorse of Commutative Algebra", see Ref. 8(p. 56). This explains the importance of depth.

It is in general not easy to determine the exact value of depth$(R)$. Therefore, the following simple result of Brodmann[6] is of great interest.

**Theorem 2.1.** [6(Theorem 2)] *Let $I \subset R$ be a proper ideal, which is assumed to be graded if $R$ is graded. Then*

*(i) depth$(R/I^n)$ is constant for all $n \gg 0$.*

*(ii) Denote the above constant by $\lim_{n \to \infty} \text{depth}(R/I^n)$. Let $\mathcal{R}(I) = \oplus_{n \geq 0} I^n t^n$ be the Rees algebra of $I$. Then*

$$\lim_{n \to \infty} \text{depth}(R/I^n) \leq \dim(R) - \ell(I),$$

*where $\ell(I) = \dim \mathcal{R}(I)/\mathfrak{m}\mathcal{R}$ is the analytic spread of $I$.*

In fact, Brodmann's result was formulated for modules. Brodmann's proof as well as a new proof by Herzog and Hibi (see Ref. 21(Theorem 1.1)) are based on the Noetherian property of the Rees algebra $\mathcal{R}(I)$. As a corollary of these proofs, one has a similar statement for the so-called integral closures of powers of an ideal. Recall that the *integral closure* of an arbitrary ideal $\mathfrak{a}$ of $R$ is the set of elements $x$ in $R$ that satisfy an integral relation

$$x^n + a_1 x^{n-1} + \cdots + a_{n-1} x + a_n = 0,$$

where $a_i \in \mathfrak{a}^i$ for $i = 1, \ldots, n$. This is an ideal of $R$ and is denoted by $\bar{\mathfrak{a}}$. In the local case, assume in addition that $R$ is complete. Then the algebra $\overline{\mathcal{R}(I)} := \oplus_{n \geq 0} \overline{I^n} t^n$ is a module-finite extension of $\mathcal{R}(I)$. So, Brodmann's result implies that depth$(R/\overline{I^n})$ also is constant for all $n \gg 0$.

**Definition 2.1.** For an ideal $I \subset R$, set

$$\text{dstab}(I) := \min\{s|\ \text{depth}(R/I^n) = \text{depth}(R/I^s)\ \forall n \geq s\}.$$

In the local case, assume in addition that $R$ is complete. Set

$$\overline{\text{dstab}}(I) := \min\{s|\ \text{depth}(R/\overline{I^n}) = \text{depth}(R/\overline{I^s})\ \forall n \geq s\}.$$

The proofs of Brodmann and Herzog-Hibi give no information on when the functions depth$(R/I^n)$ and depth$(R/\overline{I^n})$ become stable. Therefore, the following problem attracts attention of many researchers:

**Problem 1.** *Give upper bounds on* dstab$(I)$ *and* $\overline{\text{dstab}}(I)$ *in terms of other invariants of $R$ and $I$.*

Until now there is no approach to solve this problem in the general setting as above. The reason is that there is no effective way to compute depth. Therefore all known nontrivial results until now are dealing with monomial ideals in a polynomial ring. These results will be summarized in Section 4. Below we describe one of the main tools to be used.

For the moment, let $R = K[X_1, ..., X_r]$ be a polynomial ring with $r$ indeterminate $X_1, ..., X_r$. A monomial ideal $I$ of $R$ is an ideal generated by monomials $\mathbf{X}^\alpha := X_1^{\alpha_1} \cdots X_r^{\alpha_r}$, where $\alpha = (\alpha_1, ..., \alpha_r) \in \mathbb{N}^r$. In this case one can effectively describe the local cohomology module $H_\mathfrak{m}^i(R/I)$, where $\mathfrak{m} = (X_1, ..., X_r)$. Let us recall it here.

Since $R/I$ is an $\mathbb{N}^r$-graded algebra, $H_\mathfrak{m}^i(R/I)$ is an $\mathbb{Z}^r$-graded module over $R$, i.e. $H_\mathfrak{m}^i(R/I) = \oplus_{\alpha \in \mathbb{Z}^r} H_\mathfrak{m}^i(R/I)_\alpha$, such that $\mathbf{X}^\beta H_\mathfrak{m}^i(R/I)_\alpha \subseteq H_\mathfrak{m}^i(R/I)_{\alpha+\beta}$. Each $\alpha$-component $H_\mathfrak{m}^i(R/I)_\alpha$ can be computed via the reduced simplicial homology.

Recall that a simplicial complex $\Delta$ on the finite set $[r] := \{1, ..., r\}$ is a collection of subsets of $[r]$ such that $F \in \Delta$ whenever $F \subseteq F'$ for some $F' \in \Delta$. Notice that we do not impose the condition that $\{i\} \in \Delta$ for all $i \in [r]$. An element of $\Delta$ is called a *face*. A simplicial complex $\Delta$ is defined by the set of its *facets* (i.e. maximal faces) - denoted by $\mathcal{F}(\Delta)$. In this case we also write $\Delta = \langle \mathcal{F}(\Delta) \rangle$. To each monomial ideal $I$ we can associate a simplicial complex $\Delta(I)$ defined by

$$\Delta(I) = \{\{i_1, ..., i_s\} \subseteq [r] \mid X_{i_1} \cdots X_{i_s} \notin \sqrt{I}\}.$$

Thus $\Delta(I)$ is defined upto the radical $\sqrt{I}$ of $I$. This notation was first introduced for the so-called Stanley-Reisner ideals, which are generated by square-free monomials, see Ref. 56(Chapter 2).

$$
\begin{aligned}
I &= (X_1^2 X_3^3 X_4, X_2^3 X_4^2) \\
\sqrt{I} &= (X_1 X_3 X_4, X_2 X_4) \\
\mathcal{F}(\Delta(I)) &= \{\{1, 2, 3\}, \{1, 4\}, \{3, 4\}\}
\end{aligned}
$$

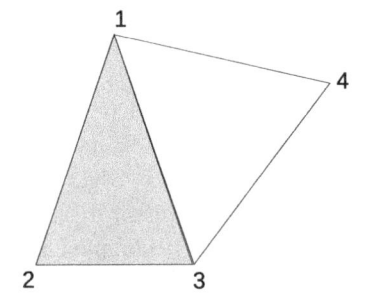

Fig. 1.

For every $\alpha = (\alpha_1, ..., \alpha_r) \in \mathbb{Z}^r$, we define the negative part of its support to be the set $CS_\alpha := \{i \mid \alpha_i < 0\}$. For a subset $F$ of $[r]$, let

$R_F := R[X_i^{-1} \mid i \in F]$ be the localization of $R$ at $F$. Set

$$\Delta_{\boldsymbol{\alpha}}(I); = \{F \subseteq [r] \setminus CS_{\boldsymbol{\alpha}} \mid \mathbf{X}^{\boldsymbol{\alpha}} \notin IR_{F \cup CS_{\boldsymbol{\alpha}}}\}. \tag{1}$$

We set $\tilde{H}_i(\emptyset; K) = 0$ for all $i$, $\tilde{H}_i(\{\emptyset\}; K) = 0$ for all $i \neq -1$, and $\tilde{H}_{-1}(\{\emptyset\}; K) = K$. Thanks to Ref. 17(Lemma 1.1) we may reformulate Takayama's result as follows.

**Theorem 2.2.** [57(Theorem 2.2)]

$$\dim_K H_{\mathfrak{m}}^i(R/I)_{\boldsymbol{\alpha}} = \dim_K \tilde{H}_{i-|CS_{\boldsymbol{\alpha}}|-1}(\Delta_{\boldsymbol{\alpha}}(I); K).$$

It was shown in Ref. 42(Lemma 1.3) that $\Delta_{\boldsymbol{\alpha}}(I)$ is a subcomplex of $\Delta(I)$. As a consequence, $H_{\mathfrak{m}}^i(R/I)_{\boldsymbol{\alpha}} = 0$ provided $CS_{\boldsymbol{\alpha}} \notin \Delta(I)$. Assume that $CS_{\boldsymbol{\alpha}} \in \Delta(I)$. Then, by Ref. 31(Lemma1.2),

$$\Delta_{\boldsymbol{\alpha}}(I) = \{F \in \mathrm{lk}_{\Delta(I)}(CS_{\boldsymbol{\alpha}}) \mid x^{\boldsymbol{\alpha}} \notin IR_{F \cup CS_{\boldsymbol{\alpha}}}\},$$

where the link of a face $F$ of a simplicial complex $\Delta$ is defined by

$$\mathrm{lk}_{\Delta}(F) = \{G \subseteq [r] \setminus F \mid F \cup G \in \Delta\}.$$

Using this remark, one can see that in the case of Stanley-Reisner ideals, Takayama's formula is exactly the famous Hochster's formula, see Ref. 56(Theorem 4.1). Hochster's formula plays crucial role in the theory of Stanley-Reisner ideals, where one can find rich interaction between commutative algebra and combinatorics (see, e.g., Ref. 8,56). We will see in this paper, that Takayama's theorem is very useful in the study of some invariants of powers of monomial ideals.

Another important notion in commutative algebra is the set of associated primes of a ring. Now we go back to an arbitrary Noetherian ring $R$. Recall that a prime ideal $\mathfrak{p} \subset R$ is called an associated prime if $\mathfrak{p}$ is the annihilator $\mathrm{ann}(x) := \{a \in R \mid ax = 0\}$ for some $x \in R$. The set of associated primes of $R$ is written as $\mathrm{Ass}(R)$. One can say that this set has a central position in commutative algebra like prime divisors of a natural number in number theory. However, it is difficult to compute $\mathrm{Ass}(R)$. Therefore the following result by Brodmann is very nice and also finds a lot of application:

**Theorem 2.3.** [5] *Let $I \subset R$ be a proper ideal. Then the set $\mathrm{Ass}(R/I^n)$ is stable for all $n \gg 0$.*

The research of Brodmann was motivated by a Conjecture of Ratliff in Ref. 47 which says that $\mathrm{Ass}(R/I^n) \subseteq \mathrm{Ass}^{\infty}(R/I)$ for all $n \geq 1$. Note that already in Ref. 40(Page 80), there is an example constructed to show that the sequence $\{\text{Ass}(R/I^n)\}$ is not monotone. For integral closures, extending a result by McAdam and Eakin[40](Props. 7 and 18), Ratliff shows that a stronger result holds

**Theorem 2.4.**[48](Theorems 2.4 and 2.7) *Let $I \subset R$ be a proper ideal. Then the sequence of sets $\{\text{Ass}(R/\overline{I^n})\}$ is increasing and becomes stable when $n \gg 0$.*

We would like to know when the sequences $\{\text{Ass}(R/I^n)\}$ and $\{\text{Ass}(R/\overline{I^n})\}$ become stable. For that, we need

**Definition 2.2.** For a proper ideal $I$ of a Noetherian ring $R$, set

$$\text{astab}(I) := \min\{s|\ \text{Ass}(R/I^n) = \text{Ass}(R/I^s)\ \forall n \geq s\},$$

and

$$\overline{\text{astab}}(I) := \min\{s|\ \text{Ass}(R/\overline{I^n}) = \text{Ass}(R/\overline{I^s})\ \forall n \geq s\}.$$

**Problem 2.** *Give upper bounds on $\text{astab}(I)$ and $\overline{\text{astab}}(I)$ in terms of other invariants of $R$ and $I$.*

In the general case this problem seems to be very hard, because there is no effective way to compute the sets $\text{Ass}(R/I^n)$ and $\text{Ass}(R/\overline{I^n})$. However, the prime divisors of a monomial ideal are easily to be found. Therefore one can solve Problem 2 for monomial ideals. This will be summarized in Section 3.

## 3. Stability of associated primes

From now on, let $R = K[X_1, ..., X_r]$, $\mathfrak{m} = (X_1, ..., X_r)$ and $I$ a proper monomial ideal of $R$. If $r \geq 2$, then for a positive integer $j \leq r$ and $\alpha = (\alpha_1, ..., \alpha_r) \in \mathbb{R}^r$, we set

$$\alpha[j] = (\alpha_1, ..., \alpha_{j-1}, \alpha_{j+1}, ..., \alpha_r).$$

Denote $\mathbf{X}^\alpha[j]$ the monomial obtained from $\mathbf{X}^\alpha$ by setting $X_j = 1$. Let $I[j]$ be the ideal of $R$ generated by all monomials $\mathbf{X}^\alpha[j]$ such that $\mathbf{X}^\alpha \in I$. Since any associated prime $\mathfrak{p}$ of a monomial ideal $\mathfrak{a}$ is generated by a subset of variables and there is a monomial $m \notin \mathfrak{a}$ such that $\mathfrak{p} = \mathfrak{a} : m$, one can easily show

**Lemma 3.1.**[61](Proposition 4, Lemma 11) *Let $\mathfrak{m} = (X_1, ..., X_r)$ and $r \geq 2$. Then for all $n \geq 1$ we have:*
  *(i) $\text{Ass}(R/I^n) = \text{Ass}(I^{n-1}/I^n)$ and $\text{Ass}(R/\overline{I^n}) = \text{Ass}(\overline{I^{n-1}}/\overline{I^n})$,*

*(ii)* $\mathrm{Ass}(I^n/I^{n+1}) \setminus \{\mathfrak{m}\} = \cup_{i=1}^r \mathrm{Ass}(I[i]^n/I[i]^{n+1})$,
*(iii)* $\mathrm{Ass}(\overline{I^n}/\overline{I^{n+1}}) \setminus \{\mathfrak{m}\} = \cup_{i=1}^r \mathrm{Ass}(\overline{I[i]^n}/\overline{I[i]^{n+1}})$.

**Remark 3.1.** On one side, Lemma 3.1 allows us to do induction on the number of variables. On the other side, in order to study the stability of the set of associated primes, it reduces to checking if $\mathfrak{m} \in \mathrm{Ass}(R/I^n)$ or $\in \mathrm{Ass}(R/\overline{I^n})$, respectively.

### 3.1. *Associated primes of integral closures of powers*

One can identify a monomial $\mathbf{X}^{\alpha}$ with the integer point $\alpha \in \mathbb{N}^r \subset \mathbb{R}^r$. For a subset $A \subseteq R$, the exponent set of $A$ is

$$E(A) := \{\alpha \mid \mathbf{X}^{\alpha} \in A\} \subseteq \mathbb{N}^r.$$

So a monomial ideal $\mathfrak{a}$ is completely defined by its exponent set $E(\mathfrak{a})$. Then, we can geometrically describe $\overline{\mathfrak{a}}$ by using its Newton polyhedron.

**Definition 3.1.** Let $\mathfrak{a}$ be a monomial ideal of $R$. The Newton polyhedron of $\mathfrak{a}$ is $NP(\mathfrak{a}) := \mathrm{conv}\{E(\mathfrak{a})\}$, the convex hull of the exponent set $E(\mathfrak{a})$ of $\mathfrak{a}$ in the space $\mathbb{R}^r$.

The following results are well-known (see Ref. 50):

$$E(\overline{I}) = NP(I) \cap \mathbb{N}^r,$$

and

$$NP(I^n) = nNP(I) = n\,\mathrm{conv}\{E(I)\} + \mathbb{R}_+^r \text{ for all } n \geq 1. \tag{2}$$

The above equalities say that (exponents of) all monomials of $\overline{I}$ form the set of integer points in $NP(I)$ (while we do not know which points among them do not belong to $I$), and the Newton polyhedron $NP(I^n)$ of $I^n$ is just a multiple of $NP(I)$.

**Remark 3.2.** By the definition of $NP(I^n)$ and (2) it follows that $\mathfrak{m} \in \mathrm{Ass}(R/\overline{I^n})$ if and only if there is $\alpha \notin nNP(I)$ and $\alpha + \mathbf{e}_i \in nNP(I)$ for all $1 \leq i \leq r$, where $\mathbf{e}_1, ..., \mathbf{e}_r$ form the canonical basis of $\mathbb{R}^r$.

Let $G(I)$ denote the minimal monomial generating system of $I$ and

$$d(I) := \max\{\alpha_1 + \cdots + \alpha_r \mid \mathbf{X}^{\alpha} \in G(I)\},$$

the maximal generating degree of $I$. Using convex analysis and lineal algebra, one can show

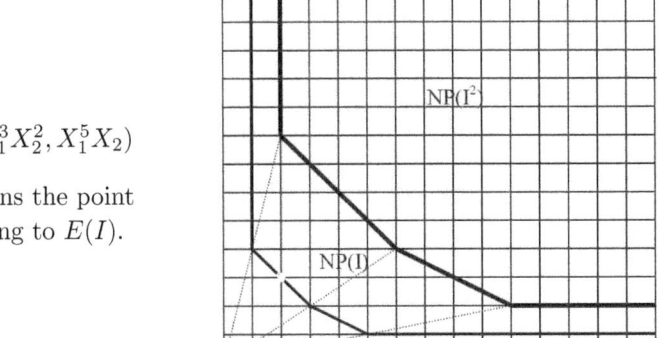

$$I = (X_1 X_2^4, X_1^3 X_2^2, X_1^5 X_2)$$

The hole means the point does not belong to $E(I)$.

Fig. 2.

**Lemma 3.2.** [61(Lemma 6), 32(Lemma 2.2)] *The Newton polyhedron $NP(I)$ is the set of solutions of a system of inequalities of the form*

$$\{\mathbf{x} \in \mathbb{R}^r \mid \langle \mathbf{a}_j, \mathbf{x} \rangle \geq b_j, \; j = 1, \ldots, q\}, \tag{3}$$

*such that each hyperplane with the equation $\langle \mathbf{a}_j, \mathbf{x} \rangle = b_j$ defines a facet of $NP(I)$, which contains $s_j$ affinely independent points of $E(G(I))$ and is parallel to $r - s_j$ vectors of the canonical basis. Furthermore, we can choose $\mathbf{0} \neq \mathbf{a}_j \in \mathbb{N}^r, b_j \in \mathbb{N}$ for all $j = 1, \ldots, q$; and if we write $\mathbf{a}_j = (a_{j1}, \ldots, a_{jr})$, then*

$$a_{ji} \leq s_j d(I)^{s_j - 1} \quad \text{for all } i = 1, \ldots, r,$$

*where $s_j$ is the number of non-zero coordinates of $\mathbf{a}_j$.*

Now one can give an effective necessary condition for $\mathfrak{m} \in \mathrm{Ass}(R/\overline{I^n})$ for some $n > 0$. It follows from Remark 3.2 and Lemma 3.2.

**Lemma 3.3.** [61(Lemma 13)] *Assume that $\mathfrak{m} \in \mathrm{Ass}(R/\overline{I^n})$ for some $n > 0$. Then there is a vector $\mathbf{a}_i$ determined in Lemma 3.2 such that $\mathbf{a}_i > 0$, that is $a_{ij} > 0$ for all $1 \leq j \leq r$.*

**Lemma 3.4.** *Let $I$ be a monomial ideal in $R$ with $r > 2$. If $\mathfrak{m} \in \mathrm{Ass}(R/\overline{I^s})$ for some $s \geq 1$, then $\mathfrak{m} \in \mathrm{Ass}(R/\overline{I^n})$ for all $n \geq (r - 1)rd(I)^{r-2}$.*

**Proof.** (Sketch): Let $m := (r - 1)rd(I)^{r-2}$. Since the sequence $\{\mathrm{Ass}(R/\overline{I^n})\}_{n \geq 1}$ is increasing by Theorem 2.4, it suffices to show that $\mathfrak{m} \in \mathrm{Ass}(R/\overline{I^m})$. As $\mathfrak{m} \in \mathrm{Ass}(R/\overline{I^s})$, by Lemma 3.3, there is a supporting hyperplane of $NP(I)$, say $H$, of the form $\langle \mathbf{a}, \mathbf{x} \rangle = b$ such that all coordinates of **a** are positive. By Lemma 3.2, this hyperplane passes through $r$ affinely independent points of $E(G(I))$, say $\alpha_1, \ldots, \alpha_r$. Denote the barycenter of the simplex $[\alpha_1, \ldots, \alpha_r]$ by $\alpha := \frac{1}{r}(\alpha_1 + \cdots + \alpha_r)$, and let $\beta := m\alpha - \mathbf{e}_r$. Then one can show that $\beta \notin mNP(I)$ and $\beta + \mathbf{e}_i \in mNP(I)$ for all $i \geq 1$. Hence the lemma follows from Remark 3.2.                                    □

The first bound on $\overline{\mathrm{astab}}(I)$ is given in Ref. 61(Theorem 16). It is then improved as follows.

**Theorem 3.1.** [32 (Theorem 2.7)] *Let $I$ be a monomial ideal of $R$. Then*

$$\overline{\mathrm{astab}}(I) \leq \begin{cases} 1 & \text{if } \ell(I) \leq 2, \\ \ell(I)(\ell(I) - 1)d(I)^{\ell(I)-2} & \text{if } \ell(I) > 2. \end{cases}$$

This theorem almost immediately follows from Lemma 3.3 and Remark 3.1 by using induction on $r$ (based on Lemma 3.1).

**Remark 3.3.** By Ref. 4(Theorem 2.3), we can compute $\ell(I)$ in terms of geometry of $NP(I)$.

$$\ell(I) = \max\{\dim F + 1 \mid F \text{ is a compact face of } NP(I)\}.$$

**Example 3.1.** [61 (Prposition 17)] Let $r \geq 4$ and $d > r - 3$. We put

$$u = X_1^{\binom{r-3}{0}} X_2^{\binom{r-3}{1}} \cdots X_{r-3}^{\binom{r-4}{0}} \text{ and } v = X_1^{\beta_1} \cdots X_{r-3}^{\beta_{r-4}} X_{r-2}^{d-r+2},$$

where

$$\beta_i = \begin{cases} 0 & \text{if } r - 3 - i \text{ is even}, \\ 2\binom{r-3}{i} & \text{if } r - 3 - i \text{ is odd}. \end{cases}$$

Let

$$I = (uX_1^d, uX_2^{d-1}X_r, \ldots, uX_{r-2}^{d-r+3}X_r^{r-3}, uX_{r-1}X_r^{d-1}, vX_r^{r-3}).$$

It is generated by monomials of the same degree $d(I) = d + 2^{r-3} - 1$. Then $\mathfrak{m} \in \mathrm{Ass}(R/\overline{I^n})$ for all $n \gg 0$ and if $\mathfrak{m} \in \mathrm{Ass}(R/\overline{I^n})$, then $n \geq n_0 := \frac{d(d-1)\cdots(d-r+3)}{r(r-3)}$. In particular,

$$\overline{\mathrm{astab}}(I) \geq n_0.$$

This shows that the bound in Theorem 3.1 is almost optimal, and *it must depend on the maximal generating degree of $I$.*

**Proof.** (Sketch): In this example, by Lemma 3.3, $\mathfrak{m} \in \mathrm{Ass}(R/\overline{I^n})$ for $n \gg 0$. On the other hand, the projection of $\mathrm{conv}(E(G(I)))$ into the hyperplane $\mathbb{R}^{r-1}$ of the first $(r-1)$ coordinates form a simplex, say $\Delta$. Using Lemma 3.2 and Remark 3.2, one can show that if $\mathfrak{m} \in \mathrm{Ass}(R/\overline{I^n})$ for a fix $n$, then the simplex $n\Delta$ must contain $r$ integer points of the form $\beta'$, $\beta' + \mathbf{e}'_1, ..., \beta' + \mathbf{e}'_{r-1} \subset \mathbb{N}^{r-1}$, where $\mathbf{e}'_1, ..., \mathbf{e}'_{r-1}$ are unit vectors of $\mathbb{R}^{r-1}$. The simplex $\Delta$ is so far from being "regular", that only its big multiples satisfy this combinatorial property. $\qquad\square$

**Question 3.** Assume that $I$ is a square-free monomial ideal. Is there a linear upper bound on $\overline{\mathrm{astab}}(I)$ in term of $r$?

### 3.2. Associated primes of powers

In the sequel, by abuse of terminology, for a linear functional

$$\varphi(\mathbf{x}) = a_1 x_1 + \cdots + a_r x_r,$$

where $a_i \in \mathbb{R}$, we say that $\varphi(\mathbf{x}) \geq 0$ is a homogeneous linear constraint, while $\varphi(\mathbf{x}) \geq b$ is a linear constraint. Unlike integer closures, it is much more difficult to describe the set of monomials in $I^n$ by linear constrains. However, we have the following observation:

Assume that the monomials $\mathbf{X}^{\alpha_1}, ..., \mathbf{X}^{\alpha_s}$ generate the ideal $I$. Then a monomial $\mathbf{X}^\alpha \in I^m$ if and only if there are nonnegative integers $a_1, ..., a_{s-1}$, such that $m \geq a_1 + \cdots + a_{s-1}$ and $\mathbf{X}^\alpha$ is divisible by

$$(\mathbf{X}^{\alpha_1})^{a_1} \cdots (\mathbf{X}^{\alpha_{s-1}})^{a_{s-1}} (\mathbf{X}^{\alpha_s})^{m-a_1-\cdots-a_{s-1}}.$$

This is equivalent to

$$\alpha_j \geq \alpha_{1j} a_1 + \cdots + \alpha_{(s-1)j} a_{s-1} + \alpha_{sj}(m - a_1 - \cdots - a_{s-1}),$$

for all $j = 1, ..., r$.

From this observation, $\alpha \in E(I^m)$ if and only is it is a part of an integer solution of a system of linear constrains in $r+s-1$ variables. Unfortunately this correspondence is not one-to-one, so that we cannot reverse a constrain in order to get a criterion for $\alpha \notin E(I^m)$. Nevertheless this observation is useful in finding an upper bound on $\mathrm{astab}(I)$ in Ref. 28.

The next observation is that in this case thanks to Lemma 3.1(i), it is easier to work with $\mathrm{Ass}(I^{n-1}/I^n)$ than with $\mathrm{Ass}(R/I^n)$, because the quotient modules $I^{n-1}/I^n$, $n \geq 1$, can be put together in the so-called associated graded ring of $I$:

$$G = \oplus_{n \geq 0} I^n/I^{n+1}.$$

Further, $\mathfrak{m} \in \mathrm{Ass}(I^n/I^{n+1})$ if and only if the local cohomology module $H^0_{\mathfrak{m}}(I^n/I^{n+1}) \neq 0$. This local cohomology can be computed as follows (see Ref. 28(Lemma 3.2))

$$H^0_{\mathfrak{m}G}(G)_{n-1} \cong H^0_{\mathfrak{m}}(I^{n-1}/I^n) \cong \frac{I^{n-1} \cap I[1]^n \cap \cdots \cap I[r]^n}{I^n}.$$

Using the above observation, one can associate the family of $E(I^{n-1} \cap I[1]^n \cap \cdots \cap I[r]^n)$ to a set $\mathcal{E} \subset \mathbb{N}^{rs+s}$ of integer solutions of linear constrains in $rs + s$ variables. If we denote the set of integer solutions of the corresponding system of homogeneous linear constrains by $\mathcal{S}$, then $\mathcal{S}$ is a semigroup, so that $K[\mathcal{S}]$ is a ring, and $K[\mathcal{E}]$ is a $K[\mathcal{S}]$-module. One can prove that $H^0_{\mathfrak{m}G}(G)$ is isomorphic to a quotient of $K[\mathcal{E}]$ (Ref. 28(Lemma 3.4)). Using linear algebra and Caratheodory's Theorem (see, e.g. Ref. 52(Corollary 7.1(i))) one can show that the maximal generating degree of $K[\mathcal{E}]$ over $K[\mathcal{S}]$ is bounded by

$$B_1 := d(rs + s + d)(\sqrt{r})^{r+1}(\sqrt{2}d)^{(r+1)(s-1)},$$

where $d = d(I)$, $s$ the number of minimal generators of $I$ (see Ref. 28(Proposition 3.1)). From that one obtains

**Proposition 3.1.** [28(Proposition 3.2)] *Let $n \geq B_1$ be an integer. Then*

$$\mathrm{Ass}(I^n/I^{n+1}) \supseteq \mathrm{Ass}(I^{n+1}/I^{n+2}).$$

In order to get the reverse inclusion we use another local cohomology module. Recall that the Rees algebra $\mathcal{R} := \mathcal{R}(I) = \oplus_{n \geq 0} I^n t^n$. Let $\mathcal{R}_+ = \oplus_{n > 0} I^n t^n$. The local cohomology module $H^0_{\mathcal{R}_+}(G)$ is also a $\mathbb{Z}$-graded $\mathcal{R}$-module. Let

$$a_0(G) = \sup\{n|\ H^0_{\mathcal{R}_+}(G)_n \neq 0\}.$$

(This number is to be taken as $-\infty$ if $H^0_{\mathcal{R}_+}(G) = 0$.) It is related to the so-called Castelnuovo-Mumford regularity of $G$ (see, e.g., Ref. 53). Then S. McAdam and P. Eakin show that $\mathrm{Ass}(I^n/I^{n+1}) \subseteq \mathrm{Ass}(I^{n+1}/I^{n+2})$ for all $n > a_0(G)$ (see Ref. 40(pp. 71, 72), and also Ref. 53(Proposition 2.4)). Now one can again use (another) system of linear constrains to show

**Proposition 3.2.** [28(Proposition 3.3)] *We have*

$$a_0(G) < B_2 := s(s + r)^4 s^{r+2} d^2 (2d^2)^{s^2 - s + 1}.$$

Putting together Propositions 3.1 and 3.2, we get

**Theorem 3.2.** [28(Theorem 3.1)] *We have*

$$\text{astab}(I) \leq \max\{d(rs + s + d)(\sqrt{r})^{r+1}(\sqrt{2}d)^{(r+1)(s-1)},$$
$$s(s+r)^4 s^{r+2} d^2 (2d^2)^{s^2-s+1}\}.$$

Of course, one can bound $s$ in terms of $d$ and $r$. But then the resulted bound would be a double exponential bound. In spite of Theorem 3.1, we would like to ask:

**Question 4.** *i) Is there an upper bound on* $\text{astab}(I)$ *of the order* $d(I)^r$ ?
*ii) Assume that* $I$ *is a square-free monomial ideal. Is there a linear upper bound on* $\text{astab}(I)$ *in term of* $r$ ?

Note that using Example 1, one can construct an example to show that in the worst case, an upper bound on $\text{astab}(I)$ much be at least of the order $d(I)^{r-2}$ (provided that $r$ is fixed), see Ref. 28(Example 3.1).

Another interesting problem is to find the stable set $\text{Ass}^\infty(R/I)$ for $n \gg 0$. There is not much progress in this direction. However, Bayati, Herzog and Rinaldo can completely solve a kind of reverse problem: In Ref. 3 they prove that any set of nonzero monomial prime ideals can be realized as the stable set of associated primes of a monomial ideal.

### 3.3. *Case of edge ideals and square-free monomial ideals*

There are some partial classes of monomial ideals where $\text{astab}(I)$ is bounded by a linear function of $r$, that gives a partial affirmative answer to Question 4. The best results are obtained for square-free monomial ideals whose generators are all of degree two. Then one can associate such an ideal to a graph. More precisely, let $G = (V, E)$ be a simple undirected graph with the vertex set $V = [r] := \{1, 2, ..., r\}$ and the edge set $E$. The ideal

$$I(G) = (x_i x_j \mid \{i, j\} \in E) \subset R,$$

is called *edge ideal*. Recall that $G$ is *bipartite* if $V = V_1 \cup V_2$ such that $V_1 \cap V_2 = \emptyset$ and there is no edge connecting two vertices of the same set $V_i$. In our terminology, one can reformulate Ref. 54(Theorem 5.9) as follows.

**Theorem 3.3.** *The graph* $G$ *is bipartite if and only if* $\text{astab}(I(G)) = 1$.

The following result is a reformulation of Ref. 9(Corollary 2.2), which reduces the problem of bounding $\text{astab}(I(G))$ to the case of connected graphs.

**Lemma 3.5.** *Assume that* $G_1, ..., G_s$ *are connected components of* $G$. *Then*

$$\text{astab}(I(G)) \leq \sum_{i=1}^{s} \text{astab}(I(G_i)) - s + 1.$$

Recall that a cycle $C$ in a graph is a sequence of different vertices $\{i_1, ..., i_s\} \subset V$ such that $\{i_j, i_{j+1}\} \in E$ for all $j \leq s$, where $i_{s+1} \equiv i_1$. The number $s$ is called the length of $C$. It is an elementary fact in the graph theory that $G$ is bipartite if and only if $G$ does not contain odd cycles.

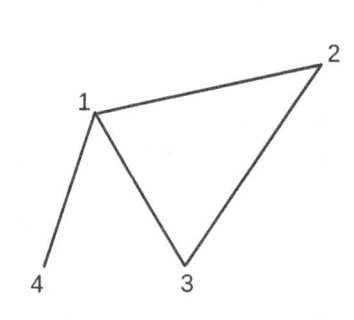
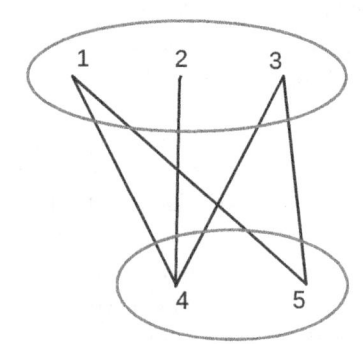

Non-bipartite graph $G_1$        Bipartite graph $G_2$

Fig. 3.

The following result together with Theorem 3.3 and Lemma 3.5 gives a good bound on $\text{astab}(I(G))$.

**Theorem 3.4.** [9(Lemma 3.1, Proposition 4.2)] *Assume that* $G$ *is non-bipartite connected graph. Let* $2k+1$ *be the smallest odd cycle contained in* $G$ $(k \geq 1)$.
  *(i) If* $G$ *is a cycle (of length* $2k+1$*), then* $\text{astab}(I(G)) = k + 1$,
  *(ii) If* $G$ *is not a cycle, then* $\text{astab}(I(G)) \leq r - k - 1$.

The proofs of the above results in Ref. 9 are quite long, but they are of combinatorial nature and do not require much from the graph theory. Even the above bound is quite good, by using other invariants, one can get a better bound. For an example, in the situation of Theorem 3.4(ii), if $G$ has $s$ vertices of degree one, then $\text{astab}(I(G)) \leq r - k - s$ (see Ref. 9(Corollary 4.3)).

If we set $\text{Min}(I)$ to be the set of minimal associated primes of $R/I$, then $\text{Min}(I) \subseteq \text{Ass}(R/I^n)$ for all $n \geq 1$. An element from $\text{Ass}(R/I^n) \setminus \text{Min}(I)$ is called an *embedded associated prime*. Therefore, in order to study the stability of $\text{Ass}(R/I^n)$, it suffices to study embedded associated primes.

In Ref. 27, using the Takayama's Theorem 2.2, Hien, Lam and N. V. Trung show that embedded primes of $R/I(G)^n$ are characterized by the existence of (vertex) weighted graphs with special matching properties, see Ref. 27(Theorem 2.4). There are (infinite) many weighted graphs with the base graph $G$. Using techniques in the graph theory, they can give necessary or sufficient conditions for an ideal generated by a subset of variables to be an embedded associated primes of $R/I(G)^n$ in terms of vertex covers of $G$ which contain certain types of subgraphs of $G$, see Ref. 27(Theorem 2.10, Theorem 3.5). From that they derive a stronger bound on $\mathrm{astab}(I(G))$, see Ref. 27(Corollary 3.7) and also Ref. 27(Example 3.8). Moreover, their method gives an algorithm to compute $\mathrm{Ass}(R/I(G)^t)$ for a fixed integer $t$. This was done for $t = 2, 3$ in Ref. 27 and for $t = 4$ in Ref. 26.

In Ref. 36, after extending some results in the graph theory, Lam and N. V. Trung are able to characterize the existence of weighted graphs with special matching properties in terms of the so-called generalized ear decompositions of the base graph $G$. Using this notion they give a new upper bound on $\mathrm{astab}(I(G))$ (see Ref. 36(Theorem 4.7)). Moreover, they can give a precise formula for $\mathrm{astab}(I(G))$. In order to formulate their results one needs to introduce some rather technical notions on graphs. Therefore we do not go to the details here. We only want to state the following nice main result of Ref. 38, that they can derive as an immediate consequence of their results.

**Theorem 3.5.** [38(Theorem 2.15), 36(Corollary 4.5)] *The sequence* $\{\mathrm{Ass}(R/I(G)^n)\}$ *is ascending, that is* $\mathrm{Ass}(R/I(G)^n) \subseteq \mathrm{Ass}(R/I(G)^{n+1})$ *for all* $n \geq 1$.

This property does not hold for square-free monomial ideals. Using a counter-example in combinatorics, Kaiser et al. constructed a square-free monomial ideal $J$ in 12 variables such that $\mathrm{Ass}(R/J^4)$ is a proper subset of $\mathrm{Ass}(R/J^3)$, see Ref. 35(Theorem 11).

For an arbitrary monomial $I$, the sequence $\{\mathrm{Ass}(R/I^n)\}$ is even not necessarily monotone. The first example is given in page 549 of Ref. 22. Recently, there is given a general construction of monomials ideals $I$ for which the non-monotonicity of $\{\mathrm{Ass}(R/I^n)\}$ can be arbitrarily long.

**Theorem 3.6.** [19(Corollary 6.8)] *Let $A$ be any finite set of positive integers. Then there exists a monomial ideal $I$ in a polynomial ring $R$ such that* $\mathfrak{m} \in \mathrm{Ass}(R/I^n)$ *if and only if $n \in A$.*

The ascending property of Ass, that means $\mathrm{Ass}(R/I^n) \subseteq \mathrm{Ass}(R/I^{n+1})$, $n \geq 1$, is also referred as the *persistence property* (with respect to associated ideals). This property is important in finding the stable set $\mathrm{Ass}^\infty(R/I)$.

For arbitrary square-free monomial ideals there are some interesting results given in Ref. 14,15,18,24, that relate $\mathrm{astab}(I)$ to combinatorics. In order to say about their study we need to introduce some notions.

The set of bases of a polymatroid of rank $d$ based on $[r]$ is a set $\mathcal{B}$ of integer points $\boldsymbol{\alpha} \in \mathbb{N}^r$ satisfying the following conditions:

- $|\boldsymbol{\alpha}| := \alpha_1 + \cdots + \alpha_r = d$ for all $\boldsymbol{\alpha} \in \mathcal{B}$,
- (Exchange property) For all $\boldsymbol{\alpha}, \boldsymbol{\beta} \in \mathcal{B}$ for which $\alpha_i > \beta_i$ for some i, there exists $j$ such that $\beta_j > \alpha_j$ and $\boldsymbol{\alpha} - \mathbf{e}_i + \mathbf{e}_j \in \mathcal{B}$.

A monomial ideal $I$ is called a *polymatroidal ideal*, if there exists a set of bases $\mathcal{B} \subset \mathbb{Z}^r$ of a polymatroid, such that $I = (\mathbf{X}^{\boldsymbol{\alpha}} | \ \boldsymbol{\alpha} \in \mathcal{B})$.

**Theorem 3.7.**[24] *(Proposition 2.4 and Theorem 4.1)* *Let $I$ be a polymatroidal ideal. Then $I$ satisfies persistence property and* $\mathrm{astab}(I) \leq \ell(I)$.

A simple hypergraph $\mathcal{H}$ is a pair of the vertex set $V = [r]$ and an edge set $\mathcal{E} = \{E_1, ..., E_t\}$, where $E_i \subseteq V$. We assume that $\mathcal{H}$ has no isolated vertices, each $E_i$ has at least two elements and that $E_i \not\subseteq E_j$ for all $i \neq j$. When the $E_i's$ all have cardinality two, then $\mathcal{H}$ is a simple graph. The ideal generated by all square-free monomials $X_{i_1} \cdots X_{i_s}$, $\{i_1, ..., i_s\} = E_i$, is called *edge ideal* of $\mathcal{H}$ and denoted by $I(\mathcal{H})$. For a very special class of $\mathcal{H}$, Ha and Morey determine the least number $k$ such that $\mathfrak{m} \in \mathrm{Ass}(R/I^k)$ and $\mathfrak{m} \notin \mathrm{Ass}(R/I^t)$ for all $t < k$, see Ref. 18(Theorem 4.6).

A vertex cover of $\mathcal{H}$ is a subset $W$ of $V$ such that if $E \in \mathcal{E}$, then $W \cap E \neq \emptyset$. A vertex cover is minimal if no proper subset is also a vertex cover. Denote $J = J(\mathcal{H})$ the *cover ideal* of $\mathcal{H}$, which is generated by the square-free monomials corresponding to the minimal vertex covers of $\mathcal{H}$. Francisco, Ha and Van Tuyl propose a conjecture related to the chromatic number of a graph $G$, and prove the persistence property of $\mathrm{Ass}(R/J(G)^n)$ provided that the conjecture holds (see Ref. 14(Theorem 2.6)). In another paper, they give an explicit description of all associated primes of $\mathrm{Ass}(R/J(\mathcal{H})^n)$, for any fixed number $n \geq 1$, in terms of the coloring properties of hypergraphs arising from $\mathcal{H}$, see Ref. 15(Corollary 4.5). From this description they give a lower bound on $\mathrm{astab}(J(\mathcal{H}))$.

Recall that a $t$-coloring of $\mathcal{H}$ is any partition of $V = C_1 \cup \cdots \cup C_t$ into $t$ disjoint sets such that for every $E \in \mathcal{E}$, we have $E \not\subseteq C_i$ for all $i = 1, ..., d$.

The $C'_i s$ are called the *color classes*. The chromatic number of $\mathcal{H}$, denoted $\chi(\mathcal{H})$, is the minimal $t$ such that $\mathcal{H}$ has a $t$-coloring.

**Proposition 3.3.** [15]*(Corollary 4.9)* $\mathrm{astab}(J(\mathcal{H})) \geq \chi(\mathcal{H}) - 1$.

Inspired by this result they pose the following question:

**Question 5** [15]*(Question 4.10)*. *For each integer $\geq 0$, does there exist a hypergraph $\mathcal{H}_n$ such that the stabilization of associated primes occurs at $a \geq \chi(\mathcal{H}_n) - 1) + n$?*

As one can see from the above discussion, all results concerning the existence of a linear bounding on $\mathrm{astab}(I)$ and the persistence property are given for very special squaree-free monomial ideals. Nevertheless, these results establish surprising relationships between some seemingly unrelated notions of commutative algebra and combinatorics and raise many more problems and questions. Thus, they will stimulate intensive investigation in the near future.

## 4. Stability of Depth

### 4.1. *Depth of powers of integral closures*

Due to some reasons, we can completely solve Problem 1 for $\overline{\mathrm{dstab}}(I)$. First, as an immediate consequence of Theorem 2.2, one can get the following "quasi-decreasing" property of the depth function $\mathrm{depth}\, R/\overline{I^n}$. We don't know if this property holds for integral closures of powers of an arbitrary homogeneous ideal.

**Lemma 4.1.** [32]*(Lemma 2.5)* *For any monomial ideal $I$ of $R$, we have*

*(1)* $\mathrm{depth}\, R/\overline{I^m} \geq \mathrm{depth}\, R/\overline{I^{mn}}$ *for all $m, n \geq 1$.*

*(2)* $\lim_{n \to \infty} \mathrm{depth}\, R/\overline{I^n} = \dim R - \ell(I)$, *where $\ell(I)$ denotes the analytic spread of $I$.*

**Proof.** (Sketch): For proving the first statement one can set $m = 1$. Then, using Theorem 2.2 and the fact that for any $\alpha \in \mathbb{Z}^r$, $CS_{n\alpha} = CS_\alpha$ and $\Delta_\alpha(\overline{I}) = \Delta_{n\alpha}(\overline{I^n})$, one can quickly show that $H^t_{\mathfrak{m}}(R/\overline{I})_\alpha \neq 0$ implies $H^t_{\mathfrak{m}}(R/\overline{I^n})_{n\alpha} \neq 0$.

The second statement follows from Ref. 11(Propostion 3.3) and the fact that $\overline{I^{r-1}}$ is torsion-free (see Ref. 64(Corollar 7.60).  $\square$

**Question 6.** *Is the depth function* depth $R/\overline{I^m}$ *decreasing?*

As we can see from Theorem 2.2, in order to study the local cohomology module, we need to have an effective description of $\Delta_{\boldsymbol{\alpha}}(\overline{I^n})$. In the case of integral closures, we do have it. Keeping the notations in Lemma 3.2, we set $\mathrm{supp}(\mathbf{a}_j) := \{i \mid a_{ji} \neq 0\}$.

**Lemma 4.2.** [32(Lemma 3.1)] *For any* $\boldsymbol{\alpha} \in \mathbb{N}^r$ *and* $n \geq 1$, *we have*

$$\Delta_{\boldsymbol{\alpha}}(\overline{I^n}) = \langle [r] \setminus \mathrm{supp}(\mathbf{a}_j) \mid j \in \{1,\ldots,q\} \text{ and } \langle \mathbf{a}_j, \boldsymbol{\alpha} \rangle < nb_j \rangle .$$

The following lemma is the main step in the proof of Theorem 4.1.

**Lemma 4.3.** [32(Lemma 3.2)] *Let* $m \geq 1$ *and* $t := \mathrm{depth}\, R/\overline{I^m}$. *Assume that* $H_{\mathfrak{m}}^t(R/\overline{I^m})_{\boldsymbol{\beta}} \neq 0$ *for some* $\boldsymbol{\beta} \in \mathbb{N}^r$. *If* $r \geq 3$, *then*

$$\mathrm{depth}\, R/\overline{I^n} \leq t \quad \text{for all } n \geq r(r^2-1)r^{r/2}(r-1)^r d(I)^{(r-2)(r+1)}.$$

**Proof.** (Sketch): Assume that

$$\langle \mathbf{a}_j, \boldsymbol{\beta} \rangle < b_j \text{ for } j = 1,\ldots,p,$$

and

$$\langle \mathbf{a}_j, \boldsymbol{\beta} \rangle \geq b_j \text{ for } j = p+1,\ldots,q,$$

for some $0 \leq p \leq q$. Then, by Lemma 4.2,

$$\Delta_{\boldsymbol{\beta}}(\overline{I^m}) = \langle [r] \setminus \mathrm{supp}(\mathbf{a}_j) \mid j = 1,\ldots,p \rangle .$$

For each $n \geq 1$, put

$$\Gamma(\overline{I^n}) := \{\boldsymbol{\alpha} \in \mathbb{N}^r \mid \Delta_{\boldsymbol{\alpha}}(\overline{I^n}) = \Delta_{\boldsymbol{\beta}}(\overline{I^m})\},$$

and

$$C_n := \{\mathbf{x} \in \mathbb{R}_+^r \mid \langle \mathbf{a}_j, \mathbf{x} \rangle < nb_j, \langle \mathbf{a}_l, \mathbf{x} \rangle \geq nb_l \text{ for } j \leq p; \ p+1 \leq l \leq q\} \subseteq \mathbb{R}_+^r. \tag{4}$$

Assume that $C_n \cap \mathbb{N}^r \neq \emptyset$. Then for any $\boldsymbol{\alpha} \in C_n \cap \mathbb{N}^r$, by Theorem 2.2, we will have $H_{\mathfrak{m}}^t(R/\overline{I^n}))_{\boldsymbol{\alpha}} \neq 0$, whence depth $R/\overline{I^n} \leq t$. It remains to show that $C_n \cap \mathbb{N}^r \neq \emptyset$ for any $n \geq r(r^2-1)r^{r/2}(r-1)^r d(I)^{(r-2)(r+1)}$.    $\square$

**Remark 4.1.** From the above sketch of proof we can see that the main technique is to find a number $n_0$ such that $C_n$ contains an integer point for all $n \geq n_0$, or equivalently that the system of linear constrains in (4) do have integer solutions. This is related to the research carried out by Ehrhart in Ref. 12,13, where he shows that the number of integer points in the closure $\overline{C_n} \subset \mathbb{R}^r$ is a quasi-polynomial!

Using Lemma 4.3, Theorem 2.2 and induction on $r$, one can prove the following main result of this subsection.

**Theorem 4.1.** [32 (Theorem 3.3)] *Let $I$ be a monomial ideal of $R$. Then*

$$\overline{\mathrm{dstab}}(I) \leq \begin{cases} 1 & \text{if } r \leq 2, \\ r(r^2-1)r^{r/2}(r-1)^r d(I)^{(r-2)(r+1)} & \text{if } r > 2. \end{cases}$$

It seems that this bound is too big. However, Example 3.1 shows that an upper bound on $\overline{\mathrm{dstab}}(I)$ must be at least of the order $d(I)^{r-2}$.

**Question 7.** *Is $\overline{\mathrm{dstab}}(I)$ bounded by a function of the order $d(I)^r$?*

### 4.2. Depth of symbolic powers

The $n$-th symbolic power of an ideal $\mathfrak{a} \subset R = K[X_1, ..., X_r]$ is the ideal

$$\mathfrak{a}^{(n)} := R \cap (\cap_{\mathfrak{p} \in \mathrm{Min}(I)} \mathfrak{a}^n R_{\mathfrak{p}}).$$

In other words, $\mathfrak{a}^{(n)}$ is the intersection of the primary components of $\mathfrak{a}^n$ associated to the minimal primes of $\mathfrak{a}$.

When $K$ is algebraically closed and $\mathfrak{a}$ is a radical ideal, Nagata and Zariski showed that $\mathfrak{a}^{(n)}$ consists of polynomials in $R$ whose partial derivatives of orders up to $n-1$ vanish on the zero set of $\mathfrak{a}$. Therefore, symbolic powers of an ideal carry richer geometric structures and more subtle information than ordinary powers!

Unlike the ordinary powers, the behavior of $\mathrm{depth}(R/I^{(n)})$ is much more mysterious. If $I$ is a monomial ideal, then the symbolic Rees algebra $\mathcal{R}_s(I) = \oplus_{n \geq 0} I^{(n)}$ is finitely generated (see [23] Theorem 3.2). Then Brodmann's Theorem 2.1 implies that $\mathrm{depth}(R/I^{(n)})$ is periodically constant for $n \gg 0$. Very recently D. H. Nguyen and N. V. Trung are able to construct monomial ideals $I$ for which $\mathrm{depth}(R/I^{(n)})$ is not constant for $n \gg 0$, see Ref. 46(Theorem 5.4). Moreover, they can show

**Theorem 4.2.** [46 (Theorem 5.1)] *Let $\varphi(t)$ be an arbitrary asymptotically periodic positive numerical function. Given a field $K$, there exist a polynomial ring $R$ over a purely transcendental extension of $K$ and a homogeneous ideal $I \subset R$ such that $\mathrm{depth}(R/I^{(t)}) = \varphi(t)$ for all $t \geq 1$.*

The construction in Ref. 46 only gives non monomial ideals. Therefore we would like to ask:

**Question 8.** *Does Theorem 4.2 hold for the class of monomial ideals?*

From now until the rest of this subsection, assume that $I$ is a square-free monomial ideal. Let $\Delta = \Delta(I)$. The correspondence $I \leftrightarrow \Delta(I)$ is one-to-one, and we also write $I_\Delta$ for $I$. Assume that $\mathcal{F}(\Delta) = \{F_1, \ldots, F_m\}$. Then

$$I_\Delta = \bigcap_{F \in \mathcal{F}(\Delta)} P_F,$$

where $P_F$ is the prime ideal of $R$ generated by variables $X_i$ with $i \notin F$, and

$$I_\Delta^{(n)} = \bigcap_{F \in \mathcal{F}(\Delta)} P_F^n.$$

In this case, it immediately follows from Ref. 30(Theorem 4.7) that $\operatorname{depth}(R/I^{(n)})$ is constant for all $n \gg 0$. Hence, for a Stanley-Reisner ideal $I_\Delta$, we can introduce the following notation.

**Definition 4.1.** Let $I$ be a square-free monomial ideal $I$. Set

$$\operatorname{dstab}^*(I) := \min\{m \geq 1 \mid \operatorname{depth}(R/I^{(n)}) = \operatorname{depth}(R/I^{(m)}) \text{ for all } n \geq m\}.$$

We can define the *symbolic analytic spread* of $I_\Delta$ by

$$\ell_s(I_\Delta) := \dim \mathcal{R}_s(I_\Delta)/\mathfrak{m}\mathcal{R}_s(I_\Delta).$$

Let $\operatorname{bight}(I_\Delta)$ be the *big height* of $I_\Delta$.

**Theorem 4.3.** [29 *(Theorem 2.4)*] *Let $I_\Delta$ be a Stanley-Reisner ideal of $R = k[X_1, \ldots, X_r]$. Then:*

*(1)* $\operatorname{depth}(R/I_\Delta^{(n)}) \geq \dim R - \ell_s(I_\Delta)$ *for all $n \geq 1$;*
*(2)* $\operatorname{depth}(R/I_\Delta^{(n)}) = \dim R - \ell_s(I_\Delta)$ *for all $n \geq r(r+1)\operatorname{bight}(I_\Delta)^{r/2}$.*

*In particular,* $\operatorname{dstab}^*(I_\Delta) \leq r(r+1)\operatorname{bight}(I_\Delta)^{r/2}$.

The idea of the proof of this theorem is similar to that of Theorem 4.1, because in this case we can also effectively compute $\Delta_\alpha(I_\Delta^{(n)})$.

**Lemma 4.4.** [42 *(Lemma 1.3)*] *For all $\alpha \in \mathbb{N}^r$ and $n \geq 1$, we have*

$$\Delta_\alpha(I_\Delta^{(n)}) = \left\langle F \in \mathcal{F}(\Delta) \mid \sum_{i \notin F} \alpha_i \leq n-1 \right\rangle.$$

We think that this bound is too big. Therefore, we would like to ask

**Question 9.** *Assume that $I$ is a square-free monomial ideal $I$.*
*(i) Is the depth function* $\operatorname{depth}(R/I^{(n)})$ *decreasing?*

*(ii) Is there a linear bound on* $\mathrm{dstab}^*(I)$ *in terms of* $r$ ?

Note that the "quasi-decreasing property" of $\mathrm{depth}(R/I^{(n)})$ can be proved similarly to Lemma 4.1. There are some partial positive answers to this question. As a corollary of Ref. 42(Theorems 2.3 and 2.4) (also see Ref. 31(Lemma 2.1)), we get

**Proposition 4.1.** *Assume that* $\dim R/I \leq 2$. *Then* $\mathrm{depth}(R/I^{(n)})$ *is decreasing and* $\mathrm{dstab}^*(I) \in \{1, 2, 3\}$.

Let $G$ be a simple graph with the vertex set $V = [r]$. Then the cover ideal

$$J(G) = \bigcap_{\{i,j\} \in E(G)} (X_i, X_j) \subset R.$$

It is clear that every unmixed squarefree monomial ideal of height two is uniquely correspondent to a cover ideal of a graph and vice versa. Using the so-called polarization technique one can show the non-increasing property of $\mathrm{depth}(R/J(G)^{(n)})$. Note that this property does not hold for the sequence $\mathrm{depth}(R/J(G)^n)$ on ordinary powers of $J(G)$ (see Ref. 35(Theorem 13)). The graph constructed there has 12 vertices and $\mathrm{depth}(R/J(G)^3) = 0$ while $\mathrm{depth}(R/J(G)^4) = 4$.

**Theorem 4.4.** [29(Theorem 3.2)] *Let* $G$ *be a simple graph. Then for* $n \geq 2$,

$$\mathrm{depth}(R/J(G)^{(n)}) \leq \mathrm{depth}(R/J(G)^{(n-1)}).$$

In order to formulate an effective bound on $\mathrm{dstab}^*(J(G))$ we recall some terminology from the graph theory. A set $M \subseteq E(G)$ is a matching of $G$ if any two distinct edges of $M$ have no vertex in common. Let $M = \{\{a_i, b_i\} \mid i = 1, \ldots, s\}$ be a nonempty matching of $G$. According to Ref. 10, we say that $M$ is an *ordered matching* if:

(1) $\{a_1, \ldots, a_s\}$ is a set of independent vertices,
(2) $\{a_i, b_j\} \in E(G)$ implies $i \leq j$.

**Example 4.1.** In the graph $G = C_4$, the subset $M = \{\{a_1 = 1, b_1 = 4\}, \{2, 3\}\}$ is a matching, but not an ordered matching, since the first property above would imply $a_2 = 3$ and $b_2 = 2$. Then, $\{a_2, b_1\} = \{3, 4\} \in E(G)$ and the second property above would not hold.

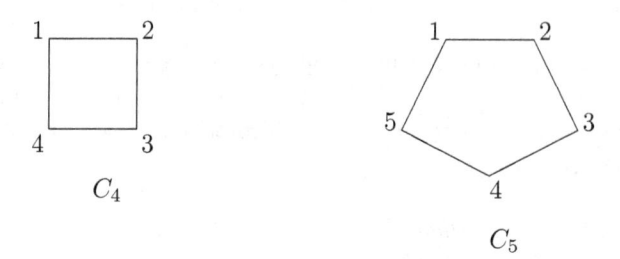

$$C_4 \qquad\qquad C_5$$

In the graph $G = C_5$, by setting $a_1 = 1$, $b_1 = 2$, $a_2 = 4$, $b_2 = 3$, $M = \{\{1,2\},\{3,4\}\}$ is an ordered matching.

**Definition 4.2.** The *ordered matching number* of $G$ is:

$$\nu_0(G) := \max\{|M| \mid M \subseteq E(G) \text{ is an ordered matching of } G\}.$$

Then we have

**Theorem 4.5.** [29(Theorem 3.4)] *Let $G$ be a simple graph with $r$ vertices. Then,*

$$\operatorname{depth} R/J(G)^{(n)} = r - \nu_0(G) - 1 \text{ for all } n \geq 2\nu_0(G) - 1.$$

*In particular* $\operatorname{dstab}^*(I) \leq 2\nu_0(G) - 1 < r$.

The proof of this theorem is based on Takayama's Theorem 2.2 and Lemma 4.4. Note that this bound is sharp, see Ref. 29(Proposition 3.6).

In Ref. 25 Herzog and Vladoiu describe some classes of square-free monomial ideals $I$ with constant depth function, i.e. their $\operatorname{dstab}(I) = 1$.

### 4.3. Depth of powers

Unlike the case of integral closures of an arbitrary monomial ideal and symbolic powers of square-free monomial ideals, the behavior of depth function of a monomial ideal is very bad until it reaches the stability. This is first observed by Herzog and Hibi[21]. A more complicated picture is given in Ref. 1. Very recently Ha et al. obtain the following surprising result, which completely solves the problem of the initial behavior of the depth function.

**Theorem 4.6.** [19(Theorem 6.7)] *Let $f(n)$ be any convergent non-negative numerical function. Then there exists a monomial ideal $I$ in $R = K[x_1, ..., x_r]$ such that $f(n) = \operatorname{depth}(R/I^n)$ for all $n$.*

This maybe is a reason why Problem 1 for $\overline{\text{dstab}}(I)$ is much more difficult than for $\text{dstab}(I)$. Example 3.1 in Ref. 28 constructed from Example 3.1 shows that the bound (if exists) must be at least of the order $O(d(I)^{r-2})$. However for the case of edge ideals there is a nice bound established by T. N. Trung[62]. Recall that a *leaf* in a graph $G$ is a vertex of degree one and a *leaf edge* is an edge incident with a leaf. For an example, in Figure 3, leafs are: the vertex 4 in $G_1$ and the vertex 2 in $G_2$, and edge leafs are: the edge $\{1,4\}$ in $G_1$ and $\{2,4\}$ in $G_2$. A connected graph is called a *tree* if it contains no cycles. We use the symbols $\varepsilon(G)$ and $\varepsilon_0(G)$ to denote the number of edges and leaf edges of $G$, respectively.

**Theorem 4.7.** [62(Theorem 4.6)] *Let $G_1, ..., G_s$ be all connected bipartite components and $G_{s+1}, ..., G_p$ all connected non-bipartite components of $G$. Let $2k_i$ be the maximal length of cycles of $G_i$, $i \le s$ ($k_i = 1$ if $G_i$ has no cycle), and let $2k_i - 1$ be the maximal length of odd cycles of $G_i$, $i > s$. Then*

$$\text{dstab}(I) \le r - \varepsilon_0(G) - \sum_{i=1}^{p} k_i + 1 \le r - p.$$

The proof is quite long and complicate. First, the author studies connected graphs. From properties of $\text{Ass}(R/I(G)^n)$ it turns out that $\text{depth}(R/I(G)^n) > 0$ for all $n \ge 1$ provided $G$ is bipartite (see Theorem 3.3) and $\text{depth}(R/I(G)^n) = 0$ for all $n \gg 0$ if $G$ is non-bipartite[9(Corollary 3.4)]. In the case of connected non-bipartite graphs, the proof intensively uses the construction developed in Ref. 9. When $G$ is a connected bipartite graph, thanks to Theorem 3.3, $I(G)^n = I(G)^{(n)}$ for all $n \ge 1$. Hence, one can apply Lemma 4.4 to describe $\Delta_\alpha(I(G)^n)$. A key point in Ref. 62 is to show that $\text{depth}(R/I(G)^n) = 1$ for all $n \gg 0$, see Ref. 62(Lemma 3.1 and 3.3). There the Takayama's Theorem 2.2 is used only to show that $\Delta_\alpha(I(G)^n))$ is disconnected, but Lemma 4.4 is very important. Some results in graph theory are also needed. Finally, the following result allows to reduce the problem to connected components of $G$.

**Theorem 4.8.** [62(Theorem 4.4)] *Keep the notation in Theorem 4.7. Then*

*(1)* $\min\{\text{depth}(R/I(G)^n)| \ n \ge 1\} = s$.
*(2)* $\text{dstab}(I(G)) = \min\{n \ge 1| \ \text{depth}(R/I(G)^n) = s\}$.
*(3)* $\text{dstab}(I(G)) = \sum_{i=1}^{p} \text{dstab}(R/I(G_i)) - p + 1$.

Below are some other partial solutions to Problem 1.

(1) Ref. 21(Corollary 3.4). The square-free Veronese ideal of degree $d$ in the variables $X_{i_1}, ..., X_{i_s}$, $s \le r$, is the ideal of $R$ generated by

all square-free monomials in $X_{i_1}, ..., X_{i_s}$ of degree $d$. Let $2 \leq d < n$ and let $I = I_{r,d}$ be the square-free Veronese ideal of degree $d$ in $r$ variables. Then

$$\text{depth}(R/I_{r,d}^n) = \max\{0, r - n(r - d) - 1\}.$$

In particular, $\text{dstab}(I_{r,d}) \leq \frac{r-1}{r-d}$.

(2) Ref. 21(Corollary 3.8). Let $P$ be a finite partially ordered set (called *poset* for short). A *poset ideal* of $P$ is a subset $I \subset P$ such that if $x \in I$, $y \in P$ and $y \leq x$, then $y \in P$. In particular, the empty set as well as $P$ itself is a poset ideal of $P$. Write $\mathcal{J}(P)$ for the finite poset which consists of all poset ideals of $P$, ordered by inclusion.

Let $P = \{p_1, ..., p_r\}$ be a finite poset and $S = K[X_1, ..., X_r, Y_1, ..., Y_r]$. Consider the square-free monomial ideal

$$H_P = (u_I := (\Pi_{p_i \in I} X_i)(\Pi_{p_i \notin I} Y_i) \mid I \in \mathcal{J}(P)) \subset S.$$

Then $\text{dstab}(H_P) = \text{rank}(P) + 1 \leq r$ and $\text{depth}(S/H_P) > \text{depth}(S/H_P^2) > \cdots > \text{depth}(S/H_P^{\text{rank}(\overline{P})+1}) = r - 1$, where $\text{rank}(P)$ is the so-called rank of $P$.

(3) Ref. 24(Theorem 4.1). Let $I$ be a polymatroidal ideal. Then $\text{dstab}(I) \leq \ell(I) \leq r$.

(4) Let $\mathcal{H} = (V, \mathcal{E})$ be a hypergraph. The incident matrix $M = (a_{ij})$ of $\mathcal{H}$ has $|V|$ rows and $|\mathcal{E}|$ columns such that $a_{ij} = 1$ if $i \in E_j$ and $a_{ij} = 0$ otherwise. A hypergraph $\mathcal{H}$ is said to be *unimodular* if every square submatrix of its incident matrix has determinant equal to $0, 1$ or $-1$. Then Ref. 20(Theorem 2.3 and Theorem 3.2) state that $\text{depth}(R/J(\mathcal{H})^n)$ is non-decreasing and $\text{dstab}(J(\mathcal{H})) \leq r$, provided $\mathcal{H}$ is an unimodular hypergraph.

### 4.4. *Cohen-Macaulay property of powers*

In this case, the depth gets its maximal value, and obtained results look nicest. Recall that a *matroid* is a simplicial complex $\Delta$ with the following property: If $F, G \in \Delta$ and $|F| > |G|$, then there is $a \in F \setminus G$ such that $G \cup \{a\} \in \Delta$.

**Theorem 4.9.** [58 (Theorem 1.1)] *Let $\Delta$ be a simplicial complex. Then the following conditions are equivalent:*

(1) $R/I_\Delta^{(n)}$ is Cohen-Macaulay for every $n \geq 1$;
(2) $R/I_\Delta^{(n)}$ is Cohen-Macaulay for some $n \geq 3$;
(3) $\Delta$ is a matroid.

The equivalence (1) $\Leftrightarrow$ (3) are independently proved by Minh and N.V. Trung[43(Theorem 3.5)] and by Varbaro[63(Theorem 2.1)]. In the approach of Ref. 43, Lemma 4.4 plays an important role. It allows them to use tool from linear programming to show that the Cohen-Macaulayness of all symbolic powers characterizes matroid complexes.

We say that $\Delta$ is a complete intersection if $I_\Delta$ is a complete intersection. This is equivalent to the property that no two minimal non-faces of $\Delta$ share a comon vertex. Since $R/I^n$ is Cohen-Macaulay if and only if $I^n = I^{(n)}$ and $R/I^{(n)}$ is Cohen-Macaulay, using Theorem 4.9, Terai and N. V. Trung can prove

**Theorem 4.10.** [58(Theorem 1.2)] *Let $\Delta$ be a simplicial complex. Then the following conditions are equivalent:*

(1) $R/I_\Delta^n$ is Cohen-Macaulay for every $n \geq 1$;
(2) $R/I_\Delta^n$ is Cohen-Macaulay for some $n \geq 3$;
(3) $\Delta$ is a complete intersection.

The idea for the proof of (2) $\Rightarrow$ (3) in both theorems above comes from the fact that matroid and complete intersection complexes can be characterized by properties of their links. The main technical result of Ref. 58 shows that a complex $\Delta$ with dim $\Delta \geq 2$ is a matroid if and only if it is connected and locally a matroid. A similar result on complete intersection are proved in Ref. 60(Theorem 1.5).

It is worth to mention that the Cohen-Macaulay property of the second (ordinary or symbolic) power of a Stanley-Resiner ideal is completely different and is still not completely understood, see Ref. 33,34,51,59,60.

The following result follows from Lemma 4.1 and Theorem 4.1.

**Theorem 4.11.** [32(Theorem 4.1)] *Let $I$ be a monomial ideal of $R$. The following conditions are equivalent*

(1) $R/\overline{I^n}$ is a Cohen-Macaulay ring for all $n \geq 1$,
(2) $R/\overline{I^n}$ is a Cohen-Macaulay ring for some $n \geq r(r^2 - 1)r^{r/2}(r - 1)^r d(I)^{(r-2)(r+1)}$,
(3) $I$ is an equimultiple ideal of $R$.

Following the idea of the proof of Theorem 4.10 we also get a similar result for the integral closures.

**Theorem 4.12.** [32](*Theorem 4.7*) *Let $\Delta$ be a simplicial complex. Then the following conditions are equivalent:*

(1) $R/\overline{I_\Delta^n}$ *is Cohen-Macaulay for every $n \geq 1$;*
(2) $R/\overline{I_\Delta^n}$ *is Cohen-Macaulay for some $n \geq 3$;*
(3) $I_\Delta$ *is a complete intersection;*
(4) $I_\Delta$ *is an equimultipe ideal.*

The property that the Cohen-Macaulayness of $R/\overline{I^n}$ for some $n \geq 3$ forces that for all $n$ is very specific for square-free monomial ideals. For an arbitrary monomial ideal, the picture is much more complicate, as shown by the following example.

**Example 4.2.** [32](Example 1) Let $d \geq 3$ and $I = (X^d, XY^{d-2}Z, Y^{d-1}Z) \subset R = K[X, Y, Z]$. Then

(1) $R/\overline{I^n}$ is Cohen-Macaulay for each $n = 1, \ldots, d-1$;
(2) $R/\overline{I^n}$ is not Cohen-Macaulay for any $n \geq d$.

Note that $\mathrm{ht}(I) = 2$ and $\ell(I) = 3$ in this case.

### *Acknowledgment*

This work is partially supported by the Project VAST.HTQT.NHAT.01/16-18. I would like to thank Prof. K. P. Shum for inviting me to be a keynote speaker at the Third International Congress in Algebras and Combinatorics (ICAC 2017), Hong Kong, where I had chance to give this lecture to a broad audience.

# References

1. S. Bandari, J. Herzog and T.Hibi, *Monomial ideals whose depth function has any given number of strict local maxima.* Ark. Mat. **52** (2014), 11-19.
2. A. Banerjee, S. Beyarslan and H. T. Ha, *Regularity of Edge Ideals and Their Powers*, Advances in alge- bra, 1752, Springer Proc. Math. Stat., 277, Springer, Cham, 2019.
3. S. Bayati, J. Herzog, and G. Rinaldo, *On the stable set of associated prime ideals of a monomial ideal.* Arch. Math. (Basel) **98** (2012), 213-217.
4. C. Bivia-Ausina, *The analytic spread of monomial ideals.* Comm. Algebra, **31** (2003), 3487-3496.
5. M. Brodmann, *Asymptotic stability of* Ass($M/I^n M$). Proc. Amer. Math. Soc. **74** (1979), no. 1, 16-18.
6. M. P. Brodmann, *The Asymptotic Nature of the Analytic Spread.* Math. Proc. Cambridge Philos Soc. **86** (1979), 35-39.
7. M. P. Brodmann and R. Y. Sharp, Local cohomology. An algebraic intro- duction with geometric applications. Second edition. Cambridge Studies in Advanced Mathematics, 136. Cambridge University Press, Cambridge, 2013.
8. W. Bruns and J. Herzog, Cohen-Macaulay rings. Cambridge Studies in Ad- vanced Mathematics, 39. Cambridge University Press, Cambridge, 1993.
9. J. Chen, S. Morey, A. Sung, *The stable set of associated primes of the ideal of a graph.* Rocky Mountain J. Math. **32**(2002), no. 1, 71-89.
10. A. Constantinescu and M. Varbaro, *Koszulness, Krull dimension, and other properties of graph-related algebras.* J. Algebr. Comb. **34** (2011), 375-400.
11. D. Eisenbud and C. Huneke, *Cohen-Macaulay Rees Algebras and their Spe- cializations.* J. Algebra **81** (1983) 202-224.
12. Ehrhart, E. *Sur un problème de géométrie diophantienne linéaire. I. Polyèdres et réseaux.* (French) J. Reine Angew. Math. **226** (1967), 1-29.
13. Ehrhart, E. *Sur un problème de géométrie diophantienne linéaire. II. Systémes diophantiens linéaires.* (French) J. Reine Angew. Math. **227** (1967), 25-49.
14. C. A. Francisco, H. T. Ha and A. Van Tuyl, *A conjecture on critical graphs and connections to the persistence of associated primes.* Discrete Math. **310** (2010), 2176-2182.
15. C. A. Francisco, H. T. Ha and A. Van Tuyl, *Colorings of hypergraphs, perfect graphs, and associated primes of powers of monomial ideals.* J. Algebra **331** (2011), 224-242.
16. L. Fouli, and S. Morey, *A lower bound for depths of powers of edge ideals.* J. Algebraic Combin. **42** (2015), 829-848.
17. D. H. Giang and L. T. Hoa, *On local cohomology of a tetrahedral curve.* Acta Math. Vietnam., **35** (2010), 229-241.
18. H. T. Ha and S. Morey *Embedded associated primes of powers of square-free monomial ideals.* J. Pure Appl. Algebra **214** (2010), 301-308.
19. H. T. Ha, D. H. Nguyen, N. V. Trung and T. N. Trung, *Symbolic powers of sums of ideals.* Preprint ArXiv:1702.01766.
20. N. T. Hang and T. N. Trung, *The behavior of depth functions of cover ideals*

*of unimodular hypergraphs.* Ark. Mat. **55** (2017), 89-104.

21. J. Herzog and T. Hibi, *The Depth of Powers of an Ideal.* J. Algebra **291** (2005), 534-550.

22. J. Herzog and T. Hibi, Monomial ideals. Graduate Texts in Mathematics, 260. Springer-Verlag London, Ltd., London, 2011.

23. J. Herzog, T. Hibi and N. V. Trung, *Symbolic powers of monomial ideals and vertex cover algebra.* Adv. Math. **210** (2007), 304-322.

24. J. Herzog and A. A. Qureshi, *Persistence and stability properties of powers of ideals.* J. Pure Appl. Algebra **219** (2015), 530-542.

25. J. Herzog and M. Vladoiu *Squarefree monomial ideals with constant depth function.* J. Pure and Appl. Algebra, **217** (2013), 1764-1772.

26. H. T. T. Hien and H. M. Lam, *Combinatorial characterizations of the saturation and the associated primes of the fourth power of edge ideals.* Acta Math. Vietnam. **40** (2015), 511-526.

27. H. T. T. Hien, H. M. Lam and N. V. Trung, *Saturation and associated primes of powers of edge ideals.* J. Algebra **439** (2015), 225-244.

28. L. T. Hoa, *Stability of associated primes of monomial ideals.* Vietnam. J. Math. **34** (2006), 473-487.

29. L. T. Hoa, K. Kimura, N. Terai and T. N. Trung, *Stability of depths of symbolic powers of Stanley-Reisner ideals.* J. Algebra **473** (2017), 307-323.

30. L. T. Hoa and T. N. Trung, *Partial Castelnuovo-Mumford regularities of sums and intersections of powers of monomial ideals.* Math. Proc. Cambridge Philos Soc. **149** (2010), 1-18.

31. L. T. Hoa and T. N. Trung, *Castelnuovo-Mumford regularity of symbolic powers of two-dimensional square-free monomial ideals.* J. Commut. Algebra **8** (2016), 77- 88.

32. L. T. Hoa and T. N. Trung, *Stability of Depth and Cohen-Macaulayness of Integral Closures of Powers of Monomial Ideals.* Acta Math. Vietnam. **43** (2018), 67-81.

33. D. T. Hoang, N. C. Minh and T. N. Trung *Combinatorial characterizations of the Cohen-Macaulayness of the second power of edge ideals.* J. Combin. Theory Ser. A **120** (2013), 1073-1086.

34. D. T. Hoang and T. N. Trung, Â characterization of triangle-free Gorenstein graphs and Cohen-Macaulayness of second powers of edge ideals. J. Algebraic Combin. **43** (2016), 325-338.

35. T. Kaiser, M. Stehlik and R. Skrekovski, *Replication in critical graphs and the persistence of monomial ideals.* J. Combin. Theory Ser. A **123** (2014), 239-251.

36. H. M. Lam and N. V. Trung, *Associated primes of powers of edge ideals and ear decompositions of graphs.* Trans. Amer. Math. Soc (to appear), arXiv:1506.01483.

37. F. S. Macaulay, *Some properties of enumeration in the theory of modular systems.* Proc. London Math. Soc. **26** (1927), 531-555.

38. J. Martinez-Bernal, S. Morey and R. H. Villarreal, *Associated primes of powers of edge ideals.* Collect. Math. **63** (2012), 361-374.

39. S. McAdam, Asymptotic prime divisors. Lecture Notes in Mathematics, 1023.

Springer-Verlag, Berlin, 1983.
40. S. McAdam and P. Eakin, *The asymptotic Ass.* J. Algebra **61** (1979), 71-81.
41. E. Miller and B. Sturmfels, Combinatorial commutative algebra. Graduate Texts in Mathematics, 227. Springer-Verlag, 2005.
42. N. C. Minh and N. V. Trung, *Cohen-Macaulayness of powers of two-dimensional square-free monomial ideals.* J. Algebra **322** (2009), 4219-4227.
43. N. C. Minh and N. V. Trung, *Cohen-Macaulayness of monomial ideals and symbolic powers of Stanley-Reisner ideals.* Adv. Math. **226** (2011), 1285-1306.
44. S. Morey, *Stability of associated primes and equality of ordinary and symbolic powers of ideals.* Comm. Algebra **27**(1999), 3221-3231.
45. S. Morey, *Depths of powers of the edge ideal of a tree.* Comm. Algebra **38** (2010), 4042-4055.
46. D. H. Nguyen and N. V. Trung, *Depth function of symbolic powers of homogeneous ideals*, Invent. Math. (to appear).
47. L. J. Ratliff, Jr., *On prime divisors of In, n large.* Michigan Math. J. **23** (1976), 337-352.
48. L. J. Ratliff, Jr., *On asymptotic prime divisors.* Pacific J. Math. **111** (1984), 395-413.
49. D. Rees, *The grade of an ideal or module.* Proc. Cambridge Philos. Soc. **53** (1957), 28-42.
50. L. Reid, L. G. Roberts and M. A. Vitulli, *Some results on normal homogeneous ideals.* Comm. Algebra **31** (2003), 4485-4506.
51. G. Rinaldo, N. Terai, and K.-I. Yoshida, *On the second powers of Stanley-Reisner ideals.* J. Commut. Algebra **3** (2011), 405-430.
52. A. Schrijver, Theory of linear and integer programming. John Wiley & Sons, 1998.
53. R. Y. Sharp, *Convergence of sequences of sets of associated primes*, Proc. Amer. Math. Soc. **131**(2003), 3009-3017.
54. A. Simis, W. V. Vasconcelos, R. H. Villarreal, *On the ideal theory of graphs.* J. Algebra **167** (1994), 389-416.
55. R. Stanley, *The upper bound conjecture and Cohen-Macaulay rings.* Studies in Appl. Math. **54** (1975), 135-142.
56. R. Stanley, Combinatorics and commutative algebra. Second edition. Progress in Mathematics, 41. Birkhuser, 1996.
57. Y. Takayama, *Combinatorial characterizations of generalized Cohen-Macaulay monomial ideals.* Bull. Math. Soc. Sci. Math. Roumanie (N.S.) **48** (2005), 327-344.
58. N. Terai and N. V. Trung, *Cohen-Macaulayness of large powers of Stanley-Reisner ideals.* Adv. Math. **229** (2012), 711-730.
59. N. Terai and N. V.Trung, *On the associated primes and the depth of the second power of square-free monomial ideals.* J. Pure Appl. Algebra **218** (2014), 1117-1129.
60. N. Terai and K. Yoshida, *Locally complete intersection Stanley-Reisner ideals.* Illinois J. Math. **53** (2009), 413-429.
61. T. N. Trung, *Stability of associated primes of integral closures of monomial ideals.* J. Combin. Ser. A., **116** (2009), 44-54.

62. T. N. Trung, *Stability of depths of powers of edge ideals.* J. Algebra **452** (2016), 157-187.

63. M. Varbaro, *Symbolic powers and matroids.* Proc. Amer. Math. Soc. **139** (2011), 2357-2366.

64. W. Vasconcelos, Integral closure: Rees algebras, multiplicities, algorithms. Springer Monographs in Mathematics. Springer, New York, 2005.

65. R. H. Villarreal, Monomial algebras. Second edition. Monographs and Research Notes in Mathematics. CRC Press, Boca Raton, FL, 2015. xviii+686 pp.

# Quantum Calculus
## An Introduction

Michel Jambu

*Professor Emeritus, Laboratoire Dieudonné,*
*University Cote d'Azur*
*Nice, France*

*jambu@unice.fr*

## 1. Introduction

Quantum physics appeared in the early beginning of the previous century. However, mathematicians, Euler, Gauss, Jacobi, created the algebraic tools more than one century before. Classical physics needs classical Newton-Leibnitz calculus, quantum physics needs quantum calculus. In classical physics, complete knowledge of the past gives knowledge of the future, dynamic variables are smoothly varying continuous values. In quantum physics, even if we have complete knowledge of the past, only probabilistic predictions of the future can be done. Quantum physics takes its name from the observation that certain quantities, most notably energy and angular momentum, are restricted to certain discrete or 'quantized' values under special circumstances. The in-between values are forbidden.

These two different approaches require different mathematical tools.

For example, the "classical" plane is the set of points of coordinates $(x, y)$ and the functions $(x, y) \longmapsto xy, (x, y) \longmapsto yx$ are the same. In the quantum plane, we have $yx = q.xy$ for some parameter $q$. Then the notion of point is not defined if $q \neq 1$.

In algebraic geometry, the algebra of functions on the space to the complex numbers, for example, are more important than points. In physics, particles may be viewed as less important than fields. In quantum mechanics, phase space appears through the algebra of observables, i.e. operators on a Hilbert space.

Let $V$ be a vector space and consider the algebra of polynomial functions on $V$, i.e. the symmetric algebra

$$SV^* = TV^*/\langle f \otimes g - g \otimes f \rangle$$

In quantum theory, we consider quantum vector spaces, basically invented by Manin, where the commutation relations are generated by the Yang-Baxter operator $R$. We get the $R$-symmetric quotient by the ideal generated by elements of the form

$$f \otimes g - (R^{-1})^*(f \otimes g)$$

where $(R^{-1})^*$ is the Yang-Baxter operator on the dual space of linear forms on $V$.

It gives the $R$-symmetric algebra $S_R V$ as $TV^*$ modulo the ideal generated by the elements $v \otimes w - R(v \otimes w)$.

Calculus is a theory in which there are a

$$\textit{differentiation } f \longmapsto f' \text{ and an } \textit{integration } f \longmapsto \int f \text{ with } \left( \int f \right)' = f$$

Quantum calculus is a version of calculus in which we do not take limits. Derivatives are differences and antiderivations are sums. So, a priori, quantum calculus seems much simpler than classical calculus. Classical calculus can be viewed as limit of quantum calculus.

Quantum calculus is of two types:

- $h$-**calculus**, where $h$ stands for Planck's constant.
- $q$-**calculus**, where $q$ stands for quantum with the relation $q = e^{ih}$.

When $h \to 0$ or $q \to 1$, we get the classical calculus.

Here are some applications of the quantum calculus:

- $q$-analogs find applications in a number of areas, including the study of fractals and multi-fractal measures, and expression for the entropy of chaotic dynamical systems.
- $q$-analogs also appear in the study of quantum groups and $q$-deformed superalgebras.
- Certain $q$-series identities have been useful in proving many combinatorial identities.
- It is known that Lie algebras play a special role in the characterization of special functions, and similarly, quantum groups play the analogous role in $q$-special functions.

- The topic of $q$-special functions is ubiquitous in mathematical physics, in particular, they play a fundamental role in statistical mechanics. In recent years, there are many new developments and applications of the $q$-calculus in mathematical physics, especially concerning special functions and quantum mechanics.

  Notice that the foundational work on $q$-special functions could be attributed to Euler (1748), more than one century before quantum mechanics.

In this paper, we will give an introduction to the quantum calculus and we only consider $q$-calculus where it is defined $q$-analogs of the objects of classical calculus. In [2], R.P. Stanley wrote: " *In general, a q-analogue of a mathematical object is an object depending on the variable q that "reduces" to (an admittedly vague term) the original object when we set $q = 1$. To be "satisfactory" q-analogue more is required, but there is no precise definition of what is meant by "satisfactory". ...*"

We will give some examples where there exists several $q$-analogs of a given mathematical expression.

## 2. q-Analogs of Numbers

### 2.1. q-*Analogs of non-negative Integers*

Let $q \in \mathbb{C}$ be a parameter. The theory of $q$-series is based on the following fact:

$$\lim_{q \to 1} \frac{1 - q^n}{1 - q} = n$$

For $n \in \mathbb{N}_{>0}$, $\dfrac{1 - q^n}{1 - q} = 1 + q + \cdots + q^{n-1}$ and for $q = 1$, we get $n$.

**Definition 2.1.** The **q-integer**, where $q \neq 0$, is

$$[n]_q := \frac{1 - q^n}{1 - q} = 1 + q + q^2 + \cdots + q^{n-1}, \ n \in \mathbb{N}_{>0}, \ [0]_q = 0$$

$[n]_q$ is a polynomial of degree $n - 1$ in the variable $q$. For example, $[1]_q = 1, [2]_q = 1 + q, [3]_q = 1 + q + q^2, \ldots$.

Notice that $[m + n]_q \neq [m]_q + [n]_q$ and $[m.n]_q \neq [m]_q.[n]_q$.

## 2.2. q-*Analogs of Factorial*

Let $n \in \mathbb{N}$ and $q \neq 0$. Then $n! = 1.2.\cdots.(n-1).n$.

**Definition 2.2.** The **q-factorial** function is

$$[n]_q! := [1]_q[2]_q \cdots [n-1]_q[n]_q = \prod_{k=1}^{n} [k]_q!, \quad \text{with } [0]_q! = 1$$

$$= \frac{(1-q)(1-q^2)\cdots(1-q^n)}{(1-q)(1-q)\cdots(1-q)} = \prod_{k=1}^{n} \frac{1-q^k}{1-q}$$

$$= (1+q)(1+q+q^2)\cdots(1+q+q^2+\cdots+q^{n-1})$$

Don't confuse $[n]_q!$ and $[n!]_q = 1 + q + q^2 + \cdots + q^{(n!-1)}$.

**Example 2.1.** $[4]_q! = (1+q)(1+q+q^2)(1+q+q^2+q^3) = 1 + 3q + 5q^2 + 6q^3 + 5q^4 + 3q^5 + q^6$
$[4!]_q = [24]_q = 1 + q + q^2 + \cdots + q^{23}$.

**Combinatorics.** Recall that a permutation of length $n$ is a bijection of the set $A$ of $n$ elements onto itself.

- $n!$ counts the number of permutations of length $n$.
- $[n]_q!$ counts permutations of length $n$ while keeping track of the number of inversions.

$$\sum_{\sigma \in \mathfrak{S}_n} q^{\text{inv}(\sigma)} = [n]_q! = (1+q)(1+q+q^2)\cdots(1+q+q^2+\cdots+q^{n-1})$$

where $\mathfrak{S}_n$ is the set of all permutations of length $n$ and $\text{inv}(\sigma)$ is the number of inversions of $\sigma \in \mathfrak{S}_n$. The permutation $\sigma = \begin{pmatrix} 1 & \cdots & i & \cdots & n \\ \sigma(1) & \cdots & \sigma(i) & \cdots & \sigma(n) \end{pmatrix}$ is denoted $(\sigma(1)\ldots\sigma(i)\ldots\sigma(n))$. An **inversion** of $\sigma$ is such that $i < j$ and $\sigma(i) > \sigma(j)$ and is denoted $(\sigma(i), \sigma(j))$.

**Example 2.2.**

- $\sigma = (\sigma(1)\sigma(2)\sigma(3)\sigma(4)\sigma(5)) = (23154)$ has three inversions $(2,1)$, $(3,1), (5,4)$. For example, $(3,1)$ is an inversion. $1 < 3$ and $\sigma(1) = 2 > 1 = \sigma(3)$. Then $q^{inv(23154)} = q^3$.
- The permutation $(12345)$ has no inversion. Then $q^{inv(12345)} = q^0 = 1$.
- The permutation $(54321)$ has 10 inversions. Then $q^{inv(54321)} = q^{10}$ is the highest degree term of $(1+q)(1+q+q^2)(1+q+q^2+q^3)(1+q+q^2+q^3+q^4)$.

## Proposition 2.1.

$$\sum_{\sigma \in \mathfrak{S}_n} q^{\text{inv}(\sigma)} = [n]_q! = (1+q)(1+q+q^2)\cdots(1+q+q^2+\cdots+q^{n-1})$$

$$= \prod_{i=1}^{n-1}(1+q+\cdots+q^i) = \prod_{i=1}^{n-1}\left(\sum_{k_i=0}^{i} q^{k_i}\right)$$

**Proof:** Let $inv(\sigma) = k = k_1 + k_2 + \cdots + k_{n-1} \le \dfrac{n(n-1)}{2}$ where $k_i \le i$, is the number of inversions $(i+1, j)$ of $\sigma$, i.e. $j < i+1$.
Then $q^{inv(\sigma)} = q^{k_1} q^{k_2} \cdots q^{k_{n-1}}$ where $q^{k_i}$ is the term of the factor $(1+q+q^2+\cdots+q^i)$.
Conversely, let $(k_1, k_2, \cdots, k_{n-1})$ where $k_i \le i$. Define $\sigma \in \mathfrak{S}_n$ as follows:

| | | | | | | | | |
|---|---|---|---|---|---|---|---|---|
| 12 | if | $k_1 = 0$ | | or | 21 | if | $k_1 = 1$ | |
| 123 | if | $k_1 = 0$ | $k_2 = 0$ | or | 213 | if | $k_1 = 1$ | $k_2 = 0$ |
| 132 | if | $k_1 = 0$ | $k_2 = 1$ | or | 231 | if | $k_1 = 1$ | $k_2 = 1$ |
| 312 | if | $k_1 = 0$ | $k_2 = 2$ | or | 321 | if | $k_1 = 1$ | $k_2 = 2$ ... |

then $inv(\sigma) = k_1 + k_2 + \cdots + k_n$ and $q^{inv(\sigma)} = q^{k_1} q^{k_2} \cdots q^{k_{n-1}}$, where $q^{k_i}$ is a term of $(1+q+q^2+\cdots+q^i)$. ∎

**Some notations.** The notations in quantum calculus are not standard, so here we choose the following notations.

$$(\alpha; q)_n := (1-\alpha)(1-\alpha q)\cdots(1-\alpha q^{n-1}), \quad (\alpha; q)_0 = 1, \quad \alpha \in \mathbb{C}$$

$$(q^a; q)_n = (1-q^a)(1-q^{a+1})\cdots(1-q^{a+n-1})$$

$$= \prod_{k=0}^{n-1}(1-q^{a+k}), \; a \ne 0, \quad (q^a; q)_0 = 1$$

$$(q; q)_n = (1-q)(1-q^2)\cdots(1-q^n)$$

$$= \prod_{k=1}^{n}(1-q^k) = (1-q)^n [n]_q!, \quad (q; q)_0 = 1.$$

and for $|q| < 1$, $(\alpha; q)_\infty = \lim_{n \to \infty}(\alpha; q)_n = \prod_{n \ge 0}(1-\alpha q^n)$.

**Infinite products.** Let $D$ be a region in the complex plane and $\mathcal{H}(D)$ denote the holomorphic functions. Let $f_n \in \mathcal{H}(D), n \in \mathbb{N}_{>0}$ where no $f_n$ is identically 0 in any component of $D$, and the series $\sum_{n>0} |1 - f_n(z)|$ converges uniformly on compact subsets of $D$. Then

$$f(z) = \prod_{n>0} f_n(z)$$

converges uniformly on compact subsets of $D$ and $f \in \mathcal{H}(D)$.

For $|q| < 1$, $(\alpha; q)_\infty = \lim_{n \to \infty} (\alpha; q)_n = \prod_{n \geq 0} (1 - \alpha q^n)$.

$$[n]_q! = \prod_{k=1}^{n} \frac{1 - q^k}{1 - q} = \frac{(q; q)_n}{(1 - q)^n} = \frac{(q; q)_\infty}{(1 - q)^n (q^{n+1}; q)_\infty}, \quad n \in \mathbb{N}, \ 0 < q < 1$$

## 2.3. q-*Analogs of complex numbers*

This is an extension of the $q$-analog of an integer to the complex numbers. For any complex number $a$, let $q^a = e^{a \log q}$, then

$$\lim_{q \to 1} \frac{1 - q^a}{1 - q} = a \quad \text{(L'Hôpital rule)}$$

The $q$-**analog** of $a \in \mathbb{C}$ is defined as

$$[a]_q := \frac{1 - q^a}{1 - q}$$

Notice that $[a]_q$ is a polynomial only if $a \in \mathbb{N}$.

## 3. q-Pochhammer Symbol

The **Pochhammer symbol** (or **shifted factorial**) is defined as

$$(a)_n := a(a + 1)(a + 2) \cdots (a + n - 1), a \in \mathbb{C}$$

Notice that if $a = 1$, $(1)_n = n!$, so the name "shifted factorial".

Recall that $(q^a; q)_n = \prod_{k=0}^{n-1} (1 - q^{a+k})$ and $[a + k]_q = \frac{1 - q^{a+k}}{1 - q}$.

So $\dfrac{(q^a; q)_n}{(1 - q)^n} = \displaystyle\prod_{k=0}^{n-1} \frac{1 - q^{a+k}}{1 - q}$ and $\displaystyle\lim_{q \to 1} \frac{(q^a; q)_n}{(1 - q)^n} = (a)_n$.

Hence, a **q-analog of the Pochhammer symbol** $(a)_n$, called **q-Pochhammer Symbol** is given by

$$\frac{(q^a; q)_n}{(1 - q)^n} = [a]_q [a + 1]_q [a + 2]_q \cdots [a + n - 1]_q$$

and for $a = 1$, then $\dfrac{(q; q)_n}{(1 - q)^n} = [n]_q!$.

The Pochhammer symbol is a generalization of the factorial function, the

$q$-Pochhammer symbol is a generalization of the $q$-factorial function. Unlike the ordinary Pochhammer symbol, the $q$-Pochhammer symbol can be extended to an infinite product.

## 4. q-Analogs and Vector spaces

### 4.1. *q-Factorial and Vector Spaces*

Another approach to $q$-analogs is through the vector spaces. The original object concerns finite sets, while its $q$-analog can be interpreted in terms of subspaces of finite-dimensional vector spaces over the finite field $\mathbb{F}_q$ where cardinality is replaced by dimension.

| Inclusion-Exclusion rule for sets | Dimension rule for vector spaces |
|---|---|
| $\mathrm{card}(A \cup B) + \mathrm{card}(A \cap B)$ $= \mathrm{card}(A) + \mathrm{card}(B)$ | $\dim(U+V) + \dim(U \cap V)$ $= \dim(U) + \dim(V)$ |

Let $q$ be a prime power, $m \in \mathbb{N}^*$, and $q = p^m \in \mathbb{N}^*$, A $n$-dimensional vector space over $\mathbb{F}_q$ is the $q$-analog of an $n$-element set.

- $n$ is the number of elements of the set $\{1, 2, \ldots, n\}$.
  Then $[n]_q$ is the number of 1-dimensional subspaces of $\mathbb{F}_q^n$.
- $n!$ is the number of sequences (strict inclusions)

$$\emptyset = S_0 \subset S_1 \subset \cdots \subset S_k \subset \cdots \subset S_{n-1} \subset S_n = S$$

  where $S_k$ has $k$ elements.
  Then $[n]_q!$ is given by the number of sequences of vector subspaces of $\mathbb{F}_q^n$.

**Proposition 4.1.** $[n]_q!$ *is the number of sequences (**flags**)*

$$0 = V_0 \subset V_1 \subset \cdots \subset V_k \subset \cdots \subset V_{n-1} \subset V_n = \mathbb{F}_q^n$$

*where $V_k$ is $k$-dimensional subspace of the $n$-dimensional vector space $\mathbb{F}_q^n$ over $\mathbb{F}_q$.*

**Proof:** $[n]_q = \dfrac{q^n - 1}{q - 1}$ choices for the subspace $V_1$ of $\mathbb{F}_q^n$: there are $q^n$ vectors, so $q^n - 1$ non zero vectors and each line is defined by any of $q - 1$ collinear vectors.
$\mathbb{F}_q^n / V_1$ is $(n-1)$-dimensional, so $[n-1]_q$ choices for $V_2, \ldots$ and the result follows by induction. $\blacksquare$

## 4.2. q-*Binomial Coefficients and Vector Spaces*

The number of subsets of $k$ elements in a set of $n$ elements is

$$\binom{n}{k} = \frac{n!}{k!(n-k)!}.$$

**Proposition 4.2.** *The number of $k$-dimensional subspaces of $\mathbb{F}_q^n$ is*

$$\begin{bmatrix} n \\ k \end{bmatrix}_q := \frac{[n]_q!}{[k]_q![n-k]_q!}.$$

**Proof:** Let $1 \le k \le n-1$. Denote this number by $C$ and let $N$ be the number of ordered $k$-tuples $(v_1, \ldots, v_k)$ of linearly independent vectors in $\mathbb{F}_q^n$.

Choose $(v_1, \ldots, v_k)$ by first choosing a $k$-dimensional subspace $V$ in $C$ ways, then choosing $v_1 \in V$ in $q^k - 1$ ways, $v_2 \in V$ in $q^k - q$ ways, and so on, yielding

$$N = C(q^k - 1)(q^k - q) \cdots (q^k - q^{k-1})$$

Hence

$$C = \frac{(q^n - 1)(q^n - q) \cdots (q^n - q^{k-1})}{(q^k - 1)(q^k - q) \cdots (q^k - q^{k-1})} = \frac{[n]_q!}{[k]_q![n-k]_q!} = \begin{bmatrix} n \\ k \end{bmatrix}_q$$

$\blacksquare$

Notice that $\begin{bmatrix} n \\ k \end{bmatrix}_q$ is a polynomial function in $q = p^m \in \mathbb{N}$.

For example,

$$\begin{bmatrix} 5 \\ 2 \end{bmatrix}_q = \frac{(q^5 - 1)(q^5 - q)}{(q^2 - 1)(q^2 - q)} = 1 + q + 2q^2 + 2q^3 + 2q^4 + q^5 + q^6 \in \mathbb{N}^*$$

is the number of 2-subspaces of $\mathbb{F}_q^5$.

## 4.3. q-*Analog of the Symmetric Groups*

A bijection $\sigma : \{1, 2, \ldots, n\} \longrightarrow \{1, 2, \ldots, n\}$ is called **permutation**.

The groups of permutations $\sigma$ called **symmetric groups**, are denoted $\mathfrak{S}_n, n \in \mathbb{N}_{>0}$.

A $q$-analog of $\sigma$ is a bijection $f : \mathbb{F}_q^n \longrightarrow \mathbb{F}_q^n$ preserving the structures of vector spaces, i.e. a linear isomorphism.

$$\mathbf{GL(n; q)} = \{\mathbf{f} : \mathbb{F}_{\mathbf{q}}^{\mathbf{n}} \longrightarrow \mathbb{F}_{\mathbf{q}}^{\mathbf{n}} \mid \mathbf{f} \text{ linear isomorphism}\} = \mathbf{q}\text{-analog of } \mathfrak{S}_{\mathbf{n}}$$

**Proposition 4.3.**

$$|GL(n;q)| = (q^n - 1)(q^n - q)(q^n - q^2) \cdots (q^n - q^{n-1}) = q^{\binom{n}{2}}(q-1)^n [n]_q!$$

**Proof:** It is clear that $|GL(n;q)| = (q^n - 1)(q^n - q)(q^n - q^2) \cdots (q^n - q^{n-1})$.

$$(q^n - 1)(q^n - q)(q^n - q^2) \cdots (q^n - q^{n-1})$$
$$= (q^n - 1)q(q^{n-1} - 1)q^2(q^{n-2} - 1) \cdots q^{n-1}(q-1)$$
$$= q^{\binom{n}{2}}(q^n - 1)(q^{n-1} - 1) \cdots (q - 1)$$
$$= q^{\binom{n}{2}}(q-1)^n [n]_q!$$
$$= q^{\binom{n}{2}}(q-1)^n \sum_{\sigma \in \mathfrak{S}_n} q^{inv(\sigma)}$$

∎

## 5. q-Weyl Calculus

The binomial formula is

$$(x+y)^n = \sum_{k=0}^{n} \frac{n!}{k!(n-k)!} x^k y^{n-k}$$

Let denote

$$\{(x+y)^n\}_q := \sum_{k=0}^{n} \frac{[n]_q!}{[k]_q! [n-k]_q!} x^k y^{n-k}$$

What is the meaning of $\{(x+y)^n\}_q$?

Let denote $\begin{bmatrix} n \\ k \end{bmatrix}_q = \dfrac{[n]_q!}{[k]_q! [n-k]_q!} = \begin{bmatrix} n \\ n-k \end{bmatrix}_q$ for $0 \le k \le n$.

When $q \to 1$, then $\begin{bmatrix} n \\ k \end{bmatrix}_q \longrightarrow \dfrac{n!}{k!(n-k)!} = \binom{n}{k}$.

**Example 5.1.** Recall that $[1]_q = 1$ and $\begin{bmatrix} 2 \\ 1 \end{bmatrix}_q = \dfrac{[2]_q}{[1]_q [1]_q} = [2]_q = \dfrac{1-q^2}{1-q} = 1 + q$.

$$\{(x+y)^2\}_q = \begin{bmatrix} 2 \\ 0 \end{bmatrix}_q x^2 + \begin{bmatrix} 2 \\ 1 \end{bmatrix}_q xy + \begin{bmatrix} 2 \\ 2 \end{bmatrix}_q y^2$$
$$= x^2 + (1+q)xy + y^2$$
$$= x^2 + xy + yx + y^2 \text{ where } yx = qxy$$

It is a **non-commutative** version of the formula $(x+y)^2$.
If $q = 1$, then we have the classical formula (commutativity $yx = xy$).

### Example 5.2. Lattice Paths

Write $(x + y)^2 = xx + xy + yx + yy$ and represent this formula with the paths going from $(0,0)$ to $(2 - k, k)$ by two horizontal or vertical steps, i.e. a **lattice path**, and more generally, from $(0,0)$ to $(n - k, k)$ when we consider $(x + y)^n$.

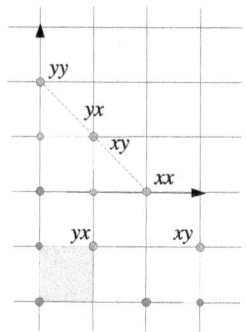

For example, there is only one path to go to $xx$, to $yy$, but two paths to go to the the third point on the diagonal line. The number of lattice paths from $(0,0)$ to $(m, n)$ is $\binom{m + n}{n}$.

Let $n = 4, k = 2$. The six distinct paths from $(0,0)$ to $(2, 2)$ are

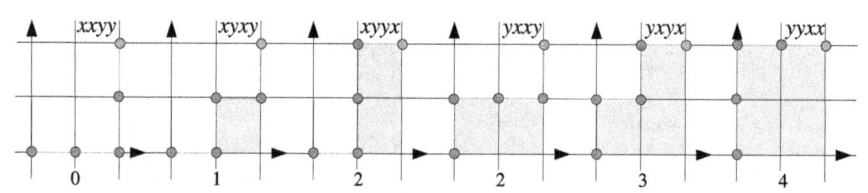

The areas below each path are **0, 1, 2, 2, 3, 4**.

$(x, y)$ is a $q$-**Weyl pair** if $yx = qxy, xq = qx, yq = qy$. So, we consider the algebra with two generators $x$ and $y$ and the relations $yx = qxy, xq = qx, yq = qy$.

**Example 5.3.** We compute $(x + y)^4 = (x + y)(x + y)(x + y)(x + y)$ in this algebra.

$$(x + y)^4 = xxxx + (xxxy + xxyx + xyxx + yxxx) + (xxyy + xyxy + xyyx$$
$$+ yxxy + yxyx + yyxx) + (xyyy + yxyy + yyxy + yyyx) + yyyy$$

$$xxyx = xxqxy = qxxxy, \qquad xyxx = xqxyx = qxxyx = qxxqxy = q^2xxxy,$$
$$yxxx = qxyxx = qxqxyx = q^2xxqxy = q^3xxxy$$

Then

$$xxxy + xxyx + xyxx + yxxx = (1 + q + q^2 + q^3)xxxy, \qquad xyxy = qxxyy,$$
$$xyyx = q^2xxyy, \qquad yxxy = q^2xxyy, \qquad yxyx = q^3xxyy,$$
$$yyxx = yqxyx = qyxyx = q^2xyyx = q^2xyqxy = q^3xyxy = q^3xqxyy = q^4xxyy,$$
$$yxyy = qxyyy, \qquad yyxy = yqxyy = q^2xyyy,$$
$$yyyx = yyqxy = q^2yxyy = q^3xyyy.$$

Then

$$xxxy + xxyx + xyxx + yxxx = (q^0 + q^1 + q^2 + q^3)xxxy$$

$$xxyy + xyxy + xyyx + yxxy + yxyx + yyxx$$
$$= (q^0 + q^1 + q^2 + q^2 + q^3 + q^4)xxyy$$
$$= (1 + q + 2q^2 + q^3 + q^4)xxyy$$
$$= (1 + q^2)(1 + q + q^2)xxyy.$$

$$xyyy + yxyy + yyxy + yyyx = (q^0 + q^1 + q^2 + q^3)xyyy.$$

$$1 + q + q^2 + q^3 = \frac{(1 - q^4)}{1 - q} = \frac{[4]_q!}{[1]_q![3]_q!}.$$

$$(1 + q^2)(1 + q + q^2) = \frac{(1 - q^4)(1 - q^3)(1 - q^2)(1 - q)}{(1 - q^2)(1 - q)(1 - q^2)(1 - q)} = \frac{[4]_q!}{[2]_q![2]_q!}.$$

$$(x + y)^4 = x^4 + \begin{bmatrix} 4 \\ 1 \end{bmatrix}_q x^3y + \begin{bmatrix} 4 \\ 2 \end{bmatrix}_q x^2y^2 + \begin{bmatrix} 4 \\ 3 \end{bmatrix}_q xy^3 + y^4$$

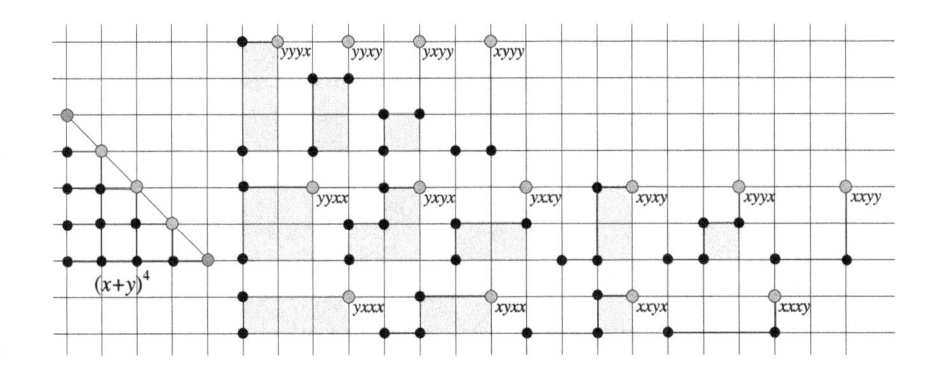

### 5.1. *What are* q-*Weyl pairs?*

What are the variables of $x$ and $y$ in a $q$-Weyl pair $(x, y)$?

- They can be viewed as **linear operators**.
  Let $\hat{x}$ and $J_q$ be the two linear operators on the space of polynomials

$$\hat{x}(f(x)) := xf(x) \quad \text{and} \quad J_q(f(x)) := f(qx)$$

  Then

$$J_q\hat{x}(f(x)) = J_q(xf(x)) = qxf(qx) = q\hat{x}J_q(f(x))$$

  so

$$J_q\hat{x} = q\hat{x}J_q \quad (yx = qxy)$$

  $J_q^2(f(x)) = J_q(f(qx)) = f(q^2x))$, and, more generally $J_q^n(f(x)) = f(q^n x))$.
  Notice that the operators $J$ and $\hat{x}$ don't commute.
- They can also be viewed as **matrices**.

$$x = \begin{pmatrix} 1 & 0 & 0 & \cdots & 0 \\ 0 & q & 0 & \cdots & 0 \\ 0 & 0 & q^2 & \cdots & 0 \\ \vdots & \vdots & \vdots & \ddots & \vdots \\ 0 & 0 & 0 & \cdots & q^{n-1} \end{pmatrix} \qquad y = \begin{pmatrix} 0 & 1 & 0 & \cdots & 0 \\ 0 & 0 & 1 & \cdots & 0 \\ \vdots & \vdots & \vdots & \ddots & \vdots \\ 0 & 0 & 0 & \cdots & 1 \\ 0 & 0 & 0 & \cdots & 0 \end{pmatrix}$$

We get the relations $qx = xq, qy = yq$ and $yx = qxy$.

## 6. q-Binomial Theorem

### 6.1. q-*Binomial Coefficients or Gaussian Polynomials*

The binomial formula is

$$(x + y)^n = \sum_{k=0}^{n} \frac{n!}{k!(n-k)!} x^{n-k} y^k$$

The binomial coefficients are given by the Pascal rule:

$$\frac{n!}{k!(n-k)!} = \frac{(n-1)!}{k!(n-k-1)!} + \frac{(n-1)!}{(k-1)!(n-k)!}$$

The $q$-binomial coefficients are also given by some Pascal rules as we will show. Let us give an explicit description of the formula, (i.e. of the $q$-binomial coefficients,)

$$\{(x+y)^n\}_q = \sum_{k=0}^n \begin{bmatrix} n \\ k \end{bmatrix}_q x^{n-k} y^k$$

in the $\mathbb{C}$-algebra generated by $x, y$ with Weyl relations, $(yx = qxy, qx = xq, qy = yq)$.

$$\{(x+y)^2\}_q = \begin{bmatrix} 2 \\ 0 \end{bmatrix}_q x^2 + \begin{bmatrix} 2 \\ 1 \end{bmatrix}_q xy + \begin{bmatrix} 2 \\ 2 \end{bmatrix}_q y^2 = x^2 + (1+q)xy + y^2.$$

$$\{(x+y)^{n+1}\}_q = \{(x+y)(x+y)^n\}_q = x\sum_{k=0}^n \begin{bmatrix} n \\ k \end{bmatrix}_q x^{n-k} y^k + y\sum_{k=0}^n \begin{bmatrix} n \\ k \end{bmatrix}_q x^{n-k} y^k$$

$$= \sum_{k=0}^n \begin{bmatrix} n \\ k \end{bmatrix}_q x^{n-k+1} y^k + \sum_{k=0}^n q^{n-k} \begin{bmatrix} n \\ k \end{bmatrix}_q x^{n-k} y^{k+1}$$

so $\quad \begin{bmatrix} n+1 \\ k \end{bmatrix}_q = \begin{bmatrix} n \\ k \end{bmatrix}_q + q^{n-k+1} \begin{bmatrix} n \\ k-1 \end{bmatrix}_q \qquad$ (**First q-Pascal rule**)

$$\{(x+y)^{n+1}\}_q = \{(x+y)^n (x+y)\}_q = \sum_{k=0}^n \begin{bmatrix} n \\ k \end{bmatrix}_q x^{n-k} y^k x + \sum_{k=0}^n \begin{bmatrix} n \\ k \end{bmatrix}_q x^{n-k} y^{k+1}$$

$$= \sum_{k=0}^n q^k \begin{bmatrix} n \\ k \end{bmatrix}_q x^{n-k+1} y^k + \sum_{k=0}^n \begin{bmatrix} n \\ k \end{bmatrix}_q x^{n-k} y^{k+1}$$

so

$$\begin{bmatrix} n+1 \\ k \end{bmatrix}_q = q^k \begin{bmatrix} n \\ k \end{bmatrix}_q + \begin{bmatrix} n \\ k-1 \end{bmatrix}_q \qquad \text{(\textbf{Second q-Pascal rule})}$$

**The q-binomial coefficient** $\begin{bmatrix} n \\ k \end{bmatrix}_q$ is a polynomial in $q$ of degree $k(n-k)$.

$$\begin{bmatrix} n \\ k \end{bmatrix}_q + q^{n-k+1} \begin{bmatrix} n \\ k-1 \end{bmatrix}_q = q^k \begin{bmatrix} n \\ k \end{bmatrix}_q + \begin{bmatrix} n \\ k-1 \end{bmatrix}_q$$

and

$$\begin{bmatrix} n \\ k \end{bmatrix}_q = \frac{1-q^{n-k+1}}{1-q^k} \begin{bmatrix} n \\ k-1 \end{bmatrix}_q$$

Finally $\begin{bmatrix} n \\ 0 \end{bmatrix}_q = \dfrac{[n]_q!}{[0]_q![n]_q!} = 1$

and if we assume that

$$\begin{bmatrix} n \\ k-1 \end{bmatrix}_q = \frac{[n]_q!}{[k-1]_q![n-k+1]_q!}$$

$$= \frac{(1-q)(1-q^2)\cdots(1-q^n)}{(1-q)(1-q^2)\cdots(1-q^{k-1})(1-q)(1-q^2)\cdots(1-q^{n-k+1})},$$

we get

$$\begin{bmatrix} n \\ k \end{bmatrix}_q = \frac{1-q^{n-k+1}}{1-q^k} \begin{bmatrix} n \\ k-1 \end{bmatrix}_q$$

$$= \frac{(1-q)(1-q^2)\cdots(1-q^n)}{(1-q)(1-q^2)\cdots(1-q^{k-1})(1-q^k)(1-q)(1-q^2)\cdots(1-q^{n-k})}$$

$$= \frac{[n]_q!}{[k]_q![n-k]_q!}$$

The **Gaussian** or **$q$-binomial coefficient** is defined for non-negative integers $n$ and $k$

$$\begin{bmatrix} n \\ k \end{bmatrix}_q = \frac{(q^n-1)(q^{n-1}-1)\cdots(q^{n-k+1}-1)}{(q^k-1)(q^{k-1}-1)\cdots(q-1)}$$

- Note that this is zero if $k > n$; so we may assume that $k \le n$.
- Note also that a Gaussian coefficient can be written out as a polynomial in $q$.

### 6.2. q-*analog of the Binomial Theorem*

The $q$-Pochhammer symbol $\dfrac{(q^a;q)_n}{(q-1)^n}$ corresponds to $(a)_n$ and $[n]_q! = \dfrac{(q;q)_n}{(q-1)^n}$ to $n! = (1)_n$.

So, the **q-analog** of the **binomial theorem** is:

$$(1-x)^{-a} = \sum_{n=0}^{\infty} \frac{(a)_n}{n!} x^n, \text{ for } |x| < 1$$

$$[(1-x)^{-a}]_q := \sum_{n=0}^{\infty} \frac{(q^a;q)_n}{(q;q)_n} x^n, \text{ for } |x| < 1$$

Recall that $(\alpha;q)_\infty = \lim_{n\to\infty} (\alpha;q)_n = \prod_{n=0}^{\infty} (1-\alpha q^n)$ when $|q| < 1$.

The $q$-analog of $(1-x)^{-a}$ is given by the infinite products

$$\frac{(q^a x; q)_\infty}{(x; q)_\infty}$$

Suppose $a = n \in \mathbb{N}$,

$$[(1-x)^{-a}]_q = [(1-x)^{-n}]_q = \frac{(q^n x; q)_\infty}{(x; q)_\infty} = \frac{(1-q^n x)(1-q^{n+1} x) \cdots}{(1-x)(1-qx) \cdots (1-q^n x) \cdots}$$

$$= \frac{1}{(1-x)(1-qx) \cdots (1-q^{n-1}x)}$$

The $q$-analog of $(1+x)^n$ is

$$[(1+x)^n]_q := (1+x)(1+qx) \cdots (1+q^{n-1}x) = (-x; q)_n$$

$$= \sum_{k=0}^{n} \frac{[n]_q!}{[k]_q![n-k]_q!} q^{k(k-1)/2} x^k$$

From the $q$-analog of $(1-x)^n$ we deduce the $q$-analog of the binomial formula

$$(x+y)^n = \sum_{k=0}^{n} \frac{n!}{k!(n-k)!} x^{n-k} y^k$$

$$[(x+y)^n]_q := (x+y)(x+qy) \cdots (x+q^{n-1}y) =$$

$$\sum_{k=0}^{n} \frac{[n]_q!}{[k]_q![n-k]_q!} q^{k(k-1)/2} x^{n-k} y^k$$

$$[(x+y)^{m+n}]_q = [(x+y)^m]_q \cdot [(x+q^m y)^n]_q$$

$$[(x+y)^{m+n}]_q = ((x+y)(x+qy) \cdots (x+q^{m-1}y))((x+q^m y) \cdots (x+q^{m+n-1}y)).$$
Don't confuse

- $[(x+y)^n]_q$ (algebra generated by $x$ and $y$ with the relation $xy = yx$)
- $\{(x+y)^n\}_q$ (algebra generated by $x$ and $y$ with the relations $qxy = yx, qx = xq, qy = yq$)

## 7. q-Derivative or Jackson's Derivative

The derivative of the function $f$ is, by definition

$$f'(x) = \lim_{h \to 0} \frac{f(x+h) - f(x)}{h}$$

Now, let us consider the following expression:

$$f'(x) = \lim_{h \to 0} \frac{f(qx) - f(x)}{qx - x}, \quad x \neq 0$$

which can be written as

$$f'(x) = \lim_{h \to 0} \frac{f(x + (q-1)x) - f(x)}{(q-1)x}, \quad (h = (q-1)x)$$

At the beginning of the XXth century, F.H. Jackson defined the **q-derivative operator**

**Definition 7.1.**

$$(D_q f)(x) = \frac{d_q(f(x))}{d_q(x)} := \begin{cases} \dfrac{f(qx) - f(x)}{qx - x} & \text{if } x \in (a,b) \setminus \{0\}, q \in \mathbb{C} \setminus \{1\} \\[2mm] \dfrac{df}{dx}(x) & \text{if } q = 1 \\[2mm] f'(0) & \text{if } x = 0 \text{ and } f'(0) \text{ defined.} \end{cases}$$

**Example 7.1.   q-Analog of $\dfrac{d}{dx}x^n$**

$$D_q x^n = \frac{(qx)^n - x^n}{qx - x} = \frac{q^n - 1}{q - 1} x^{n-1} = [n]_q x^{n-1}$$

which is the $q$-analog of

$$\frac{d}{dx}x^n = nx^{n-1}$$

**Properties of the q-Derivative Operator.** One can easily check that the $q$-derivative operator is linear:

$$D_q(f + g) = D_q(f) + D_q(g)$$

$$D_q(\lambda f) = \lambda D_q(f)$$

The product rule is slighlty modified but it approaches the usual product rule when $q$ goes to 1:

$$(D_q(fg))(x) = f(qx)(D_q)g(x) + (D_q f)(x)g(x)$$

## 8. q-Binomial Theorem

This series $\displaystyle\sum_{n=0}^{\infty} \frac{(q^a; q)_n}{(q; q)_n} x^n$ can be summed and it is shown that its evaluation in terms of infinite products is given by the **q-binomial theorem**

$$\sum_{n=0}^{\infty} \frac{(\alpha; q)_n}{(q; q)_n} x^n = \frac{(\alpha x; q)_\infty}{(x; q)_\infty}, \quad \text{for } |x| < 1, |q| < 1$$

Recall how we got the formula $(1-x)^{-a} = \sum_{n=0}^{\infty} \frac{(a)_n}{n!} x^n$, for $|x| < 1$

Denote $g_a(x) = \sum_{n=0}^{\infty} \frac{(a)_n}{n!} x^n$, then $g_a'(x) = \sum_{n=1}^{\infty} \frac{(a)_n}{(n-1)!} x^{n-1} = a g_{a+1}(x)$

$$g_a(x) - g_{a+1}(x) = \sum_{n=1}^{\infty} \frac{(a)_n - (a+1)_n}{n!} x^n = -x g_{a+1}(x)$$

$$g_a(x) = (1-x) g_{a+1}(x) \quad \text{and} \quad \frac{g_a'(x)}{g_a(x)} = \frac{a}{1-x}$$

so

$$g_a(x) = (1-x)^{-a}.$$

The proof of the $q$-binomial theorem is similar and requires $q$-difference operator.

$$\sum_{n=0}^{\infty} \frac{(\alpha; q)_n}{(q; q)_n} x^n = \frac{(\alpha x; q)_\infty}{(x; q)_\infty}$$

Consider $f_\alpha(x) = \sum_{n=0}^{\infty} \frac{(\alpha; q)_n}{(q; q)_n} x^n$ and apply the $q$-difference operator $D_q$ to both sides. Then

$$\frac{f_\alpha(x) - f_\alpha(qx)}{x} = \sum_{n=0}^{\infty} \frac{(\alpha; q)_n}{(q; q)_n} (1 - q^n) x^{n-1} = (1 - \alpha) f_{\alpha q}(x)$$

$$f_\alpha(x) - f_{\alpha q}(x) = \sum_{n=0}^{\infty} \frac{(\alpha q; q)_{n-1}}{(q; q)_n} (1 - \alpha - 1 + \alpha q^n) x^n = -\alpha x f_{\alpha q}(x)$$

$$f_\alpha(x) = (1 - \alpha x) f_{\alpha q}(x) \text{ and } f_\alpha(x) - f_\alpha(qx) = x(1 - \alpha) f_{\alpha q}(x),$$

$$\text{thus } f_\alpha(x) = \frac{1 - \alpha x}{1 - x} f_\alpha(qx)$$

Then iterate this relation $n$ times and let $n \to \infty$, to get

$$f_\alpha(x) = \frac{(\alpha x; q)_n}{(x; q)_n} f_\alpha(q^n x) = \frac{(\alpha x; q)_\infty}{(x; q)_\infty} f_\alpha(0) = \frac{(\alpha x; q)_\infty}{(x; q)_\infty}$$

## 9. q-Trigonometry

### 9.1. q-*Exponentials*

In the standard approach to the $q$-calculus, two $q$-exponential functions are used:

$$[e^z]_q = \sum_{n=0}^{\infty} \frac{z^n}{[n]_q!} \quad \text{and} \quad [E^z]_q = \sum_{n=0}^{\infty} \frac{z^n}{[\tilde{n}]_q!}$$

where $[\tilde{n}]_q = 1 + \dfrac{1}{q} + \dfrac{1}{q^2} + \cdots + \dfrac{1}{q^{n-1}}$ and $|q| > 1$ or $0 < q < 1$ and $|z| < |1 - q|^{-1}$.

Both exponential functions can be represented by infinite products

$$[e^z]_q = \prod_{n=0}^{\infty} (1 - (1-q)q^n z)^{-1} \quad \text{and} \quad [E^z]_q = \prod_{n=0}^{\infty} (1 + (1-q)q^n z)$$

The existence of two representations of $q$-exponential functions (infinite series and infinite product) is related to well known formulae for the usual exponential function

$$e^z = \sum_{n=0}^{\infty} \frac{z^n}{n!} = \lim_{n \to \infty} \left(1 + \frac{1}{z^n}\right)^n$$

The classical exponential function is also defined as solution of the differential equation

$$df(z) = f(z)dz$$

The $q$-analog $[e^z]_q$ is solution of the $q$-difference equation

$$D_q f(z) = f(z)$$

$$D_q [e^z]_q = [e^z]_q$$

and the $q$-analog $[E^z]_q$ is solution of the $q$-difference equation

$$D_q f(z) = f(qz)$$

$$D_q [E^z]_q = [E^{qz}]_q$$

**Another q-Exponential.** A few years ago, a new definition of the $q$-exponential, motivated by recent developments in the time scale calculus, was introduced. This $q$-exponential function maps the imaginary axis into the unit circle and the resulting $q$-trigonometric functions are bounded and satisfy the Pythagorean identity.

$$\mathcal{E}_q^z = [e^{\frac{z}{2}}]_q - [E^{\frac{z}{2}}]_q$$

$$\mathcal{E}_q^z = \prod_{n=0}^{\infty} \frac{1 + q^n (1-q)\frac{z}{2}}{1 - q^n (1-q)\frac{z}{2}}$$

**Properties**

$$\mathcal{E}_q^{-z} = (\mathcal{E}_q^z)^{-1}, \qquad |\mathcal{E}_q^{ix}| = 1$$

$$\mathcal{E}_q^z = \mathcal{E}_{\frac{1}{q}}^z \quad , \quad D_q \mathcal{E}_q^z = \frac{\mathcal{E}_q^z + \mathcal{E}_q^{qz}}{2}, \quad z \in \mathbb{C},\ x \in \mathbb{R}$$

## 9.2. q-*cosine and* q-*sine functions*

Two exponential functions of the quantum calculus generate two pairs of the $q$-trigonometric functions.

$$\sin_q x = \frac{[e^{ix}]_q - [e^{-ix}]_q}{2i} \qquad \mathrm{Sin}_q x = \frac{[E^{ix}]_q - [E^{-ix}]_q}{2i}$$

$$\cos_q x = \frac{[e^{ix}]_q + [e^{-ix}]_q}{2} \qquad \mathrm{Cos}_q x = \frac{[E^{ix}]_q + [E^{-ix}]_q}{2}$$

$$\sin_q x\ \mathrm{Sin}_q x + \cos_q x\ \mathrm{Cos}_q x = 1$$

$$\sin_q x\ \mathrm{Cos}_q x = \cos_q x\ \mathrm{Sin}_q x$$

$$D_q \cos_q x = -\sin_q x \qquad D_q \mathrm{Cos}_q x = -\mathrm{Sin}_q(qx)$$
$$D_q \sin_q x = \cos_q x \qquad D_q \mathrm{Sin}_q x = \mathrm{Cos}_q(qx)$$

We also have another $q$-cosine and $q$-sine functions.

$$\mathcal{S}in_q(x) = \frac{\mathcal{E}_q^{iz} - \mathcal{E}_q^{-iz}}{2i}$$

$$\mathcal{C}os_q(x) = \frac{\mathcal{E}_q^{iz} + \mathcal{E}_q^{-iz}}{2}$$

**Properties**

$$Cos_q^2(x) + Sin_q^2(x) = 1 \quad \text{(Pythagorean identity)}$$
$$D_q Sin_q(x) = \quad \langle Cos_q(x) \rangle$$
$$D_q Cos_q(x) = -\langle Sin_q(x) \rangle$$

where $\langle f(z) \rangle = \dfrac{f(z) + f(qz)}{2}$.

## 10. Jackson's Integral

While the discrete analogue of the derivative is a difference, we will see that the discrete analogue of the integral is a sum. $F(x)$ is a $q$-antiderivative of $f(x)$ if $D_q F(x) = f(x)$ where $D_q F(x) = \dfrac{F(qx) - F(x)}{(q-1)x}$.

In ordinary calculus, the uniqueness is up to adding a constant, in quantum calculus, it is more subtle.

$D_q F(x) = 0$ iff $F(qx) = F(x)$. However, if $F(x)$ is a formal power series, $F(qx) = F(x)$ implies $F = Cst$.

Let $J_q$ and $\hat{x}$ be the operators defined by $J_q(F(x)) = F(qx), \hat{x}(f(x)) = xf(x)$. Then, formally

$$\frac{1}{(q-1)\hat{x}}(J_q - 1)F(x) = f(x)$$

$$F(x) = \frac{1}{1 - J_q}\left((1-q)\hat{x}f(x)\right) = (1-q)\sum_{n=0}^{\infty} J_q^n(xf(x))$$

$$= (1-q)x\sum_{n=0}^{\infty} q^n f(q^n x)$$

Let $(D_q F) = f$ be a continuous function, $0 < q < 1$, the function $F$ is called the q-antiderivative or **q-integral** or **Jackson's integral** of $f$, and it is denoted

$$\int f(x)d_q(x) = (1-q)x\sum_{n\geq 0} f(q^n x)q^n$$

Suppose $0 < a < b$, the definite q-integral is defined as

$$\int_0^b f(x)d_q(x) = (1-q)b\sum_{n\geq 0} f(q^n b)q^n$$

$$\int_a^b f(x)d_q(x) = \int_0^a f(x)d_q(x) - \int_0^b f(x)d_q(x)$$

The $q$-integral is a Riemann-Stieltjes integral with respect to a step function having infinitely many points. The height of each strip is $f(aq^k)$, making the rectangle
$$(aq^k - aq^{k+1})f(aq^k) = a(1-q)q^k f(aq^k).$$
If $f$ is continuous in $[0, b]$, $\displaystyle\lim_{q\to 1}\int_0^b f(x)d_q x = \int_0^b f(x)dx$

**Proposition 10.1.** *Let $0 < q < 1$. Then, up to adding constants, any function has at most one $q$-antiderivative that is continuous.*

**Theorem 10.1 (Fundamental Theorem of Quantum Calculus).**
*If $F$ is an $q$-antiderivative of the function $f$, i.e. $D_q F = f$, continuous at $x = a$, then*
$$\int_a^b f(x)d_q x = F(b) - F(a)$$

**Properties**

- For any function $f$, $D_q\left(\displaystyle\int_a^x f(t)d_q t\right) = f(x)$.
- $q$-analog of Leibniz's rule:
$$D_q(f(x)g(x)) = f(x)D_q(g(x)) + g(x)D_q(f(x))$$

**Improper Jackson's Integral.** Notice that $\displaystyle\int_a^b f(x)d_q x$ does not converge when $b \to \infty$.

$$\int_{q^{j+1}}^{q^j} f(x)d_q x = \int_0^{q^j} f(x)d_q x - \int_0^{q^{j+1}} f(x)d_q x$$
$$= (1-q)\sum_{k=0}^{\infty} q^{j+k} f(q^{j+k}) - (1-q)\sum_{k=0}^{\infty} q^{j+k+1} f(q^{j+k+1})$$
$$= (1-q)q^j f(q^j)$$

So the definition of the improper $q$-integral,

$$\int_0^{\infty} f(x)d_q(x) = \sum_{j=-\infty}^{\infty} \int_{q^{j+1}}^{q^j} f(x)d_q x, \quad 0 < q < 1$$
$$\int_0^{\infty} f(x)d_q(x) = \sum_{j=-\infty}^{\infty} \int_{q^j}^{q^{j+1}} f(x)d_q x, \quad q > 1$$

**Examples of q-Integral**

$$\int (1+x)_q^n d_q x = \frac{(1+x)_q^{n+1}}{[n+1]_q} + C$$

$$\int [e]_q(x) d_q x = [e]_q(x) + C, \quad \int [E]_q(x) d_q x = [E]_q\left(\frac{x}{q}\right) + C$$

$$\int \frac{d_q x}{x} = \frac{q-1}{\ln q} \ln x + C$$

**q-Gamma Function**

Let us define $\displaystyle\int_0^{\infty/A} f(x) d_q x := (1-q) \sum_{k=-\infty}^{\infty} \frac{q^k}{A} f\left(\frac{q^k}{A}\right).$

$$\Gamma_q(x) := \int_0^{\infty/1-q} t^{x-1} [E]_q^{-qt} d_q t = \int_0^{\frac{1}{1-q}} t^{x-1} [E]_q^{-qt} d_q t$$

$$= \frac{(q;q)_\infty}{(q^x;q)_\infty} (1-q)^{1-x}, \quad 0 < |q| < 1$$

## 11. Jacobi Elliptic Functions & Theta Functions

There is a close connection between $q$-calculus and both elliptic functions and theta functions. The theory of elliptic functions has been studied for two centuries and it was initiated by Euler, Gauss, Abel and Jacobi. These functions play a fundamental role in the theory of elliptic curves and they were used by A. Wiles to prove Fermat's theorem.

### 11.1. *Elliptic Functions*

The rational functions on the Riemann sphere form a field.
The meromorphic functions on the torus are the doubly periodic elliptic functions on $\mathbb{C}$, and they form a field.
Let $L = \mathbb{Z}\omega_1 + \mathbb{Z}\omega_2 \subset \mathbb{C}$ be a lattice, i.e. an additive subgroup, $(\omega_1, \omega_2)$ linearly independent over $\mathbb{R}$. Let $\mathbb{C}/L$ be a torus.
An **elliptic function** with respect to $L$ is a meromorphic function $f \in \mathcal{M}(\mathbb{C}/L)$.

$$\mathbb{C}/L \xrightarrow{\sim} \mathbb{C}/(\mathbb{Z}\tau + \mathbb{Z}) \qquad z \mapsto z/\omega_2$$

where $\tau = \omega_1/\omega_2, Im(\tau) > 0$.

$$\mathbb{C}/(\mathbb{Z}\tau + \mathbb{Z}) \xrightarrow{\sim} \mathbb{C}/q^{\mathbb{Z}} \qquad z \mapsto t = e^{2\pi i z}, q = e^{2\pi i \tau}, 0 < |q| < 1$$

The exponential map $\mathbb{C}/\mathbb{Z} \xrightarrow{\sim} \mathbb{C}^*$, $z \mapsto e^{2\pi i z}$ replaces the additive periodicity with respect to $\tau$ by the multiplicative periodicity with respect to $q$. In terms of the multiplicative variable $t = e^{2\pi i z}$, an elliptic function $f \in \mathcal{M}(\mathbb{C}^*/q^{\mathbb{Z}})$ is a meromorphic function $f \in \mathcal{M}(\mathbb{C}^*)$ satisfying

$$f(qt) = f(t) \qquad t \in \mathbb{C}^*, |q| < 1$$

## 11.2. Theta Functions

A **theta function** with respect to a lattice $L$, is a holomorphic function $F(z)$ satisfying

$$F(z + u) = e^{a(u)z + b(u)} F(z) \qquad z \in \mathbb{C}, u \in L$$

where $a(u), b(u) \in \mathbb{C}$.
For example, $L = \mathbb{Z}\tau + \mathbb{Z}, q = e^{2i\pi\tau}, \tau > 0$, and $t = e^{2i\pi z}$ the function

$$A(z) = (1 - t) \prod_{n \geq 1} (1 - q^n t)(1 - q^n t^{-1})$$

is a theta function.
Theta functions are the elliptic analogs of the exponential function. The Jacobi elliptic functions are formed as quotients of the four **theta functions**

$$\theta_1(z; q) = 2 \sum_{n=0}^{\infty} (-1)^n q^{\left(\frac{n+1}{2}\right)^2} \sin(2n + 1)z$$

$$= -i q^{1/4} e^{iz} (q^2; q^2)_\infty (q^2 e^{2iz}; q^2)_\infty (e^{-2iz}; q^2)_\infty$$

$$\theta_2(z; q) = 2 \sum_{n=0}^{\infty} q^{\left(\frac{n+1}{2}\right)^2} \cos(2n + 1)z$$

$$= q^{1/4} e^{iz} (q^2; q^2)_\infty (-q e^{2iz}; q^2)_\infty (-e^{-2iz}; q^2)_\infty$$

$$\theta_3(z; q) = 1 + 2 \sum_{n=1}^{\infty} q^{n^2} \cos 2nz = (q^2; q^2)_\infty (-q e^{2iz}; q^2)_\infty (-q e^{-2iz}; q^2)_\infty$$

$$\theta_4(z; q) = 1 + 2 \sum_{n=1}^{\infty} (-1)^n q^{n^2} \cos 2nz = (q^2; q^2)_\infty (q e^{2iz}; q^2)_\infty (q e^{-2iz}; q^2)_\infty$$

When $z = 0$ we denote $\theta_i = \theta_i(0; q), i = 1, 2, 3, 4$. Then we obtain the following Jacobi elliptic functions where we set $k^{1/2} = \theta_2/\theta_3, K = (\pi/2)\theta_3^2, u = 2Kz/\pi$.

$$sn(u; k) = \frac{\theta_3 \theta_1 \left(u/\theta_3^2; q\right)}{\theta_2 \theta_4 \left(u/\theta_3^2; q\right)} = \frac{2\pi}{Kk} \sum_{n=0}^{\infty} \frac{q^{n+1/2} \sin(2n + 1)z}{1 - q^{2n+1}}$$

$$cn(u;k) = \frac{\theta_2 \theta_2 \left(u/\theta_3^2;q\right)}{\theta_3 \theta_4 \left(u/\theta_3^2;q\right)} = \frac{\pi}{2K} + \frac{2\pi}{K} \sum_{n=0}^{\infty} \frac{q^n \cos 2nz}{1 - q^{2n}}$$

$$dn(u;k) = \frac{\theta_4 \theta_3 \left(u/\theta_3^2;q\right)}{\theta_3 \theta_4 \left(u/\theta_3^2;q\right)} = \frac{2\pi}{Kk} \sum_{n=0}^{\infty} \frac{q^{n+1/2} \cos(2n+1)z}{1 - q^{2n+1}}$$

# References

1. V. Kac and P. Cheung, Quantum Calculus. Universitext, Springer, 2001.
2. R.P. Stanley, Enumerative Combinatorics, Cambridge University Press in April, 1997.
3. T. Ernst, A Comprehensive Treatment of $q$-Calculus, Birkhäuser, 2012.
4. T. Ernst, The History of $q$-Calculus and a new Method (Licentiate Thesis). U. U. D. M. Report 2000:16.

# Tree Breadth of the Continued Fractions Root Finding Method

K. Kalorkoti

*School of Informatics, University of Edinburgh, 10 Crichton Street*
*Edinburgh EH8 9LE, U.K.*

*kk@inf.ed.ac.uk*

In honour of the 80th birthday of Professor Leonid Arkad'evich Bokut'

The continued fractions algorithm for isolating the real roots of polynomials with certainty is one of the most efficient known and widely used. It can be viewed as exploring a tree created by a sequence of simple transformations. In this paper we produce new upper bounds for the breadth of the tree that are significantly smaller than the degree of the input polynomial. We also consider the expected breadth under a reasonable distribution and derive a bound, subject to a plausible assumption, that grows logarithmically with the degree and coefficient size.

*2010 Mathematics subject classification:* 13Y05.

*Keywords:* Polynomial roots, continued fractions, tree breadth.

## 1. Introduction

The continued fractions algorithm for isolating the (positive) real roots of polynomials is one of the most efficient known and widely used, excluding algorithms that are subject to numerical instability. Its ultimate source is a result of Vicent[21] which guarantees the termination of a sequence of transformations that enable us to explore the real roots in $(0, 1)$ and $(1, \infty)$. Alesina and Galuzzi[4] give a modern proof of the result as well as historical information. Rather than reiterate the extensive bibliography we will refer the reader to a few papers that cover it in great detail (in particular Krandick and Mehlhorn[15], Tsigaridas and Emiris[20]).

Krandick and Mehlhorn[15] analyse a version that uses homothetic transformations given by Collins and Akritas[7], obtaining new bounds on the number of recursive subdivisions. They also prove that the breadth of the recursion tree is bounded by the degree of the input square free polynomial.

By constrast, Tsigaridas and Emiris[20] give complexity and implementation results for the method without homothetic transformations (see below for further details on this method). Their analysis relies on a conjecture regarding the continued fractions expansions of non-quadratic algebraic irrationals (see p.161 of Ref. 20). They also discuss an earlier analysis by Akritas[2,3].

The original continued fractions method (without homothetic transformations) can be seen as exploring a tree of transformations with vertices labeled by polynomials and edges by one of two transformations. In this paper we provide bounds for the tree breadth of individual polynomials as well as for the average tree breadth (subject to an assumption) under a reasonable distribution.

For individual polynomials, Lemma 2.4 provides a bound in terms of the number of certain types of roots, which immediately implies that the degree is an upper bound. Theorem 3.2 provides a bound in terms of the degree and the size of coefficients which is good for cases of sequences where the coefficients have sufficiently controlled growth, the bound is even better if the number of non-zero coefficients is bounded.

For the average case we consider drawing uniformly at random square free polynomials of degree $n$ with integer coefficients in $[-B, B]$ where $B \geq 1$, we also consider primitive square free polynomials. Theorem 4.1 provides an upper bound that is logarithmic in $n$ and $B$ but this is subject to an assumption on the distribution of roots in the unit disc (discussed at the start of §4.1, see also §4.2).

## 2. Definitions

Throughout we consider non-zero polynomials in $z$ with coefficients from $\mathbb{R}$ (in practice the coefficients are from $\mathbb{Q}$ or, equivalently for root finding, from $\mathbb{Z}$ and we will assume this in some places). If

$$f = a_n z^n + a_{n-1} z^{n-1} + \cdots + a_0$$

then we define $\mathrm{vc}(f)$, the *variation of coefficients*, to be the number of sign changes in the sequence $a_n, a_{n-1}, \ldots, a_0$ of coefficients (ignoring as usual any occurrences of 0). The continued fractions method of finding the non-negative real roots of a square free $f$ is based on the transformations

$$f^T = f(z+1) = a_n(z+1)^n + a_{n-1}(z+1)^{n-1} + \cdots + a_0,$$

and

$$f^I = f(1/(z+1))(z+1)^n = a_n + a_{n-1}(z+1) + \cdots + a_0(z+1)^n,$$

where we have assumed that $n$ is the degree of $f$ (for the sake of completeness we can set $0^T = 0^I = 0$). For simplicity later on, if $f^T(0) = 0$ then we replace $f^T(z)$ by $f^T(z)/z$ and similarly for $f^I(z)$. This ensures that 0 is not a root of $f^T$ of of $f^I$ and so it has non-zero constant coefficient enabling us to express various bounds using this (the alternative is to use the trailing coefficient). For the same reasons we will assume that $a_0 \neq 0$.

In practice transformations of the first type are replaced by $z \mapsto z + c$ where $c$ is a good integer lower bound on the smallest positive root of $f$, see Ref. 2,3. Without this optimisation the algorithm is provably exponential, see Ref. 7. For the purposes of this paper it makes no difference to the overall situation so we will stay with the definition as given (our analysis holds unchanged no matter which version is used). Vincent's Theorem[21] shows that if $f$ is square free then for all sufficiently long sequences of transformations involving $T$ and $I$ we produce a polynomial with variation of coefficients either 0 or 1. This is false if the polynomial is not square free, for example if $f = (2z^2 - 1)^2$ then $f^{IIT} = f^I = z^4 + 4z^3 + 2z^2 - 4z + 1$. In connection with non-square free polynomials see the result cited by Ref. 20 as Theorem 5 of that paper. A polynomial $f$ is called *terminal* if and only if $\mathrm{vc}(f) \leq 1$.

In order to avoid confusion we note here that we will employ two forms of notation for transforms. The exponent form is compact and makes for ease of readability, but it must be borne in mind that it conforms to the algebraic notation of writing the argument on the left with order of application being as shown. Thus $f^{IT}$ means that we apply $I$ first and then $T$. In some situations it is more convenient to use the standard function notation with the argument on the right so that the order of application is the reverse of the written one. Thus if we wish to represent $z^{IT}$ in standard notation it becomes $T(I(z))$, which can be abbreviated to $TI(z)$; this will be significant when considering sequences of transformations as single Moebius transforms.

Given a polynomial $f$ we associate with it a binary tree, denoted by tree$(f)$, of the possible sequences of transformations of $f$ as follows. If $f$ is terminal the tree is empty otherwise the root is labeled with $f$. At any vertex $v$ labeled with the polynomial $g$ if $g^T$ not terminal then there is a right child with the edge labeled $T$ and the vertex labeled with $g^T$. Similarly if $g^I$ is not terminal there is a left child with the edge labeled $I$ and the vertex labeled with $g^I$.

We define the depth of a vertex to be the number of edges from the unique path to it starting at the root. Thus the root has depth 0 and its children, if any, have depth 1. The breath at depth $d$ of the tree is the number of vertices of depth $d$. The breadth of a tree is 0 if it is empty otherwise it is the maximum breadth over all depths (if this is unbounded we take the breadth to be $\infty$, this does not happen in our case). We denote the breadth of tree($f$) by br($f$).

We define $\mathbf{R}^+(f)$ to be the set of strictly positive real roots of $f$. We will say that $\alpha$ is complex to mean that $\alpha \in \mathbb{C} - \mathbb{R}$. Let $\mathbf{C}(f)$ denote the set of complex roots $\alpha$ of $f$ such that $\mathrm{Re}(\alpha) > 0$ and $\mathrm{Re}(\alpha) - \lfloor \mathrm{Re}(\alpha) \rfloor > |\alpha - \lfloor \mathrm{Re}(\alpha) \rfloor|^2$. We also define $\mathbf{c}(f)$ to be 0 if all complex roots $\alpha$ of $f$ that do not belong to $\mathbf{C}(f)$ have $\mathrm{Re}(\alpha) \le 0$ and to be 1 otherwise. Note that $\mathrm{vc}((z - \alpha)(z - \overline{\alpha})) = 0$ if and only if $\mathrm{Re}(\alpha) \le 0$ since $(z - \alpha)(z - \overline{\alpha}) = z^2 - 2\,\mathrm{Re}(\alpha)z + |\alpha|^2$. It follows that if $|\mathbf{R}^+(f)| + \mathbf{c}(f) = 0$ then $\mathrm{vc}(f) = 0$.

**Lemma 2.1.** *For all polynomials $f$*

> *(1) $\mathbf{C}(f)$ consists of all complex roots $\alpha$ of $f$ with $\mathrm{Re}(\alpha) > 0$ that are in the open disc $|z - \lfloor \mathrm{Re}(z) \rfloor - 1/2| < 1/2$.*
> *(2) If $\alpha \notin \mathbf{C}(f)$ is complex then $\mathrm{Re}(\alpha) \le |\alpha|^2$.*

**Proof.** Let $\alpha$ be a root of $f$ and set $m = \lfloor \mathrm{Re}(\alpha) \rfloor$ throughout. For the first part, we have $\alpha$ with $\mathrm{Re}(\alpha) > 0$ and $\alpha$ is in the given disc if and only if

$$(\alpha - m - 1/2)(\overline{\alpha} - m - 1/2) < 1/4$$
$$\Longleftrightarrow (\alpha - m)(\overline{\alpha} - m) - (\alpha - m)/2 - (\overline{\alpha} - m)/2 + 1/4 < 1/4$$
$$\Longleftrightarrow |\alpha - m|^2 - (\mathrm{Re}(\alpha) - m) < 0$$

which completes the proof.

For the second part, if $\mathrm{Re}(\alpha) \le 0$ the claim is trivial. So assume that $\mathrm{Re}(\alpha) > 0$. Since $\alpha \notin \mathbf{C}(f)$ we have

$$\mathrm{Re}(\alpha) - m \le |\alpha - m|^2$$
$$= (\alpha - m)(\overline{\alpha} - m)$$
$$= |\alpha|^2 - 2m\,\mathrm{Re}(\alpha) + m^2$$

Thus $\mathrm{Re}(\alpha) \le |\alpha|^2 - m(2\,\mathrm{Re}(\alpha) - m - 1)$. If $m = 0$ the claim follows immediately. Otherwise if $m > 0$ then $2\,\mathrm{Re}(\alpha) - m - 1 \ge 2m - m - 1 = m - 1 \ge 0$ and hence $\mathrm{Re}(\alpha) \le |\alpha|^2$. $\qquad\square$

We note that the preceding lemma remains true for the set $\widehat{\mathbf{C}}(f)$ that is defined is the same way as $\mathbf{C}(f)$ but without the condition that its members must be complex. This will be useful in §4.1.

Define $\mathbf{C}_m(f)$ to be the set of points of $\mathbf{C}(f)$ satisfying $|z - m - 1/2| < 1/2$ for $m = 0, 1, 2, \ldots$. Note that $\mathbf{C}_0(f)$ is the same as the set $C$ of Definition 19 in Ref. 15. It follows from Lemma 2.1 that $\mathbf{C}(f) = \cup_{m=0}^{\infty}\mathbf{C}_m(f)$. We define $\widehat{\mathbf{C}}_m(f)$ similarly. Clearly the sets $\mathbf{C}_i(f)$ are disjoint as are the sets $\widehat{\mathbf{C}}_i(f)$. The next Lemma follows from Lemma 2.6, the proof given here is more direct.

**Lemma 2.2.** $C(f^T) = \cup_{m=1}^{\infty}T^{-1}(\mathbf{C}_m(f))$ *and* $\mathbf{C}(f^I) \subseteq I^{-1}(\mathbf{C}_0(f))$. *In particular* $\mathbf{C}_m(f^T) = T^{-1}(\mathbf{C}_{m+1}(f))$. *Furthermore a root* $\alpha$ *of* $f$ *cannot both be mapped by* $T$ *a root in* $\mathbf{C}(f^T)$ *and mapped by* $I$ *to a root in* $\mathbf{C}(f^I)$.

**Proof.** We have for a complex $\beta$ that $\beta \in \mathbf{C}_m(f^T)$ if and only if

$$f^T(\beta) = 0 \,\&\, \mathrm{Re}(\beta) > 0 \,\&\, |\beta - m - 1/2| < 1/2$$
$$\Leftrightarrow f(\alpha) = 0 \,\&\, \beta = \alpha - 1 \,\&\, \mathrm{Re}(\alpha) > 1 \,\&\, |\alpha - 1 - m - 1/2| < 1/2$$
$$\Leftrightarrow \alpha \in \mathbf{C}_{m+1}(f) \,\&\, \beta = T^{-1}(\alpha).$$

This proves the claims regarding $f^T$. For the claim regarding $f^I$, suppose $\beta \in \mathbf{C}(f^I)$. Then $\beta = 1/\alpha - 1 = \overline{\alpha}/|\alpha|^2 - 1$ where $f(\alpha) = 0$. If $\alpha \notin \mathbf{C}_0(f)$ then one of three possibilities holds: (i) $\mathrm{Re}(\alpha) \leq 0$, (ii) $0 < \mathrm{Re}(\alpha) < 1$ and $\mathrm{Re}(\alpha) \leq |\alpha|^2$ or (iii) $\mathrm{Re}(\alpha) \geq 1$. If any of these conditions hold then $\mathrm{Re}(\beta) \leq 0$ which contradicts the assumption that $\beta \in \mathbf{C}(f^I)$.

The final claim follows from the description of $\mathbf{C}(f^T)$ and $\mathbf{C}(f^I)$ given by the first part. $\qquad\square$

Note that we cannot strengthen the second containment to an equality, e.g., if $f = 8z^2 - 4z + 1$ then $\mathbf{C}_0(f) = \{1/4 + i/4, 1/4 - i/4\}$. However $f^I = z^2 - 2z + 5$ and $\mathbf{C}(f^I) = \emptyset$ since the roots are $1 \pm 2i$. Indeed it follows from Lemma 2.6 that $\mathbf{C}_m(f^I)$ consists of all $I^{-1}(\alpha)$ where $\alpha$ is a root of $f$ that is in the open disc $|z - (2m+3)/2(m+1)(m+2)| < 1/2(m+1)(m+2)$. Just as above, the preceding lemma holds for $\widehat{\mathbf{C}}$ and $\widehat{\mathbf{C}}_m$.

**Lemma 2.3.** *For all polynomials* $f$ *we have*

    *(1)* If $\alpha \in \mathbf{R}^+(f)$ but $T^{-1}(\alpha) \notin \mathbf{R}^+(f^T)$ and $I^{-1}(\alpha) \notin \mathbf{R}^+(f^I)$ then $\alpha = 1$. Moreover $T(\mathbf{R}^+(f^T)) \cap I(\mathbf{R}^+(f^I)) = \emptyset$
    *(2)* $|\mathbf{R}^+(f)| \geq |\mathbf{R}^+(f^T)| + |\mathbf{R}^+(f^I)|$.
    *(3)* $|\mathbf{C}(f)| \geq |\mathbf{C}(f^T)| + |\mathbf{C}(f^I)|$.
    *(4)* $|\mathbf{C}_0(f)| \geq |\mathbf{C}(f^I)| + c(f^I)$ and $c(f) \geq c(f^T)$.

**Proof.** For the first claim note that

$$T^{-1}(1) = I^{-1}(1) = 0 \notin T^{-1}(\mathbf{R}^+(f^T)) \cup I^{-1}(\mathbf{R}^+(f^I)).$$

If $\alpha > 1$ then $T^{-1}(\alpha) = \alpha - 1 > 0$ while $I^{-1}(\alpha) = 1/\alpha - 1 < 0$ and so $T^{-1}(\alpha) \in \mathbf{R}^+(f^T)$ but $I^{-1}(\alpha) \notin \mathbf{R}^+(f^I)$. If $\alpha < 1$ the same argument shows that $T^{-1}(\alpha) \notin \mathbf{R}^+(f^T)$ but $I^{-1}(\alpha) \in \mathbf{R}^+(f^I)$.

The second claim follows from the first.

For the third claim, suppose a complex root $\alpha \notin \mathbf{C}(f)$ then under $I$ it is mapped to $1/\alpha - 1 = \overline{\alpha}/|\alpha|^2 - 1$. From Lemma 2.1 we have $\mathrm{Re}(\alpha) \leq |\alpha|^2$. It follows that $\mathrm{Re}(1/\alpha - 1) \leq 0$ and so $1/\alpha - 1 \notin \mathbf{C}(f^T) \cup \mathbf{C}(f^I)$. Under $T$ the root is mapped to $\alpha - 1$ and again this cannot be in $\mathbf{C}(f^T) \cup \mathbf{C}(f^I)$, e.g., by the second part of the preceding Lemma. It follows from Lemma 2.2 that if $\alpha \in \mathbf{C}(f)$ then we cannot have both $1/\alpha - 1 \in \mathbf{C}(f^I)$ and $\alpha - 1 \in \mathbf{C}(f^T)$. The claim now follows.

We now deal with the first part of the fourth claim. By Lemma 2.2 we have $\mathbf{C}(f^I) \subseteq I^{-1}(\mathbf{C}_0(f))$ and so if $\mathbf{c}(f^I) = 0$ the claim follows immediately. Suppose now that $\mathbf{c}(f^I) = 1$ so there is a complex root $\beta$ of $f^I$ with $\mathrm{Re}(\beta) > 0$ and $\beta \notin \mathbf{C}(f^I)$. It follows that $\beta = 1/\alpha - 1$ for some complex root $\alpha$ of $f$. Thus $\mathrm{Re}(\beta) = \mathrm{Re}(\alpha)/|\alpha|^2 - 1$ and so $\mathrm{Re}(\alpha) > |\alpha|^2$. It follows that $0 < \mathrm{Re}(\alpha) < 1$ and hence $\alpha \in \mathbf{C}_0(f)$. Since $\mathbf{C}(f^I) \subseteq I^{-1}(\mathbf{C}_0(f))$ and $\beta \notin \mathbf{C}_0(f^I)$ the containment is strict. The claim now follows. The second part of fourth claim follows immediately from Lemma 2.2. $\qquad \square$

**Lemma 2.4.** $br(f) \leq |\boldsymbol{R}^+(f)| + |\boldsymbol{C}(f)| + \boldsymbol{c}(f)$ *and hence* $br(f) \leq \deg(f)$.

**Proof.** If $\mathrm{tree}(f)$ is empty the claim is trivial. We now assume that $\mathrm{tree}(f)$ is not empty and use induction on the depth $d$. If $d = 0$ then $|\mathbf{R}^+(f)| + |\mathbf{C}(f)| + \mathbf{c}(f) \geq 1$ since $f$ is not terminal and so $br(f) = 1 \leq |\mathbf{R}^+(f)| + |\mathbf{C}(f)| + \mathbf{c}(f)$. Assume now that $d > 0$. By induction $br(f^T) \leq |\mathbf{R}^+(f^T)| + |\mathbf{C}(f^T)| + \mathbf{c}(f^T)$ and $br(f^I) \leq |\mathbf{R}^+(f^I)| + |\mathbf{C}(f^I)| + \mathbf{c}(f^I) \leq |\mathbf{R}^+(f^I)| + |\mathbf{C}_0(f)|$ by the fourth part of Lemma 2.3. Now, using Lemma 2.2 in the third line below and Lemma 2.3 in the last line,

$$
\begin{aligned}
br(f) &\leq br(f^T) + br(f^I) \\
&\leq |\mathbf{R}^+(f^T)| + |\mathbf{C}(f^T)| + \mathbf{c}(f^T) + |\mathbf{R}^+(f^I)| + |\mathbf{C}_0(f)| \\
&\leq |\mathbf{R}^+(f^T)| + |\mathbf{R}^+(f^I)| + \sum_{m=1}^{\infty} |\mathbf{C}_m(f)| + |\mathbf{C}_0(f)| + \mathbf{c}(f) \\
&\leq |\mathbf{R}^+(f)| + |\mathbf{C}(f)| + \mathbf{c}(f),
\end{aligned}
$$

which establishes the main inequality. The consequence is immediate since $f$ has at least $|\mathbf{R}^+(f)| + |\mathbf{C}(f)| + \mathbf{c}(f)$ distinct roots. $\qquad \square$

Krandick and Mehlhorn[15] prove that the breadth is bounded by the degree of the input square free polynomial for the variant method that also employs homothetic transformations (see their Theorem 29).

## 2.1. *Effect of Möbius Transforms*

For the reader's convenience we collect together some simple results on the effect of Möbius transforms on certain discs.

**Lemma 2.5.** *Assume that* $(ck - cr - a)(ck + cr - a) > 0$. *Then the set of values satisfying* $|(az + b)/(cz + d) - k| < r$ *is given by the open disc*

$$\left| z - \frac{(ad + bc)k - ab - cd(k^2 - r^2)}{(ck - cr - a)(ck + cr - a)} \right| < \frac{|ad - bc|r}{(ck - cr - a)(ck + cr - a)}$$

*In particular if* $k > r \geq 0$ *then under the transform* $z \mapsto 1/z$ *the disc* $|z - k| < r$ *goes to* $|z - k/(k^2 - r^2)| < r/(k^2 - r^2)$.

**Proof.** The given disc is the same as $|(az+b) - k(cz+d)| < r|cz+d|$. Setting $z = u+iv$ and squaring both sides, the disc is given by $((a-kc)u+b-kd)^2 + (a - kc)^2 v^2 - r^2(cu + d)^2 - r^2 c^2 v^2 < 0$, since $(ck - cr - a)(ck + cr - a) > 0$ the derived inequality is equivalent to $(u - C)^2 + v^2 - R^2 < 0$ where

$$C = \frac{(ad + bc)k - ab - cd(k^2 - r^2)}{(ck - cr - a)(ck + cr - a)} \text{ and } R = \frac{|ad - bc|r}{(ck - cr - a)(ck + cr - a)},$$

which is the claimed disc.

The rest follows by noting that $a = 0$, $b = 1$, $c = 1$, $d = 0$ so that the condition $(ck - cr - a)(ck + cr - a) > 0$ reduces to $(k - c)(k + c) > 0$ and since $k > c \geq 0$ it is satisfied. Substituting the values of $a, b, c, d$ into the general derived disc we obtain claimed disc. □

We note that if $M$ consists of a sequence of $T^{-1}$ and $I^{-1}$ transformations then $ad - bc = (-1)^s$ where $s$ is the number of occurrences of $I^{-1}$, cf. Theorem 8 of Collins and Krandick[6]. This can be shown by a straightforward induction. Thus, in this situation, the disc in the preceding lemma can be written as

$$\left| z - \frac{(ad + bc)k - ab - cd(k^2 - r^2)}{(ck - cr - a)(ck + cr - a)} \right| < \frac{r}{(ck - cr - a)(ck + cr - a)}$$

**Lemma 2.6.** *Let* $M(z)$ *be a Möbius transform composed of* $T$ *and* $I$ *and set* $M^{-1}(z) = (az + b)/(cz + d)$. *Then* $C_m(f^M)$ *consists of all* $M^{-1}(\alpha)$ *where* $\alpha$ *is a complex root of* $f$ *that is in the open disk*

$$\left| z - \frac{(ad + bc)(m + 1/2) - ab - cdm(m + 1)}{(cm - a)(c(m + 1) - a)} \right| < \frac{1}{2(cm - a)(c(m + 1) - a)}$$

**Proof.** The roots $\beta$ of $f^M$ are precisely all $\beta = M^{-1}(\alpha)$ where $\alpha$ is a root of $f$. Now a root $\beta$ belongs to $\mathbf{C}_m(f^M)$ if and only $|\beta - (m + 1/2)| < 1/2$, i.e., $|M^{-1}(\alpha) - (m + 1/2)| < 1/2$. Since $M$ is a composition of $T$ and $I$ we have $M(z) = (Az + B)/(Cz + D)$ with $A, B, C, D \geq 0$ and not both $C, D$ are 0 (similarly for $A, B$). We have $M^{-1}(z) = (Dz - B)/(-Cz + A)$. Thus the inequality $(ck - cr - a)(ck + cr - a) > 0$ of Lemma 2.5 becomes $(-C(m+1/2)+C/2-D)(-C(m+1/2)-C/2-D) > 0$ which is equivalent to $(Cm + D)(C(m + 1) + D) > 0$. Since $C, D \geq 0$ and at least one is non-zero the inequality holds. The result now follows from Lemma 2.5. $\square$

## 2.2. *Bounds on the number of roots*

This section summarises some well known results for the reader's convenience. The Mahler measure of a polynomial $f = a_n z^n + a_{n-1} z^{n-1} + \cdots + a_0$ with roots $\alpha_1, \ldots, \alpha_n$ is $M(f) = |a_n| \prod_{j=1}^n \max\{1, |\alpha_j|\}$. As is well known, Jensen's formula (see, e.g., Ahlfors[1]) yields the bound $M(f) \leq \sum_{j=0}^n |a_j|$. Suppose $\rho > 1$ and set $\epsilon(f, \rho) = |\{\alpha_j \mid 1 \leq j \leq n \ \& \ |\alpha_j| > \rho\}|$. Then

$$\epsilon(f, \rho) \leq \frac{1}{\log \rho} \left( \log \sum_{j=0}^n |a_j| - \log |a_n| \right). \tag{1}$$

This follows from the simple observation that $\rho^{\epsilon(f,\rho)} \leq |a_n|^{-1} M(f) \leq |a_n|^{-1} \sum_{j=0}^n |a_j|$. Now suppose that $\rho < 1$ and set $\mu(f, \rho) = |\{\alpha_j \mid 1 \leq j \leq n \ \& \ |\alpha_j| < \rho\}|$. It follows from (1) and the transformation $z \mapsto 1/z$ that

$$\mu(f, \rho) \leq \frac{1}{\log(1/\rho)} \left( \log \sum_{j=0}^n |a_j| - \log |a_0| \right). \tag{2}$$

Hughes and Nikeghbali[11] give the bounds

$$\epsilon(f, 1/(1 - \rho)) \leq \frac{1}{\rho} \left( \log \sum_{i=0}^n |a_i| - \log |a_n| \right),$$

$$\mu(f, 1 - \rho) \leq \frac{1}{\rho} \left( \log \sum_{i=0}^n |a_i| - \log |a_0| \right),$$

These are slightly weaker than the ones above since $\log(1/(1 - \rho)) = \rho + \rho^2/2 + \rho^3/3 + \cdots$.

It will be helpful to set

$$L(f) = \frac{1}{\sqrt{|a_0||a_n|}} \sum_{j=0}^n |a_j|.$$

**Lemma 2.7.** *Let* $f = a_n z^n + \cdots + a_0$ *where* $a_n a_0 \neq 0$ *and assume that* $\rho > 1$. *Then*

$$\epsilon(f, \rho) + \mu(f, 1/\rho) \leq \frac{2}{\log \rho} L(f).$$

**Proof.** Straightforward application of (1) and (2). □

**Lemma 2.8.** *Let* $S$ *be a set of univariate polynomials of degree* $n$ *with non-zero constant term and coefficients bounded in absolute value from above by* $B$. *Consider drawing polynomials at random from* $S$ *(using any probability distribution). Then*

$$\mathbb{E}[\epsilon(f, \rho)] \leq \frac{\log(n+1)B}{\log \rho},$$

*when* $\rho > 1$ *and*

$$\mathbb{E}[\mu(f, \rho)] \leq \frac{\log(n+1)B}{\log(1/\rho)},$$

*when* $\rho < 1$.

**Proof.** The inequalities are an immediate consequence of (1) and (2). □

**Lemma 2.9.** *Let* $f = a_n z^n + \cdots + a_0$ *be square free where* $f \in \mathbb{Z}[z]$ *and* $a_n a_0 \neq 0$, *then the number of positive integer roots of* $f$ *is at most* $1 + \log |a_0| / \log 2$.

**Proof.** Suppose $m_1, \ldots, m_r$ are integer roots of $f$, necessarily non-zero. A simple argument based on Gauss's Lemma for the content and primitive part of polynomials shows that $m_1 \cdots m_r \mid a_0$. Consider now the positive integer roots of $f$ other than 1 and assume there are $s$ of them. Since each root is an integer which is at least 2, it follows that $s \leq \log |a_0| / \log 2$. Taking the possibility that 1 is a root into account now yields the result. □

## 3. Bound on the tree breadth for a polynomial

We will use a famous result of Erdős and Turàn[9] in the form given by Rahman and Schmeisser[18], Theorem 11.6.4: denote by $n_f[\theta, \phi]$ the number of zeros in the sector $\{z \mid \theta \leq \arg z < \phi\}$, where $0 < \phi - \theta \leq 2\pi$. Then

$$\left| n_f[\theta, \phi] - \frac{\phi - \theta}{2\pi} n \right| \leq C \sqrt{n \log L(f)},$$

where $C = \sqrt{2\pi/G} < 2.62$ and $G = \sum_{m=0}^{\infty}(-1)^{m-1}(2m+1)^{-2}$ is Catalan's constant. The original paper[9] had $C = 16$, the improved constant is due to Ganelius[10].

We are interested in the number of roots $N_f(\phi)$ within a wedge defined by the angles $-\phi$, $\phi$. Note that in this we include those roots $\alpha$ with $\arg(\alpha) = \pm\phi$. It follows that

$$N_f(\phi) < \sqrt{\frac{2\pi}{G}}\sqrt{n \log L(f)} + \frac{\epsilon\phi}{\pi}n \tag{3}$$

for all $\epsilon > 1$.

Suppose now that $f$ has no more than $k$ non-zero coefficients. Proposition 11.2.4 of Ref. 18 states: for $0 < \phi - \theta < 2\pi$, denote by $n_f(\alpha, \beta)$ the number of zeros of $f$ in the sector $\{z \mid \theta < \arg z < \phi\}$ then

$$\left| n_f(\theta, \phi) - \frac{\phi - \theta}{2\pi}n \right| \leq k.$$

It follows that

$$N_f(\phi) < k + \frac{\epsilon\phi}{\pi}n \tag{4}$$

for all $\epsilon > 1$.

**Theorem 3.1.** *Let $f$ be any polynomial of degree $n$ with real coefficients and non-zero constant term. For $m \geq 0$ define $\phi = \arctan\left(1/2(m+2)\sqrt{(m+1)(m+3)}\right)$. Then*

*(1) the breadth of tree$(f)$ satisfies*

$$br(f) \leq \frac{2}{\log\left(\frac{m+2}{m+1}\right)} \log L(f) + \sqrt{\frac{2\pi}{G}}\sqrt{n \log L(f)} + \frac{\phi}{\pi}n + 4.$$

*Suppose $a$ satisfies $0 < a < 4$ and $b \geq 0$. Choose $m_0$ such that $a(m+b)^4 < 4(m+1)(m+2)^2(m+3)$ for all $m \geq m_0$. If $\sqrt{a}(m_0 + b)^2 \geq 2/\pi$, then*

$$br(f) < \frac{1}{\sqrt[4]{a}\log^2(2)\sqrt{\phi}} \log L(f) + \frac{\phi}{\pi}n + \sqrt{\frac{2\pi}{G}}\sqrt{n \log L(f)}$$
$$- \frac{1}{\log 2}\left(\frac{b}{\log 2} - 2\right)\log L(f) + 4.$$

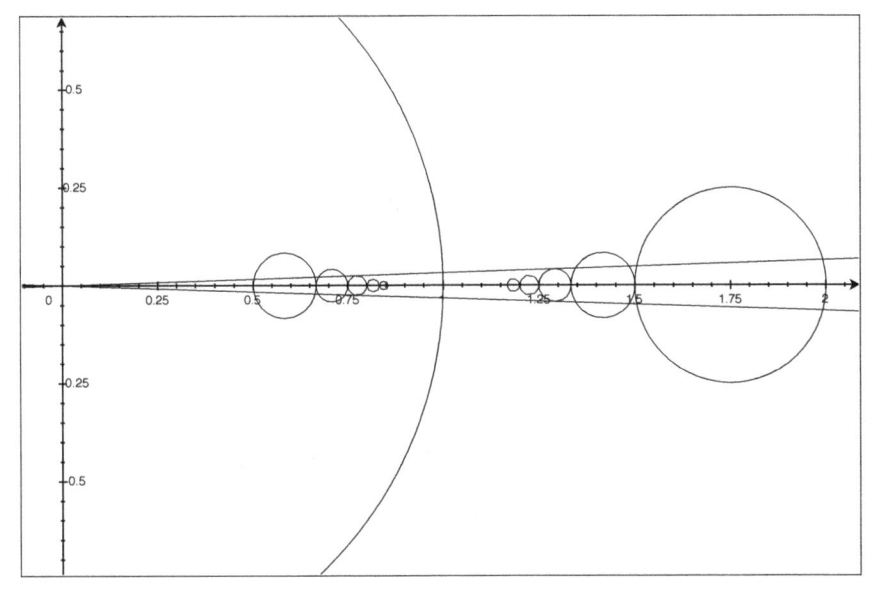

Fig. 1. The discs corresponding to $\mathbf{C}_m(f^M)$ for $M(z) = II(z) = z^{II}$ (left of unit circle) and $M(z) = IT(z) = z^{TI}$ (right of unit circle) and $m = 0, 1, 2, 3, 4$. Also shown is the wedge enclosing the discs corresponding to $\mathbf{C}_m(f^M)$ for $m \geq 2$

(2) *Suppose that the number of non-zero coefficients of $f$ is no more than $k$. Then*

$$br(f) \leq \frac{2}{\log\left(\frac{m+2}{m+1}\right)} \log L(f) + \frac{\epsilon\phi}{\pi} n + k + 4.$$

*Let $a$, $b$, $\phi$ and $m_0$ be as above. If $\sqrt{a}(m_0 + b)^2 \geq 2/\pi$, then*

$$br(f) < \frac{1}{\sqrt[4]{a}\log^2(2)\sqrt{\phi}} \log L(f) + \frac{\epsilon\phi}{\pi} n +$$

$$k - \frac{1}{\log 2}\left(\frac{b}{\log 2} - 2\right)\log L(f) + 4.$$

**Proof.** We look first at item 1. Consider tree$(f)$. If it is not complete at depth 2 (i.e., one of $f^I$, $f^T$, $f^{II}$, $f^{IT}$, $f^{TI}$, $f^{TT}$ is terminal) we may extend it artificially by adding the appropriate vertices and paths. It follows from Lemma 2.4 that br$(f)$ is bounded from above by the maximum of 2 and

$$\mathrm{br}(f^{II}) + \mathrm{br}(f^{IT}) + \mathrm{br}(f^{TI}) + \mathrm{br}(f^{TT})$$
$$\leq |\mathbf{C}(f^{II})| + |\mathbf{C}(f^{IT})| + |\mathbf{C}(f^{TI})| + |\mathbf{C}(f^{TT})| +$$
$$\mathbf{R}^+(f^{II}) + \mathbf{R}^+(f^{IT}) + \mathbf{R}^+(f^{TI}) + \mathbf{R}^+(f^{TT}) +$$
$$\mathbf{c}(f^{II}) + \mathbf{c}(f^{IT}) + \mathbf{c}(f^{TI}) + \mathbf{c}(f^{TT})$$
$$\leq |\mathbf{C}(f^{II})| + |\mathbf{C}(f^{IT})| + |\mathbf{C}(f^{TI})| +$$
$$|\mathbf{C}(f^{TT})| + \mathbf{R}^+(f) + 4$$

where the final line is justified by the second part of Lemma 2.3. Thus the preceding expression acts as an upper bound for $\mathrm{br}(f)$ in any case.

Using Lemmas 2.2 and 2.6 we have the following cases.

(1) $\mathbf{C}_m(f^{II})$ is a subset of the set of all $I^{-1}I^{-1}(\alpha)$ where $\alpha$ ranges over all roots $\alpha$ of $f$ in $\mathbf{C}(f)$ that lie in the open disc

$$\left| z - \frac{2m^2 + 8m + 7}{2(m+2)(m+3)} \right| < \frac{1}{2(m+2)(m+3)}.$$

(2) $\mathbf{C}_m(f^{IT})$ is a subset of the set of all $I^{-1}T^{-1}(\alpha)$ where $\alpha$ ranges over all roots $\alpha$ of $f$ in $\mathbf{C}(f)$ that lie in the open disc

$$\left| z - \frac{2m + 5}{2(m+2)(m+3)} \right| < \frac{1}{2(m+2)(m+3)}.$$

(3) $\mathbf{C}_m(f^{TI})$ is a subset of the set consists of all $T^{-1}I^{-1}(\alpha)$ where $\alpha$ ranges over all roots $\alpha$ of $f$ in $\mathbf{C}(f)$ that lie in the open disc

$$\left| z - \frac{2m^2 + 8m + 7}{2(m+1)(m+2)} \right| < \frac{1}{2(m+1)(m+2)}.$$

(4) $\mathbf{C}(f^{TT})$ consists of all $T^{-1}T^{-1}(\alpha)$ where $\alpha$ ranges over all roots $\alpha$ of $f$ in $\cup_{m=2}^{\infty}\mathbf{C}_m(f)$ that lie in any of the open discs

$$\left| z - \frac{2m + 5}{2} \right| < \frac{1}{2}.$$

Note that, by Lemma 2.5, cases 1 and 3 are dual to each other via the transformation $z \mapsto 1/z$ and similarly for 2 and 4; however this observation does not lead to any advantage over the proof below. We can deal with the second and fourth cases easily. For the second case the disc for $m$ is contained in a disc $D(0, 1/(m+2))$ and hence $\mathbf{C}(f^{IT})$ is contained in the disc $D(0, 1/2)$. For the fourth case, all the roots $\alpha$ in question satisfy $|\alpha| > 2$. Thus, by Lemma 2.7, $|\mathbf{C}(f^{IT})| + |\mathbf{C}(f^{TT})| \leq (2/\log 2) \log L(f)$.

For the other two cases we consider a wedge that encloses $C_m(f^{II})$. Suppose the lines $y = \pm \tan(\phi)x$, where $\phi > 0$, are tangent to the boundary of $\mathbf{C}_m(f^{II})$ with centre $c$ and radius $r$, see Figure 1. It follows easily that

$$\tan^2(\phi) = \frac{r^2}{c^2 - r^2} = \frac{1}{4(m+1)(m+2)^2(m+3)}. \tag{5}$$

Clearly such a wedge will enclose $\mathbf{C}_s(f^{II})$ for all $s \geq m$. Furthermore it does the same for $\mathbf{C}_s(f^{TI})$, since the ratio of the centre to the radius of the disc for $\mathbf{C}_m(f^{TI})$ is the same as that for $\mathbf{C}_m(f^{II})$ and $\tan^2(\alpha) = 1/(c^2/r^2 - 1)$.

The boundary of the disc corresponding to $\mathbf{C}_m(f^{II})$ crosses the $x$-axis at $x = (m+1)/(m+2)$, nearest to the origin, while the boundary for the disc corresponding to $\mathbf{C}_m(f^{TI})$ crosses the $x$-axis at $x = (m+2)/(m+1)$, furthest from the origin. Thus $|\cup_{i=0}^{m-1} \mathbf{C}_m(f^{II})|$ is bounded from above by the number of roots of $f$ in the interior of the disc with radius $\sigma = (m+1)/(m+2)$. Similarly $|\cup_{i=0}^{m-1} \mathbf{C}_m(f^{TI})|$ is bounded from above by the number of roots of $f$ in the exterior of the disc with radius $1/\sigma$. By Lemma 2.7 the number of such roots is bounded from above by $2\log L(f)/\log((m+2)/(m+1))$.

Since the wedge includes $\mathbf{R}^+(f)$ it follows from (3) that for all $\epsilon > 1$.

$$|\mathbf{R}^+(f)| + |\mathbf{C}(f^{II})| + |\mathbf{C}(f^{IT})| + |\mathbf{C}(f^{TI})| + |\mathbf{C}(f^{TT})| <$$

$$\frac{2}{\log\left(\frac{m+2}{m+1}\right)} \log L(f) + \sqrt{\frac{2\pi}{G}} \sqrt{n \log L(f)} + \frac{\epsilon\phi}{\pi} n.$$

The first claim of item 1 now follows from (5) and the observation regarding $\mathrm{br}(f)$ at the start of this proof by taking $\epsilon$ arbitrarily close to 1.

For the second claim, It is easily seen that $\tan(\phi) > \phi$ for $0 < \phi < \pi/2$ and so, by (5), we have $4(m+1)(m+2)^2(m+3) < 1/\phi^2$. For all large enough $m$, say $m \geq m_0 \geq 0$, we have $a(m+b)^4 \leq 4(m+1)(m+2)^2(m+3)$. Thus it suffices to have $m < 1/\sqrt[4]{a}\sqrt{\phi} - b$ in order to ensure the condition $4(m+1)(m+2)^2(m+3) < 1/\phi^2$. So for a given $\phi > 0$ the corresponding value of $m$ satisfies $m < 1/\sqrt[4]{a}\sqrt{\phi} - b$ . In order to have $m \geq m_0$ we need $\phi < 1/\sqrt{a}(m_0 + b)^2 \leq \pi/2$.

We claim that $\log((m+2)/(m+1)) \geq 2\log^2 2/(m+2\log 2)$. To see this consider the function $h(x) = \log((x+2)/(x+1)) - 2\log^2 2/(x+2\log 2)$ defined over the non-negative reals. The denominator of $dh/dx$ is positive and the numerator is $(c_1 x + c_0)x$ where $c_1 = 2\log^2 2 - 1 = -0.390\ldots$ and $c_0 = 6\log^2 2 - 4\log 2 = 0.110\ldots$. Hence $h$ is increasing for $0 \leq x \leq -c_0/c_1 \approx 2.817$ and decreasing for $x \geq -c_0/c_1$. Now $h(0) = 0$ and then becomes positive while $\lim_{x\to\infty} h(x) = 0$. Hence $h(x) \geq 0$ for all $x \geq 0$ and the claim is established. By the first part of item 1 we have

$$
\begin{aligned}
\mathrm{br}(f) &\leq 2\left(\frac{1}{\log 2} + \frac{m}{2\log^2 2}\right)\log L(f) + \sqrt{\frac{2\pi}{G}}\sqrt{n\log L(f)} + \frac{\phi}{\pi}n + 4 \\
&< \frac{2}{\log 2}\log L(f) + \frac{1}{\log^2 2}\left(\frac{1}{\sqrt[4]{a}\sqrt{\phi}} - b\right)\log L(f) + \\
&\quad \sqrt{\frac{2\pi}{G}}\sqrt{n\log L(f)} + \frac{\phi}{\pi}n + 4 \\
&= \frac{1}{\sqrt[4]{a}\log^2(2)\sqrt{\phi}}\log L(f) + \frac{\phi}{\pi}n + \sqrt{\frac{2\pi}{G}}\sqrt{n\log L(f)} - \\
&\quad \frac{1}{\log 2}\left(\frac{b}{\log 2} - 2\right)\log L(f) + 4
\end{aligned}
$$

The claims of item 2 follow by using (4) instead of (3) in the derivation above. $\qquad\square$

**Theorem 3.2.** *Let $f$ be any polynomial of degree $n$ with real coefficients and non-zero constant term. Suppose $a$ satisfies $0 < a < 4$ and $b \geq 0$. Choose $m_0$ such that $a(m+b)^4 \leq 4(m+1)(m+2)^2(m+3)$ and set $c = (b/\log 2 - 2)/\log 2$. If $\sqrt{a}(m_0 + b)^2 \geq 2/\pi$ and $\log L(f) \leq 2n\log^2(2)/\pi\sqrt{a}(m_0 + b)^3$ then*

$$
\begin{aligned}
\mathrm{br}(f) &< \frac{3}{\sqrt[3]{4\pi\sqrt{a}\log^4 2}}\sqrt[3]{n\log^2 L(f)} + \sqrt{\frac{2\pi}{G}}\sqrt{n\log L(f)} - c\log L(f) + 4 \\
&< \frac{2.11}{\sqrt[6]{a}}\sqrt[3]{n\log^2 L(f)} + 2.62\sqrt{n\log L(f)} - c\log L(f) + 4
\end{aligned}
$$

*Suppose that the number of non-zero coefficients of $f$ is no more than $k$. Then*

$$
\begin{aligned}
\mathrm{br}(f) &< \frac{3}{\sqrt[3]{4\pi\sqrt{a}\log^4 2}}\sqrt[3]{n\log^2 L(f)} - c\log L(f) + k + 4 \\
&< \frac{2.11}{\sqrt[6]{a}}\sqrt[3]{n\log^2 L(f)} - c\log L(f) + k + 4
\end{aligned}
$$

**Proof.** The function $g(\phi) = u/\sqrt{\phi} + v\phi$ has its minimum at $\phi = (u/2v)^{2/3}$. As seen in the proof of Theorem 3.1 the r.h.s. needs to be bounded from above by $1/\sqrt{a}(m_0 + b)^2$ and this holds provided that $u < 2v/\sqrt[4]{a^3}(m_0 + b)^3$. Thus the minimum value of $g(\phi)$ is $(2u^2v)^{1/3} + (u^2v/4)^{1/3} = 3(u^2v/4)^{1/3}$. Setting $u = \log L(f)/\sqrt[4]{a}\log^2 2$ and $v = n/\pi$

the inequality is $\log L(f) < 2n \log^2 2/\pi \sqrt{a}(m_0 + b)^3$ and the value of $g(\phi)$ is $3(n \log^2 L(f)/4\pi \sqrt{a} \log^4 2)^{1/3}$. Substituting into the second inequality of item 1 of Theorem 3.1 yields the desired bound. The second bound follows by substituting into the second inequality of item 2 of Theorem 3.1. $\square$

The assumption that $\log L(f) \leq 2n \log^2(2)/\pi \sqrt{a}(m_0 + b)^3$ ensures that the bounds of the preceding lemmas are strictly positive; a simple argument shows that under the assumption

$$c \log L(f) < 3 \sqrt[3]{n \log^2 L(f)} / \sqrt[3]{4\pi \sqrt{a} \log^4 2}.$$

If the condition on $L(f)$ is satisfied then, since $\sqrt{a}(m_0 + b)^2 \geq 2/\pi$, we have $\log L(f) \leq n \log^2(2)/(m_0 + b)$ . Thus $L(f) \leq 2^{n \log(2)/(m_0+b)}$. If $f$ has coefficients bounded in absolute value by $B$ (the situation of the next section) than $\log L(f) \leq \log(n + 1)B$ and the condition on $L(f)$ is satisfied if $B \leq 2^{n \log(2)/(m_0+b)}/(n + 1)$, i.e., for all polynomials of sufficiently high degree. As a simple illustration we may take $a = 9/8$, $b = 3$ and $m_0 = 1$. The condition $\sqrt{a}(m_0 + b)^2 \geq 2/\pi$ is satisfied while the condition on $L(f)$ is satisfied if $\log L(f) \leq 0.05n$. The second bound of the preceding lemma becomes

$$\mathrm{br}(f) < 2.07 \sqrt[3]{n \log^2 L(f)} - 3.35 \log L(f) + k + 4.$$

## 4. Bound on the expected breadth of the transformation tree

Let $B > 0$ be an integer and let $P(n, B)$ be the set of non-zero polynomials from $\mathbb{Z}[z]$ of degree $n$ with non-zero constant term and coefficients from $[-B, B]$. It will be convenient to set $M = 2B + 1$ in various places below. We set

$$P^+(n, B) = \{f \in P(n, B) \mid f \text{ is square free and}$$
$$\text{has positive leading coefficient}\},$$
$$P^\circ(n, B) = \{f \in P^+(n, B) \mid f \text{ is primitive}\},$$
$$\mathrm{sq}(n, B) = |P^+(n, B)|,$$
$$\mathrm{psq}(n, B) = |P^\circ(n, B)|.$$

This situation is typical of experiments to measure statistics for algorithms on polynomials, we generate them by choosing the coefficients uniformly at random from the chosen range $[-B, B]$ and accept the polynomial if it satisfies any further conditions (for efficiency we draw the leading coefficient from $[1, B]$). The requirement in $P^+(n, B)$ that $f$ has positive leading co-efficient and $f(0) \neq 0$ is clearly reasonable in the context of root finding. Insisting that such a polynomial is primitive is also well motivated, but note that the procedure of first generating coefficients does not guarantee uniform sampling, for example all 8 members of $P^+(1, 2)$ are square free and all but $2z \pm 2$ are primitive. We can correct this as follows. For a polynomial $f$ define $|f|$ to be the maximum absolute value of its coefficients. A generated polynomial $f$ is accepted with probability $1/c$ where $c = \lfloor B/|\operatorname{pp}(f)| \rfloor$ and $\operatorname{pp}(f)$ is the primitive part of $f$. The reason for this is that a primitive polynomial $f$ of degree $n$ and non-zero constant term accounts for $\lfloor B/|f| \rfloor$ members of $P^+(n, B)$.

**Lemma 4.1.** *Let $p$ be any prime that divides $M = 2B + 1$. Then*

*(1) $\operatorname{sq}(n, B) \geq (1 - 1/p)^2 B(2B + 1)^n$.*
*(2) $\operatorname{psq}(n, B) \geq (1 - 1/p)^2 (2B + 1)^n$.*

**Proof.** It suffices to show that $|P(n, B)| \geq (1 - 1/p)^2 (M - 1)M^n$ since clearly $\operatorname{sq}(n, B) = |P(n, B)|/2$.

Define $\phi : P(n, B) \to \mathbb{F}_p[z]$ by $f \mapsto f \bmod p$, where $\mathbb{F}_p$ is the field of integers modulo $p$. If $g \in \mathbb{F}_p[z]$ is square free and has degree $n$ then the same holds for every $f \in \phi^{-1}(g)$. For if $f = h^2 q$ where $h$ is not a constant then $p \nmid \operatorname{lc}(h)$ since $p \nmid \operatorname{lc}(f)$ and so $\phi(h)$ is not a constant. Hence $\phi(f) = \phi(h)^2 \phi(q)$ is not square free, which is a contradiction.

Clearly the image of $\phi$ contains all polynomials in $\mathbb{F}_p[z]$ of degree $n$. Furthermore $|\phi^{-1}(g)| = (M/p)^{n+1}$ for all $g \in \mathbb{F}_p[z]$ with $n = \deg g$ and $g(0) \neq 0$. For if $g(0) \neq 0$ then $f(0) \neq 0$, for all $f \in \phi^{-1}(g)$. The cardinality claim follows from the fact that if $r$ is a residue modulo $p$ then there are $M/p$ numbers in $[-B, B]$, equivalently in $[0, 2B]$, with residue $r$; namely $r + qp$ for $q = 0, 1, \ldots M/p - 1$. Similarly $|\phi^{-1}(g)| = (M/p)^n(M/p - 1)$ for all $g \in \mathbb{F}_p[z]$ with $n = \deg g$ and $g(0) = 0$. For if $f \in \phi^{-1}(g)$ then $f(0) \neq 0$ provided the constant term is non-zero and this is so for $(M/p)^n(M/p - 1)$ members of $\phi^{-1}(g)$.

As is well known, the number of monic square free polynomials in $\mathbb{F}_p[z]$ of degree $n \geq 2$ is $p^n - p^{n-1}$, e.g., see Mignotte[16]. The number of these that have 0 as a root is the number of square free monic polynomials of degree $n - 1$. Hence, for $n \geq 3$, the number of monic square free polynomials that do not have 0 as a root is $(p^n - p^{n-1}) - (p^{n-1} - p^{n-2}) = (1 - 1/p)^2 p^n$. For $n = 2$ the number that have 0 as a root is the number of those of form $z^2 + az$ with $a \neq 0$, i.e., there are $p - 1$ of them. Hence the number of

monic square free polynomials of degree 2 that do not have 0 as a root
is $(p^2 - p) - (p - 1) = (1 - 1/p)^2 p^2$. It follows that the formula $(1 - 1/p)^2 p^n$ applies to the case $n = 2$ as well. Hence the number of square
free such polynomials is $(p - 1)(1 - 1/p)^2 p^n$. These correspond to $(p - 1)(1 - 1/p)^2 p^n (M/p)^{n+1} = (1 - 1/p)^3 M^{n+1}$ members of $P(n, B)$. Similarly
the number of square free polynomials in $\mathbb{F}_p[z]$ of degree $n \geq 2$ with 0
as a root is $(p - 1)(p^{n-1} - p^{n-2})$. These correspond to $(p - 1)(p^{n-1} - p^{n-2})(M/p)^n (M/p - 1) = (1 - 1/p)^2 M^n (M/p - 1)$ members of $P(n, B)$.
Thus the number polynomials in $P(n, B)$ with $n \geq 2$ is at least

$$(1 - 1/p)^3 M^{n+1} + (1 - 1/p)^2 M^n (M/p - 1) = (1 - 1/p)^2 (M - 1)M^n.$$

For $n = 1$ we have $|P(1, B)| = (M - 1)^2$. Since $p \leq M$ we have $(1 - 1/p)^2 (M - 1)M < (M - 1)^2$. For $n = 0$ we have $|P(0, B)| = M - 1 > (1 - 1/p)^2 (M - 1)$.

For the second claim suppose that $f \in P^\circ(n, B)$. This corresponds to
all $cf \in P^+(n, B)$ for all $c$ with $1 \leq c \leq \lfloor B/|f| \rfloor$, i.e., to $\lfloor B/|f| \rfloor \leq B$
distinct members of $P^+(n, B)$. Since different $f$ give rise to different sets
of members of $P^+(n, B)$ it follows that $|P^\circ(n, B)| \geq |P^+(n, B)|/B$ and the
claim follows from part 1.                                              □

We note that the assumption that $p$ divides $M$ is for the sake of simplicity,
if $p \leq M$ is any prime then the preceding argument shows that $\mathrm{sq}(n, B) \geq (1 - 1/p)^2 p^n \lfloor M/p \rfloor^n (p \lfloor M/p \rfloor - 1)$.

An alternative approach to a (weaker) lower bound for $\mathrm{sq}(n, B)$ is to
observe that $f \in P(n, B)$ is square free provided that the leading coefficient
or the trailing coefficient is a square free integer. Using $Q(B)$ to denote the
number of square free integers between 1 and $B$ we then have that

$$|P(n, B)| \geq 2Q(B)M^n + 4Q(B)(B - Q(B))M^{n-1}$$

Now $6B/\pi^2 \leq Q(B) < 6B/\pi^2 + \sqrt{B}$, see Moser and McLeod[17], so that

$$|P(n, B)| > \frac{12}{\pi^2} B M^n + \frac{24}{\pi^2} B^2 \left(1 - \frac{6}{\pi^2} - \frac{1}{\sqrt{B}}\right) M^{n-1}$$

$$= \frac{6}{\pi^2}(M - 1)M^n + \frac{6}{\pi^2}\left(1 - \frac{6}{\pi^2} - \frac{1}{\sqrt{B}}\right)(M - 1)^2 M^{n-1}$$

Since $\mathrm{sq}(n, B) = |P(n, B)|/2$ we have

$$\mathrm{sq}(n, B) \geq \frac{6}{\pi^2} B(2B + 1)^n + \frac{12}{\pi^2}\left(1 - \frac{6}{\pi^2} - \frac{1}{\sqrt{B}}\right) B^2 (2B + 1)^{n-1}.$$

### 4.1. Expected breadth

We define two open discs parametrised by $m$. $S_m$ is the disc given by

$$\left| z - \frac{2m^2 + 8m + 7}{2(m+2)(m+3)} \right| < \frac{1}{2(m+2)(m+3)}.$$

This is the disc in Case 1 of Theorem 3.1. $L_m$ is the disc given by

$$|z| < \frac{m+2}{m+3}.$$

Note that $L_m$ is the smallest disc centred at the origin that encloses $S_m$.

**Root Distribution Assumption:** Let $N_m$ denote the expected number of roots in $L_m$ of members of $P^+(n, B)$ drawn uniformly at random. There is a constant $m_0$ such that for all $m \geq m_0$ the expected number of roots in $S_m$ is at most $N_m(s_m/l_m)^2$ where $s_m$ is the radius of $S_m$ and $l_m$ is the radius of $L_m$. The same applies to the situation in which members of $P^\circ(n, B)$ are drawn uniformly at random.

It follows from the second bound in Lemma 2.8 that

$$N_m \leq \log((n+1)B)/\log((m+3)/(m+2)).$$

Shepp and Vanderbei[19] study the expected number of complex roots in a region for the case of coefficients drawn from a normal distribution while Ibragimov and Zeitouni[12] prove a more general result. Unfortunately it does not seem possible to deduce the assumption stated above form these results. An intuitive justification is that roots cluster around the unit disc and are distributed fairly uniformly by angle, see the experimental evidence presented in §4.4.

**Lemma 4.2.** *Assume that members of $P^+(n, B)$ are drawn uniformly at random and that the root distribution assumption holds. Then*

$$\mathbb{E}[|\widehat{C}(f^{II})|] + \mathbb{E}[|\widehat{C}(f^{IT})|] + \mathbb{E}[|\widehat{C}(f^{TI})|] + \mathbb{E}[|\widehat{C}(f^{TT})|]$$
$$< (4.44 + m_0)\log(n+1)B.$$

*The same inequality holds for $P^\circ(f)$.*

**Proof.** As in Theorem 3.1, we consider the four cases of $\widehat{C}_m(f^{II})$, $\widehat{C}_m(f^{IT})$, $\widehat{C}_m(f^{TI})$ and $\widehat{C}_m(f^{TT})$, but with $\widehat{C}$ instead of $C$, using the same enumeration as there. It is shown there that $|\widehat{C}(f^{IT})| + |\widehat{C}(f^{TT})| \leq (2/\log 2)\log L(f)$; the inequality there is observed for $C$ but it is clearly valid for $\widehat{C}$.

For the first case, the disc for $m$ is contained in the disc $L_m$ of the root distribution assumption. The discs for $m < m_0$ are all contained in the disc centred at the origin with radius $(m_0 + 1)/(m_0 + 2)$. Thus the total number of roots of any polynomial is no more than $\log(n+1)B/\log((m_0+2)/(m_0+1))$. The transformation $f(z) \mapsto (-1)^{\deg f} f(-z)$ preserves $P^+(n, B)$ as well as $P^\circ(n, B)$ and shows that the number of roots with strictly positive real part is no more than half the total number of roots. Thus $|\cup_{0 \le i < m_0} \widehat{\mathbf{C}}_m(f^{II})| \le \log(n+1)B/2\log((m_0+2)/(m_0+1))$. Assuming that $m_0 > 0$,

$$\mathbb{E}[|\widehat{\mathbf{C}}(f^{II})|] = \mathbb{E}\left[\sum_{m=0}^{m_0-1} |\widehat{\mathbf{C}}_m(f)|\right] + \mathbb{E}\left[\sum_{m=m_0}^{\infty} |\widehat{\mathbf{C}}_m(f)|\right]$$

$$\le \frac{\log(n+1)B}{2\log\left(\frac{m_0+2}{m_0+1}\right)} + \log(n+1)B \times$$

$$\sum_{m=m_0}^{\infty} \frac{1}{4(m+2)^2(m+3)^2} \left(\frac{m+3}{m+2}\right)^2 \frac{1}{\log\left(\frac{m+3}{m+2}\right)}.$$

A simple argument shows that $\log(1 + x) > 2x/(x + 2)$ for all $x > 0$, hence $1/\log(1 + x) < 1/x + 1/2$. Now

$$\mathbb{E}[|\widehat{\mathbf{C}}(f^{II})|] < \log(n+1)B \left(\frac{2m_0+3}{4} + \sum_{m=m_0}^{\infty} \frac{m+5/2}{4(m+2)^4}\right)$$

$$\le \log(n+1)B \left(\frac{2m_0+3}{4} + \frac{m_0+5/2}{4(m_0+2)^4} + \int_{m_0}^{\infty} \frac{m+5/2}{4(m+2)^4} dm\right)$$

$$= \log(n+1)B \left(\frac{2m_0+3}{4} + \frac{m_0+5/2}{4(m_0+2)^4} + \left[\frac{-3m-7}{24(m+2)^3}\right]_{m_0}^{\infty}\right)$$

$$= \left(\frac{m_0+2}{2} - \frac{1}{4} + \frac{1}{8(m_0+2)^2} + \frac{7}{24(m_0+2)^3} + \frac{1}{8(m_0+2)^4}\right) \log(n+1)B.$$

As pointed out in the proof of Theorem 3.1, the first and third cases transform to each other by the transform $f(z) \mapsto (-z)^{\deg(f)} f(1/z)$. Hence $|\widehat{\mathbf{C}}(f^{II})| \le |\widehat{\mathbf{C}}(f^{TI})|$ and $|\widehat{\mathbf{C}}(f^{TI})| \le |\widehat{\mathbf{C}}(f^{II})|$, hence $|\widehat{\mathbf{C}}(f^{TI})| = |\widehat{\mathbf{C}}(f^{II})|$. Thus bound above applies to $\mathbb{E}[|\widehat{\mathbf{C}}(f^{TI})|]$ and so

$$\mathbb{E}[|\widehat{\mathbf{C}}(f^{II})|] + \mathbb{E}[|\widehat{\mathbf{C}}(f^{IT})|] + \mathbb{E}[|\widehat{\mathbf{C}}(f^{TI})|] + \mathbb{E}[|\widehat{\mathbf{C}}(f^{TT})|]$$

$$\le 2\left(\frac{1}{\log 2} - \frac{1}{4} + \frac{m_0+2}{2} + \frac{1}{8(m_0+2)^2} + \frac{7}{24(m_0+2)^3} + \frac{1}{8(m_0+2)^4}\right) \times$$

$$\log(n+1)B$$

$$\leq 2 \left( \frac{1}{\log 2} - \frac{1}{4} + \frac{m_0 + 2}{2} + \frac{1}{8 \cdot 3^2} + \frac{7}{24 \cdot 3^3} + \frac{1}{8 \cdot 3^4} \right) \log(n+1)B$$

$$= \left( \frac{2}{\log 2} + \frac{503}{324} + m_0 \right) \log(n+1)B$$

$$< (4.44 + m_0) \log(n+1)B.$$

If $m_0 = 0$ the argument above becomes simpler and leads to

$$\mathbb{E}[|\widehat{\mathbf{C}}(f^{II})|] + \mathbb{E}[|\widehat{\mathbf{C}}(f^{IT})|] + \mathbb{E}[|\widehat{\mathbf{C}}(f^{TI})|] + \mathbb{E}[|\widehat{\mathbf{C}}(f^{TT})|]$$

$$\leq 2 \left( \frac{1}{\log 2} + \frac{29}{384} \right) \log(n+1)B < 3.037 \log(n+1)B.$$

$\square$

**Theorem 4.1.** *Assume that members $f$ of $P^+(n, B)$ or of $P^\circ(f)$ are drawn uniformly at random and that the root distribution assumption holds. Then*

$$\mathbb{E}[br(f)] < (4.44 + m_0) \log(n+1)B + 1.45 \log B + 5.$$

**Proof.** As in Theorem 3.1 we have

$$\mathbb{E}[br(f)] \leq \mathbb{E}[|\operatorname{br}(f^{II})|] + \mathbb{E}[|\operatorname{br}(f^{IT})|] + \mathbb{E}[|\operatorname{br}(f^{TI})|] + \mathbb{E}[|\operatorname{br}(f^{TT})|].$$

It follows from the first part of Lemma 2.3 that $\mathbb{E}[|\widehat{\mathbf{C}}(f^{II})|] + \mathbb{E}[|\widehat{\mathbf{C}}(f^{IT})|] + \mathbb{E}[|\widehat{\mathbf{C}}(f^{TI})|] + \mathbb{E}[|\widehat{\mathbf{C}}(f^{TT})|]$ accounts for all the real roots except for the integer roots of $f$. By Lemmas 2.4 and 2.9 we now have

$$\mathbb{E}[br(f)] \leq \mathbb{E}[|\widehat{\mathbf{C}}(f^{II})|] + \mathbb{E}[|\widehat{\mathbf{C}}(f^{IT})|] + \mathbb{E}[|\widehat{\mathbf{C}}(f^{TI})|] + \mathbb{E}[|\widehat{\mathbf{C}}(f^{TT})|]$$

$$+ 1 + \log B / \log 2 + 4.$$

The result follows by Lemma 4.2.

$\square$

### 4.2. *Alternative hypothesis*

The experimental evidence of §4.4 suggests that the Root Distribution Assumption is true, with one exception, for all $m \geq 4$ provided we replace the upper bound estimate $N_m(s_m/l_m)^2$ by $N_m(s_m/l_m)^{1+\epsilon}$ for some $\epsilon > 0$. Under this assumption the first bound in Lemma 4.2 becomes

$$\mathbb{E}[|\widehat{\mathbf{C}}(f^{II})|] \leq \log(n+1)B \left( \sum_{m=0}^{\infty} \frac{m+5/2}{2^{1+\epsilon}(m+2)^{2(1+\epsilon)}} \right)$$

$$\leq \log(n+1)B \left( \sum_{m=0}^{4} \frac{m+5/2}{2(m+2)^2} + \int_{m=4}^{\infty} \frac{m+5/2}{2^{1+\epsilon}(m+2)^{2(1+\epsilon)}} dm \right)$$

$$= \log(n+1)B\left(\frac{12209}{14400} + \left[\frac{-((1+2\epsilon)m+2+5\epsilon)}{\epsilon(1+2\epsilon)2^{2+\epsilon}(m+2)^{1+2\epsilon}}\right]_4^\infty\right)$$

$$< \log(n+1)B\left(\frac{12209}{14400} + \frac{6+13\epsilon}{\epsilon(1+2\epsilon)2^{2+\epsilon}6^{1+2\epsilon}}\right)$$

A simple argument shows that $(6+13\epsilon)/\epsilon(1+2\epsilon)2^{2+\epsilon}6^{1+2\epsilon} \le 1/2\epsilon\log 2$ and so

$$\mathbb{E}[|\widehat{\mathbf{C}}(f^{II})|] < \log(n+1)B\left(\frac{12209}{14400} + \frac{1}{2\epsilon\log 2}\right).$$

Thus

$$\mathbb{E}[|\widehat{\mathbf{C}}(f^{II})|] + \mathbb{E}[|\widehat{\mathbf{C}}(f^{IT})|] + \mathbb{E}[|\widehat{\mathbf{C}}(f^{TI})|] + \mathbb{E}[|\widehat{\mathbf{C}}(f^{TT})|]$$

$$< 2\left(\frac{1}{\log 2} + \frac{12209}{14400} + \frac{1}{2\epsilon\log 2}\right)\log(n+1)B.$$

The bound of Theorem 4.1 now becomes

$$\mathbb{E}[\mathrm{br}(f)] < 2\left(\frac{1}{\log 2} + \frac{12209}{14400} + \frac{1}{2\epsilon\log 2}\right)\log(n+1)B + 1.45\log B + 5$$

$$< \left(4.59 + \frac{1}{\epsilon\log 2}\right)\log(n+1)B + 1.45\log B + 5.$$

### 4.3. *Dependence of expected breadth on the real roots*

Lemma 2.4 shows that $\mathbb{E}[\mathrm{br}(f)] \le \mathbb{E}[\mathbf{R}^+(f)] + \mathbb{E}[\mathbf{C}(f)] + 1$. In this section we show that $\mathbb{E}[\mathbf{R}^+(f)]$ is in line with the bound of Theorem 4.1 but without any assumptions.

Consider the situation in which we draw the coefficients of non-zero polynomials of degree at *most* $n$ uniformly at random from $[-B, B]$, where $B \ge 1$. Thus the probability of drawing any given polynomial is $1/(M^{n+1} - 1)$ since the zero polynomial is excluded. Let $N(n, B)$ be the expected number of real roots. It is shown by Ibragimov and Maslova[13] that if we draw the coefficients independently from the same distribution that has mean zero and which belongs to the domain of attraction of the normal law then the expected number of real roots is equal to

$$\frac{2}{\pi}\log n + o(\log n).$$

This generalises the corresponding result of Kac[14] for the standard normal distribution. The situation of Ref. 13 covers our case, as observed by Cucker and Roy[8]. Since the transformation $z \mapsto (-1)^{\deg(f)}z$ swaps positive and negative roots while keeping the class of polynomials invariant it follows that the number of expected positive real roots, $N^+(n, B)$, is equal to

$$\frac{1}{\pi} \log n + o(\log n).$$

Consider now drawing the members of $P^+(n, B)$, respectively $P^\circ(n, B)$, uniformly at random and denote the expected number of real roots by $N^+(n, B)$, respectively $N^\circ(n, B)$.

**Lemma 4.3.** *Let $p$ be a prime that divides $M = 2B + 1$. Then*

$$N^+(n, B) < \frac{(2 + 1/B)}{\pi(1 - 1/p)^2} \log n + o(\log n) < 2.13 \log n + o(\log n),$$

*and*

$$N^\circ(n, B) < \frac{(2B + 1)}{\pi(1 - 1/p)^2} \left(\log n + o(\log n)\right) < 0.72(2B+1)\left(\log n + o(\log n)\right).$$

**Proof.** Using the first part of Lemma 4.1 in the fourth line below, we have

$$
\begin{aligned}
N^+(n, B) &= \sum_{f \in P^+(n,B)} \frac{\mathbf{R}^+(f)}{\mathrm{sq}(n, B)} \\
&= \frac{M^{n+1} - 1}{\mathrm{sq}(n, B)} \sum_{f \in P^+(n,B)} \frac{\mathbf{R}^+(f)}{M^{n+1} - 1} \\
&\leq \frac{M^{n+1} - 1}{\mathrm{sq}(n, B)} \sum_{f \in P(n,B)} \frac{\mathbf{R}^+(f)}{M^{n+1} - 1} \\
&< \frac{M^{n+1}}{(1 - 1/p)^2 B M^n} N^+(n, B) \\
&= \frac{(2 + 1/B)}{\pi(1 - 1/p)^2} \log n + o(\log n).
\end{aligned}
$$

The numerical constant follows from the fact that $M \geq 3$ is odd and so $p \geq 3$. The second inequality follows similarly by using the second part of Lemma 4.1. $\square$

### 4.4. *Experimental evidence*

The results presented here are based on data produced by Zhao Zheng, a student at the School of Informatics in Edinburgh working under the supervision of the author. The root finding package used from Ref. 5 is based on Laguerre's method (org.apache.commons.math3.analysis.solvers.Lague rreSolver).

Table 1. Comparison of acoustic for frequencies for piston-cylinder problem.

| $n$ | $B$ | Value of $m$ when $u_m \geq \mathbb{E}[S_m]$ | Value of $m$ when $e_m > 1$ | Sample size |
|---|---|---|---|---|
| 10 | 1 | $m \geq 32$ | $m \geq 4$ | 162803 |
| | 10 | $m \geq 256$ | $m \geq 1$ | 150895 |
| | 100 | $m \geq 512$ | $m \geq 1$ | 162179 |
| 20 | 1 | $m \geq 128$ | $m \geq 4$ | 161519 |
| | 10 | $m \geq 512$ | $m \geq 1$ | 155512 |
| | 100 | $m \geq 512$ | $m \geq 1$ | 154807 |
| 30 | 1 | $m \geq 256$ | $m \geq 4$ | 152639 |
| | 10 | $m \geq 512$ | $m \geq 1$ | 152852 |
| | 100 | $m = 256, 512$ | $m \geq 2$ | 143495 |

In Table 1 we present a summary of the results of the experiments in which square free and primitive polynomials were generated. The values of $m$ used were all $2^i$ for $0 \leq i \leq 10$. The estimates of the expected number of roots in the discs $S_m$ and $L_m$ is denoted by $\mathbb{E}[S_m]$ and $\mathbb{E}[L_m]$ respectively. In the third column $u_m = \mathbb{E}[L_m](s_m/l_m)^2$ and the values of $m$ for which the Root Distribution Assumption holds are given. In the fourth column $e_m = \log(\mathbb{E}[S_m]/\mathbb{E}[L_m])/\log(s_m/l_m)$ and the values of $m$ for which the modified conjecture of §4.2 holds, i.e., $e_m > 1$. A comparison was made with purely square free polynomials (for two cases), these showed no significant difference with the corresponding cases of square free and primitive polynomials.

In each experiment polynomials with the appropriate degree and coefficient bound were generated uniformly at random then accepted with appropriate probability if they were square free and after taking the primitive part. Each experiment was run for a maximum time, this is the reason that the sample sizes vary somewhat. Once the roots were found for each polynomial the polynomial was evaluated at each root and this sample was rejected if any value was not close to 0 (within $10^{-8}$). As can be seen from Table 1 the two hypotheses hold after an initial period and stay this way except for for $n = 30$ and $B = 100$ where the result for $m = 1024$ is somewhat anomalous. It is not clear if this is a genuine trend or a software issue, unfortunately further investigation was not possible due to time constraints but it will be investigated in the future*. It is also clear that for the Root Distribution Assumption it is likely that the initial value of $m$ from which it holds depends on $B$, assuming it is indeed true.

---

*Note added in proof: further experiments by David Grilec, a student at the School of Informatics in Edinburgh working under the supervision of the author, have shown that this anomaly was indeed a software issue.

Table 2. Ratio of estimate for
$\mathbb{E}[|\widehat{\mathbf{C}}|]$ and $\log(n+1)B$.

| $n$ | $B$ | $\mathbb{E}[|\widehat{\mathbf{C}}|]/\log(n+1)B$ |
|---|---|---|
| 10 | 1 | 1.047349 |
| | 10 | 0.642732 |
| | 100 | 0.435704 |
| 20 | 1 | 1.334795 |
| | 10 | 0.833969 |
| | 100 | 0.586072 |
| 30 | 1 | 1.547431 |
| | 10 | 1.016662 |
| | 100 | 0.740089 |

The bounds obtained in §§4.1, 4.2 are dominated by $c\log(n+1)B$ for some constant $c$. By Lemmas 2.3 and 2.4 $\mathbb{E}[\mathrm{br}(f)]$ is dominated by $\mathbb{E}[|\widehat{\mathbf{C}}|]$. Table 2 shows the ratio of $\mathbb{E}[|\widehat{\mathbf{C}}|]$ to $\log(n+1)B$. In all cases this is well below the theoretical constant, however there is an increase as $n$ increases for any given $B$.

## Acknowledgments

The author is grateful to an anonymous referee for helpful comments.

## References

1. L. AHLFORS, *Complex Analysis*, McGraw-Hill Book Company, second ed., 1966.
2. A. G. AKRITAS, *An implementation of Vincent's theorem*, Numerische Mathematik, 36 (1980), pp. 53–62.
3. A. G. AKRITAS, *Elements of Computer Algebra with Applications*, J. Wiley & Sons, 1989.
4. A. ALESINA AND M. GALUZZI, *A new proof of Vincent's theorem*, L'Enseignement Matheématique, 44 (1998).
5. APACHE COMMONS LIBRARY, *Laguerre solver*, http://commons.apache.org/proper/commons-math/.
6. G. COLLINS AND W. KRANDICK, *On the computing time of the continued fractions method*, Journal of Symbolic Computation, 47 (2012), pp. 1372–1412.
7. G. E. COLLINS AND A. G. AKRITAS, *Polynomial real root isolation using Descarte's rule of signs*, in Proceedings of the 1976 Symposium on Symbolic and Algebraic Computation, R. D. Jenks, ed., ACM Press, 1976, pp. 272–275.
8. F. CUCKER AND M.-F. ROY, *A theorem on random polynomials and some consequences in average complexity*, J. Symbolic Computation, 10 (1990), pp. 405–409.

9. P. ERDŐS AND P. TURÁN, *On the distribution of roots of polynomials*, Ann. Math., 51 (1950), pp. 105–119.

10. T. GANELIUS, *Sequences of analytic functions and their zeros*, Ark. Math., 3 (1954), pp. 1–50.

11. C. P. HUGHES AND A. NIKEGHBALI, *The zeros of random polynomials cluster uniformly near the unit circle*, Composition Math., 144 (2008), pp. 734–746.

12. I. IBRAGIMOV AND O. ZEITOUNI, *On roots of random polynomials*, Trans. of the American Math. Soc., 349 (1997), pp. 2427–2441.

13. I. A. IBRAGIMOV AND N. B. MASLOVA, *The mean number of real zeros of random polynomials. I. Coefficients with zero mean*, Theor. Probability Appl., 16 (1971), pp. 228–248.

14. M. KAC, *On the average number of real roots of a random algebraic equation*, Bull. Amer. Math. Soc., 49 (1943), pp. 314–320 and 938.

15. W. KRANDICK AND K. MEHLHORN, *New bounds for the Descartes method*, J. Symbolic Computation, 41 (2006).

16. M. MIGNOTTE, *Mathematics for Computer Algebra*, Springer, 1992.

17. L. MOSER AND R. A. MCLEOD, *The error term for square free integers*, Canadian Mathematics Bulletin, 9 (1966), pp. 303–306.

18. Q. I. RAHMAN AND G. SCHMEISSER, *Analytic Theory of Polynomials*, no. 26 in London Mathematical Society Monographs, New Series, Clarendon Press, Oxford, 2002.

19. L. SHEPP AND R. VANDERBEI, *The complex zeros of random polynomials*, Trans. of the American Math. Soc., 347 (1995), pp. 4365–4384.

20. E. P. TSIGARIDAS AND I. Z. EMIRIS, *On the complexity of real root isolation using continued fractions*, Theoretical Computer Science, 392 (2008), pp. 158–173.

21. A. J. H. VINCENT, *Sur la résolution des équations numériques*, Journal de mathématiques pures et appliquées, 1 (1836), pp. 341–372.

# Combinatorial rank of quantum groups
## of infinite series

V. K. Kharchenko

*FES-Cuautitlán Universidad Nacional Autónoma de México;*
*Sobolev Institute of Mathematicas, Novosibirsk 630090, Russia*

*vlad@unam.mx*

M. L. Díaz Sosa

*FES-Cuautitlán Universidad Nacional Autónoma de México;*

*mlds@apolo.acatlan.unam.mx*

We demonstrate that the combinatorial rank of the multiparameter version of the Lusztig "small" quantum group $u_q(\mathfrak{sp}_{2n})$, or, equivalently, of the Frobenius-Lusztig kernel of type $C_n$, equals $\lfloor \log_2(n-1) \rfloor + 2$ provided that $q$ has a finite multiplicative order $t > 3$. It is known that the combinatorial rank of the multiparameter version of the Frobenius-Lusztig kernel of type $A_n$ equals $\lfloor \log_2 n \rfloor + 1$, whereas in case $B_n$ it is equal to $\lfloor \log_2(n-1) \rfloor + 2$, and in case $D_n$ it is $\lfloor \log_2(n-2) \rfloor + 2$.

*2010 Mathematics subject classification:* 16W30, 16W35, 17B37.

*Keywords*: Hopf algebra, primitive generators, PBW basis.

## 1. Introduction

In this article, we continue investigation of the combinatorial rank of Lusztig "small" quantum groups $u_q(\mathfrak{g})$ started in Ref. 4–6. The combinatorial rank of a character Hopf algebra $H$ generated by skew primitive elements $a_1, a_2 \ldots, a_n$ is the length of the sequence of bi-ideals

$$0 = J_0 \subset J_1 \subset J_2 \subset \cdots \subset J_i \subset \ldots \subset J_\kappa = \ker \varphi,$$

where $\varphi$ is the natural epimorphism $G\langle X \rangle \to H$, $x_i \to a_i$ of the free character Hopf algebra, and $J_{i+1}/J_i$ is an ideal generated by all skew primitive elements of $\ker \varphi / J_i$.

We show that the combinatorial rank of the multiparameter version of the Lusztig "small" quantum group $u_q(\mathfrak{sp}_{2n})$, equals $\lfloor \log_2(n-1) \rfloor + 2$

provided that $q$ has a finite multiplicative order $t > 3$. It is known from Ref. 5 that the combinatorial rank of $u_q(\mathfrak{sl}_{n+1})$ equals $\lfloor \log_2 n \rfloor + 1$, whereas in case $\mathfrak{g} = \mathfrak{so}_{2n+1}$ it has the same value $\lfloor \log_2(n-1) \rfloor + 2$ (see Ref. 6), and in case $\mathfrak{g} = \mathfrak{so}_{2n}$ it equals $\lfloor \log_2(2n-3) \rfloor + 1 = \lfloor \log_2(n-2) \rfloor + 2$ (see Ref. 4).

The method of calculation is based on the recently discovered explicit coproduct formula (see Ref. 7), and it is very similar to that of previous papers[4-6]. An important difference is that in case $C_n$, the PBW generators $v[k, \phi(k)]$ of the explicit coproduct formula are not defined by Lyndon–Shirshov words. At the same time, the subalgebra of constants, which is very important for calculations, is generated by powers of bracketed Lyndon–Shirshov words. In fact, when $t$ is even, the elements $v[k, \phi(k)]^{t/2}$ not always are constants. By this reason, in Section 5, we have to develop specific calculations for bracketed Lyndon-Shirshov words $[v_k]$ that do not appear in the coproduct formula.

In the second, third and fourth sections, we briefly recall the main concepts and basic statements which are of use later. In the fifth section we make specific calculations and demonstrate that $U_q^+(\mathfrak{sp}_{2n})$ and $u_q^+(\mathfrak{sp}_{2n})$ have the same bracketed Lyndon-Shirshov words that define the PBW bases. Then, in sixth section, we prove that every intermediate biideal is generated by powers of that bracketed words.

In the seventh section, in perfect analogy with Ref. 4(Section 8), we demonstrate that $U_q^+(\mathfrak{so}_{2n})$ is a finite extension of a skew central Hopf subalgebra $GC$ of quantum polynomials in fixed powers of the PBW generators $y_i = [w]^{t_w}$. The elements $y_i$ are precisely the generators of the subalgebra $C$ of constants with respect to the non-commutative differential calculus that naturally arise on the subalgebra generated by $x_i$, $1 \le i \le n$. In the one parameter case considered by C. De Concini, V.G. Kac, and C. Procesi in Ref. 2,3, the subalgebra $C$ is central, and $GC = GZ_0$, where $Z_0$ is the smallest subalgebra invariant with respect to Lusztig braid group action containing all $x_i^t$, $g_i^t$, $1 \le i < n$, $x_n^{t'}$, $g_n^{t'}$. Here $t' = t$ if the parameter $q$ has odd multiplicative order, and $t' = t/2$ otherwise. We show that each Hopf ideal $J_j$, $j > 1$ is generated by some of the elements $y_i$. This allows us to find the combinatorial rank in Sections 8 using the explicit coproduct formula.

## 2. Preliminaries

In this section, we collect some known results on the structure of an arbitrary character Hopf algebra. Recall that a Hopf algebra $H$ is referred to as a *character* Hopf algebra if the group $G$ of all grouplike elements is commutative and $H$ is generated over $\mathbf{k}[G]$ by skew primitive semi-invariants $a_i$, $i \in I$:

$$\Delta(a_i) = a_i \otimes 1 + g_i \otimes a_i, \quad g^{-1}a_i g = \chi^i(g)a_i, \quad g, g_i \in G, \qquad (1)$$

where $\chi^i$, $i \in I$ are characters of the group $G$.

## 2.1. *Skew brackets*

Let us associate a "quantum" variable $x_i$ to $a_i$. For each word $u$ in $X = \{x_i \mid i \in I\}$, let $g_u$ or $\mathrm{gr}(u)$ denote an element of $G$ that appears from $u$ by replacing each $x_i$ with $g_i$. In the same way, $\chi^u$ denotes a character that appears from $u$ by replacing each $x_i$ with $\chi^i$. We define a bilinear skew commutator on homogeneous linear combinations of words in $a_i$ or in $x_i$, $i \in I$ by the formula

$$[u, v] = uv - \chi^u(g_v)vu, \qquad (2)$$

where we sometimes use the notation $\chi^u(g_v) = p_{uv} = p(u, v)$. Of course, $p(u, v)$ is a bimultiplicative map:

$$p(u, vt) = p(u, v)p(u, t), \quad p(ut, v) = p(u, v)p(t, v). \qquad (3)$$

The brackets satisfy the following Jacobi identity:

$$[[u, v], w] = [u, [v, w]] + p_{wv}^{-1}[[u, w], v] + (p_{vw} - p_{wv}^{-1})[u, w] \cdot v. \qquad (4)$$

The Jacobi identity (4) implies the following conditional identity

$$[[u, v], w] = [u, [v, w]], \quad \text{provided that } [u, w] = 0. \qquad (5)$$

The brackets are related with the product by the following ad-identities

$$[u \cdot v, w] = p_{vw}[u, w] \cdot v + u \cdot [v, w], \qquad (6)$$

$$[u, v \cdot w] = [u, v] \cdot w + p_{uv}v \cdot [u, w]. \qquad (7)$$

## 2.2. *Radford biproduct and the ideal* $\Lambda$

The group $G$ acts on the free algebra $\mathbf{k}\langle X \rangle$ by $g^{-1}ug = \chi^u(g)u$, where $u$ is an arbitrary monomial in $X$. The skew group algebra $G\langle X \rangle$ has the natural Hopf algebra structure

$$\Delta(x_i) = x_i \otimes 1 + g_i \otimes x_i, \quad i \in I, \quad \Delta(g) = g \otimes g, \quad g \in G.$$

We fix a Hopf algebra homomorphism

$$\xi : G\langle X \rangle \to H, \quad \xi(x_i) = a_i, \quad \xi(g) = g, \quad i \in I, \quad g \in G. \tag{8}$$

If the kernel of $\xi$ is contained in the ideal $G\langle X \rangle^{(2)}$ generated by $x_i x_j$, $i, j \in I$, then there exists a Hopf algebra projection $\pi : H \to \mathbf{k}[G]$, $a_i \to 0$, $g_i \to g_i$. Hence, by the Radford theorem[14], we have a decomposition in a biproduct, $H = A \star \mathbf{k}[G]$, where $A$ is a subalgebra generated by $a_i$, $i \in I$ (see Ref. 1(§1.5, §1.7)).

**Definition 2.1.** In what follows, $\mathbf{\Lambda}$ denotes the biggest Hopf ideal in $G\langle X \rangle^{(2)}$, where, as above, $G\langle X \rangle^{(2)}$ is the ideal of $G\langle X \rangle$ generated by $x_i x_j$, $i, j \in I$. The ideal $\mathbf{\Lambda}$ is homogeneous in each $x_i \in X$ (see Ref. 5(Lemma 2.2)).

The algebra $A$ has the structure of a *braided Hopf algebra*[17], with a braiding $\tau(u \otimes v) = p(v, u)^{-1} v \otimes u$. If $\operatorname{Ker} \xi = \mathbf{\Lambda}$ then $A$ is a Nichols algebra[1](§1.3, Section 2) or, equivalently, a quantum symmetric algebra.

### 2.3. *Differential calculi*

The free algebra $\mathbf{k}\langle X \rangle$ has two closely related differential calculi defined by the following Leibniz rules:

$$\partial_j(x_i) = \delta_i^j, \quad \partial_i(uv) = \partial_i(u) \cdot v + \chi^u(g_i) u \cdot \partial_i(v). \tag{9}$$

$$\partial_j^*(x_i) = \delta_i^j, \quad \partial_i^*(uv) = p(x_i, v) \partial_i^*(u) \cdot v + u \cdot \partial_i^*(v). \tag{10}$$

**Lemma 2.1.** *(Ref. 12(Lemma 2.10)). Let $u \in \mathbf{k}\langle X \rangle$ be an element homogeneous in each $x_i$, $1 \leq i \leq n$. If $p_{uu}$ is a primitive $t^{\text{th}}$ root of 1, then*

$$\partial_i(u^t) = p(u, x_i)^{t-1} \underbrace{[u, [u, \ldots [u, \partial_i(u)] \ldots]].}_{t-1} \tag{11}$$

The partial derivatives and coproduct are related by

$$\Delta(\partial_i(u)) = \sum_{(u)} g_i^{-1} u^{(1)} \otimes \partial_i(u^{(2)}), \tag{12}$$

where we use the Sweedler notations $\Delta(u) = \sum_{(u)} u^{(1)} \otimes u^{(2)}$. Recall that the *braided antipode* $\sigma^b : \mathbf{k}\langle X \rangle \to \mathbf{k}\langle X \rangle$ acts on a word $u$ as follows

$$\sigma^b(u) = \operatorname{gr}(u) \sigma(u),$$

where $\sigma$ is the antipode of $G\langle X \rangle$. The braided antipode satisfies

$$\sigma^b([u, v]) \sim [\sigma^b(v), \sigma^b(u)] \tag{13}$$

and

$$\sigma^b(\partial_i^*(u)) \sim \partial_i(\sigma^b(u)), \tag{14}$$

where $\sim$ is the projective equality: $a \sim b \iff a = \alpha b,\ 0 \neq \alpha \in \mathbf{k}$.

**Lemma 2.2.** *(Milinski—Schneider criterion). Suppose that $Ker\xi = \Lambda$. If a polynomial $f \in \mathbf{k}\langle X \rangle$ is a constant in $A$ (that is, $\partial_i(f) \in \Lambda,\ i \in I$), then there exists $\alpha \in \mathbf{k}$ such that $f - \alpha = 0$ in $A$.*

See details in Ref. 4(Section 2).

## 3. Combinatorial representation

Recall that the quantum groups $U_q(\mathfrak{sp}_{2n})$, $u_q(\mathfrak{sp}_{2n})$, are generated as $\mathbf{k}$-algebras by the grouplike elements

$$g_1, g_2, \ldots, g_n, f_1, f_2, \ldots, f_n,\quad \Delta(g_i) = g_i \otimes g_i,\quad \Delta(f_i) = f_i \otimes f_i,$$

by their inverses, and by the skew primitive elements $x_1, x_2, \ldots, x_n$, $x_1^-, x_2^-, \ldots, x_n^-$,

$$\Delta(x_i) = x_i \otimes 1 + g_i \otimes x_i;\quad \Delta(x_i^-) = x_i^- \otimes 1 + f_i \otimes x_i^-.$$

Whereas all the grouplike elements commute each with other, the skew primitive generators commute with the grouplikes via

$$x_i g_j = p_{ij} g_j x_i,\ x_i^- g_j = p_{ij}^{-1} g_j x_i^-,\ x_i f_j = p_{ji} f_j x_i,\ x_i^- f_j = p_{ji}^{-1} f_j x_i^-,$$

where $p_{ij}$ are arbitrary parameters satisfying the following relations

$$p_{ii} = q,\ 1 \leq i < n;\quad p_{i\,i-1} p_{i-1\,i} = q^{-1},\ 1 < i < n;\quad p_{ij} p_{ji} = 1,\ j > i+1;$$

$$p_{nn} = q^2 \quad p_{n-1\,n}\, p_{n\,n-1} = q^{-2}.$$

The group $G$ of the grouplike elements may satisfy arbitrary additional relations that are compatible with the above commutation rules (for example, if $p_{ij} = p_{ji}$ for all $i, j$ then one may additionally suppose that $f_i = g_i$).

Let $G\langle X \cup X^- \rangle$ be the Hopf algebra defined as above with free skew primitives $x_i, x_i^-$. Then we have Hopf algebra morphisms

$$\varphi : G\langle X \cup X^- \rangle \to U_q(\mathfrak{sp}_{2n}),\quad \varphi_1 : G\langle X \cup X^- \rangle \to u_q(\mathfrak{sp}_{2n}).$$

By its definition, the ideal $\ker \varphi$ is generated by the elements

$$\mu_{ij} \overset{df}{=} x_i x_j^- - p_{ji} x_j^- x_i - \delta_i^j (1 - g_i f_i)$$

and the quantum Serre polynomials

$$S_{ij}(x_i, x_j), \quad S_{ij}(x_i^-, x_j^-), \quad 1 \leq i, j \leq n.$$

It is important here that all of these elements are skew primitive in $G\langle X \cup X^- \rangle$; see Ref. 11(Theorem 6.1). In the explicit form the polynomials $S_{ij}(x_i, x_j)$ are:

$$[x_i, [x_i, x_{i+1}]] = [[x_i, x_{i+1}], x_{i+1}] = 0, \ 1 \leq i < n - 1;$$
$$[x_i, x_j] = 0, \quad j > i + 1; \tag{15}$$

$$[[x_{n-1}, x_n], x_n] = [x_{n-1}, [x_{n-1}, [x_{n-1}, x_n]]] = 0. \tag{16}$$

The ideal $\ker \varphi_1$ is generated by $\mu_{ij}$ and the two sets $\Lambda$, $\Lambda^-$, see Subsection 2.2.

## 4. Hard super-letters and PBW basis

We shall concentrate on the positive quantum Borel subalgebra, the subalgebra generated over $G$ by by the $x_i$'s.

On the set of all words in the $x_i$'s we fix the lexicographical order with the priority from the left to the right considering $x_1 > x_2 > \ldots > x_n$, where a proper beginning of a word is considered to be greater than the word itself.

A non-empty word $u$ is called a *standard* word (Lyndon word, Lyndon-Shirshov word) if $vw > wv$ for each decomposition $u = vw$ with non-empty $v, w$. A *nonassociative* word is a word where brackets $[,]$ somehow arranged to show how multiplication applies. If $[u]$ denotes a nonassociative word, then $u$ denotes an associative word obtained from $[u]$ by removing the brackets. The set of *standard nonassociative* words is the biggest set $SL$ that contains all variables $x_i$ and satisfies the following properties.

1) If $[u] = [[v][w]] \in SL$, then $[v], [w] \in SL$, and $v > w$ are standard.

2) If $[u] = [[[v_1][v_2]][w]] \in SL$, then $v_2 \leq w$.

Every standard word has only one arrangement of brackets such that the appeared nonassociative word is standard (Shirshov theorem [15]). In order to find this arrangement one may use the following inductive procedure:

**Algorithm.** The factors $v, w$ of the nonassociative decomposition $[u] = [[v][w]]$ are the standard words such that $u = vw$ and $v$ has the minimal length (Ref. 16, see also Ref. 13).

**Definition 4.1.** A *super-letter* is a polynomial that equals a nonassociative standard word where the brackets are defined as follows $[u, v] = uv - p(u, v)vu$, while $p(u, v)$ is a bimultiplicative map, see (3), defined on words so that $p(x_i, x_j) = p_{ij}$. A *super-word* is a word in super-letters.

By Shirshov's theorem every standard word $u$ defines only one super-letter, in what follows we shall denote it by $[u]$. The order on the super-letters is defined in the natural way: $[u] > [v] \iff u > v$.

**Definition 4.2.** In what follows we fix a homogeneous in each $x_i$ bi-ideal $J$ of $G\langle X \rangle$ such that $\ker \varphi \cap G\langle X \rangle \subseteq J \subseteq \Lambda$.

By Ref. 5(Lemma 2.2) the bi-ideal $\Lambda$ itself is homogeneous.

**Definition 4.3.** A super-letter $[u]$ is called *hard in* $G\langle X \rangle / J$ provided that its value in $G\langle X \rangle / J$ is not a linear combination of values of super-words in smaller than $[u]$ super-letters.

**Definition 4.4.** We say that a *height* of a hard super-letter $[u]$ in $G\langle X \rangle / J$ equals $h = h([u])$ if $h$ is the smallest number such that: first, $p(u, u)$ is a primitive $s$-th root of 1 and either $h = s$ or $h = sl^r$, where $l =$char($\mathbf{k}$); and next the value of $[u]^h$ in $G\langle X \rangle / J$ is a linear combination of super-words in less than $[u]$ super-letters. If there exists no such a number, the height is infinite.

**Theorem 4.1.** (*Ref. 10(Theorem 2)*). *The values of all hard in* $G\langle X \rangle / J$ *super-letters with the above defined height function form a set of PBW generators for* $G\langle X \rangle / J$ *over* $\mathbf{k}[G]$; *that is, the set of all products*

$$g[u_1]^{n_1}[u_2]^{n_2} \cdots [u_k]^{n_k}, \quad [u_1] < [u_2] < \ldots < [u_k], \ n_i < h([u_i]), g \in G$$

*form a basis of* $G\langle X \rangle / J$.

The hard super-letters for $U_q^+(\mathfrak{sp}_{2n})$ are described in Ref. 9 (Theorem $C_n$).

**Definition 4.5.** In what follows, $x_i$, $n < i < 2n$ denotes the generator $x_{2n-i}$. Respectively, $v(k, m)$, $1 \leq k \leq m < 2n$ is the word $x_k x_{k+1} \cdots x_{m-1} x_m$. If $1 \leq i < 2n$, then $\phi(i)$ denotes the number $2n - i$, so that $x_i = x_{\phi(i)}$.

The word $v(k, m)$ is standard if and only if $k \leq m < \phi(k)$ or $k = m = n$. The standard arrangement of brackets, $[v(k, m)]$, is described in

Ref. 9(Lemma 7.18). In Ref. 7(Proposition 4.1), it is shown that the value of $[v(k,m)]$ in $U_q^+(\mathfrak{sp}_{2n})$ coincides with the value of the long skew commutator

$$v[k,m] = [[\ldots[[x_k, x_{k+1}], x_{k+2}]\ldots x_{m-1}], x_m], \quad k \le m < \phi(k).$$

Following Ref. 7(p. 21) we define the bracketing of $v(k,m)$, $k \le m < 2n$ as follows:

$$v[k,m] = \begin{cases} [[[\ldots[x_k, x_{k+1}], \ldots], x_{m-1}], x_m], & \text{if } m < \phi(k); \\ [x_k, [x_{k+1}, [\ldots, [x_{m-1}, x_m]\ldots]]], & \text{if } m > \phi(k); \\ [v[k, m-1], x_m], & \text{if } m = \phi(k), \end{cases} \quad (17)$$

where in the latter term, $[\![u, v]\!] \overset{df}{=} uv - q^{-1}p(u,v)vu$.

By Ref. 7(Theorem 5.1) the coproduct on the elements $v[k,m]$, $k \le m < 2n$ has the following explicit form:

$$\Delta(v[k,m]) = v[k,m] \otimes 1 + g_{km} \otimes v[k,m] \quad (18)$$

$$+ \sum_{i=k}^{m-1} \tau_i(1 - q^{-1})g_{ki}\, v[i+1, m] \otimes v[k, i],$$

where $\tau_i = 1$ with two exceptions, being $\tau_{n-1} = 1 + q^{-1}$ if $m = n$, and $\tau_n = 1 + q^{-1}$ if $k = n$. Here $g_{ki} = \mathrm{gr}(v(k,i)) = g_k g_{k+1} \cdots g_i$.

Conditional identity (5) implies that the value of $v[k,m]$ in $U_q^+(\mathfrak{sp}_{2n})$ is independent of the precise arrangement of brackets, provided that $m \le n$ or $k \ge n$. In general, the value of bracketed word $v[k,m]$ is almost independent of the precise arrangement of brackets.

**Lemma 4.1.** (*Ref. 7(Lemma 3.6)*). *If $k \le n \le m < \phi(k)$, then the value in $U_q^+(\mathfrak{sp}_{2n})$ of the bracketed word $[y_k x_n x_{n+1} \cdots x_m]$, where $y_k = v[k, n-1]$, is independent of the precise arrangement of brackets.*

**Lemma 4.2.** (*Ref. 7(Lemma 3.7)*). *If $k \le n$, $\phi(k) < m$, then the value in $U_q^+(\mathfrak{sp}_{2n})$ of the bracketed word $[x_k x_{k+1} \cdots x_n y_m]$, where $y_m = v[n+1, m]$, is independent of the precise arrangement of brackets.*

**Definition 4.6.** We define the words $v_k$, $1 \le k \le n$ as follows:

$$v_k = v(k, n-1)v(k, n) = x_k x_{k+1} \ldots x_{n-1} x_k x_{k+1} \ldots x_{n-1} x_n, \quad v_n = x_n.$$

Certainly, these are standard words and the standard arrangement of brackets is $[v_k] = [[v(k, n-1)][v(k, n)]]$.

**Proposition 4.1.** (*Ref. 9(Theorem $C_n$)*). *If $q^3 \ne 1$, $q \ne -1$, then the set*

$$\mathfrak{C} = \{[v(k,m)], [v_s] \mid k \le m < \phi(k), 1 \le s \le n\} \quad (19)$$

*is the set of all hard super-letters for $U_q^+(\mathfrak{sp}_{2n})$.*

## 5. PBW generators for $u_q^+(\mathfrak{sp}_{2n})$

Now we are going to demonstrate that the set $\mathfrak{C}$ are hard super-letters for $u_q^+(\mathfrak{sp}_{2n})$ as well.

Recall that $\deg_i(w)$, $1 \le i \le n$ denotes a degree in $x_i$ of the word $w$, the number of occurrences of $x_i$ in $w$. Respectively, $D(w) = (\deg_1(w), \deg_2(w), \ldots, \deg_n(w))$ is a *multidegree* of $w$.

**Lemma 5.1.** *If $s$ is an arbitrary natural number and $[w] \in \mathfrak{C}$, then $sD(w)$ is not a sum of multidegrees $D(u)$ of lesser than $[w]$ elements $[u] \in \mathfrak{C}$.*

**Proof.** Let $w = v(k, m)$, $k \le m < \phi(k)$. If $[u] \in \mathfrak{C}$ and $u < v(k, m)$, then either $\deg_k(u) \le 1$ or $u$ contains a letter that does not occur in $w$. Certainly, in the latter case $D(u)$ may not appear in the decomposition of $sD(w)$ in the sum.

Hence, at least $s$ super-letters $[u]$ with $\deg_k(u) = 1$ appear in the decomposition. At the same time, all such super-letters are in the list

$$[v(k, m+1)], \ [v(k, m+2)], \ldots, \ [v(k, \phi(k) - 1)]. \tag{20}$$

If $m < n$, then $[v(k, m)]$ belongs to the Hopf subalgebra generated by the elements $x_1, x_2, \ldots, x_{n-1}$. In this case, the above list of hard super-letters reduces to

$$[v(k, m+1)], \ [v(k, m+2)], \ldots, \ [v(k, n-1)].$$

All elements in this list, if any, depend on $x_{m+1}$, whereas $[v(k, m)]$ is independent of it.

If $m \ge n$, then all super-letters from the list (20) are of degree 2 in $x_{m+1}$. Therefore, the $(m+1)$th component of the sum of $s$ multidegrees of that type is $2s$. However, the $(m+1)$th component of $sD(v(k, m))$ equals $s$. Thus, the decomposition is still impossible.

Assume $w = v_k$, $1 \le k < n$. In this case the list (20) take the form

$$[v(k, n)], \ [v(k, n+1)], \ldots, \ [v(k, \phi(k) - 1)].$$

All words in this list are linear in $x_k$ and in $x_n$. Hence the $k$th component of $\sum D(u)$ is less than or equal to the $n$th component of it. At the same time, the $k$th component of $sD(v_k)$ equals $2s$, and the $n$th component is $s$. $\qquad \square$

**Lemma 5.2.** *If a standard word $w$ is independent of $x_n$, then either $w = v(k, m)$, $k \le m < n$ or $[w] = 0$ in $U_q^+(\mathfrak{sp}_{2n})$.*

**Proof.** Subalgebra generated by $x_1, x_2, \ldots x_{n-1}$ is the quantization of type $A_{n-1}$. Hence Ref. 9 (Theorem $A_n$ case 3) applies. $\qquad \square$

**Lemma 5.3.** *If $k + 1 < i \leq n$, then $[v[k, \phi(i) - 1], x_{i-1}] = 0$.*

**Proof.** If $i = n$, then

$$[v[k, \phi(i) - 1], x_{i-1}] = [[v[k, n - 2], x_{n-1}], x_{n-1}] = 0$$

due to the relations $[[x_{n-2}, x_{n-1}], x_{n-1}] = 0$ and $[x_j, x_{n-1}] = 0$, $j < n - 2$.

Assume $i < n$. By Lemma 4.2, we have $v[k, \phi(i) - 1] = [v[k, n - 1], y]$, where $y = v[n + 1, \phi(i) - 1]$. In this case, $[y, x_{i-1}] = 0$ due to the defining relations $[x_j, x_{i-1}] = 0$, $i < j \leq n$, whereas

$$[v[k, n - 1], x_{i-1}] = [v[k, i - 3], [v[i - 2, n - 1], x_{i-1}]] = 0$$

because the standard word $v(i - 2, n - 1)x_{i-1}$ is independent of $x_n$ and the standard bracketing is precisely $[v[i - 2, n - 1], x_{i-1}]$. $\square$

**Lemma 5.4.** *If $i \neq k$, then $\partial_i(v[k, m]) = 0$, $k \leq m < \phi(k)$.*

**Proof.** If an element $u$ is independent of $x_i$ then, of course, $\partial_i(u) = 0$. The Leibniz formula (9) implies

$$\partial_i([u, x_i]) = \partial_i(ux_i) - p(u, x_i)\partial_i(x_iu) = p(u, x_i)u \cdot \partial_i(x_i) - p(u, x_i)\partial_i(x_i) \cdot u = 0,$$

provided that $\partial_i(u) = 0$. It remains to apply definition (17) of the bracketed word $v[k, m]$. $\square$

**Lemma 5.5.** *We have $\partial_k(v[k, m]) \sim v[k + 1, m]$ provided that $k < m < \phi(k)$.*

**Proof.** If $m = k + 1$, then

$$\partial_k([x_k, x_{k+1}]) = \partial_k(x_k x_{k+1}) - p(x_k, x_{k+1})\partial_k(x_{k+1}x_k)$$

$$= x_{k+1} - p(x_k, x_{k+1})p(x_{k+1}, x_k)x_{k+1} = (1 - q^{-\varepsilon})x_{k+1},$$

where $\varepsilon = 1$ or $\varepsilon = 2$. In the general case, we may perform the evident induction:

$$\partial_k(v[k, m]) = \partial_k([v[k, m - 1], x_m])$$

$$= \partial_k(v[k, m - 1]x_m) - p(v(k, m - 1), x_m)\partial_k(x_m v[k, m - 1])$$

$$= \alpha v[k + 1, m - 1]x_m - p(v(k, m - 1), x_m)p(x_m, x_k)x_m(\alpha v[k + 1, m - 1]).$$

If $k + 1 < m < \phi(k) - 1$, then $p(x_k, x_m)p(x_m, x_k) = 1$ and the above linear combination equals $\alpha[v[k + 1, m - 1], x_m] = \alpha v[k + 1, m]$, which is required. If $m = \phi(k) - 1$, then $p(x_k, x_m)p(x_m, x_k) = q^{-1}$ and we arrive to $\alpha v[k + 1, \phi(k + 1)]$ due to (17). $\square$

**Lemma 5.6.** *In the algebra* $U_q^+(\mathfrak{sp}_{2n})$, *we have*

$$[\, v[k,m], [v[k,m], \partial_k(v[k,m])]\,] = 0, \quad k \le m \le \phi(k) - 2. \tag{21}$$

**Proof.** Consider the word $w = v(k,m)v(k,m)v(k+1,m)$. This is a standard word, and the standard arrangement of brackets given by the Algorithm p. 233 is precisely

$$[\,[v(k,m)]\,[\,[v(k,m)]\,[v(k+1,m)]\,]\,]\,.$$

By Proposition 4.1, all hard super-letters in $U_q^+(\mathfrak{sp}_{2n})$ are

$$[v(s,r)], \ 1 \le s \le r < \phi(s); \quad [v_s], \ 1 \le s \le n. \tag{22}$$

In particular, $[w]$ is not hard. If $m < n$, then Lemma 5.2 applies. Assume $m \ge n$. The multiple use of Definition 4.3 demonstrates that the value of $[w]$ is a linear combination of super-words in smaller than $[w]$ hard super-letters. Because $U_q^+(\mathfrak{sp}_{2n})$ is homogeneous, the hard super-letters that may appear in the linear combination are

$$[v(k,r)], \ m < r < \phi(k); \quad [v(s,r)], \ k < s \le r < \phi(s); \quad [v_r], \ k < r \le n.$$

In the above list, all super-letters that depend on $x_k$ have degree 2 in $x_{m+1}$ and degree 1 in $x_k$. At the same time $w$ has degree 2 in $x_k$ and degree 3 in $x_{m+1}$. Therefore the linear combination is empty, $[w] = 0$. $\qquad\square$

**Lemma 5.7.** *In the algebra* $U_q^+(\mathfrak{sp}_{2n})$, *we have*

$$[v[k,\phi(k)-1], \partial_k(v[k,\phi(k)-1])] = 0, \quad 1 \le k < n. \tag{23}$$

**Proof.** If $k = n - 1$, then the required relation reduces to the defining relation $[[[x_{n-1}]x_n], x_n] = 0$.

Assume $k < n - 1$. By Lemma 5.5 the element $\partial_k([v(k,\phi(k)-1)])$ is proportional to

$$v[k+1,\phi(k+1)] = [\![v[k+1,\phi(k)-2], x_{k+1}]\!].$$

Hence, it suffices to prove two equalities:

$$[v[k,\phi(k)-1], x_{k+1}] = 0, \quad [v[k,\phi(k)-1], v[k+1,\phi(k)-2]] = 0.$$

Consider the word $w = v(k,\phi(k)-1)x_{k+1}$. This is a standard word, and it does not appear in the list (22). Therefore the value of $[w]$ in $U_q^+(\mathfrak{sp}_{2n})$ is a linear combination of super-words in smaller than $w$ hard super-letters. Since $w$ is linear in $x_k$, it follows that each term of the linear combination has a hard super-letter linear in $x_k$ which is less than $[w]$. But in the list (22) there does not exist such a super-letter. Thus $[w] = 0$ in $U_q^+(\mathfrak{sp}_{2n})$. According to the Algorithm p. 233, the standard arrangement of brackets in $w$ is precisely $[[v(k, \phi(k) - 1)]x_{k+1}]$.

Consider the word $w_1 = v(k, \phi(k) - 1)v(k + 1, \phi(k) - 2)$. This is also a standard word. The super-letter $[w_1]$ is not in the list (22). Therefore, the value of $[w_1]$ in $U_q^+(\mathfrak{sp}_{2n})$ is a linear combination of super-words in smaller than $w_1$ hard super-letters. Since $w$ is linear in $x_k$, it follows that each term of the linear combination has a hard super-letters linear in $x_k$ which is less than $[w_1]$. In the list (22), there does not exist such a super-letter, hence $[w_1] = 0$ in $U_q^+(\mathfrak{sp}_{2n})$.

According to the Algorithm, the standard arrangement of brackets in $w$ is

$$[v(k, \phi(k) - 2)][x_{k+1}[v(k + 1, \phi(k) - 2)]]. \tag{24}$$

To correct the arrangement of brackets, we may use the conditional identity (5). To this end, consider the word

$$w_2 = v(k, \phi(k) - 2)v(k + 1, \phi(k) - 2).$$

The super-letter $[w_2]$ is not in the list (22). Therefore the value of $[w_2]$ in $U_q^+(\mathfrak{sp}_{2n})$ is a linear combination of super-words in smaller than $w_2$ hard super-letters.

Since in the list (22) there does not exist a linear in $x_k$ super-letter which is less than $[w_2]$, it follows that $[w_2] = 0$ in $U_q^+(\mathfrak{sp}_{2n})$. It remains to note that according to the Algorithm, the standard arrangement of brackets in $w_2$ has the required form $[[v(k, \phi(k) - 2)][v(k + 1, \phi(k) - 2)]]$. $\qquad\square$

**Lemma 5.8.** *In the algebra* $U_q^+(\mathfrak{sp}_{2n})$, *we have*

$$[[v_k], \partial_k([v_k])] = 0, \quad 1 \le k < n. \tag{25}$$

**Proof.** Denote for short $u = [v(k, n - 1)]$, $v = [v(k, n)]$, $u' = \partial_k(u)$, $v' = \partial_k(v)$. In this case by definition $[v_k] = [u, v]$. By means of Leibniz formula (9), we have

$$\partial_k([v_k]) = \partial_k(uv) - p(u, v)\partial_k(vu)$$
$$= u'v + p(u, x_k)uv' - p(u, v)v'u - p(u, v)p(v, x_k)vu'. \tag{26}$$

First, we note that $[v_k]$ skew commutes with the first an the last terms of the above linear combination. To this end, it suffices to check the equalities

$$[[v_k], v] = [[v_k], u'] = 0. \tag{27}$$

Consider the word $v_k v = v(k, n-1)v(k, n)v(k, n)$. This is a standard word, and the standard arrangement of brackets (see Algorithm p. 233) is precisely

$$[[v(k, n-1)v(k, n)]v(k, n)].$$

This word does not belong to the list (22), hence $[[v_k]v]$ is not hard. The multiple use of Definition 4.3 demonstrates that the value of $[v_k v]$ is a linear combination of super-words in smaller than $[v_k v]$ hard super-letters. Because $U_q^+(\mathfrak{sp}_{2n})$ is homogeneous, the hard super-letters that may appear in the linear combination are

$$[v(k, m)], \ n \le m < \phi(k);$$
$$[v(s, m)], \ k < s \le m < \phi(s); \quad [v_r], \ k < r \le n. \tag{28}$$

Each of that super-letter is either independent of $x_k$ or linear both in $x_k$ and $x_n$. Since $v_k v$ is of degree 3 in $x_k$ and of degree 2 in $x_n$, it follows that the linear combination is empty; that is, $[[v_k]v] = 0$ in $U_q^+(\mathfrak{sp}_{2n})$.

Consider the word $w_1 = v_k v(k+1, n-1)$. This is also a standard non hard word, therefore its value is a linear combination of super-words in smaller than $w_1$ hard super-letters. The super-letters that may appear in the linear combination are the same (28). As we have noted before, each of that super-letter is either independent of $x_k$ or linear both in $x_k$ and $x_n$. Since $v_k v(k+1, n-1)$ is of degree 2 in $x_k$ and of degree 1 in $x_n$, it follows that the linear combination is empty; that is, $[v_k v(k+1, n-1)] = 0$ in $U_q^+(\mathfrak{sp}_{2n})$.

In this case, the standard arrangement of brackets due to the Algorithm p. 233 is $[v(k, n-1)[v(k, n)v(k+1, n-1)]]$ but not the required

$$[[v(k, n-1)v(k, n)] [v(k+1, n-1)]].$$

Nevertheless, we may apply the conditional identity (5) because

$$[[v(k, n-1)], [v(k+1, n-1)]] = 0$$

due to Lemma 5.2. Since $\partial_k(u) \sim [v(k+1, n-1)]$, it follows that $[[v_k], u'] = 0$.

Let us turn to the second and the third terms of (26). We have

$$p(u, x_k)uv' - p(u, v)v'u = p(u, x_k)[u, v'].$$

Therefore, it remains to prove the equality $[[v_k], [u, v']] = 0$.

Consider the word $w_2 = v(k, n-1)v(k, n)v(k, n-1)v(k+1, n)$. This is a standard word. Therefore the value of $[w_2]$ in $U_q^+(\mathfrak{sp}_{2n})$ is a linear combination of super-words in smaller than $w_2$ hard super-letters. The super-letters that may appear in the linear combination are precisely (28), that are either independent of $x_k$ or linear both in $x_k$ and $x_n$. Since $w_2$ is of degree 3 in $x_k$ and of degree 2 in $x_n$, it follows that the linear combination is empty; that is, $[w_2] = 0$ in $U_q^+(\mathfrak{sp}_{2n})$.

The standard arrangement of brackets $[w_2]$ is

$$[[v(k, n-1)v(k, n)][x_k[v(k+1, n-1)v(k+1, n)]]]$$

that differs from the required

$$[[v(k, n-1)v(k, n)][[x_k v(k+1, n-1)][v(k+1, n)]]].$$

Using the quantum Jacobi identity (4) with

$$u \leftarrow x_k, \quad v \leftarrow [v(k+1, n-1)], \quad w \leftarrow [v(k+1, n)],$$

we see that

$$[[x_k, [v(k+1, n-1)]], [v(k+1, n)]]$$

is a linear combination of the following three terms

$$[x_k[v(k+1, n-1)v(k+1, n)]],$$
$$[v(k, n)] \cdot [v(k+1, n-1)], \quad [v(k+1, n-1)] \cdot [v(k, n)].$$

Equality $[w_2] = 0$ implies that the element $[v_k]$ skew commutes with the first term. Relations (27) demonstrate that $[v_k]$ skew commutes with $[v(k, n)]$ and also with $[v(k+1, n-1)]$. $\qquad\square$

**Lemma 5.9.** *If $1 \leq i \leq n$, $k < m < \phi(k)$, then we have*

$$\partial_i^*(v[k, m]) \sim \begin{cases} v[k, m-1], & \text{if } m = i \text{ or } m = \phi(i); \\ 0, & \text{otherwise.} \end{cases}$$

**Proof.** If an element $u$ is independent of $x_i$, then, of course, $\partial_i^*(u) = 0$. If $\partial_i^*(u) = 0$, then the Leibniz formula (10) implies

$$\partial_i^*([u, x_i]) = \partial_i^*(ux_i) - p(u, x_i)\partial_i^*(x_i u) = (1 - p(u, x_i)p(x_i, u))u. \quad (29)$$

In particular, taking $u = v[k, i-1]$, we obtain the required formula for $m \leq i$.

If $x_j \neq x_i$, then again by the Leibniz formula (10), we have

$$\partial_i^*([u, x_j]) = \partial_i^*(ux_j) - p(u, x_j)\partial_i^*(x_j u) = p(x_i, x_j)[\partial_i^*(u), x_j]. \quad (30)$$

In particular, taking $u = v[k, i]$, $j = i+1$ and using the evident relation

$$[v[k, i-1], x_{i+1}] = 0, \ i \leq n,$$

we obtain the required $\partial_i^*(v[k, i+1]) = 0$, and then step-by-step the relations $\partial_i^*(v[k, m]) = 0$, $i < m < \phi(i)$.

Further, the substitution $u \leftarrow v[k, \phi(k) - 1]$ in (29) implies the required formula with $m = \phi(i)$. Then, substitution $u \leftarrow v[k, \phi(i)]$, $j = i - 1$ in (30) demonstrate that $\partial_i^*(v[k, \phi(i) + 1])$ is proportional to $[v[k, \phi(i) - 1], x_{i-1}]$, which is equal to zero due to Lemma 5.3 because condition $\phi(i) + 1 < \phi(k)$ implies $k + 1 < i$. Thus the required formula is valid for $m = \phi(i) + 1 < \phi(k)$. Now, step-by-step, using (30), we have $\partial_i^*(v[k, m]) = 0$, $\phi(i) < m < \phi(k)$. □

**Lemma 5.10.** *If* $1 \le k < n$, *then*

$$\partial_i^*([v_k]) \sim \begin{cases} v[k, n-1]^2, & \text{if } i = n; \\ 0, & \text{otherwise.} \end{cases}$$

**Proof.** Denote $u = v[k, n-1]$, and $v = v[k, n]$. In this case, $[v_k] = [u, v]$. By Lemma 5.9 and the Leibniz formula (10), we have $\partial_i^*([v_k]) = 0$ if $i < n - 1$.

If $i = n - 1$, then $\partial_i^*(u) \sim v[k, n-2]$ and $\partial_i^*(v) = 0$. The Leibniz formula (10) yields

$$\partial_i^*([v_k]) \sim p(x_{n-1}, v)v[k, n-2] \cdot v - p(u, v)v \cdot v[k, n-2]$$
$$= p(x_n, v)[v[k, n-2], v].$$

In this formula, if $k = n - 1$, one has to replace $v[k, n-2]$ with 1, and the resulting expression becames 0. In general case, still

$$[v[k, n-2], v] = [v[k, n-2], [v[k, n-1], x_n]]$$
$$= [[v[k, n-2], v[k, n-1]], x_n] = 0$$

because, first, $[[v[k, n-2], x_n] = 0$, and, next, $[v[k, n-2], v[k, n-1]] = 0$ as the word $v(k, n-2)v(k, n-1)$ is standard and independent of $x_n$ (see, Lemma 5.2).

Assume $i = n$. In this case, $\partial_i^*(u) = 0$ and $\partial_i^*(v) \sim u$. Using the Leibniz formula (10), we obtain the required relation $\partial_i^*([v_k]) \sim u^2$. □

**Lemma 5.11.** *If* $1 \le i \le n$, $k < m < \phi(k)$, $m \ne n$, *then*

$$[\partial_i^*(v[k, m]), v[k, m]] = 0.$$

**Proof.** By Lemma 5.9, we have to demonstrate that $[v[k, m-1], v[k, m]] = 0$. The word $w = v(k, m-1)v(k, m)$ is standard with standard arrangement of brackets $[w] = [[v(k, m-1)][v(k, m)]]$. If $m < n$, then $w$ is independent of $x_n$ and Lemma 5.2 implies $[w] = 0$.

If $m > n$, then $[w]$ still is not a hard super-letter. Therefore its value is a linear combination of super-words in smaller than $[w]$ hard super-letters. The smaller hard super-letters are:

$$[v(k, r)], \ m \leq r < \phi(k); \quad [v(s, r)], \ k < s < r < \phi(s); \quad [v_r], \ k < r < n.$$

Each of them is either independent of $x_k$ or linear in $x_k$ and of degree 2 in $x_m$. At the same time $w$ has degree 2 in $x_k$ and degree 3 in $x_m$. Hence the linear combination is empty, and $[w] = 0$. □

**Lemma 5.12.** *If $1 \leq i \leq n$, then*

$$[[\partial_i^*(v[k, n]), v[k, n]], v[k, n]] = 0.$$

**Proof.** By Lemma 5.9, we have to demonstrate that $[[v[k, n-1], v[k, n]], v[k, n]] = 0$. This equality is already proven, see (27). □

**Lemma 5.13.** *If $1 \leq i \leq n$, then*

$$[\partial_i^*([v_k]), [v_k]] = 0.$$

**Proof.** By Lemma 5.10, we have to consider just one case $i = n$. In this case, $\partial_i^*([v_k]) \sim v[k, n-1]^2$. Therefore it suffices to demonstrate that $[v[k, n-1], [v_k]] = 0$.

Consider the word $w = v(k, n-1)v(k, n-1)v(k, n)$. This is a standard word with the following standard arrangement of brackets:

$$[w] = [v(k, n-1)[v(k, n-1)v(k, n)]].$$

Since $[w]$ is not a hard super-letter, it follows that its value is a linear combination of super-words in smaller than $[w]$ hard super-letters:

$$[v(k, r)], \ n \leq r < \phi(k); \quad [v(s, r)], \ k < s < r < \phi(s); \quad [v_r], \ k \leq r < n.$$

The word $w$ is linear in $x_n$ and of degree 3 in $x_k$, whereas all words in the above list which depend on $x_k$ have degree 1 or 2 in $x_2$ and degree 1 in $x_n$. Thus, the linear combination is empty, and $[w] = 0$. □

**Theorem 5.1.** *If the multiplicative order $t$ of $q$ is finite, $t > 3$, then the values of*

$$v[k, m], \ k \leq m < \phi(k), \quad [v_s], \ 1 \leq s \leq n$$

*form a set of PBW generators for $u_q^+(\mathfrak{sp}_{2n})$ over $k[G]$. The height of $v[k, m]$, equals $t$. The height $h$ of $[v_s]$, $1 \leq s \leq n$ equals $t$ if $t$ is odd, otherwise $h = t/2$. In all cases $v[k, m]^t = 0$, $[v_s]^h = 0$ in $u_q^+(\mathfrak{sp}_{2n})$.*

**Proof.** First, we note that Definition 4.3 implies that a non-hard super-letter in $U_q^+(\mathfrak{sp}_{2n})$ remains non-hard in $u_q^+(\mathfrak{sp}_{2n})$. Hence, all hard super-letters for $u_q^+(\mathfrak{sp}_{2n})$ are in the list (19).

Next, if $[v(k,m)]$, $k \le m < \phi(k)$ is not hard in $u_q^+(\mathfrak{sp}_{2n})$, then by the multiple use of Definition 4.3, the value of $[v(k,m)]$ is a linear combination of super-words in hard super-letters smaller than given $v(k,m)$. Because $u_q^+(\mathfrak{sp}_{2n})$ is homogeneous, each of the super-words $M$ in that decomposition has a hard super-letter smaller than $v(k,m)$ and of degree 1 in $x_k$. At the same time, all such super-letters are in the list

$$[v(k,m+1)], \ [v(k,m+2)], \ldots, \ [v(k,\phi(k)-1)].$$

Each of them has degree 2 in $x_{m+1}$ if $m \ge n$, and at least 1 if $m < n$. Hence the super-word $M$ has degree of at least 2 if $m \ge n$, and at least 1 if $m < n$. However $u(k,m)$ is of degree 1 in $x_{m+1}$ if $m \ge n$, and is independent of $x_{m+1}$ if $m < n$. Therefore the decomposition is empty, and $[v(k,m)] = 0$ in $u_q^+(\mathfrak{sp}_{2n})$. Nevertheless, all elements $v[k,m]$ are nonzero in $u_q^+(\mathfrak{sp}_{2n})$ because by Lemma 5.9 we have $\partial_m^*(v[k,m]) \sim v[k,m-1]$, and evident induction applies.

Similarly, if $[v_k]$ is not hard in $u_q^+(\mathfrak{sp}_{2n})$, then its value is a linear combination of super-words in hard super-letters smaller than $v_k$. Each of the super-words $M$ in that decomposition has a hard super-letter of degree 2 in $x_k$ or at least two hard super-letters of degree 1 in $x_k$ because the degree of $v_k$ in $x_k$ equals 2 (unless $k = n$). All possible super-letters of $M$ are in the list

$$[v(k,n)], \ [v(k,n+1)], \ldots, \ [v(k,\phi(k)-1)].$$

No one of them has degree 2 in $x_k$, and each of them has degree 1 in $x_n$. Hence $M$ is of degree at least 2 in $x_n$. However $[v_k]$ is of degree 1 in $x_n$. Therefore the decomposition is empty, and $[v_k] = 0$ in $u_q^+(\mathfrak{sp}_{2n})$.

By Lemma 5.10 we obtain $v[k,n-1] \sim \partial_n^*([v_k]) = 0$. Using Leibniz formula (10), we have $0 = (\partial_{n-1}^*)^2(v[k,n-1]^2) \sim v[k,n-2]^2$ unless $q = -1$. Applying operator $(\partial_{n-2}^*)^2$ we find $v[k,n-3]^2 = 0$, and so on. Finally, $x_k^2 \ne 0$ in $u_q^+(\mathfrak{sp}_{2n})$ because $\partial_k(x_k^2) \sim x_k \ne 0$.

Let us find the heights of hard super-letters in $u_q^+(\mathfrak{sp}_{2n})$.

For short we put $v = v[k,m]$, $u = [v_k]$. Applying Ref. 7(Lemma 3.5), we have $p_{vv} = q$ and $p_{uu} = q^2$. By Definition 4.4 the minimal possible value for the height is precisely the $h$ given in the proposition. It remains to show that $v^t = 0$, $u^h = 0$ in $u_q^+(\mathfrak{sp}_{2n})$.

By Lemma 2.2 it suffices to prove that all partial derivatives of the related elements are zero. Lemma 2.1 yields

$$\partial_i(v^t) = p(v, x_i)^{t-1} \underbrace{[v, [v, \ldots [v, \partial_i(v)] \ldots ]]}_{t-1}.$$

$$\partial_i(u^h) = p(u, x_i)^{h-1} \underbrace{[u, [u, \ldots [u, \partial_i(u)] \ldots ]]}_{h-1}.$$

It remains to apply Lemma 5.4, Lemma 5.6, Lemma 5.7 and Lemma 5.8. □

## 6. Defining relations for $G\langle X \rangle / J$

We are reminded that an ideal $J$ is given in Definition 4.2 as an arbitrary homogeneous intermediate ideal, whereas the set $\mathfrak{C}$ of hard in $U_q^+(\mathfrak{sp}_{2n})$ and $u_q^+(\mathfrak{sp}_{2n})$ super-letters is defined in (19).

**Proposition 6.1.** *The values in $G\langle X \rangle / J$ of the set $\mathfrak{C}$ form a set of PBW generators for $G\langle X \rangle / J$ over $\mathbf{k}[G]$. The ideal $J$ is uniquely defined by the heights of $\mathfrak{C}$. More precisely, $J$ is generated by relations of $U_q^+(\mathfrak{sp}_{2n})$ and $[w]^h$, where $[w] \in \mathfrak{C}$ and $h$ is the height of $[w]$ in $G\langle X \rangle / J$.*

**Proof.** The definition of the hard super-letter implies that a hard in $G\langle X \rangle / \Lambda = u_q^+(\mathfrak{sp}_{2n})$ super-letter is hard in $G\langle X \rangle / J$, whereas a hard in $G\langle X \rangle / J$ one is hard in $U_q^+(\mathfrak{sp}_{2n})$. Hence, $\mathfrak{C}$ is the set of all hard in $G\langle X \rangle / J$ super-letters.

By Theorem 4.1, the values of $\mathfrak{C}$ form a set of PBW generators (with some height function) for $G\langle X \rangle / J$ over $\mathbf{k}[G]$; that is, the set of all products

$$g[w_1]^{n_1}[w_2]^{n_2} \cdots [w_k]^{n_k}, \quad [w_1] < [w_2] < \ldots < [w_k], \ n_i < h([w_i]), \ g \in G$$

form a basis of $G\langle X \rangle / J$.

If $h = h([w])$ is the height of $[w]$ in $G\langle X \rangle / J$, then by definition, the value of $[w]^h$ in $G\langle X \rangle / J$ is a linear combination of super-words in hard super-letters smaller than $[w]$. Because $J$ is homogeneous, the multidegree of each super-word $M$ in that decomposition equals the multidegree of $w^h$. Lemma 5.1 states that this is impossible. Therefore the decomposition is empty, and $[w]^h = 0$ in $G\langle X \rangle / J$; that is, $[w]^h \in J$.

Finally, let $J'$ be the ideal generated by relations of $U_q^+(\mathfrak{so}_{2n})$ and $[w]^h$, where $[w] \in \mathfrak{C}$, while $h$ is the height of $[w]$ in $G\langle X \rangle / J$. By Theorem 4.1 applied to $U_q^+(\mathfrak{sp}_{2n})$, the set of all products

$$g[w_1]^{n_1}[w_2]^{n_2} \cdots [w_k]^{n_k}, \quad [w_1] < [w_2] < \ldots < [w_k], \ [w_i] \in \mathfrak{C}, \ g \in G$$

form a basis of $U_q^+(\mathfrak{sp}_{2n})$. Hence values of that elements with additional restriction, $n_i < h([w_i])$, span $G\langle X \rangle / J'$. Since the products with that restriction are linearly independent in $G\langle X \rangle / J$, and $J' \subseteq J$, it follows that they are still linearly independent in $G\langle X \rangle / J'$. In other words, $G\langle X \rangle / J'$ and $G\langle X \rangle / J$ have the same basis, therefore $J' = J$.    $\square$

**Corollary 6.1.** *Each homogeneous skew-primitive in $G\langle X \rangle / J$ element $u$ of total degree $> 1$ has the form $u = \alpha[w]^h$, $\alpha \in \mathbf{k}$, where $[w] \in \mathfrak{C}$. If $t$ is odd or $w = v(k, m)$, then $h = t$ or $h = tl^s$ in the case of characteristic $l > 0$. If $t$ is even and $w = v_k$, then $h = t/2$ or $h = (t/2)l^s$ in the case of characteristic $l > 0$.*

**Proof.** By Ref. 9(Lemma 4.9) (essentially proven in Ref. 10(Lemmas 12, 13)), the decomposition of $u$ in the PBW basis has the form $u = \alpha[w]^h + \sum_i \alpha_i W_i$, where $W_i$ are basis super-words in less than $[w]$ hard super-letters. In Proposition 5.1, it is proven that the multidegree of $[w]^h$ is not a sum of multidegrees of lesser than $[w]$ super-letters from $\mathfrak{C}$. Therefore, $u = \alpha[w]^h$.

If $t$ is odd or $w = v(k, m)$, then $p(w, w) = q$ (see Ref. 7(Lemma 3.5)) and due to sited above[9](Lemma 4.9) the exponent $h$ is $t$, or 1, or $tl^s$ in the case of characteristic $l > 0$. If $t$ is even and $w = v_k$, then $p(w, w) = q^2$ which implies that the exponent $h$ may have only values $t/2$, $(t/2)l^s$, or 1. No one of $[w]$, $w \neq x_i$ is skew-primitive in $G\langle X \rangle / \Lambda$ because each of them has nonzero partial derivatives; that is, $h \neq 1$.    $\square$

## 7. Constants of differential calculi

Let $\mathfrak{A}$ be a subalgebra of $U_q^+(\mathfrak{sp}_{2n})$ generated by the $x_i$'s, and let

$$C = \{u \in \mathfrak{A} \mid \partial_i(u) = 0, 1 \leq i \leq n\}$$

be the subalgebra of constants for calculus (9), whereas

$$C^* = \{u \in \mathfrak{A} \mid \partial_i^*(u) = 0, 1 \leq i \leq n\}$$

be the subalgebra of constants for calculus (10). Because the operators $\partial_i$, $\partial_i^*$ diminish degree in $x_i$ of every monomial by one and do not change the degree in other variables, both subalgebras are homogeneous in each variable. By means of the substitution $u \leftarrow C^*$ in (14) we have $\sigma^b(C^*) \subseteq C$. Similarly the substitution $u \leftarrow (\sigma^b)^{-1}(C)$ implies $(\sigma^b)^{-1}(C) \subseteq C^*$; that is, $C = \sigma^b(C^*)$.

**Theorem 7.1.** *The following statements are valid.*

1. *The algebra $C$ is generated by the elements $v[k,m]^t$, $k \leq m < \phi(k)$, $[v_k]^h$, $1 \leq k < n$, where $h = t$ if $t$ is odd and $h = t/2$ if $t$ is even.*

2. *We have $[f, v[k,m]^t] = [v[k,m]^t, f] = 0$ and $[f, [v_k]^h] = [[v_k]^h, f] = 0$ for all homogeneous $f \in \mathfrak{A}$.*

3. *$C = C^*$.*

4. *A subalgebra $GC$ of $U_q^+(\mathfrak{sp}_{2n})$ generated by $G$ and $C$ is a Hopf subalgebra.*

5. *$U_q^+(\mathfrak{sp}_{2n})$ is a free finitely generated module over $GC$ of rank $t^{n^2}$ if $t$ is odd and of rank $t^{n^2-n}(t/2)^n$ if $t$ is even.*

**Proof.** The proof is very similar to that of Ref. 4(Theorem 2). Let us note, first, that $v[k,m]^t, [v_k]^h \in C \cap C^*$. By Lemma 2.1 we have

$$\partial_i([w]^{t'}) \sim \underbrace{[[w], [[w], \ldots [[w]}_{t'-1}, \partial_i([w])] \ldots ]],$$

where $t' = t$ if $w = v(k,m)$ or $t$ is odd, and $t' = t/2$ if $w = v_k$ and $t$ is even. Applying Lemma 5.4, Lemma 5.6, and Lemma 5.7, we obtain $\partial_i(v[k,m]^t) = 0$, whereas Lemma 5.8 implies $\partial_i([v_k]^h) = 0$.

Similarly, Lemma 2.1 and (13), (14) imply

$$\partial_i^*([w]^{t'}) \sim [\ldots [[\partial_i^*([w]), [w]], [w]], \ldots [w]].$$
$$\underbrace{\phantom{xxxxxxxxxxxxxxxxxxxxxxx}}_{t'-1}$$

Applying Lemma 5.9, Lemma 5.11, and Lemma 5.12, we have $\partial_i^*(v[k,m]^t) = 0$. Lemma 5.13 implies $\partial_i^*([v_k]^h) = 0$. Thus $[w]^{t'} \in C \cap C^*$ for all $[w] \in \mathfrak{C}$.

Using (12) let us note that $C$ is a left coideal: $\Delta(C) \subseteq G\mathfrak{A} \otimes C$. If $c \in C$, then

$$0 = \Delta(\partial_i(c)) = \sum_{(c)} g_i^{-1} c^{(1)} \otimes \partial_i(c^{(2)}).$$

Because $g_i^{-1} c^{(1)}$ are linearly independent, we have $\partial_i(c^{(2)}) = 0$, and $c^{(2)} \in C$.

This implies that $GC^* = G(\sigma^b)^{-1}(C) = G\sigma^{-1}(C)$ is a right coideal subalgebra which contains the coradical $\mathbf{k}[G]$. By Ref. 8(Theorem 4.1), the subalgebra $GC^*$ has a PBW basis that can be extended to a PBW basis of $U_q^+(\mathfrak{sp}_{2n})$.

The PBW generators may be chosen in the following way. For every $[w] \in \mathfrak{C}$, we choose an arbitrary element, if any, with the minimal possible $a$ of the form

$$c_w = [w]^a + \sum_i \alpha_i W_i R_i \in GC^*, \quad \alpha_i \in \mathbf{k}, \tag{31}$$

where the $W_i$'s are nonempty basis super-words in less than $[w]$ super-letters, whereas the $R_i$'s are basis super-words in greater than or equal to $[w]$ super-letters. According to Ref. 8(Lemma 4.3), the number $a$ either equals 1, or $p(w, w)$ is a primitive $r$th root of 1 and $a = r$ or (in the case of positive characteristic) $a = r(\text{char } \mathbf{k})^s$.

In our case, $p(w, w) = q$ is a primitive $t$th root of 1 if $t$ is odd or $w = v(k, m)$, and $p(w, w) = q^2$ is a primitive $(t/2)$th root of 1 if $t$ is even and $w = v_k$ (see Ref. 7(Lemma 3.5)). In both cases, $a \neq 1$ because otherwise due to Milinski–Schneider criterion (Lemma 2.2), we have $c_w - \alpha \in \Lambda$, $\alpha \in \mathbf{k}$. However, this is impossible as (31) with $m = 1$ is a linear combination of elements from PBW basis of $u_q^+(\mathfrak{sp}_{2n})$ with the leading term $[w]$.

Thus, we can choose $c_w = [w]^h$ because we already know that $[w]^h \in C^*$. In particular, $C^* = GC^* \cap \mathfrak{A}$ as an algebra is generated by the elements $c_w = [w]^h$ with $[w] \in \mathfrak{C}$. Since all of that elements belong to $C$, it follows that $C^* \subseteq C$.

The map $\sigma^2$ is an automorphism such that $\sigma^2(x_i) = p_{ii}x_i$, $\sigma^2(g_i) = g_i$. This implies $\sigma^2(f) \sim f$ for every homogeneous polynomial. Moreover,

$$(\sigma^b)^2(f) = \text{gr}(f)\sigma(\text{gr}(f)\sigma(f)) = \text{gr}(f)f\text{gr}(f)^{-1} \sim f$$

as well. Applying this proportion to $f = [w]^h$, we have $(\sigma^b)^2(C^*) = C^*$. Let us apply $\sigma^b$ to the the already proven relation $C^* \subseteq C = \sigma^b(C^*)$. We obtain

$$C = \sigma^b(C^*) \subseteq \sigma^b(C) = (\sigma^b)^2(C^*) = C^*.$$

This completes the proof of 3 and 1.

Let $a \in C$. Using Leibniz rule (9), we have $\partial_i(x_i a) = a$, whereas $\partial_i(a x_i) = p(a, x_i)a$. Hence, $\partial_i([a, x_i]) = p(a, x_i)a - p(a, x_i)a = 0$. Because certainly $\partial_k([a, x_i]) = 0$, $x_k \neq x_i$, we get $[a, x_i] \in C$ and also $[x_i, a] = \sigma^b([a, x_i]) \in C$. At the same time, the degree in $x_i$ of each generator $[w]^h$ is a multiple of $h$ (which is $t$ or $t/2$). Therefore the degree in $x_i$ of each homogeneous constant is a multiple of $t$ or $t/2$ also. However, $\deg_i([a, x_i]) = \deg_i([x_i, a]) \equiv 1(\text{mod } t/2)$. This is possible only if $[a, x_i] = [x_i, a] = 0$. Hence, all constants, particularly $[w]^h$, $[w] \in \mathfrak{C}$, are skew central. This proves 2.

We have seen above that $C$ is a left coideal, and $GC^*$ is a right coideal. Since $C^* = C = \sigma^b(C)$, it follows that the subalgebra $GC$ is both a left and a right coideal and it is invariant with respect to the antipode; that is, $GC$ is a Hopf subalgebra, which completes 4.

Finally, each element $[w]^r$ has a decomposition $[w]^r = [w]^{r_0} \cdot ([w]^{h_w})^g$, $0 \leq r_0 < h_w$. Hence, the products $\prod_{[w] \in \mathfrak{C}} [w]^{k_w}$, $0 \leq k_w < h_w$, form a basis of $U_q^+(\mathfrak{sp}_{2n})$ over $GC$. The total number of such products is $t^{n^2}$ if $t$ is odd and it is $t^{n^2-n}(t/2)^n$ if $t$ is even. $\qquad\square$

**Corollary 7.1.** *If $[w] \in \mathfrak{C}$, then the ideal $J_w$ of $U_q^+(\mathfrak{sp}_{2n})$ generated by all elements $[u]^{h_u}$ with $[u] \in \mathfrak{C}$, $u \neq w$ contains no one of the elements $[w]^b$, $b \geq 1$.*

The proof literally coincides with the proof of Ref. 4(Corolary 2).

## 8. Combinatorial rank

Consider the chain that defines the combinatorial rank

$$J_0^+ = G\langle X\rangle \cap \ker \varphi \subset J_1^+ \subset J_2^+ \subset \ldots \subset J_\kappa^+ = \Lambda.$$

**Proposition 8.1.** *All $J_i^+$ are homogeneous Hopf ideals. Let $[w] \in \mathfrak{C}$.*
*If $w = v(k, m)$, then $[w]^h \in J_i^+$ if and only if $h \geq t$ and $m - k < 2^i - 1$.*
*If $w = v_k$, and $t$ is odd, then $n - k < 2^{i-1}$ implies $[w]^t \in J_i^+$.*
*If $w = v_k$, and $t$ is even, then $n - k < 2^i - 1$ implies $[w]^{t/2} \in J_i^+$.*

**Proof.** We perform induction on $i$. Let $t' = t/2$ if $t$ is even, and $t' = t$ otherwise.

Theorem $C_n$ from Ref. 9 describes all skew primitive elements of $U_q^+(\mathfrak{sp}_{2n})$. They are $x_i$, $x_i^t$, $x_i^{t_i l^r}$, $1 \leq i < n$, $x_n$ $x_n^{t'}$, $x_n^{t'l^r}$, $1 - g$, $g \in G$ and possibly some linear combinations of these elements. Because all that elements are skew-primitive in $G\langle X\rangle$, the ideal $J_1^+$ is generated by $x_n^{t'}$, $x_i^t$, $1 \leq i < n$ and quantum Serre relations $S_{ij}(x_i, x_j)$ given in (15) and (16). As a result, Corollary 7.1 implies that $[v(k, m)]^h \in J_1^+$ if and only if $k = m < n$, $h \geq t$, whereas $[v_k]^h \in J_1^+$ if and only if $k = n$, $h \geq t'$. At the same time, $m - k < 2^1 - 1$ & $m < \phi(k)$ is equivalent to $k = m < n$; and $n - k < 2^{1-1}$ means $k = n$. If $t$ is even and $w = v_k$, then $n - k < 2^{1-2}$ implies $k = n$, and $[v_n]^{t/2} = x_n^{t/2} \in J_1$.

Suppose that the statement is valid for $J_{i-1}^+$. Corollary 6.1 implies that every homogeneous skew primitive element of $G\langle X\rangle/J_{i-1}^+$ is proportional to $[w]^h$ with $[w] \in \mathfrak{C}$, $h = t$ or $h = t/2$. Because $J_{i-1}^+$ is a homogeneous Hopf ideal, each homogeneous component of a skew primitive element of $G\langle X\rangle/J_{i-1}^+$ is again skew primitive. Thus, $J_i^+$ is generated by both $J_{i-1}^+$ and all elements $[w]^h$ that are skew primitive in $G\langle X\rangle/J_{i-1}^+$. In particular, $J_i^+$ is a homogeneous Hopf ideal.

Moreover, Corollary 7.1 implies that $[w]^h \in J_i^+$ if and only if $[w]^{h'}$, $h' \le h$ is in the list of the skew primitives of $G\langle X \rangle / J_{i-1}^+$.

Let us demonstrate, first, that if $m - k < 2^i - 1$ then $[w]^t$ with $w = v(k, m)$ is skew primitive in $G\langle X \rangle / J_{i-1}^+$. By Theorem 7.1 the subalgebra $GC$ generated over $G$ by the elements $T_u = [u]^{t_u}$, $[u] \in \mathfrak{C}$ is a Hopf subalgebra (here $t_u = t$ if $u = v(k, m)$ or $t$ is odd, and $t_u = t/2$ othrwise). Therefore there exists a decomposition

$$\Delta([w]^t) = \sum_{(c)} \mathrm{gr}(c^{(2)}) c^{(1)} \otimes c^{(2)}, \tag{32}$$

such that $c^{(1)}, c^{(2)}$ are words (products) in $T_u$.

Assume that $[w]^t$ is not skew primitive in $G\langle X \rangle / J_{i-1}^+$. Let us fix a tensor $c^{(1)} \otimes c^{(2)}$ with nonempty $c^{(1)}$ and $c^{(2)}$ which is not zero in $G\langle X \rangle / J_{i-1}^+ \otimes G\langle X \rangle / J_{i-1}^+$. Certainly, no one of the factors in $c^{(1)} \sim \prod_{\mu \in M_1} [u_\mu]^{t_\mu}$ and $c^{(2)} \sim \prod_{\mu \in M_2} [u_\mu]^{t_\mu}$ is zero in $G\langle X \rangle / J_{i-1}^+$. Hence, by the inductive supposition we have $m_\mu - k_\mu \ge 2^{i-1} - 1$, if $u_\mu = v(k_\mu, m_\mu)$, and $n - k_\mu \ge 2^{i-2}$ if $u_\mu = v_{k_\mu}$ and $t$ is odd, whereas $n - k_\mu \ge 2^{i-1} - 1$ if $u_\mu = v_{k_\mu}$ and $t$ is even.

The total degree of the tensor equals the total degree of $[w]^t$. At the same time, the total degree of $[u_\mu]^{t_\mu}$ equals $(m_\mu - k_\mu + 1)t$ if $u_\mu = v(k_\mu, m_\mu)$, and it is

$$(2n - 2k_\mu + 1)t' = (m_\mu - k_\mu + 1)t',$$

where by definition we put $m_\mu = \phi(k_\mu)$ provided that $u_\mu = v_{k_\mu}$. Hence, we have

$$(m - k + 1)t = \sum_{\mu \in M_1 \cup M_2} (m_\mu - k_\mu + 1)t_\mu. \tag{33}$$

If $t$ is odd, then $t_\mu = t$, and the above equality with conditions on $m - k$ and $m_\mu - k_\mu$ imply

$$2^i > m - k + 1 \ge |M_1 \cup M_2| \cdot 2^{i-1}. \tag{34}$$

This is a contradiction because no one of the sets $M_1$, $M_2$ is empty.

If $t$ is even and no one of $u_\mu$ has the form $v_{k_\mu}$, then we arrive to the same contradiction (34). If $u_\mu = v_\mu$, then the degree in $x_n$ of $[u_\mu]^{t/2}$ equals $t/2$. Since degree in $x_n$ of $[w]^t$ either is zero or equals $t$, it follows that there exists a unique $\nu \in M_1 \cup M_2$, $\nu \ne \mu$, such that $u_\nu = v_{k_\nu}$. In this case, the equality (33) implies

$$2^i t > (m - k + 1)t \ge (2n - 2k_\mu + 1)\frac{t}{2} + (2n - 2k_\nu + 1)\frac{t}{2} \ge (2^i - 1)t, \tag{35}$$

and hence $2^i > m - k + 1 \geq 2^i - 1$. This inequality means $m - k + 1 = 2^i - 1$, whereas (35) becomes the equality

$$(2n - 2k_\mu + 1)\frac{1}{2} + (2n - 2k_\nu + 1)\frac{1}{2} = 2^i - 1,$$

which is possible only if $n - k_\mu$ and $n - k_\nu$ take minimal possible value (that is, $k_\mu = k_\nu = n - 2^{i-1} + 1$) and $M_1 \cup M_2 = \{\mu, \nu\}$. Since $w = v(k, m)$ depends on $x_k$, and is independent of $x_i$, $i < k$, it follows that $k_\mu = k_\nu = k$. But this is still impossible because $[v(k, m)]^t$ is of degree $t$ in $x_k$, whereas $[v_k]^{t/2} \otimes [v_k]^{t/2}$ is of degree $2t$ in $x_k$.

Let $n - k < 2^{i-1}$ and $t$ is odd. We have to demonstrate that $[v_k]^t$ is skew primitive in $G\langle X\rangle/J_{i-1}^+$. The decomposition (32) with $w = v_k$ implies

$$(2n - 2k + 1)t = \sum_{\mu \in M_1 \cup M_2} (m_\mu - k_\mu + 1)t \geq |M_1 \cup M_2| 2^{i-1} t \geq 2^i t. \quad (36)$$

Inequality $2^i > 2n - 2k$ implies $2^i > 2n - 2k + 1$ because $2^i$ is even, and we obtain a contradiction $2^i > 2^i$.

Let $n - k < 2^i - 1$ and $t$ is even. We shall prove that $[v_k]^{t/2}$ is skew primitive in $G\langle X\rangle/J_{i-1}^+$. Consider the decomposition (32) with $w = v_k$. The degree of $[v_k]^{t/2}$ in $x_n$ equals $t/2$, therefore there exists one and only one $\mu$, such that $u_\mu = v_{k_\mu}$. Since no one of sets $M_1$, $M_2$ is empty, it follows that there exists at least one $\nu$, such that $u_\nu = v(k_\nu, m_\nu)$. In this case the equality (33) takes the form

$$(2n - 2k + 1)\frac{t}{2} = (2n - 2k_\mu + 1)\frac{t}{2} + (m_\nu - k_\nu + 1)t + \cdots.$$

By the induction hypothesis, we have $n - k_\mu \geq 2^{i-1} - 1$ and $m_\nu - k_\nu \geq 2^{i-1} - 1$. This implies a contradiction:

$$\frac{2^{i+1} - 1}{2} > \frac{2n - 2k + 1}{2} \geq \frac{2^i - 1}{2} + 2^{i-1}.$$

Next, we show that if $m - k \geq 2^i - 1$, $k \leq m < \phi(k)$, then $[w]^t$ with $w = v(k, m)$ is not skew primitive in $G\langle X\rangle/J_{i-1}^+$. Let $s$ be an arbitrary number less than $n$. We shall analyze all tensors of the decomposition

$$\Delta([w]^t) = (\Delta([w]))^t = \sum_{(c)} c^{(1)} \otimes c^{(2)}$$

such that $\deg_s(c^{(2)}) = t$, $\deg_{s+1}(c^{(2)}) = 0$. By the coproduct formula (18) each tensor of that decomposition has the form

$$\alpha g a_1 a_2 \cdots a_t \otimes b_1 b_2 \cdots b_t,$$

where $a_\lambda = v[1 + i_\lambda, m]$, $b_\lambda = v[k, i_\lambda]$. Because $\deg_{s+1}(b_\lambda) = 0$, we have $i_\lambda \leq s$. Therefore the inequality $s < n$ implies $\deg_s(b_\lambda) \leq 1$. At the same time,

$$\sum_{\lambda=1}^{t} \deg_s(b_\lambda) = \deg_s(c^{(2)}) = t.$$

Hence, $\deg_s(b_\lambda) = 1$, all $\lambda$. In particular $i_\lambda \geq s$. Thus, $i_\lambda = s$ for all $\lambda$, and there is only one tensor of the required degrees in the decomposition:

$$\alpha g_k^t g_{k+1}^t \cdots g_s^t v[s+1, m]^t \otimes v[k, s]^t, \ \alpha \neq 0. \tag{37}$$

By the inductive supposition $v[k, s]^h \notin J_{i-1}^+$ if $s - k \geq 2^{i-1} - 1$. At the same time, either $v[s+1, m]$ or $\sigma^b(v[s+1, m]) = v[\phi(m), \phi(s+1)]$ belongs to $\mathfrak{C}$, unless $m = \phi(s+1)$. Hence, again by the inductive supposition, $v[s+1, m]^t \notin J_{i-1}^+$ provided that $m - s - 1 \geq 2^{i-1} - 1$, $m \neq \phi(s+1)$. Denote $s_{\min} = 2^{i-1} - 1 + k$, $s_{\max} = m - 2^{i-1}$ for short.

To show that $[w]^t$ is not skew primitive in $G\langle X \rangle / J_{i-1}^+$, it suffices to find at least one point $s$ satisfying $s < n$, $m \neq \phi(s+1)$ in the interval $[s_{\min}, s_{\max}]$.

This interval is not empty: $s_{\max} - s_{\min} = m - k - 2^i + 1 \geq 0$. We have $s_{\min} + s_{\max} = k + m - 1 \leq 2n - 2$ because $m < \phi(k)$.

If the interval contains at least two points, $s_{\min} \leq s_{\max} - 1$, then $2s_{\min} \leq s_{\min} + s_{\max} - 1 \leq 2n - 3$. Hence, $s_{\min} \leq n - 2$; that is, the interval contains at least two points satisfying $s < n$. One of them satisfies $m \neq \phi(s+1)$.

If the interval contains just one point, $s = s_{\min} = s_{\max}$, then $m + s = m + s_{\max} = 2m - 2^{i-1}$ is an even number (here of course $i > 1$). At the same time, $m = \phi(s+1)$ is equivalent to $m + s + 1 = 2n$. Hence, $m \neq \phi(s+1)$.  $\square$

**Theorem 8.1.** *The combinatorial rank of $u_q^+(\mathfrak{sp}_{2n})$ equals $\lfloor \log_2(n-1) \rfloor + 2$.*

**Proof.** First, we note that $J_\kappa^+$ with $\kappa = \lfloor \log_2(n-1) \rfloor + 2$ contains all elements $[v(k, m)]^t$, and $[v_k]^{t'}$.

Using the evident inequality $a < 1 + \lfloor a \rfloor$, we have

$$m - k \leq (\phi(1) - 1) - 1 = 2n - 3 = 2^{1+\log_2(n-1)} - 1 < 2^{2+\lfloor \log_2(n-1) \rfloor} - 1,$$

and Proposition 8.1 implies that $[v(k, m)]^t \in J_\kappa^+$.

Similarly,

$$n - k \leq n - 1 = 2^{\log_2(n-1)} < 2^{1+\lfloor \log_2(n-1) \rfloor} = 2^{\kappa-1} \leq 2^\kappa - 1.$$

Hence, Proposition 8.1 implies $[v_k]^t \in J_\kappa^+$ if $t$ is odd, and $[v_k]^{t/2} \in J_\kappa^+$ if $t$ is even.

Next, we note that $[v(1, 2n-2)]^t \notin J^+_{\kappa-1}$. Using inequality $a \geq \lfloor a \rfloor$, we have

$$(2n-2) - 1 = 2^{1+\log_2(n-1)} - 1 \geq 2^{\kappa-1} - 1,$$

and Proposition 8.1 applies. $\qquad\square$

**Theorem 8.2.** *The combinatorial rank of* $u_q(\mathfrak{sp}_{2n})$ *is* $\lfloor \log_2(n-1) \rfloor + 2$.

The proof almost literally coincides with the proof of Ref. 6(Theorem 8.1) or Ref. 4(Theorem 4).

## Acknowledgments

The authors were supported by PAPIIT grant IN 105219 UNAM and PI-API1824 of the FES-Cuautitlán UNAM. The first author express his deep gratitude to the Max Planck Institute for Mathematics in Bonn for the financial grant and the hospitality.

## References

1. N. Andruskiewitsch, H.-J. Schneider, Pointed Hopf algebras, in: S. Montgomery, H.-J. Schneider (Eds.) New Directions in Hopf Algebras, MSRI Publications, 43(2002), 1–68.
2. C. De Concini and V. G. Kac, *Representations of quantum groups at roots of 1*, Colloque Dixmier 1989, pp. 471-506; Progress in Math., vol. 92, Birkhauser, Boston, MA, 1990.
3. C. De Concini, V.G. Kac, C. Procesi, *Quantum coadjoint action*, Journal of AMS, 5(1992), 151–189.
4. M.L. Díaz Sosa, V. Kharchenko, Combinatorial rank of $u_q(\mathfrak{so}_{2n})$, Journal of Algebra, 448(2016), 48–73.
5. V. K. Kharchenko, A. Andrade Álvarez, *On the combinatorial rank of Hopf algebras*, Contemporary Mathematics, v. 376(2005), 299–308.
6. V. K. Kharchenko, M. L. Diaz Sosa, *Computing of the combinatorial rank of* $u_q(\mathfrak{so}_{2n+1})$, Communications in Algebra, v. 39(2011) 4705–4718.
7. V. K. Kharchenko, *Explicit coproduct formula for quantum groups of infinite series*, Israel Journal of Mathematics, 208(2015), 13-43.
8. V. K. Kharchenko, *PBW-bases of coideal subalgebras and a freeness theorem*, TAMS, v. 300, N10(2008), 5121–5143.
9. V. K. Kharchenko, *A combinatorial approach to the quantification of Lie algebras*, Pacific Journal of Mathematics 203 N1(2002) 191–233.
10. V. K. Kharchenko, *A quantum analogue of the Poincaré-Birkhoff-Witt theorem*, Algebra i Logika 38 N4(1999) 476–507; English translation: Algebra and Logic, 38, N4(1999), 259-276; arXiv: math. QA/0005101.
11. V. K. Kharchenko, *An algebra of skew primitive elements*, Algebra i Logica, 37, N2(1998),181–224, English translation: Algebra and Logic, 37, N2(1998),101–127, arXiv: math. QA/0006077.

12. V. K. Kharchenko, *Right coideal subalgebras of* $U_q^+(\mathfrak{so}_{2n+1})$, Journal of the European Mathematical Society, 13(2011), 1677–1735.
13. M. Lothaire, Algebraic Combinatorics on Words, Cambridge Univ. Press, 2002.
14. D.E. Radford, *The structure of Hopf algebras with projection*, Journal of Algebra, 92(1985), 322–347.
15. A.I. Shirshov, *On free Lie rings*, Matem. Sbornic 45, 87(2)(1958), 113–122.
16. A.I. Shirshov, *Some algorithmic problems for Lie algebras*, Sibirskii Math. J., 3(2)(1962), 292–296.
17. M. Takeuchi, *Survey of braided Hopf algebras,* in: New Trends in Hopf Algebra Theory, Contemp. Math., vol. 267, AMS, Providence RI, 2000, pp. 301-324.

# Gröbner–Shirshov bases for associative conformal algebras with arbitrary locality function[*]

P. Kolesnikov

*Sobolev Institute of Mathematics,*
*Akad. Koptyug prosp., 4,*
*Novosibirsk 630090, Russia*

*pavelsk@math.nsc.ru*

We present an approach to the computation of confluent systems of defining relations in associative conformal algebras based on the similar technique for modules over ordinary associative algebras.

*2010 Mathematics subject classification:* 16S15, 13P10, 17A32.

## 1. Introduction

Gröbner–Shirshov bases (GSB) is a useful tool for solving theoretical problems in algebra (see, e.g., Ref. 1). It is not so common in applications as Gröbner bases in commutative algebra since the construction of a confluent rewriting system for a given ideal may not be done in a finite number of steps.

In order to establish a version of GSB theory for a particular class of algebraic systems we usually need to determine the structure of free objects (determine what is a normal word), define a linear order on the set of normal words compatible with algebraic structure (i.e., monomial order), define a sort of elimination procedure, and state an appropriate version of the Composition-Diamond Lemma (CD-Lemma). The latter is the key statement of the theory, a criterion for a set of defining relations to be confluent.

However, there are many cases when realization of the entire program described above is excessive. For example, GSB theory for modules over associative algebras[2] can be explicitly deduced from the classical CD-Lemma for associative algebras (see Theorem 2.2 below), GSB theory for

---

[*]Supported by SB RAS project 0314-2016-0001.

di-algebras[3,4] may also be considered as an application of the same statement[5]. In a similar way, GSB theory for Leibniz algebras may be derived from the Shirshov's CD-Lemma for Lie algebras.

Conformal algebras appeared in mathematical physics as "singular parts" of vertex algebras[6]. Combinatorial study of associative conformal algebras was started in Ref. 11. In a series of papers (see Refs. 7–9), different versions of the CD-Lemma for associative conformal algebras was stated. In this paper, we show how to derive GSB theory for associative conformal algebras from CD-Lemma for modules over (ordinary) associative algebras. This approach is technically simpler than and it is more general: we do not assume the locality of generators is bounded. Moreover, in this way one may use existing computer algebra systems to calculate compositions in conformal algebras. As an application, we calculate GSB for universal associative envelopes of some Lie conformal algebras.

Throughout the paper, $\Bbbk$ is a base field of characteristic zero, $\mathbb{Z}_+$ is the set of non-negative integers. Given a set $X$, $X^*$ stands for the set of nonempty words in the alphabet $X$, $X^\#$ denotes the set $X^*$ together with the empty word. We will use notation like $x^{(s)}$ for $\frac{1}{s!}x^s$ for $s \in \mathbb{Z}_+$.

## 2. Preliminaries: CD-Lemma for modules

Let us recall the classical CD-lemma for associative algebras in the form of Ref. 10.

Suppose $O$ is a set of generators and $\preceq$ is a monomial order on $O^*$. As usual, $\bar{f}$ denotes the principal monomial of a nonzero polynomial from the free associative algebra $\Bbbk\langle O \rangle$ relative to the order $\preceq$.

If $f$ and $g$ are monic polynomials in $\Bbbk\langle O \rangle$ such that $w = \bar{f} = u\bar{g}v$ for some $u, v \in O^\#$, then

$$(f, g)_w = f - ugv$$

is said to be a composition of inclusion.

If $\bar{f} = u_1 u_2$, $\bar{g} = u_2 v_2$ for some $u_1, u_2, v_2 \in O^*$ then

$$(f, g)_w = f v_2 - u_1 g, \quad w = u_1 u_2 v_2,$$

is said to be a composition of intersection.

Given a polynomial $f \in \Bbbk\langle O \rangle$, a set of polynomials $\Sigma \subset \Bbbk\langle O \rangle$, and a word $w \in O^*$, we say $f$ is trivial modulo $\Sigma$ and $w$ if $f$ can be presented in the form

$$f = \sum_i \alpha_i u_i s_i v_i, \quad \alpha_i \in \Bbbk, \; u_i, v_i \in O^\#, \; s_i \in \Sigma,$$

where $u_i \bar{s}_i v_i \prec w$. We will use the following notation:

$$f \equiv g \pmod{\Sigma, w}$$

if $f - g$ is trivial modulo $\Sigma$ and $w$.

A word $u \in O^*$ is said to be $\Sigma$-reduced if $u \neq v_1 \bar{s} v_2$, $v_1, v_2 \in O^{\#}$, $s \in \Sigma$.

A set of monic polynomials $\Sigma \subset \Bbbk\langle O \rangle$ is said to be a Gröbner–Shirshov basis (GSB) if $(f, g)_w \equiv 0 \pmod{\Sigma, w}$ for every pair of polynomials $f, g \in \Sigma$.

By abuse of terminology, we say $\Sigma$ is a GSB of the ideal $I = (\Sigma)$ in $\Bbbk\langle O \rangle$.

Note that a pair of polynomials in $\Bbbk\langle O \rangle$ may have more than one composition. In order to make sure that $\Sigma$ is a GSB we have to check if all compositions are trivial in the above-mentioned sense.

The following statement is known as the CD-Lemma for associative algebras [10].

**Theorem 2.1.** *For a set $\Sigma$ of monic polynomials in $\Bbbk\langle O \rangle$ the following statements are equivalent:*

*(1) $\Sigma$ is a GSB;*
*(2) If $f \in (\Sigma)$ then $\bar{f} = u \bar{s} v$ for appropriate $u, v \in O^{\#}$, $s \in \Sigma$;*
*(3) The set of $\Sigma$-reduced words forms a linear basis of $\Bbbk\langle O \mid \Sigma \rangle$.*

The same approach works for modules over associative algebras. Suppose $O$ is a set of generators of an associative algebra $A$, $X$ is a set of generators of a left $A$-module $M$. Let $\Sigma \subset \Bbbk\langle O \rangle$ be a set of defining relations of $A$ and let $S \subset \Bbbk\langle O \rangle \otimes \Bbbk X$ be a set of defining relations of $M$. We will identify $\Bbbk\langle O \rangle \otimes \Bbbk X$ with a subset $\Bbbk\langle O \dot{\cup} X \rangle_1$ of $\Bbbk\langle O \dot{\cup} X \rangle$ that consist of polynomials

$$\sum_i \alpha_i u_i x_i, \quad \alpha_i \in \Bbbk, \ u_i \in O^{\#}, \ x_i \in X.$$

Consider the alphabet $Z = O \dot{\cup} X$ and a monomial order on $Z^*$. Obviously, $M$ is a subspace of the associative algebra $A(M)$ generated by $Z$ relative to the defining relations $\Sigma \cup S$ along with $xa, xy$ $x, y \in X$, $a \in O$. Indeed, $A(M)$ is the split null extension of $A$ by means of $M$.

**Definition 2.1.** Let $\Sigma$ be a GSB in $\Bbbk\langle O \rangle$, $A = \Bbbk\langle O \mid \Sigma \rangle$, $F(X)$ is the free $A$-module generated by $X$. A set $S$ of monic polynomials in $\Bbbk\langle O \dot{\cup} X \rangle_1$ is said to be a Gröbner–Shirshov basis in $F(X)$ if $S \cup \Sigma \cup \{xa, xy \mid x, y \in X, a \in O\}$ is a GSB in $\Bbbk\langle O \dot{\cup} X \rangle$.

The following statement is equivalent to the CD-lemma for modules[2].

**Theorem 2.2.** *For a set of monic polynomials $S \subset \Bbbk\langle O \dot\cup X\rangle_1$ the following statements are equivalent:*

(1) *$S$ is a GSB in $F(X)$;*
(2) *If $f \in \Bbbk\langle O \dot\cup X\rangle_1$ belongs to the $A$-submodule generated by $S$ then either $\bar f = u\bar svx$, $u, v \in O^\#$, $s \in \Sigma$, $x \in X$, or $\bar f = u\bar g$, $u \in O^\#$, $g \in S$;*
(3) *The set of $(\Sigma \cup S)$-reduced set of words in $\Bbbk\langle O \dot\cup X\rangle_1$ forms a linear basis of the $A$-module $M$ generated by $X$ relative to the defining relations $S$.*

## 3. Conformal algebras

Conformal algebra[6] is a linear space $C$ equipped with a linear map $\partial : C \to C$ and a family of bilinear products $(\cdot_{(n)} \cdot)$, $n \in \mathbb{Z}_+$, such that the following axioms hold: for every $a, b \in C$

$$a_{(n)} b = 0 \quad \text{for almost all } n \in \mathbb{Z}_+; \tag{1}$$

$$\partial a_{(n)} b = -na_{(n-1)} b, \quad a_{(n)} \partial b = \partial(a_{(n)} b) + na_{(n-1)} b. \tag{2}$$

Locality function on $C$ is a map $N_C : C \times C \to \mathbb{Z}_+$, where $N_C(a, b)$ is equal to the minimal $N$ such that $a_{(n)} b = 0$ for all $n \geq N$.

If

$$a_{(n)} (b_{(m)} c) = \sum_{s \geq 0} (-1)^s \binom{n}{s} (a_{(n-s)} b)_{(m+s)} c$$

for all $a, b, c \in C$ and $n, m \in \mathbb{Z}_+$ then $C$ is said to be associative.

Axiom (1) shows the class of associative conformal algebras is not a variety in the sense of universal algebra. However, for a given set $X$ of generators and for a fixed function $N : X \times X \to \mathbb{Z}_+$ there exists a free object in the class of associative conformal algebras $C$ generated by $X$ such that $N_C(a, b) \leq N(a, b)$ for $a, b \in X$ (see Ref. 11). This free algebra is unique up to isomorphism, let us denote it $\mathrm{Conf}(X, N)$. As shown in Ref. 11, $\mathrm{Conf}(X, N)$ has a linear basis which consists of words

$$\partial^s(a_1{}_{(n_1)} (a_2{}_{(n_2)} \ldots (a_k{}_{(n_k)} a_{k+1}) \ldots)),$$

$$s \geq 0, \ k \geq 1, \ a_i \in X, \ 0 \leq n_i < N(a_i, a_{i+1}).$$

Following Ref. 12, introduce the family of operations

$$\{a_{(n)} b\} = \sum_{s \geq 0} (-1)^{n+s} \partial^{(s)}(a_{(n+s)} b), \quad a, b \in C, \ n \in \mathbb{Z}_+.$$

As shown in Ref. 12, in an associative conformal algebra we have

$$\{\partial a \;_{(n)}\; b\} = \partial\{a \;_{(n)}\; b\} + n\{a \;_{(n-1)}\; b\}, \quad \{a \;_{(n)}\; \partial b\} = -n\{a \;_{(n-1)}\; b\},$$
$$a \;_{(n)}\; \{b \;_{(m)}\; c\} = \{(a \;_{(n)}\; b) \;_{(m)}\; c\}.$$

## 4. Module construction of the free associative conformal algebra

Let $(X, \leq)$ be a well-ordered set, and $N : X \times X \to \mathbb{Z}_+$ be an arbitrary function. Consider the set of symbols

$$O = \{\partial, L_n^a, R_n^a \mid n \in \mathbb{Z}_+, \; a \in X\}$$

Define a monomial order $\preceq$ on $O^*$ in the following way: compare two words first by their degree in the variables $R_n^a$, then by deg-lex order assuming $\partial < L_n^a < R_m^b$, $L_n^a < L_m^b$ if $n < m$ or $n = m$ and $a < b$; similarly for $R_n^a$, $R_m^b$.

Denote by $A(X)$ the associative algebra generated by $O$ relative to the following defining relations:

$$L_n^a \partial - \partial L_n^a - n L_{n-1}^a, \tag{3}$$
$$R_n^a \partial - \partial R_n^a - n R_{n-1}^a, \tag{4}$$
$$R_m^b L_n^a - L_n^a R_m^b, \tag{5}$$

$a, b \in X$, $n, m \in \mathbb{Z}_+$.

**Lemma 4.1.** *Polynomials* (3)–(5) *form a GSB with respect to the order* $\preceq$ *on* $O^*$.

**Proof.** The only composition here is the composition of intersection of (3) and (5). It is straightforward to check that this composition is trivial. $\square$

Denote by $F(X)$ the free (left) $A(X)$-module generated by $X$. Lemma 4.1 implies that the linear base of $F(X)$ consists of words

$$\partial^s L_{n_1}^{a_1} \ldots L_{n_k}^{a_k} R_{m_1}^{b_1} \ldots R_{m_p}^{b_p} c, \quad s, k, p, n_i, m_j \in \mathbb{Z}_+, \; a_i, b_j, c \in X. \tag{6}$$

**Definition 4.1.** Let us call a word of type (6) *normal* if $p = 0$, $n_i < N(a_i, a_{i+1})$ for $i = 1, \ldots, k-1$, and $n_k < N(a_k, c)$. Denote by $B(X, N)$ the set of all normal words, and use $B_0(X, N)$ for the set of $\partial$-free normal words.

Let $M(X, N)$ stand for the $A(X)$-module generated by $X$ relative to the following relations:

$$L_n^a b, \quad a, b \in X, \; n \geq N(a, b), \tag{7}$$

$$R_n^b a - (-1)^n \sum_{s=0}^{N(a,b)-n-1} \partial^{(s)} L_{n+s}^a b, \quad a, b \in X, \ n \in \mathbb{Z}_+. \tag{8}$$

This module is the main object of study in this section. The following statement explains the origin of $A(X)$ and $M(X, N)$.

**Proposition 4.1.** *The free associative conformal algebra* $\mathrm{Conf}(X, N)$ *is an $A(X)$-module which is a homomorphic image of $M(X, N)$.*

**Proof.** The action of $A(X)$ on $\mathrm{Conf}(X, N)$ is given by

$$L_n^a u = a_{(n)} u, \quad R_n^a u = \{u_{(n)} a\}$$

for $a \in X$, $u \in \mathrm{Conf}(X, N)$, $n \in \mathbb{Z}_+$, $\partial$ acts naturally. It is easy to see that these rules are compatible with the defining relations of $A(X)$, as well as (7), (8) hold. Obviously, $\mathrm{Conf}(X, N)$ as an $A(X)$-module is generated by $X$. $\qquad\square$

Let us extend the order $\preceq$ to the set of words $O^* \cup O^\# X$ which occur in the computation of compositions in the free $A(X)$-module generated by $X$. For $ux, vy \in O^\# X$, where $u, v \in O^\#$, $a, y \in X$, assume $ux \prec vy$ if and only if $u \prec v$ or $u = v$ and $x < y$. Moreover, set $O^* \prec O^\# X$.

**Theorem 4.1.** *Gröbner–Shirshov basis of the $A(X)$-module $M(X, N)$ consists of (7), (8) and*

$$L_n^a L_m^b u + \sum_{q \geq 1} (-1)^q \binom{n}{q} L_{n-q}^a L_{m+q}^b u, \tag{9}$$

*where $a, b \in X$, $n \geq N(a, b)$, $m \in \mathbb{Z}_+$, $u \in B_0(X, N)$.*

**Proof.** Let us show that relations (9) follow from the defining relations (7), (8). For $u = c \in X$, consider the compositions of intersection of

$$f = R_m^c L_n^a - L_n^a R_m^c, \quad g = L_n^a b$$

for $n \geq N(a, b)$, $m \in \mathbb{Z}_+$:

$$(f, g)_{R_m^c L_n^a b} = (R_m^c L_n^a - L_n^a R_m^c) b - R_m^c (L_n^a b)$$

$$\equiv (-1)^{m+1} \sum_{s=0}^{N(b,c)-1} (-1)^s L_n^a \partial^{(s)} L_{m+s}^b c$$

$$\equiv (-1)^{m+1} \sum_{s=0}^{N(b,c)-1} \sum_{t \geq 0} (-1)^s \binom{n}{t} \partial^{(s-t)} L_{n-t}^a L_{m+s}^b c.$$

These compositions are trivial for $m \geq N(b,c)$, for $m = N(b,c) - 1$ we obtain $(f,g)_{R_m^c L_n^a b}$ to be a multiple of $L_n^a L_m^b c$. For smaller $m$, proceed by induction. Suppose (9) holds for $u = c$ and all $m > p$. Then

$$(f,g)_{R_p^c L_n^a b} \equiv (-1)^{p+1} \sum_{s=0}^{N(b,c)-1} \sum_{t\geq 0} (-1)^s \binom{n}{t} \partial^{(s-t)} L_{n-t}^a L_{p+s}^b c$$

$$= (-1)^{p+1} \left( L_n^a L_p^b c + \sum_{s=1}^{N(b,c)-1} (-1)^s \left( \sum_{t\geq 0} \binom{n}{t} \partial^{(s-t)} L_{n-t}^a L_{p+s}^b c \right) \right)$$

$$\equiv (-1)^{p+1} \left( L_n^a L_p^b c + \sum_{s=1}^{N(b,c)-1} \left( \partial^{(s)} \sum_{q\geq 1} (-1)^{q+s+1} \binom{n}{q} L_{n-q}^a L_{p+s+q}^b c \right. \right.$$

$$\left. \left. + \sum_{t\geq 1} (-1)^s \binom{n}{t} \partial^{(s-t)} L_{n-t}^a L_{p+s}^b c \right) \right)$$

$$= (-1)^{p+1} \left( L_n^a L_p^b c + \sum_{s,q\geq 1} (-1)^{q+s+1} \binom{n}{q} \partial^{(s)} L_{n-q}^a L_{p+s+q}^b c \right.$$

$$\left. + \sum_{r\geq 0, t\geq 1} (-1)^{r+t} \binom{n}{t} \partial^{(r)} L_{n-t}^a L_{p+r+t}^b c \right)$$

$$= (-1)^{p+1} \left( L_n^a L_p^b c + \sum_{q\geq 1} (-1)^q \binom{n}{q} L_{n-q}^a L_{p+q}^b c \right). \quad (10)$$

Here we have applied substitution $r = s-t$ for the last sum in the right-hand side of (10). Therefore, relation (9) holds for $m = p$ as well.

Denote by $\Sigma(X,N)$ the set of relations (7), (8), and (9). Note that the set of normal words is exactly the set of $\Sigma(X,N)$-reduced words. Images of these words under the homomorphism from Proposition 4.1 are linearly independent in $\mathrm{Conf}(X,N)$ as shown in Ref. 11. Hence, $B(X,N)$ is a linear basis of $M(X,N)$ and thus $\Sigma(X,N)$ is a GSB. $\qquad\square$

**Remark 4.1.** It is not hard to check in a straightforward way that $\Sigma(X,N)$ is closed with respect to compositions.

**Corollary 4.1.** *Free associative conformal algebra* $\mathrm{Conf}(X,N)$ *as* $A(X)$-*module is isomorphic to* $M(X,N)$. *The set of normal words* $B(X,N)$ *is a linear basis of the module* $M(X,N)$.

**Corollary 4.2.** *Ideals of* $\mathrm{Conf}(X,N)$ *are exactly* $A(X)$-*submodules of* $M(X,N)$.

**Proof.** Every (two-sided) ideal of $\text{Conf}(X, N)$ is obviously closed with respect to the action of $\partial$, $L_n^a$, and $R_n^a$. Conversely, every $A(X)$-submodule of $M(X, N)$ is $\partial$-invariant and closed relative to left and right conformal multiplications on $a \in X$. $\qquad\qquad\qquad\qquad\qquad\qquad\qquad\qquad\qquad\qquad\qquad\square$

Let us summarize the results above to state the CD-Lemma for associative conformal algebras.

**Definition 4.2.** Suppose $S$ is a set of elements in $\text{Conf}(X, N)$. Identify $S$ with a set of linear combinations of normal words $B(X, N)$ in $F(X)$. If $S$ together with $\Sigma(X, N)$ is a Gröbner–Shirshov basis (GSB) in the $A(X)$-module $F(X)$ then we say that $S$ is a GSB in $\text{Conf}(X, N)$.

By abuse of terminology, we say a GSB $S$ in $\text{Conf}(X, N)$ is a GSB of ideal generated by $S$ in $\text{Conf}(X, N)$.

**Theorem 4.2.** *The following statements are equivalent:*

- *$S$ is a GSB in $\text{Conf}(X, N)$;*
- *The set of $S$-reduced normal words forms a linear basis of the associative conformal algebra $\text{Conf}(X, N \mid S)$.*

## 5. Applications

One of the most interesting questions in the theory of conformal algebras is to determine if every torsion-free finite Lie conformal (super)algebra $L$ with operation $[\cdot\ _{(\lambda)}\ \cdot]$ is special, i.e., whether it can be embedded into an associative conformal algebra $C$ in such a way that

$$[x\ _{(\lambda)}\ y] = (x\ _{(\lambda)}\ y) - \{y\ _{(\lambda)}\ x\}, \quad x, y \in L. \qquad (11)$$

One of the most natural ways to resolve the speciality problem is to apply GSB theory. Namely, suppose $X$ is a basis of $L$ as of $\Bbbk[\partial]$-module. The multiplication table in $L$ is given by

$$[a\ _{(\lambda)}\ b] = \sum_{c \in X} f_c^{a,b}(\partial, \lambda)c, \quad f_c^{a,b}(\partial, \lambda) \in \Bbbk[\partial, \lambda], \ a, b \in X. \qquad (12)$$

Obviously, $L$ is special if and only if there exists a function $N : X \times X \to \mathbb{Z}_+$ such that the GSB of the ideal generated by the set

$$a\ _{(n)}\ b - \{b\ _{(n)}\ a\} - g_n^{a,b}, \quad a, b \in X, \ n \in \mathbb{Z}_+, \qquad (13)$$

where $g_n^{a,b} = [a\ _{(n)}\ b]$ is the coefficient of the right-hand side of (12) at $\lambda^{(n)}$, does not contain nonzero $\Bbbk[\partial]$-linear combinations of elements from $X$.

The corresponding quotient conformal algebra is the universal associative envelope of $L$ relative to the locality function $N$ on generators $X$ [13]. Let us denote it by $U(L; X, N)$.

As an $A(X)$-module, $U(L; X, N)$ is generated by $X$ relative to the defining relations (7), (8), and (13). Sesqui-linearity of the conformal $\lambda$-product implies that if $[x_{(\lambda)} y]$ is given by (11) in an associative conformal algebra $C$ then

$$[x_{(\lambda)} y]_{(\lambda+\mu)} z = x_{(\lambda)} (y_{(\mu)} z) - y_{(\mu)} (x_{(\lambda)} z)$$

for $x, y, z \in C$. Therefore, the following relations hold on $U(L; X, N)$:

$$L_n^a L_m^b u - L_m^b L_n^a u - \sum_{s \geq 0} \binom{n}{s} L_{n+m-s}^{g_s^{a,b}} u, \quad a, b \in X, \ n, m \in \mathbb{Z}_+$$

for every normal word $u \in B(X, N)$ (in the last summand, we assume $L_n^{\partial x} = -n L_{n-1}^x$). Therefore, in order to analyze the structure of $U(L; X, N)$ we may replace $A(X)$ with the algebra $A(L; X)$ generated by $X$ relative to the defining relations (3)–(5) along with

$$L_n^a L_m^b - L_m^b L_n^a - \sum_{s \geq 0} \binom{n}{s} L_{n+m-s}^{g_s^{a,b}}, \quad a, b \in X, \ n, m \in \mathbb{Z}_+, \ L_n^a > L_m^b. \quad (14)$$

**Proposition 5.1.** *Relations (3)–(5) and (14) form a GSB in the free associative algebra generated by $O$ with respect to the order $\preceq$.*

**Proof.** A relation of type (14) has compositions of intersection with (3) and (5). It is straightforward to check these compositions are trivial. Two relations of type (14) may also have a composition of intersection. Such a composition is also trivial since (14) is nothing but the multiplication table of a Lie algebra spanned by operators $L_n^a$, $n \in \mathbb{Z}_+$, $a \in X$. To be more precise, $A(L; X)$ is the universal associative envelope of the Lie algebra $\Bbbk \partial \ltimes (\mathcal{A}(L) * \mathcal{R}(X)/([\mathcal{A}(L), \mathcal{R}(X)]))$, where $\mathcal{A}(L)$ is the annihilation algebra of $L$ (positive part of the coefficient algebra), $\mathcal{R}(X)$ is the free Lie algebra generated by $R_n^a$, $*$ denotes free product of Lie algebras, and $\partial$ acts as described by (3), (4).

The classical Poincaré–Birkhoff–Witt Theorem states that the multiplication table of a Lie algebra is a GSB relative to the deg-lex order based on an arbitrary ordering of the generators. The order $\preceq$ we use is not exactly deg-lex, but all relations (3)–(5), (14) are homogeneous relative to the variables $R_n^a$, so we may still conclude that this is a GSB. □

**Remark 5.1.** The last statement in the proof of Proposition 5.1 is equivalent to so called 1/2-PBW Theorem for conformal algebras[8].

Therefore, in order to determine if $L$ is embedded into its universal envelope $U(L; X, N)$ we have to find GSB of the $A(L; X)$-module generated by $X$ relative to the defining relations (8), (7), and

$$L_n^a b - R_n^a b - g_n^{a,b}, \quad a, b \in X, \ n \in \mathbb{Z}_+. \tag{15}$$

Let us consider two examples of such a computation.

**Example 5.1.** Lie conformal algebra $L = \text{Vir} \ltimes \text{Cur} \, \mathbb{k}$ is generated as a $\mathbb{k}[\partial]$-module by $X = \{v, h\}$, where

$$[v_{(0)} v] = \partial v, \quad [v_{(1)} v] = 2v,$$
$$[v_{(0)} h] = \partial h, \quad [v_{(1)} h] = h, \quad [h_{(1)} v] = h,$$

other products are zero.

For the Heisenberg–Virasoro Lie conformal algebra $L$, $A(L; X)$ is generated by $O = \{\partial, L_n^h, R_n^h, L_n^v, R_n^v \mid n \in \mathbb{Z}_+\}$ relative to the following defining relations:

$$L_n^a \partial - \partial L_n^a - n L_{n-1}^a, \quad n \geq 0, \ a \in X, \tag{16}$$
$$R_n^a \partial - \partial R_n^a - n R_{n-1}^a, \quad n \geq 0, \ a \in X, \tag{17}$$
$$R_n^b L_m^b - L_m^b R_n^a, \quad n, m \geq 0, \ a, b \in X, \tag{18}$$
$$L_n^v L_m^v - L_m^v L_n^v - (n - m) L_{n+m-1}^v, \quad n > m \geq 0, \tag{19}$$
$$L_n^h L_m^v - L_m^v L_n^h - n L_{n+m-1}^h, \quad n, m \geq 0, \tag{20}$$
$$L_n^h L_m^h - L_m^h L_n^h, \quad n > m \geq 0. \tag{21}$$

We will use an order on $O^*$ which is slightly different from $\preceq$ used in Section 4: assume $L_n^h > L_m^v$ for all $n, m \geq 0$. Obviously, (16)–(21) remains a GSB relative to this modified order.

Let us fix the following locality function on $X$:

$$N(v, v) = N(h, v) = 2, \quad N(v, h) = 1, \quad N(h, h) = 0.$$

Then relations (7) and (8) turn into

$$L_n^v h, \ n \geq 1, \quad L_m^h h, \ m \geq 0, \tag{22}$$
$$L_n^v v, \ L_n^h v, \ n \geq 2, \tag{23}$$
$$R_0^v h - L_0^h v + \partial L_1^h v, \quad R_1^v h + L_1^h v, \quad R_n^v h, \ n \geq 2, \tag{24}$$
$$R_0^v v - L_0^v v + \partial L_1^v v, \quad R_1^v v + L_1^v v, \quad R_n^v v, \ n \geq 0. \tag{25}$$

$$R_0^h v - L_0^v h, \quad R_n^h v, \ n \geq 1, \quad R_m^h h, \ m \geq 0. \tag{26}$$

Commutation relations (15) turn into

$$L_1^v v - v, \quad L_1^h v - h, \quad L_0^h v - L_0^v h. \tag{27}$$

It is straightforward to check that the set $S$ of relations (16)–(26) form a GSB in the free $A(L; X)$-module generated by $X$. For example, let us show that the composition of intersection of $f = R_0^v L_0^h - L_0^h R_0^v$ and $g = L_0^h v - L_0^v h$ is trivial:

$$\begin{aligned}
(f, g)_{R_0^v L_0^h v} &= (R_0^v L_0^h - L_0^h R_0^v)v - R_0^v(L_0^h v - L_0^v h) = R_0^v L_0^v h - L_0^h R_0^v v \\
&\equiv L_0^v R_0^h h - L_0^h R_0^v v \equiv L_0^v(L_0^v v - \partial L_1^h v) - L_0^h(L_0^v v - \partial L_1^v v) \\
&\equiv (L_0^v L_0^h - L_0^h L_0^v)v - \partial L_0^v L_1^h v + \partial L_0^h L_1^v v \equiv \partial L_0^h L_1^v v - \partial L_0^v L_1^h v \\
&\equiv L_0^h v - L_0^v h \equiv 0 \pmod{\Sigma \cup S, R_0^v L_0^h v}.
\end{aligned}$$

The set of $(\Sigma \cup S)$-reduced normal words consists of

$$\partial^s (L_0^v)^k a, \quad a \in X, \ s, k \geq 0.$$

These words form a linear basis of $U(L; X, N)$.

**Example 5.2.** Let us evaluate the GSB of the universal associative envelope of the Virasoro conformal algebra relative to $N(v, v) = 3$.

Here we will use a different ordering of generators $O$. Denote $L_n = L_n^v$, $R_n^v = R_n$, suppose

$$L_0 < L_1 < \partial < L_2 < \cdots < R_0 < R_1 < \ldots,$$

and compare monomials in $O^*$ first by their degree in $R_n$, next by deg-lex rule.

In these settings, $A(L; X)$ is generated by $O$ relative to the following defining relations:

$$\partial L_0 - L_0 \partial, \tag{28}$$
$$\partial L_1 - L_1 \partial + L_0, \tag{29}$$
$$L_n \partial - \partial L_0 - n L_{n-1}, \quad n \geq 2, \tag{30}$$
$$R_n \partial - \partial R_n - n R_{n-1}, \quad n \geq 0, \tag{31}$$
$$R_n L_m - L_m R_n, \quad n, m \geq 0, \tag{32}$$
$$L_n L_m - L_m L_n - (n - m) L_{n+m-1}, \quad n > m \geq 0. \tag{33}$$

Polynomials (28)–(33) form a GSB in $\Bbbk \langle O \rangle$ by Proposition 5.1.

As an $A(L; X)$-module, $U(\mathrm{Vir}; v, 3)$ is generated by $X = \{v\}$ with the following relations:

$$L_n v, \quad n \geq 3, \tag{34}$$

$$R_0 v - 2L_0 v + L_1 \partial v - \partial^{(2)} L_2 v, \tag{35}$$

$$R_1 v + L_1 v - \partial L_2 v, \tag{36}$$

$$R_2 v - L_2 v, \tag{37}$$

$$R_n v, \quad n \geq 3, \tag{38}$$

$$\partial L_2 v - 2L_1 v + 2v. \tag{39}$$

Relations (28)–(39) have the following compositions of intersection:

- (32) with (34), $w = R_m L_n v$, $n \geq 3$;
- (33) with (34), $w = L_m L_n v$, $m > n \geq 3$;
- (30) with (39), $w = L_n \partial L_2 v$, $n \geq 2$;
- (31) with (39), $w = R_m L_n v$, $n \geq 0$.

Calculation of these compositions gives only one new defining relation

$$L_2 L_2 v. \tag{40}$$

Denote by $S$ the set of relations (28)–(40) Relation (40) has a composition of intersection with (33), $w = L_n L_2 L_2 v$, $n \geq 3$, which is trivial modulo $S$ and $w$. Yet another composition is obtained from (32) and (40) relative to $w = R_n L_2 L_2 v$, $n \geq 0$. This composition is obviously trivial for $n \geq 2$. Let us show as an example the computation of that composition for $n = 0$:

$$
\begin{aligned}
R_0 L_2 L_2 v \equiv L_2 L_2 R_0 v &\equiv L_2 L_2 (L_0 v - \partial L_1 v + \partial^{(2)} L_2 v) \\
&\equiv L_2 L_2 L_0 v - L_2 \partial L_2 L_1 v - 2L_2 L_1 L_1 v + 2L_2 \partial L_1 L_2 v + L_2 L_0 L_2 v \\
&\equiv L_2 L_2 L_0 v - 2L_1 L_2 L_1 v - 2L_2 L_1 L_1 v + 4L_1 L_1 L_2 v + L_2 L_0 L_2 v \\
&\equiv 2L_2 L_0 L_2 v + 2L_2 L_1 v - 2L_1 L_2 v - 2L_2 L_1 L_1 v + 2L_1 L_1 L_2 v \\
&\equiv 4L_1 L_2 v + 2L_1 L_2 v + 2L_2 v + 2L_1 L_1 L_2 v - 2L_1 L_2 v - 2L_2 L_1 L_1 v \\
&\equiv 4L_1 L_2 v + 2L_2 v + 2L_1 L_1 L_2 v - 2L_1 L_1 L_2 v - 4L_1 L_2 v - 2L_2 v \\
&\equiv 0 \quad (\mathrm{mod}\ S, R_0 L_2 L_2 v).
\end{aligned}
$$

As a result, $S$ is a GSB of $U(\mathrm{Vir}; v, 3)$.

**Remark 5.2.** Note that Proposition 5.1 remains valid for conformal superalgebras provided that we add the appropriate parity to (14):

$$L_n^a L_m^b - (-1)^{p(a)p(b)} L_m^b L_n^a - \sum_{s \geq 0} \binom{n}{s} L_{n+m-s}^{g_s^{a,b}},$$

$$a, b \in X, \ n, m \in \mathbb{Z}_+, \ L_n^a > L_m^b.$$

Therefore, in order to find a GSB for the universal envelope $U(L; X, N)$ of a Lie conformal superalgebra, we have to find a GSB of the $A(L; X)$-module generated by $X$ relative to the defining relations (8), (7), and (15) with appropriate parity:

$$L_n^a b - (-1)^{p(a)p(b)} R_n^a b - g_n^{a,b}, \quad a, b \in X, \ n \in \mathbb{Z}_+.$$

## References

1. L. A. Bokut, Y.-Q. Chen. Gröbner-Shirshov bases and their calculation, *Bull. Math. Sci.* **4**, 325–395 (2014).
2. S.-J. Kang, K.-H. Lee, Gröbner–Shirshov bases for representation theory, *J. Korean Math. Soc.* **37**, 55–72 (2000).
3. L. A. Bokut, Y.-Q. Chen, C. Liu, Gröner–Shirshov bases for dialgebras, *Internat. J. Algebra Comput.* **20**, 391–415 (2010).
4. G. Zhang, Y.-Q. Chen, A new Composition-Diamond lemma for dialgebras, *J. Algebra Appl.* **16**, 1–28 (2017).
5. P. S. Kolesnikov, Gröbner–Shirshov bases for replicated algebras, *Algebra Colloquium* **24**, 563–576 (2017).
6. V. G. Kac, *Vertex algebras for beginners*, University Lecture Series, vol. 10, AMS, Providence, RI, 1996.
7. L. A. Bokut, Y. Fong, W.-F. Ke, Gröbner–Shirshov bases and composition lemma for associative conformal algebras: an example, *Contemp. Math.* **264**, 63–90 (2000).
8. L. A. Bokut, Y. Fong, W.-F. Ke, Composition-Diamond lemma for associative conformal algebras, *J. Algebra* **272**, 739–774 (2004).
9. L. Ni, Y.-Q. Chen, A new Composition-Diamond lemma for associative conformal algebras, *J. Algebra App.* **16**, 1750094-1–1750094-28 (2017).
10. L. A. Bokut, Imbeddings into simple associative algebras (Russian), *Algebra i Logika* **15**, 117–142 (1976).
11. M. Roitman, On free conformal and vertex algebras, *J. Algebra* **217**, 496–527 (1999).
12. V. G. Kac, Formal distribution algebras and conformal algebras. In: *XII-th International Congress in Mathematical Physics (ICMP97)*, Internat. Press, Cambridge, MA, pp. 80–97 (1999).
13. M. Roitman, Universal enveloping conformal algebras, *Sel. Math. New Ser.* **6**, 319–345 (2000).

# The groups $G_{k+1}^k$ and fundamental groups of configuration spaces

V. O. Manturov *

*Bauman Moscow State Technical University and Novosibirsk State University*
*vomanturov@yandex.ru*

To Leonid Arkadievich Bokut' on the occasion of his 80th birthday

In Ref. 6, the author has constructed natural maps from fundamental groups of topological spaces (restricted configuration spaces) to the groups $G_n^k$. In the present paper, we show that in the case of $n = k+1$, the group $G_{k+1}^k$ is isomorphic to the fundamental group of some (quotient space of) some configuration space up to finite index subgroups. In particular, this leads to the solution of word and conjugacy problems in $G_4^3$ and sheds light on $G_{k+1}^k$ for higher $k$.

*2010 Mathematics subject classification:* 57M25, 57M27.

*Keywords:* group, configuration space, braid, word problem, conjugacy problem

## 1. Introduction

In Ref. 4, the author defined a family of groups $G_n^k$ depending on two natural numbers $n > k$, and formulated the following principle: *if a dynamical system describing a motion of n particles, admits some "good" codimension one property governed by exactly k particles, then this dynamical system has a topological invariant valued in* $G_n^k$. These groups are related to many problems in topology and combinatorial group theory, see, e.g., Ref. 3,7,8.

For $n \in N$, let $[n] = \{1, \cdots, n\}$.

The groups $G_n^k$ are defined as follows.

$$G_n^k = \langle a_m | (1), (2), (3) \rangle,$$

*The author is supported by the Laboratory of Topology and Dynamics, Novosibirsk State University (grant No. 14.Y26.31.0025 of the government of the Russian Federation).

where the generators $a_m$ are indexed by all $k$-element subsets of $[n]$, the relation (1) means

$$(a_m)^2 = 1 \text{ for any unordered sets } m \subset \{1, \cdots, n\}, Card(m) = k; \qquad (1)$$

(2) means

$$a_m a_{m'} = a_{m'} a_m, \text{ if } Card(m \cap m') < k - 1; \qquad (2)$$

and, finally, the relations (3) look as follows. For every set $U \subset [n]$ of cardinality $(k + 1)$, let us order all its $k$-element subsets arbitrarily and denote them by $m^1, \cdots, m^{k+1}$. Then (3) looks as

$$(a_{m^1} \cdots a_{m^{k+1}})^2 = 1. \qquad (3)$$

In view of (1), we can rewrite (3) in the form

$$a_{m^1} \cdots a_{m^{k+1}} = a_{m^{k+1}} \cdots a_{m^1}. \qquad (3')$$

Note that in the case $n = k+1$, the relations (2) are void, since any two distinct subsets $m, m'$ of $[n]$ of cardinality $k$ have intersection of cardinality $k - 1$.

In Ref. 6, groups $G^k_n$ are related to fundamental groups of the following (reduced) configuration spaces. We take $n$-point sets in the $(k - 1)$-dimensional real projective plane $\mathbb{R}P^{k-1}$, such that no $(k - 1)$ of these points belong to the same (projective) $(k - 3)$-plane. In particular, for $k = 3$ we deal with $n$-strand $\mathbb{R}P^2$-braids. There is a map from the fundamental group of the (ordered) configuration space to $G^k_n$.

These maps correspond to the good property "some $k$ points are not in general position" (i.e. belong the same $(k-2)$-projective plane). Note that the we have imposed the restriction that any $(k - 1)$ points should be in general position.

The maps constructed in Ref. 6 have some obvious kernel corresponding to rotations of $\mathbb{R}P^{k-1}$. It turns out that in the lowest level (for $n = k+1$), after factorisation by this kernel, we get an isomorphism of groups.

The kernel looks as follows: identifying $\mathbb{R}P^{k-1}$ with the quotient space of $S^{k-1}$ by the involution, we can just rotate $S^{k-1}$ about some axis by $\pi$; the corresponding motion of points is homotopically non-trivial; on the other hand, if points are in general position from the very beginning, they remain in the general position during the rotation. Hence, no singular moment occurs and the corresponding word is empty.

To remedy this, we define the configuration space $\tilde{C}'_n(\mathbb{R}P^{k-1})$ of all unordered $n$-tuples of points where the first $(k-1)$ points are fixed and any

$k-1$ points are in general position. For example, for $\mathbb{R}P^2$, we can consider the path when one point $x_1$ is fixed an three other points lie in the small neighbourhood of $x_1$; when rotating them around $x_1$, we get a non-trivial element of the fundamental group of the corresponding configuration space, and during the whole path, there are no moments when any three points are collinear. Then every closed path in the configuration space giving rise to the empty word is homotopic to the trivial path. See Lemma 3.3.

More precisely, we shall prove the following

**Theorem 1.1.** *There is a subgroup $\tilde{G}^k_{k+1}$ of the group $G^k_{k+1}$ of index $2^{k-1}$ isomorphic to $\pi_1(\tilde{C}'_{k+1}(\mathbb{R}P^{k-1}))$.*

The space $\tilde{C}'_n$ and the group $\tilde{G}^k_{k+1}$ will be defined later.
The simplest case of the above Theorem is

**Theorem 1.2.** *The group $\tilde{G}^3_4$ (which is a finite index subgroup) is isomorphic to $\pi_1(FBr_4(\mathbb{R}P^2))$, the 4-strand braid group on $\mathbb{R}P^2$ with two point fixed.*

The paper is organized as follows. In the next section, we give all necessary definitions and construct maps from configuration spaces of points in $\mathbb{R}P^{k-1}$ to $G^k_{k+1}$. In Section 3, we prove Theorem 1.1. In Section 4, we shall discuss an algebraic lemma about reduction of the word problem in $G^k_{k+1}$ to the word problem in some subgroup of it.

We conclude the paper by Section 4 by discussing some open problems for further research.

## 2. Basic definitions

Let us now pass to the definition of spaces $C'_n(\mathbb{R}P^{k-1})$ and maps from the corresponding fundamental groups to the groups $G^k_n$.

Let us fix a pair of natural number $n > k$. A point in $C'_n(\mathbb{R}P^{k-1})$ is an ordered set of $n$ pairwise distinct points in $\mathbb{R}P^{k-1}$, such that any $(k-1)$ of them are in general position. Thus, for instance, if $k = 3$, then the only condition is that these points are pairwise distinct. For $k = 4$ for points $x_1, \ldots, x_n \in \mathbb{R}P^3$ we impose the condition that no three of them belong to the same line (though some four are allowed to belong to the same plane), and for $k = 5$ a point in $C'_n(\mathbb{R}P^4)$ is a set of ordered $n$ points in $\mathbb{R}P^4$, with no four of them belonging to the same 2-plane.

Let us use the projective coordinates $(a_1 : a_2 : \cdots : a_k)$ in $\mathbb{R}P^{k-1}$ and let us fix the following $k-1$ points in general position, $y_1, y_2, \cdots, y_{k-1} \in$

$\mathbb{R}P^{k-1}$, where $a_i(y_j) = \delta_i^j$. Let us define the subspace $\tilde{C}_n'(\mathbb{R}P^{k-1})$ taking those $n$-tuples of points $x_1, \cdots, x_n \in \mathbb{R}P^{k-1}$ for which $x_i = y_i$ for $i = 1, \cdots, k-1$.

We say that a point $x \in C_n'(\mathbb{R}^{k-1})$ is *singular*, if the set of points $x = (x_1, \ldots, x_n)$, corresponding $x$, contains some subset of $k$ points lying on the same $(k-2)$-plane. Let us fix two non-singular points $x, x' \in C_n'(\mathbb{R}P^{k-1})$.

We shall consider smooth paths $\gamma_{x,x'} : [0,1] \to C_n'(\mathbb{R}^{k-1})$. For each such path there are values $t$ for which $\gamma_{x,x'}(t)$ is not in the general position (some $k$ of them belong to the same $(k-2)$-plane). We call these values $t \in [0,1]$ *singular*.

On the path $\gamma$, we say that the moment $t$ of passing through the singular point $x$, corresponding to the set $x_{i_1}, \cdots, x_{i_k}$, is *transverse* (or stable) if for any sufficiently small perturbation $\tilde{\gamma}$ of the path $\gamma$, in the neighbourhood of the moment $t$ there exists exactly one time moment $t'$ corresponding to some set of points $x_{i_1}, \cdots, x_{i_k}$ non in general position.

**Definition 2.1.** We say that a path is *good and stable* if the following holds:

(1) The set of singular values $t$ is finite;
(2) For every singular value $t = t_l$ corresponding to $n$ points representing $\gamma_{x,x'}(t_l)$, there exists only one subset of $k$ points belonging to a $(k-2)$-plane;
(3) Each singular value is *stable*.

**Definition 2.2.** We say that the path without singular values is *void*.

We shall call such paths *braids*. We say that a braid whose ends $x, x'$ coincide with respect to the order, is *pure*. We say that two braids $\gamma, \gamma'$ with endpoints $x, x'$ are *isotopic* if there exists a continuous family $\gamma_{x,x'}^s, s \in [0,1]$ of smooth paths with fixed ends such that $\gamma_{x,x'}^0 = \gamma, \gamma_{x,x'}^1 = \gamma'$. By a small perturbation, any path can be made good and stable (if endpoints are generic, we may also require that the endpoints remain fixed).

**Definition 2.3.** A path from $x$ to $x'$ is called *a braid* if the points representing $x$ are the same as those representing $x'$ (possibly, in different orders); if $x$ coincides with $x'$ with respect to order, then such a braid is called *pure*.

There is an obvious concatenation structure on the set of braids: for paths $\gamma_{x,x'}$ and $\gamma_{x',x''}'$, the concatenation is defined as a path $\gamma_{x,x''}''$ such that $\gamma''(t) = \gamma(2t)$ for $t \in [0, \frac{1}{2}]$ and $\gamma''(t) = \gamma'(2t - 1)$ for $t \in [\frac{1}{2}, 1]$; this path can be smoothed in the neighbourhood of $t = \frac{1}{2}$; the result of such smoothing is well defined up to isotopy.

Thus, the sets of braids and pure braids (for fixed $x$) admit a group structure. This latter group is certainly isomorphic to the fundamental group $\pi_1(C'_n(\mathbb{R}^{k-1}))$. The former group is isomorphic to the fundamental group of the quotient space by the action of the permutation group.

## 3. The realisability of $G^k_{k+1}$

The main idea of the proof of Theorem 1.1 is to associate with every word in $G^k_{k+1}$ a braid in $\tilde{C}'_{k+1}(\mathbb{R}P^{k-1})$.

Let us start with the main construction from Ref. 6.

With each good and stable braid from $PB_n(\mathbb{R}P^2)$ we associate an element of the group $G^k_n$ as follows. We enumerate all singular values of our path $0 < t_1 < \cdots < t_l < 1$ (we assume than 0 and 1 are not singular). For each singular value $t_p$ we have a set $m_p$ of $k$ indices corresponding to the numbers of points which are not in general position. With this value we associate the letter $a_{m_p}$. With the whole path $\gamma$ (braid) we associate the product $f(\gamma) = a_{m_1} \cdots a_{m_l}$.

**Theorem 3.1.**[6] *The map $f$ takes isotopic braids to equal elements of the group $G^k_n$. For pure braids, the map $f$ is a homomorphism $f$ :* $\pi_1(C'_n(\mathbb{R}P^2)) \to G^3_n$.

Now we claim that

*Every word from $G^k_{k+1}$ can be realised by a path of the above form.*

Note that if we replace $\mathbb{R}P^{k-1}$ with $\mathbb{R}^{k-1}$, the corresponding statement will fail. Namely, starting with the configuration of four points, $x_i, i = 1, \cdots, 4$, where $x_1, x_2, x_3$ form a triangle and $x_4$ lies inside this triangle, we see that any path starting from this configuration will lead to a word starting from $a_{124}, a_{134}, a_{234}$ but not from $a_{123}$. In some sense the point 4 is "locked" and the points are not in the same position.

The following well known theorem (see, e.g., Ref. 1) plays a crucial role in the construction

**Theorem 3.2.** *For any two sets of $k + 1$ points in general position in $\mathbb{R}P^{k-1}$, $(x_1, \cdots, x_{k+1})$ and $(y_1, \cdots, y_{k+1})$, there is an action of $PGL(k, \mathbb{R})$ taking all $x_i$ to $y_i$.*

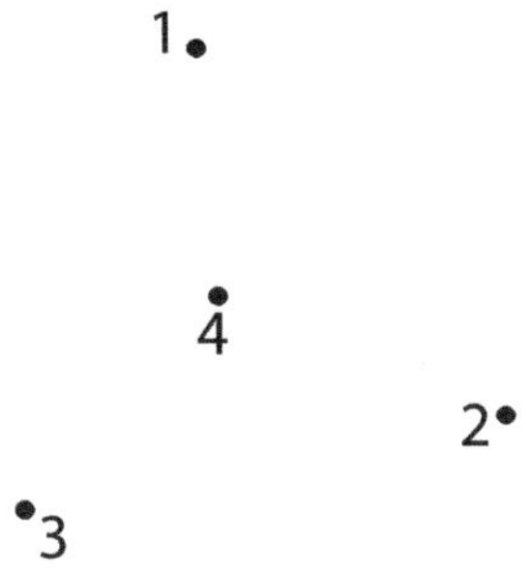

Fig. 1. The "locked" position for the move $a_{123}$

For us, this will mean that there is no difference between all possible "non-degenerate starting positions" for $k + 1$ points in $\mathbb{R}P^k$.

We shall deal with paths in $\tilde{C}'_{k+1}(\mathbb{R}P^{k-1})$ similar to braids. Namely, we shall fix a set of $2^{k-1}$ points such that all paths will start and end at these points.

We shall denote homogeneous coordinates in $\mathbb{R}P^{k-1}$ by $(a_1 : \cdots : a_k)$ in contrast to points (which we denote by $(x_1, \cdots, x_{k+1})$).

## 3.1. *Constructing a braid from a word in $G^k_{k+1}$*

Our main goal is to construct a braid by a word. To this end, we need a base point for the braid. For the sake of convenience, we shall use not one, but rather $2^{k-1}$ reference points. For the first $k$ points $y_1 = (1 : 0 : \cdots : 0), \cdots, y_k = (0 : \cdots : 0 : 1)$ fixed, we will have $2^{k-1}$ possibilities for the choice of the last point. Namely, let us consider all possible strings of length $k$ of $\pm 1$ with the last coordinate $+1$: $(1, 1, \cdots, 1, 1), (1, \cdots, 1, -1, 1), \cdots, (-1, -1, \cdots, -1, 1)$ with $a_k \sim 1$. We shall denote these points by $y_s$ where $s$ records the first $(k-1)$ coordinates of the point.

Now, for each string $s$ of length $k$ of $\pm 1$, we set $z_s = (y_1, y_2, \cdots, y_k, y_s)$. The following lemma is evident.

**Lemma 3.1.** *For every point $z \in \mathbb{R}P^{k-1}$ with projective coordinates $(a_1(z) : \cdots : a_{k-1}(z) : 1)$, let $\tilde{z} = (sign(a_1(z)) : sign(a_2(z)) : \cdots : sign(a_{k-1}(z)) : 1)$. Then there is a path between $(y_1, \cdots, y_k, z)$ and $(y_1, \cdots, y_k, \tilde{z})$ in $\tilde{C}'_{k+1}(\mathbb{R}P^k)$ with the first points $y_1, \cdots, y_k$ fixed, such that the corresponding path in $\tilde{C}'_{k+1}$ is void.*

**Proof.** Indeed, it suffices just to connect $z$ to $\tilde{z}$ by a geodesic. $\square$

From this we easily deduce the following

**Lemma 3.2.** *Every point $y \in \tilde{C}'_{k+1}(\mathbb{R}P^{k-1})$ can be connected by a void path to some $(y_1, \cdots, y_k, y_s)$ for some $s$.*

**Proof.** Indeed, the void path can be constructed in two steps. At the first step, we construct a path which moves both $y_k$ and $y_{k+1}$, so that $y_k$ becomes $(0 : \cdots 0 : 1)$, and at the second step, we use Lemma 3.1. To realise the first step, we just use linear maps which keep the hyperplane $a_k = 0$ fixed. □

The lemma below shows that the path mentioned in Lemma 3.2 is unique up to homotopy.

**Lemma 3.3.** *Let $\gamma$ be a closed path in $\tilde{C}'_{k+1}(\mathbb{R}P^{k-1})$ such that the word $f(\gamma)$ is empty. Then $\gamma$ is homotopic to the trivial braid.*

**Proof.** In $\tilde{C}'_{k+1}$, we deal with the motion of points, where all but $x_k, x_{k+1}$ are fixed.

Consider the projective hyperplane $\mathcal{P}_1$ passing through $x_1, \cdots, x_{k-1}$ given by the equation $a_k = 0$.

We know that none of the points $x_k, x_{k+1}$ is allowed to belong to $\mathcal{P}_1$. Hence, we may fix the last coordinate $a_k(x_k) = a_k(x_{k+1}) = 1$.

Now, we may pass to the affine coordinates of these two points (still to be denoted by $a_1, \cdots, a_k$).

Now, the condition $\forall i = 1, \cdots, k - 1 : a_j(x_k) \neq a_j(x_{k+1})$ follows from the fact that the points $x_1, \cdots, \hat{x}_j, x_{k+1}$ are generic.

This means that $\forall i = 1, \cdots, k - 1$ the sign of $a_i(x_k) - a_i(x_{k+1})$ remains fixed.

Now, the motion of points $x_k, x_{k+1}$ is determined by their coordinates $a_1, \cdots, a_j$, and since their signs are fixed, the configuration space for this motions is simply connected.

This means that the loop $\gamma$ is described by a loop in a two-dimensional simply connected space. □

Our next strategy is as follows. Having a word in $G^k_{k+1}$, we shall associate with this word a path in $\tilde{C}'_{k+1}(\mathbb{R}P^{k-1})$. After each letter, we shall get to $(y_1, \cdots, y_k, y_s)$ for some $s$.

Let us start from $(y_1, \cdots, y_k, y_{1,\cdots,1})$.

After making the final step, one can calculate the coordinate of the $(k+1)$-th points. They will be governed by Lemma 3.6 (see ahead). As we shall see later, those words we have to check for the solution of the word problem in $G_{k+1}^k$, will lead us to closed paths, i.e., pure braids.

Let us be more detailed.

**Lemma 3.4.** *Given a non-singular set of points $y$ in $\mathbb{R}P^{k-1}$. Then for every set of $k$ numbers $i_1, i_2, \cdots, i_k \in [n]$, there exists a path $y_{i_1 \cdots i_k}(t) = y(t)$ in $C_n'(\mathbb{R}P^{k-1})$, having $y(0) = y$ as the starting point and the final point and with only one singular moment corresponding to the numbers $i_1, \cdots, i_k$ which encode the points not in general position; moreover, we may assume that at this moment all points except $i_1$, are fixed during the path.*

*Moreover, the set of paths possessing this property is connected: any other path $\tilde{y}(t)$, possessing the above properties, is homotopic to $y(t)$ in this class.*

**Proof.** Indeed, for the first statement of the Lemma, it suffices to construct a path for some initial position of points and then apply Theorem 3.2.

For the second statement, let us take two different paths $\gamma_1$ and $\gamma_2$ satisfying the conditions of the Lemma. By a small perturbation, we may assume that for both of them, $t = \frac{1}{2}$ is a singular moment with the same position of $y_{i_1}$.

Now, we can contract the loop formed by $\gamma_1|_{t \in [\frac{1}{2}, 1]}$ and the inverse of $\gamma_2|_{t \in [\frac{1}{2}, 1]}$ by using Lemma 3.3 as this is a small perturbation of a void braid. We are left with $\gamma_1|_{t \in [0, \frac{1}{2}]}$ and the inverse of $\gamma_2|_{t \in [0, \frac{1}{2}]}$ which is contractible by Lemma 3.3 again. $\qquad \square$

**Remark 3.1.** Note that in the above lemma, we deal with the space $C_n'(\mathbb{R}P^{k-1})$, not with $\tilde{C}_n'(\mathbb{R}P^{k-1})$. On the other hand, we may always choose $i_1 \in \{k, k+1\}$; hence, the path in question can be chosen in $\tilde{C}'(\mathbb{R}P^{k-1})$.

Now, for every subset $m \subset [n], Card(m) = k+1$ we can create a path $p_m$ starting from any of the base points listed above and ending at the corresponding basepoints.

Now, we construct our path step-by step by applying Lemma 3.4 and returning to some of base points by using Lemma 3.2.

From Ref. 6, we can easily get the following

**Lemma 3.5.** *Let $i_1, \cdots, i_{k+1}$ be some permutation of $1, \cdots, k+1$. Then the concatenation of paths $p_{i_1 i_2 \cdots i_k} p_{i_1 i_3 i_4 \cdots i_{k+1}} \cdots p_{i_2 i_3 \cdots i_k}$ is homotopic to the concatenation of paths in the inverse order $p_{i_2 i_3 \cdots i_k} \cdots p_{i_1 i_3 i_4 \cdots i_{k+1}} p_{i_1 i_2 \cdots i_k}$.*

**Proof.** Indeed, in Ref. 6, some homotopy corresponding to the above mentioned relation corresponding to *some* permuation is discussed. However, since all basepoints are similar to each other as discussed above, we can transform the homotopy from Ref. 6 to the homotopy for any permutation. □

**Lemma 3.6.** *For the path starting from the point* $(y_1, \cdots, y_k, y_s)$ *constructed as in Lemma 3.4 for the set of indices $j$, we get to the point* $(y_1, \cdots, y_k, y_{s'})$ *such that:*

  (1) *if $j = 1, \cdots, k$, then $s'$ differs from $s$ only in coordinate $a_j$;*
  (2) *if $j = k + 1$, all coordinates of $s'$ differ from those coordinates of $s$ by sign.*

Denote the map from words in $G_{k+1}^k$ to paths between basepoints by $g$.

By construction, we see that for every word $w$ we have $f(g(w)) = w \in G_{k+1}^k$.

Now, we define the group $\tilde{G}_{k+1}^k$ as the subgroup of $G_{k+1}$ which is taken by $g$ to *braids*, i.e., to those paths with coinciding initial and final points. From Lemma 3.6, we see that this is a subgroup of index $(k-1)$: there are exactly $(k-1)$ coordinates.

## 3.2. *Equal words lead to homotopic paths*

Let us pass to the proof of Theorem 1.2. Our next goal is to see that equal words can originate only from homotopic paths.

To this end, we shall first show that the map $f$ from Theorem 3.1 is an isomorphism for $n = k + 1$. To perform this goal, we should construct the inverse map $g : \tilde{G}_{k+1}^k \to \pi_1(\tilde{C}'_{k+1}(\mathbb{R}P(k-1)))$.

Note that for $k = 3$ we deal with the pure braids $PB_4(\mathbb{R}P^2)$.

Let us fix a point $x \in C'_4(\mathbb{R}P^2)$. With each generator $a_m, m \subset [n], Card(m) = k$ we associate a path $g(m) = y_m(t)$, described in Lemma 3.4. This path is not a braid: we can return to any of the $2^{k-1}$ base points. However, once we take the concatenation of paths correspoding to $\tilde{G}_{k+1}^k$, we get a braid.

By definition of the map $f$, we have $f(g(a_m)) = a_m$. Thus, we have chosen that the map $f$ is a surjection.

Now, let us prove that the kernel of the map $f$ is trivial. Indeed, assume there is a pure braid $\gamma$ such that $f(\gamma) = 1 \in G_{k+1}^k$. We assume that $\gamma$ is good and stable. If this path has $l$ critical points, then we have the word corresponding to it $a_{m_1} \cdots a_{m_l} \in G_{k+1}^k$.

Let us perform the transformation $f(\gamma) \to 1$ by applying the relations of $G_{k+1}^k$ to it and transforming the path $\gamma$ respectively. For each relation of the sort $a_m a_m = 1$ for a generator $a_m$ of the group $G_{k+1}^k$, we see that the path $\gamma$ contains two segments whose concatenation is homotopic to the trivial loop (as follows from the second statement of Lemma 4).

Whenever we have a relation of length $2k+2$ in the group $G_{k+1}$, we use the Lemma 3.5 to perform the homotopy of the loops.

Thus, we have proved that if the word $f(\gamma)$ corresponding to a braid $\gamma \in G_{k+1}^k$ is equal to 1 in $G_{k+1}^k$ then the braid $\gamma$ is isotopic to a braid $\gamma'$ such that the word corresponding to it is empty. Now, by Lemma 3.3, this braid is homotopic to the trivial braid.

## 4. The group $H_k$ and the algebraic lemma

The aim of the present section is to reduce the word problem in $G_{k+1}^k$ to the word problem in a certain subgroup of it, denoted by $H_k$.

Let us rename all generators of $G_{k+1}^k$ lexicographically: $b_1 = a_{1,2,\cdots,k}, \cdots, b_{k+1} = a_{2,3,\cdots,k+1}$.

Let $H_k$ be the subgroup of $G_{k+1}^k$ consisting of all elements $x \in G_{k+1}^k$ that can be represented by words with no occurencies of the last letter $b_{k+1}$.

Our task is to understand whether a word in $G_{k+1}^k$ represents an element in $H_k$. To this end, we recall the map from Ref. 7. Consider the group $F_{k-1} = \mathbb{Z}_2^{*2^{k-1}} = \langle c_m | c_m^2 = 1 \rangle$, where all generators $c_m$ are indexed by $(k-1)$-element strings of 0 and 1 with only relations being that the square of each generator is equal to 1. We shall construct a map[†] from $G_{k+1}^k$ to $F_{k-1}$ as follows.

Take a word $w$ in generators of $G_{k+1}^k$ and list all occurencies of the last letter $b_{k+1} = a_{2,\cdots,k+1}$ in this word. With any such occurency we first associate the string of indices $0, 1$ of length $k$. The $j$-th index is the number of letters $b_j$ preceding this occurency of $b_{k+1}$ modulo 2. Thus, we get a string of length $k$ for each occurency.

Let us consider "opposite pairs" of strings $(x_1, \cdots, x_k) \sim (x_1 + 1, \cdots, x_k + 1)$ as equal. Now, we may think that the last ($k$-th) element of our string is always 0, so, we can restrict ourselves with $(x_1, \cdots, x_{k-1}, 0)$. Such a string of length $k-1$ is called the *index* of the occurency of $b_{k+1}$.

Having this done, we associate with each occurency of $b_{k+1}$ having index $m$ the generator $c_m$ of $F_{k-1}$. With the word $w$, we associate the word $f(w)$

---

[†]This map becomes a homomorphism when restricted to a finite index subgroup.

equal to the product of all generators $c_m$ in order.

In Ref. 7, the following Lemma is proved:

**Lemma 4.1.** *The map* $f : G^k_{k+1} \to F_{k-1}$ *is well defined.*

Now, let us prove the following

**Lemma 4.2.** *If* $f(w) = 1$ *then* $w \in H_k$.

In other words, the free group $F_{k-1}$ yields the only obstruction for an element from $G^k_{k+1}$ to have a presentation with no occurency of the last letter.

**Proof.** Let $w$ be a word such that $f(w) = 1$. If $f(w)$ is empty, then there is nothing to prove. Otherwise $w$ contains two "adjacent" occurencies of the same index. This means that $w = Ab_{k+1}Bb_{k+1}C$, where $A$ and $C$ are some words, and $B$ contains no occurencies of $b_{k+1}$ and the number of occurencies of $b_1, b_2, \cdots, b_k$ in $B$ are of the same parity.

Our aim is to show that $w$ is equal to a word with smaller number of $b_{k+1}$ in $G^k_{k+1}$. Then we will be able to induct on the number of $b_{k+1}$ until we get a word without $b_{k+1}$.

Thus, it suffices for us to show that $b_{k+1}Bb_{k+1}$ is equal to a word from $H_k$. We shall induct on the length of $B$. Without loss of generality, we may assume that $B$ is reduced, i.e., it does not contain adjacent $b_jb_j$.

Let us read the word $B$ from the very beginning $B = b_{i_1}b_{i_2}\cdots$ If all elements $i_1, i_2, \cdots$ are distinct, then, having in mind that the number of occurencies of all generators in $B$ should be of the same parity, we conclude that $b_{k+1}B = B^{-1}b_{k+1}$, hence $b_{k+1}Bb_{k+1} = B^{-1}b_{k+1}b_{k+1} = B^{-1}$ is a word without occurencies of $b_{k+1}$.

Now assume $i_1 = i_p$ (the situation when the first repetition is for $i_j = i_p, 1 < j < p$ is handled in the same way). Then we have $b_{k+1}B = b_{k+1}b_{i_1}\cdots b_{i_{p-1}}b_{i_1}B'$. Now we collect all indices distinct from $i_1, \cdots, i_{p-1}, k+1$ and write the word $P$ containing exactly one generator for each of these indices (the order does not matter). Then the word $W = Pb_{k+1}b_{i_1}\cdots b_{i_{p-1}}$ contains any letter exactly once and we can invert the word $W$ as follows: $W^{-1} = b_{i_{p-1}}\cdots b_{i_1}b_{k+1}P^{-1}$. Thus, $b_{k+1}B = P^{-1}(Pb_{k+1}b_{i_1}\cdots b_{i_{p-1}})b_{i_1}B' = P^{-1}b_{i_{p-1}}\cdots b_{i_1}b_{k+1}P^{-1}b_{i_1}B'$.

We know that the letters in $P$ (hence, those in $P^{-1}$) do not contain $b_{i_1}$. Thus, the word $P^{-1}b_{i_1}$ consists of distinct letters. Now we perform the same trick: create the word $Q = b_{i_2}b_{i_3}\cdots b_{i_{p-1}}$ consisting of remaining letters from $\{1, \cdots, k\}$, we get:

$$P^{-1}b_{i_{p-1}}\cdots b_{i_1}b_{k+1}P^{-1}b_{i_1}B' = P^{-1}b_{i_{p-1}}\cdots b_{i_1}QQ^{-1}b_{k+1}P^{-1}b_{i_1}B'$$
$$= P^{-1}b_{i_{p-1}}\cdots b_{i_1}Qb_{i_1}Pb_{k+1}QB'.$$

Thus, we have moved $b_{k+1}$ to the right and between the two occurencies of the letter $b_{k+1}$, we replaced $b_{i_1}\cdots,b_{i_{p-1}}b_{i_1}$ with just $b_{i_2}\cdots b_{i_{p-1}}$, thus, we have shortened the distance between the two adjacent occurencies of $b_{k+1}$.

Arguing as above, we will finally cancel these two letters $b_{k+1}$ and perform the induction step. $\qquad\square$

**Theorem 4.1.** *In the group $G_{k+1}^k$, the problem whether $x \in G_{k+1}^k$ belongs to $H_k$, is solvable.*

**Proof.** Indeed, having a word $w$ in generators of $G_{k+1}^k$, we can look at the image of this word by the map $f$. If it is not equal to 1, then, from Ref. 7, it follows that $w$ is non-trivial, otherwise we can construct a word $\tilde{w}$ in $H_k$ equal to $w$ in $G_{k+1}^k$. $\qquad\square$

**Remark 4.1.** A shorter proof of Theorem 4.1 based on the same ideas was communicated to the author by A.A.Klyachko. We take the subgroup $K_k$ of $G_{k+1}^k$ generated by products $B_\sigma = b_{\sigma_1}\cdots b_{\sigma_k}$ for all permutations $\sigma$ of $k$ indices. This group $K_k$ contains the commutor of $H_k$ and is a normal subgroup in $G_{k+1}^k$.

Moreover, the quotient group $G_{k+1}^k/K_k$ is naturally isomorphic to the free product $(H_k/K_k) * \langle b_{k+1} \rangle$. Hence, the problem whether an element of $G_{k+1}^k/K_k$ belongs to $H_k/K_k$ is trivially solved, which solves the initial problem because of the normality of $K_k$ in $G_{k+1}^k$.

Certainly, to be able to solve the word problem in $H_k$, one needs to know a presentation for $H_k$. It is natural to take $b_1,\cdots,b_k$ for generators of $H_k$. Obviously, they satisfy the relations $b_j^2 = 1$ for every $j$.

To understand the remaining relations for different $k$, we shall need geometrical arguments.

## 5. Concluding remarks

We have completely constructed the isomorphism between the (finite index subgroup) of the group $G_{k+1}^k$ and a fundamental group of some configuration space.

This completely solves the word problem in $G_4^3$ for braid groups in projective spaces are very well studied. The same can be said about the conjugacy problem in $\tilde{G}_4^3$.

Besides, we have seen that the word problem for the case of general $G_{k+1}^k$ can be reduced to the case of $H_k$.

In a subsequent paper, we shall completely describe the relations in $H_k$ by geometric reasons and apply it to the word problem in $G_5^4$. Here we just mention that it was proved by A.B.Karpov (unpubished) that the only relations in $H_3$ are $a^2 = b^2 = c^2$ which also follows from the geometrical techniques of the present paper.

The main open question which remains unsolved is how to construct a configuration space which can realise $G_n^k$ for $n > k + 1$. Even in the case of $G_4^2$ this problem seems very attractive though the word and conjugacy problems for $G_n^2$ can be solved by algebraic methods, see Ref. 3.

The word problem and the partial case of conjugacy problem in $G_4^3$ was first solved by A.B.Karpov, but the author has not yet seen any complete text of it.

It would be very interesting to compare the approach of $G_n^k$ with various generalizations of braid groups, e.g., Manin-Schechtmann groups [2,9].

The author is very grateful to I.M.Nikonov, L.A.Bokut', Jie Wu and A.A.Klyachko for extremely useful discussions and comments.

## References

1. A.J. Berrick, F.R. Cohen, Y.L. Wong, J. Wu, Configurations, braids, and homotopy groups, *Journal AMS*, **19**, 2, Pp. 265-326 (2005).
2. M. Kapranov, V. Voevoodsky, Braided Monoidal 2-Categories and Manin-Schechtmann Higher Braid Groups, Journal of Pure and Applied Mathematics, N. 92, 1994, P. 241-167.
3. V. O. Manturov, On the Groups $G_n^2$ and Coxeter Groups, Uspekhi Mat. Nauk, 72:2(434) (2017), 193–194; Russian Math. Surveys, 72:2 (2017), 378–380. http://arxiv.org/abs/1512.09273.
4. V. O. Manturov, Non-Reidemeister Knot Theory and Its Applications in Dynamical Systems, Geometry, and Topology, http://arxiv.org/abs/1501.05208.
5. V. O. Manturov, Invariants of Classical Braids Valued in $G_n^2$, http://arxiv.org/abs/1611.07434v2.
6. V. O. Manturov, The Groups $G_n^k$ and fundamental groups of configuration spaces, *J. Knot Theory & Ramifications, 2015, Vol.26.*
7. V. O. Manturov, I.M. Nikonov, On braids and groups $G_n^k$ *J. Knot Theory & Ramifications, 2015, Vol. 24, No.13.*
8. S. Kim, V. O. Manturov, The Braid Groups and Imaginary Generators, http://arxiv.org/abs/1612.03486v1.
9. Y.I. Manin, V.V. Schechtmann, Arrangements of Hyperplanes, higher braid groups and higher Bruhat orders, Adv. Stud. Pure Math., 17 (1990), P. 289-308

# Presentations of inverse semigroups: progress, problems and applications

John Meakin

*Department of Mathematics*
*University of Nebraska*
*Lincoln, NE 68588, USA*

*jmeakin@math.unl.edu*

Dedicated to Professor Leonid Bokut on his 80'th birthday

This paper surveys some recent developments in the theory of presentations of inverse semigroups by generators and relations. Particular attention is paid to the word problem for one-relator inverse monoids and for inverse semigroups admitting a cycle-free presentation. Some applications of the theory to operator algebras and topology are discussed.

## 1. Introduction

Inverse semigroups were introduced into the literature independently by V.V. Wagner[80] and G. B. Preston[59] in the early 1950's. The original impetus came from differential geometry and topology, where inverse semigroups were introduced as algebraic objects to study pseudogroups of local diffeomorphisms of a differentiable manifold or homeomorphisms between open subsets of a topological space. Inverse semigroups are essentially algebraic objects that are used to study local symmetry in mathematical objects in much the same way as groups are used to study symmetry. This point of view is exposited clearly in the book by Lawson[34], which is a good source of much basic information about the structure of inverse semigroups and their relationship to other fields.

Combinatorial inverse semigroup theory traces its roots back to the work of Gluskin[20], who described the structure of free monogenic inverse semigroups. The structure of arbitrary free inverse semigroups was described independently by Scheiblich[69] and Munn[55] in the 1970's. Scheiblich's work was an influential factor in the development of McAlister's theory of $E$-unitary inverse semigroups[44,45], and Munn's work sheds light on the solution to the word problem for free inverse semigroups and led to the introduction of a full-fledged theory of presentations of inverse semigroups by generators and relations in the work of Stephen[77,78]. Subsequently, the theory of presentations of inverse semigroups has developed into a substantial part of semigroup theory, with significant connections with geometric group theory, topology and operator algebras.

Section 2 of this paper contains some of the basic concepts and structure theory of inverse semigroups that is needed throughout the paper. It also outlines an automata-theoretic proof of Munn's theorem on the structure of free inverse semigroups and discusses Stephen's approach to the general study of presentations of inverse semigroups. Section 3 of the paper discusses several interesting applications of Stephen's approach that have appeared in the literature and highlights some of the distinctions between this theory and the theory of presentations of groups. In Section 4, we discuss various approaches to the study of the word problem for one-relator inverse monoids, one of the outstanding unsolved problems in the field. Section 5 of the paper focuses on the class of inverse semigroups admitting a cycle-free presentation and relates the word problem for this class to the corresponding problems for cycle-free groups and semigroups. In section 6 of the paper we highlight some areas where inverse semigroups play an important role in operator algebras, particularly in the theory of amalgams of $C^*$-algebras and in the study of certain maximal abelian subalgebras of von Neumann algebras. The final section of the paper discusses some recent work on the use of inverse semigroups to study immersions between cell complexes, in much the same way as groups are used to study covers of topological spaces.

## 2. Preliminaries

A semigroup $S$ is called an *inverse semigroup* if for each $a \in S$ there exists a unique inverse (denoted by $a^{-1}$) in $S$ such that

$$a = aa^{-1}a \quad \text{and} \quad a^{-1} = a^{-1}aa^{-1}.$$

Inverse monoids are inverse semigroups that have an identity. Most (but not all) of the inverse semigroups that we will consider in this paper are monoids. Of course it is always possible to obtain an inverse monoid from an inverse semigroup $S$ simply by adjoining an identity to $S$, but for many inverse monoids the group of units is non-trivial and may play an important role in the structure of the monoid. Inverse semigroups and inverse monoids arise naturally in the study of *partial symmetry* in mathematics, in much the same way as groups arise in the study of symmetry. For each non-empty set $X$, the *symmetric inverse monoid* $SIM(X)$ consists of all partial one-to-one maps between subsets of $X$ with respect to composition of partial maps. This is in a sense the "canonical" example of an inverse monoid as the following theorem, known as the *Wagner-Preston Theorem* shows.

**Theorem 2.1.** *Every inverse monoid embeds in a suitable symmetric inverse monoid.*

For a proof of this theorem, together with much basic information about the structure of inverse monoids and their connection with other fields, see the book of Lawson[34]. In particular, some standard properties and concepts involving inverse monoids are assembled in the following proposition: the proofs of all of these statements are either explicitly given in Ref. 34 or are easily deduced from facts that are proved in Ref. 34.

**Proposition 2.1.** *Let $M$ be an inverse monoid and let $E(M)$ denote the set of idempotents of $M$. Then*

*(a) The idempotents of $M$ commute. Thus $E(M)$ is a submonoid of $M$ that forms a (lower) semilattice with respect to the meet operation $e \wedge f = ef$.*

*(b) The Green's $\mathcal{R}$ and $\mathcal{L}$ relations on $M$ may be characterized by $a \mathcal{R} b$ iff $aa^{-1} = bb^{-1}$ and $a \mathcal{L} b$ iff $a^{-1}a = b^{-1}b$. As usual, the $\mathcal{R}$-class [resp. $\mathcal{L}$-class] containing an element $a \in M$ is denoted by $R_a$ [resp. $L_a$] and the $\mathcal{D}$-class [resp. $\mathcal{H}$-class] of $a$ is denoted by $D_a$ [resp. $H_a$]. Each $\mathcal{R}$-class [resp. $\mathcal{L}$-class] of $M$ contains exactly one idempotent.*

*(c) For each $e \in E(M)$, the map $a \to a^{-1}$ (for $a \in R_e$) is a bijection from $R_e$ to $L_e$. Thus the $\mathcal{D}$-classes of $M$ are "square" and may be visualized by "egg boxes" with idempotents arranged down the diagonal.*

*(d) If $a \in M$ and $e \in E(M)$ then $aea^{-1}, a^{-1}ea \in E(M)$. If $a, b \in M$ and $e \in E(M)$ then there exist $f, g \in E(M)$ such that $aeb = fab = abg$.*

*(e) The relation defined by $a \leq b$ iff $a = eb$ for some $e \in E(M)$ (equivalently $a = bf$ for some $f \in E(M)$; equivalently $a = aa^{-1}b$; equivalently $a = ba^{-1}a$) is a partial order on $M$, called the natural partial order on $M$. It is compatible with respect to the product in $M$ and with respect to taking inverses of elements in $M$. The restriction of this partial order to the semilattice $E(M)$ is the natural partial order on $E(M)$, namely $e \leq f$ iff $e = ef = fe$.*

*(f) If $e$ and $f$ are idempotents of $M$ with $e \leq f$, then for each $a \in R_f$ [resp. $a \in L_f$] there is a unique element $b \in R_e$ [resp. $c \in L_e$] such that $b \leq a$ [resp. $c \leq a$], namely $b = ea$ [resp. $c = ae$]. The corresponding maps $\phi_{f,e} : R_f \rightarrow R_e$ and $\psi_{f,e} : L_f \rightarrow L_e$ defined by $a \rightarrow ea$ [resp. $a \rightarrow ae$] are called the structure maps of $M$.*

*(g) For $a, b \in M$, we have $ab \in R_a \cap L_b$ iff there is an idempotent $e$ in $L_a \cap R_b$: such products are referred to as "trace products". The trace products and the structure maps completely determine the structure of $M$; namely for any $a, b \in M$, $ab = \psi_{a^{-1}a,e}(a).\phi_{bb^{-1},e}(b)$ where $e = a^{-1}abb^{-1}$.*

Item (g) of the proposition above forms the basis for the "inductive groupoid" approach to the structure of inverse semigroups (see Ref. 34 for details and references), or equivalently for Meakin's "structure mapping" approach [47].

For each subset $N$ of an inverse monoid $M$, we denote by $N^\omega$ the set of all elements $m \in M$ such that $m \geq n$ for some $n \in N$. The subset $N$ of $M$ is called *closed* if $N = N^\omega$. Closed inverse submonoids of an inverse monoid $M$ arise naturally in the representation theory of $M$ by partial injections on a set developed by Schein [70]. An inverse monoid $M$ acts (on the right) by injective partial functions on a set $Q$ if there is a homomorphism from $M$ to SIM$(Q)$. Denote by $q.m$ the image of $q$ under the action of $m$ if $q$ is in the domain of the action by $m$.

If an inverse monoid $M$ acts on $Q$ by injective partial functions, then for every $q \in Q$, $Stab(q) = \{m \in M : qm = q\}$ is a closed inverse submonoid of $M$. Conversely, given a closed inverse submonoid $H$ of $M$, we can construct a transitive representation of $M$ as follows. A subset of $M$ of the form $(Hm)^\omega$ where $mm^{-1} \in H$ is called a *right $\omega$-coset* of $H$. Let $X_H$ denote the set of right $\omega$-cosets of $H$. If $m \in M$, define an action on $X_H$ by $Y.m = (Ym)^\omega$ if $(Ym)^\omega \in X_H$ and undefined otherwise. This defines a transitive action of $M$ on $X_H$. Conversely, if $M$ acts transitively on $Q$, then this action is equivalent in the obvious sense to the action of $M$ on the right $\omega$-cosets of $Stab(q)$ in $M$ for any $q \in Q$. See Ref. 70 for details.

We call two closed inverse submonoids $H_1, H_2$ of an inverse monoid $M$ *conjugate* if there exists $m \in M$ such that $mH_1m^{-1} \subseteq H_2$ and $m^{-1}H_2m \subseteq H_1$. It is clear that conjugacy is an equivalence relation on the set of closed inverse submonoids of $M$: however, conjugate closed inverse submonoids of an inverse monoid are not necessarily isomorphic. For example, the closed inverse submonoids $\{1, aa^{-1}, a^2a^{-2}\}$ and $\{1, aa^{-1}, a^{-1}a, aa^{-2}a\}$ of the free inverse monoid on the set $\{a\}$ are conjugate but not isomorphic.

The relation $\sigma$ on an inverse monoid $M$ defined by $a\,\sigma\,b$ iff there exists $c \in M$ such that $c \leq a$ and $c \leq b$ is a congruence on $M$ and the quotient $M/\sigma$ is a group, the *maximum group homomorphic image of $M$*. The congruence $\sigma$ is called the *minimum group congruence* on $M$. An inverse monoid $M$ is called *E-unitary* if its minimum group congruence is *idempotent-pure*, i.e. the inverse image in $M$ of the identity of $G = M/\sigma$ under the natural map from $M$ onto $G$ is $E(M)$. Equivalently, $M$ is $E$-unitary iff, whenever $a \geq e$ for some idempotent $e \in E(M)$, then $a \in E(M)$; or equivalently $M$ is $E$-unitary iff all structure maps of $M$ are one-to-one. We refer the reader to Ref. 34 for much information about the structure of $E$-unitary inverse monoids and the important role that they play in the theory of inverse monoids. In particular McAlister's "$P$-theorem"[45] describes how $E$-unitary inverse monoids may be built from groups and semilattices, while McAlister's "covering theorem"[44] shows that every inverse monoid is an idempotent-separating homomorphic image of a suitable $E$-unitary inverse monoid.

An alternative description of how to build inverse semigroups from groups and semilattices is provided by Munn[54] who shows that every inverse semigroup admits an idempotent-separating homomorphism onto a "fundamental" inverse semigroup. An inverse semigroup $M$ is called "fundamental" if the only congruence on $M$ that separates idempotents into distinct congruence classes is the trivial congruence. In his paper[54], Munn showed that every fundamental inverse semigroup is isomorphic to a full inverse subsemigroup of a semigroup $T_E$, where $T_E$ is the inverse semigroup of all isomorphisms between principal ideals of the semilattice $E$. The *Munn representation* of an inverse semigroup $S$ is the homomorphism $\mu : S \to T_{E(S)}$ defined by sending $s \in S$ to the principal ideal isomorphism $\mu(s)$ that maps the set $\{e \in E : e \leq s^{-1}s\}$ onto the set $\{e \in E : e \leq ss^{-1}\}$, via $e \mapsto ses^{-1}$. The image $\mu(S)$ is called the *fundamental image* of $S$.

It is not difficult to show that inverse monoids form a variety of algebras of type $< 2, 1, 0 >$ (in the sense of universal algebra) defined by the identities

$$a(bc) = (ab)c, \quad 1.a = a.1 = a, \quad aa^{-1}a = a, \quad (a^{-1})^{-1} = a,$$
$$(ab)^{-1} = b^{-1}a^{-1}, \quad aa^{-1}bb^{-1} = bb^{-1}aa^{-1}.$$

(This last identity expresses the fact that idempotents commute.) It follows that free inverse monoids exist. We will denote the free inverse monoid on a set $X$ by $FIM(X)$. This is the quotient of the free monoid with involution $(X \cup X^{-1})^*$ by the *Vagner congruence*, i.e. the congruence that forces all of the identities above to hold. Similarly, the free inverse semigroup on $X$ exists and is denoted by $FIS(X)$. It is not difficult to see that $FIM(X) = FIS(X) \cup \{1\}$.

The structure of free inverse monoids was determined independently by Scheiblich[69] and Munn[55]. Scheiblich's description for elements of $FIM(X)$ is in terms of rooted Schreier subsets of the free group $FG(X)$, while Munn's description is in terms of birooted edge-labeled trees. Scheiblich's description provides an important example of a McAlister triple, in the spirit of the McAlister $P$-theorem, while Munn's description lends itself most directly to a solution to the word problem for $FIM(X)$. It is not difficult to see the equivalence of the two descriptions. Some variations on Scheiblich's description, obtaining a "linear" canonical form for elements of $FIM(X)$ are provided by Schein[71] and Preston[60]. A variation on Scheiblich's approach, motivated somewhat by Schein's description, is provided in Lawson's book[34]. The version below is a slight variation on Munn's approach, the essential difference being that for some purposes it is somewhat more convenient to regard Munn's birooted trees as subtrees of the Cayley graph of the free group $FG(X)$, with the initial root identified with the vertex 1 in the Cayley graph.

Denote by $\Gamma(X)$ the Cayley graph of the free group $FG(X)$ with respect to the usual presentation, $FG(X) = Gp\langle X : \emptyset \rangle$. Thus $\Gamma(X)$ is an infinite tree whose vertices correspond to the elements of $FG(X)$ (in reduced form) and with a directed edge labeled by $x \in X$ from $g$ to $gx$ (and an inverse edge labeled by $x^{-1}$ from $gx$ to $g$). For each word $w \in (X \cup X^{-1})^*$, denote by $MT(w)$ the finite subtree of the tree $\Gamma(X)$ obtained by reading the word $w$ as the label of a path in $\Gamma(X)$, starting at 1. Thus, for example, if $w = aa^{-1}bb^{-1}ba^{-1}abb^{-1}$, then $MT(w)$ is the tree pictured below.

One may view $MT(w)$ as a birooted tree, with initial root 1 and terminal root $r(w)$, the reduced form of the word $w$ in the usual group-theoretic sense. Munn's solution[55] to the word problem in $FIM(X)$ may be stated in the following form.

**Theorem 2.2.** *If $u, v \in (X \cup X^{-1})^*$, then $u = v$ in $FIM(X)$ iff $MT(u) = MT(v)$ and $r(u) = r(v)$.*

Thus elements of $FIM(X)$ may be viewed as pairs $(MT(w), r(w))$ (or as birooted edge-labelled trees, which was the way that Munn described his results). Multiplication in $FIM(X)$ is performed as follows. If $u, v \in (X \cup X^{-1})^*$, then $MT(uv) = MT(u) \cup r(u).MT(v)$ (just translate $MT(v)$ so that its initial root coincides with the terminal root of $MT(u)$ and take the union of $MT(u)$ and the translated copy of $MT(v)$: the terminal root is of course $r(uv)$).

We may regard $MT(w)$ as an automaton with initial state 1 and terminal state $r(w)$ (the reduced form of $w$). From this point of view, $MT(w)$ is actually an *inverse automaton*, i.e. an edge labeled by a letter $x$ from $g$ to $gx$ has a unique inverse edge labeled by $x^{-1}$ from $gx$ to $g$ and we permit edges to be read in either direction. The *language* accepted by the automaton $MT(w)$ is then $L(MT(w)) = \{u \in (X \cup X^{-1})^* : u$ labels a path in $MT(w)$ from 1 to $r(w)\}$. Then $MT(w)$ is the minimal automaton accepting this language. With this notation, we outline an automata-theoretic proof of Munn's theorem below.

**Proof of Munn's theorem**

We show in fact that $L(MT(w)) = \{u \in (X \cup X^{-1})^* : w \leq u$ in the natural partial order on $FIM(X)\}$. This clearly gives Munn's theorem since we then have $u = w$ in $FIM(X)$ iff $u \in L(MT(w))$ and $w \in L(MT(u))$, i.e. iff $MT(u) = MT(w)$ and $r(u) = r(w)$.

Start with the "linear automaton" $Lin(w)$ of the word $w$: i.e. start with a linear directed path labeled by the word $w$ and view this as an automaton with initial state the initial vertex of the path and terminal state the terminal vertex of the path. The language accepted by this automaton consists of all words $u \in (X \cup X^{-1})^*$ that label paths that can be read (reading edges in either direction) from the initial state to the terminal state of the automaton. It is clear that $w$ is in the language accepted by $Lin(w)$ and it is also the case that any word in the language accepted by $Lin(w)$ is equal to $w$ in $FIM(X)$. However, there may be words that are equal to $w$ in $FIM(X)$ that are not accepted by $Lin(w)$. For example, if $w = aa^{-1}bb^{-1}$ then $bb^{-1}aa^{-1} = aa^{-1}bb^{-1}$ in $FIM(X)$ but $bb^{-1}aa^{-1}$ is not accepted by $Lin(w)$.

If at any vertex $\alpha$ of this automaton, we see an edge with label $x$ coming in to $\alpha$ and an edge with label $x^{-1}$ leaving $\alpha$, then we perform a "Stallings folding" by identifying these two edges, obtaining a new automaton $\mathcal{A}_1$.

Notice that $w \in L(Lin(w)) \subseteq L(\mathcal{A}_1) \subseteq \{u \in (X \cup X^{-1})^* : u \geq w\}$ (i.e. the Stallings folding has the effect of increasing the language recognized by the resulting automaton). Continue applying Stallings foldings until no more foldings can be applied. The process clearly yields the Munn tree $MT(w)$ after finitely many steps. Thus we have a sequence of automata $Lin(w), \mathcal{A}_1, ..., \mathcal{A}_k = MT(u)$, with $w \subseteq L(Lin(w)) \subseteq L(\mathcal{A}_1) \subseteq ... \subseteq L(MT(w)) \subseteq \{u \in (X \cup X^{-1})^* : u \geq w\}$. Conversely, if $u \geq w$ in the natural partial order on $FIM(X)$, then $w = ww^{-1}u$. Since $u$ labels a path in $MT(w)$ from 1 to $r(w)$ iff $ww^{-1}u$ also labels a path in $MT(w)$ from 1 to $r(w)$, without loss of generality we may assume that $w = u$ in $FIM(X)$. Then it is straightforward to see by induction on the number of elementary transitions needed to get from $w$ to $u$ under the Vagner congruence that $u$ must label a path in $MT(w)$ from the initial point 1 to the terminal point $r(w)$; that is, $u \in L(MT(w))$.  ∎

Munn's theorem and in fact the proof of his theorem outlined above has been greatly extended by Stephen[77,78] in his study of presentations of inverse monoids by generators and relations.

Denote by $Inv\langle X : u_i = v_i, i \in I \rangle$ the inverse monoid presented by set $X$ of generators and set $R = \{(u_i, v_i) : i \in I\}$ of relations. Here the $u_i, v_i$ are words in $(X \cup X^{-1})^*$ and $Inv\langle X : u_i = v_i \rangle$ is the image of $FIM(X)$ obtained by imposing the relations in $R$. Clearly the group $G = Gp\langle X : u_i = v_i \rangle$ is the maximum group homomorphic image of the inverse monoid $M = Inv\langle X : u_i = v_i \rangle$. We will use the same notation for the inverse semigroup presented by a set $X$ of generators and relations $u_i = v_i$ (of course all words $u_i, v_i$ are in $(X \cup X^{-1})^+$ in that case). No confusion should arise and the context will make it clear whether the semigroup is in fact a monoid.

For example, $FIM(X) = Inv\langle X : \emptyset \rangle$ and its maximum group image is the free group $FG(X)$. The free group may be presented as an inverse monoid as $FG(X) = Inv\langle X : xx^{-1} = x^{-1}x = 1, x \in X \rangle$. The bicyclic monoid is $B = Inv\langle a : aa^{-1} = 1 \rangle$ when presented as an inverse monoid. Its maximum group image is $\mathbb{Z}$ of course.

More generally we may associate inverse semigroups with directed graphs in several ways: the construction below is standard in the operator algebra literature. Let $\Gamma$ be a directed graph with set $\Gamma^0$ of vertices and set $\Gamma^1$ of edges: view an edge $e \in \Gamma^1$ as starting at $s(e)$ and ending at $r(e)$. For each edge $e$ there is an inverse edge $e^*$ from $r(e)$ to $s(e)$. For each vertex $v \in \Gamma^0$, define $s(v) = r(e) = v$. Define $I(\Gamma)$ as the inverse semigroup generated by $\Gamma^0 \cup \Gamma^1$ together with a with a zero $z$ subject to the relations

(1) $s(e)e = er(e) = e$ for all $e \in \Gamma^1 \cup \{e^* : e \in \Gamma^1\}$
(2) $a\, b = z$ if $a, b \in \Gamma^0 \cup \Gamma^1 \cup \{e^* : e \in \Gamma^1\}$ and $r(a) \neq s(b)$
(3) $a^* b = z$ if $a, b \in \Gamma^1$ and $a \neq b$
(4) $b^* b = r(b)$ if $b \in \Gamma^1$.

If $\Gamma$ is the graph with one vertex and one edge, then $I(\Gamma)$ is the bicyclic monoid with a (removable) zero. If $\Gamma$ is the graph with one vertex and $n$ edges, then $I(\Gamma)$ is the polycyclic monoid $P_n$. The monoid $P_n$ was introduced by Nivat and Perrot [57] as the syntactic monoid of the "correct parenthesis" language with $n$ sets of parentheses. It was rediscovered in the operator algebra literature where it is referred to as the Cuntz monoid (see Paterson's book [58] for details).

If $M = Inv\langle X : u_i = v_i \rangle$ and $w \in (X \cup X^{-1})^*$, then the *Schützenberger graph* $S\Gamma(w)$ is the strongly connected component of the Cayley graph of $M$ (with respect to the generating set $X$) that contains the image of $w$ in $M$. Alternatively, $S\Gamma(w)$ is the restriction of the Cayley graph of $M$ to the $\mathcal{R}$-class of (the image of) $w$ in $M$. That is, the vertices of $S\Gamma(w)$ consist of those elements $u \in M$ such that $uu^{-1} = ww^{-1}$ in $M$, and there is an edge labeled by $x \in X$ from such a vertex $u$ to the vertex $ux$ if $uxx^{-1}u^{-1} = uu^{-1} = ww^{-1}$ (and an inverse edge labeled by $x^{-1}$ from $ux$ to $u$). Note that there is a natural edge-labeled graph morphism from $S\Gamma(w)$ to the Cayley graph of $G = Gp\langle X : u_i = v_i \rangle$. This morphism takes an edge labeled by $x$ from $u$ to $ux$ to the edge labeled by $x$ from $u\sigma$ to $(ux)\sigma$. We may regard $S\Gamma(w)$ as an inverse automaton with initial state $ww^{-1}$ and terminal state $w$. With this notation, we have the following theorem of Stephen [78].

**Theorem 2.3.** *Let* $M = Inv\langle X : u_i = v_i \rangle$ *and* $w \in (X \cup X^{-1})^*$. *Then* $L(S\Gamma(w)) = \{u \in (X \cup X^{-1})^* : u \geq w$ *in the natural partial order on* $M\}$. *Thus* $u = w$ *in* $M$ *iff* $u \in L(S\Gamma(w))$ *and* $w \in L(S\Gamma(u))$, *i.e. iff* $S\Gamma(u)$ *and* $S\Gamma(w)$ *are isomorphic as inverse automata.*

Stephen's theorem implies that the word problem is decidable for the inverse monoid $M = Inv\langle X : u_i = v_i \rangle$ iff there is an algorithm to effectively calculate $L(S\Gamma(w))$ for each word $w \in (X \cup X^{-1})^*$. In his thesis [77] and his paper [78], Stephen provided the following iterative procedure for constructing the Schützenberger automaton $S\Gamma(w)$ for each word $w$ corresponding to a presentation $M = Inv\langle X : u_i = v_i \rangle$.

Start with the linear automaton $Lin(w)$ and close under successive applications of the operations of *Stallings foldings* and *expansions*. Here by an "expansion" we mean the following. If in some automaton $\mathcal{A}$ we can read a path from a vertex $\alpha$ of $\mathcal{A}$ to a vertex $\beta$ of $\mathcal{A}$ labeled by one side (say $u_i$) of one of the defining relations $u_i = v_i$ for $M$ then we sew on a new path from $\alpha$ to $\beta$ labeled by the other side $v_i$ of the corresponding relation if no such path exists in $\mathcal{A}$. Of course after such a path has been added it may be necessary to apply additional Stallings foldings etc, so this process may or may not stop after a finite number of steps. Stephen shows that this process is confluent and then shows that the process "limits" in an appropriate sense to the Schützenberger automaton $S\Gamma(w)$. The proof is a language-theoretic argument that is essentially a generalization of the proof of Munn's theorem outlined above.

For example, if $M = FIM(X) = Inv\langle X : \emptyset \rangle$, then it is clear that $S\Gamma(w) = MT(w)$ for each word $w$ since no expansions apply: in this case Stephen's iterative process stops after a finite number of steps (the $\mathcal{R}$-classes of $FIM(X)$ are all finite). On the other hand, if $M = B = Inv\langle a : aa^{-1} = 1 \rangle$ (the bicyclic monoid) then all Schützenberger graphs are infinite. Stephen's iterative process does not stop but it is nevertheless easy to understand what the Schützenberger graphs look like. For example, if we start with the word $w = 1$, then we can read one side (namely 1) of the defining relation at this vertex so we sew on a path labeled by the other side $(aa^{-1})$ at this vertex and fold, to obtain a graph with two vertices (which we denote by 1 and $a$) and an edge labeled by $a$ from 1 to $a$. But then at the vertex $a$ we see the side of the relation labeled by 1 so we sew on another path labeled by $aa^{-1}$ at this vertex and fold, yielding a graph with three vertices (1, $a$ and $a^2$) and two edges etc. We can continue adding a new edge labeled by $a$ to any finite approximate of $S\Gamma(1)$ obtained this way. The process does not stop but it clearly limits to a graph that may be identified with $\mathbb{N} = \{1, 2, ...\}$ with an edge labeled by $a$ from $i$ to $i+1$ for each positive integer $i$. Note that in this case, each graph $S\Gamma(w)$ naturally embeds in the Cayley graph of $\mathbb{Z}$, the maximal group image of $B$. In fact this property characterizes $E$-unitary inverse monoids, as the following simple proposition shows.

**Proposition 2.2.** *Let $M = Inv\langle X : u_i = v_i \rangle$. Then $M$ is E-unitary if and only each Schützenberger graph of $M$ naturally embeds in the Cayley graph of the maximal group image $G = Gp\langle X : u_i = v_i \rangle$ of $M$.*

**Proof** This follows easily from the fact that $M$ is $E$-unitary if and only if all of its structure maps are injective; i.e. whenever $u \, \mathcal{R} \, v$ and $u \, \sigma \, v$ in $M$, then $u = v$ in $G$. ∎

This proposition has the consequence that if we know that $M$ is $E$-unitary, then we may carry out Stephen's iterative process for constructing $S\Gamma(w)$ inside the Cayley graph $\Gamma(G, X)$ of $G$ relative to the set $X$ of generators. That is, we start with the image of the linear automaton $Lin(w)$ in $\Gamma(G, X)$ and then proceed to do all expansions inside $\Gamma(G, X)$: no folding is required if we work inside the Cayley graph at each step.

Stephen's iterative process described above may be extended somewhat. Start with an automaton $\mathcal{A}$ with initial state $\alpha$ and terminal state $\omega$ and input alphabet $X$ (the set from which the labels on the edges of $\mathcal{A}$ are taken) and assume that each edge labeled by $x \in X$ has an associated inverse edge labeled by $x^{-1}$, as above. Then the language $L(\mathcal{A})$ accepted by this automaton is the set of words $w \in (X \cup X^{-1})^*$ that label paths from $\alpha$ to $\omega$ in $\mathcal{A}$. Iteratively apply Stallings foldings and expansions (relative to an inverse monoid $M = Inv\langle X : u_i = v_i \rangle$) to the automaton $\mathcal{A}$, obtaining a sequence of new automata. The following result was proved by Meakin and Szakács[50].

**Theorem 2.4.** *Let $M = Inv\langle X : u_i = v_i \rangle$ and let $\mathcal{A}$ be an automaton (as described above) with input alphabet $X$. Iteratively apply Stallings foldings and expansions to obtain a sequence of intermediate automata. Then there is an appropriate notion of a "limit" $\mathcal{A}^\omega$ of the sequence of automata obtained by applying these operations iteratively, that is independent of how they are applied. Furthermore, $L(\mathcal{A}^\omega) = \{w \in (X \cup X^{-1})^* : w \geq s$ for some $s \in L(\mathcal{A})\}$.*

Of course if $\mathcal{A}$ is the linear automaton $Lin(w)$ of some word $w$, then $\mathcal{A}^\omega$ is the Schützenberger automaton $S\Gamma(w)$. The proof of Theorem 2.4 closely follows Stephen's proof of the validity of his iterative construction of Schützenberger graphs. Meakin and Szakács[50,51] made use of this more general result to study immersions between cell complexes (see a brief discussion of this later in the present paper).

## 3. Some applications of Stephen's procedure

We remark that in general there is no relationship between decidability of the word problem for $M = Inv\langle X : u_i = v_i, \ i = 1, ..., n \rangle$ and decidability of the word problem for its maximal group image $G = Gp\langle X : u_i = v_i \rangle$.

In one direction this is easy to see. If $G$ is a finitely presented group with undecidable word problem, and if we let $M$ be the inverse monoid $M = G \cup \{0\}$, then we may use the presentation for $G$ to regard $M$ as a finitely presented inverse monoid which clearly has undecidable word problem since $G$ does. But the maximum group homomorphic image of $M$ is the trivial group, which has decidable word problem.

Conversely, suppose again that $G = Gp\langle X : u_i = 1, i = 1, ..., n \rangle$ is a group with undecidable word problem and let $M = Inv\langle X : u_i^2 = u_i, i = 1, ..., n \rangle$. Clearly $G$ is the maximum group image of $M$ and $G$ has undecidable word problem. But we claim that $M$ has decidable word problem! To see this, construct the Munn tree of a word $w$. If we see in this graph a path labeled by $u_i$ for some $i$, then we sew on a path labeled by $u_i^2$. After folding, the path labeled by $u_i$ becomes a loop, so the resulting graph has fewer vertices than the original graph. Iterating this, we see that the Schützenberger graph of $w$ is in fact finite (with no more that $|V(MT(w))|$ vertices), so the word problem for $M$ is decidable. This was observed by Stephen[77].

While this latter result seems counterintuitive, we note that the construction outlined above essentially takes the Cayley graph of $G$ and breaks it into finite pieces that represent the $\mathcal{R}$-classes of $M$. The undecidability in $G$ comes from the fact that even though we can decide the word problem for $M$ we cannot in general decide when two elements of $M$ are related via the minimum group congruence since this involves an additional existential quantifier. Given elements $a, b$ and $c$ of $M$ we can decide whether $c \leq a, b$ but we may not be able to decide whether there *exists* an element $c$ such that $c \leq a, b$.

However, this difficulty does not arise if $M$ is $E$-unitary, as the following simple proposition shows.

**Proposition 3.1.** *If $M = Inv\langle X : u_i = v_i \rangle$ is $E$-unitary and $M$ has decidable word problem, then $G = Gp\langle X : u_i = v_i \rangle$ has decidable word problem.*

**Proof** If $w \in (X \cup X^{-1})^*$ then $w = 1$ in $G$ iff $w = w^2$ in $M$ since $M$ is $E$-unitary, but this is decidable since $M$ has decidable word problem. ∎

While the word problem for an inverse monoid presented by generators and relations is in general undecidable, there are many special cases in which the techniques discussed above yield positive solutions as well as some structural information. We mention below some interesting examples that have appeared in the literature.

Let $C_n = Inv\langle a_1, a_2, ..., a_n : a_i a_j = a_j a_i \rangle$. It is clear that the maximum group homomorphic image of $C_n$ is the free abelian group $\mathbb{Z}^n$. However, the inverse monoid $C_n$ is not commutative. For example $a_1(a_1)^{-1} \neq a_1^{-1} a_1$ in $C_n$. This is because although these elements have the same Munn tree, they have different Schützenberger automata since the initial and terminal states of the Munn tree of the two elements do not coincide, and no expansions are involved in constructing the Schützenberger automata from the corresponding Munn trees. McAlister and McFadden[46] showed that $C_2$ is $E$-unitary and solved the word problem for $C_2$. However $C_n$ is not $E$-unitary if $n > 2$. To see this, consider the element $w = a_1 a_2^{-1} a_3 a_1^{-1} a_2 a_3^{-1}$ in $C_n$. Clearly the image of $w$ in the free abelian group $\mathbb{Z}^n$ is the identity of the group. But $w$ is not an idempotent of $C_n$ since $w$ and $w^2$ have different Schützenberger graphs (which coincide with the corresponding linear automata since no Stallings folding or application of an expansion is possible to these linear automata). Stephen[77] solved the word problem for $C_n$ for all $n$ by showing that all Schützenberger graphs of $C_n$ are finite.

As another example, consider an inverse monoid $M = Inv\langle X : e_i = f_i, i = 1, ..., n \rangle$ where the $e_i$ and $f_i$ are Dyck words, i.e. $r(e_i) = r(f_i) = 1$. This means that the $e_i, f_i$ are idempotents in $FIM(X)$, and the terminal root of $MT(e_i)$ [resp. $MT(f_i)$] is 1. Clearly the maximum group image of such a monoid $M$ is the free group $FG(X)$. It is not too difficult to see that $M$ is $E$-unitary. Thus we may apply Stephen's iterative procedure inside the Cayley tree $\Gamma(X)$ of $FG(X)$ to calculate the Schützenberger graph of a word $w$. We start with the Munn tree $MT(w)$, regarded as a subtree of $\Gamma(X)$, and if at some vertex $\alpha$ in this tree we see a copy of the Munn tree $MT(e_i)$ (or $MT(f_i)$) rooted at that vertex, and contained in $MT(w)$, then we augment $MT(w)$ by adding a copy of $MT(f_i)$ [resp. $MT(e_i)$] rooted at $\alpha$ inside $\Gamma(X)$. This possibly enlarges the subtree of $\Gamma(X)$. Iterate the procedure, building an increasing sequence of subtrees of $\Gamma(X)$ whose union is $S\Gamma(w)$. In Ref. 39, Margolis and Meakin showed that every such monoid $M$ has decidable word problem. The original proof in Ref. 39 makes use of Rabin's "tree theorem"[61] to decide membership in the set of vertices of $S\Gamma(w)$: of course since in this case $S\Gamma(w)$ is a subtree of $\Gamma(X)$, we know the tree if we know its vertices. An alternative language-theoretic proof that avoids the use of Rabin's tree theorem was given by Silva[73]: another proof, involving an algorithm of much lower complexity, was subsequently provided by Lohrey and Ondrusch[36].

Many additional applications of Stephen's approach to the theory of presentations of inverse monoids (or inverse semigroups) may be found in the literature. For example, Jones, Margolis, Meakin and Stephen[33] used this approach to study free products of inverse semigroups, Margolis, Meakin and Stephen[41] and Meakin and Sapir[49] used it to study varieties of inverse semigroups, Bennett[4-7] as well as Cherubini, Meakin and Piochi[9,10] and Cherubini and Rodaro[11] used these techniques to study certain kinds of amalgamated free products of inverse semigroups, Jajcayova[31,32] and Yammamura[82,83] used it to study HNN extensions of inverse semigroups, several papers (Ivanov, Margolis and Meakin[30], Birget, Margolis and Meakin[8], Hermiller, Lindblad and Meakin[26]) have used these techniques to study one relator inverse monoids, and these methods have played a role in Steinberg's topological theory of inverse semigroups[75,76]. Some additional information may be found in the survey papers by Meakin[48] and Margolis, Meakin and Sapir[40]. Further information about one-relator inverse monoids is provided in the next section of this paper.

## 4. One-relator inverse monoids

It is well known that the word problem for a one-relator group $G = Gp\langle X : w = 1\rangle$ is decidable. This, together with Magnus' "Freiheitssatz" was proved by Magnus in the early 1930's. The proof that is included in the book by Lyndon and Schupp[37] is phrased in terms of $HNN$ extensions and induction on the length of the relator. Magnus' algorithm has high computational complexity and it is not known whether there is a more computationally efficient algorithm.

In his paper[1], Adian proved decidability of the word problem for all one-relator monoids of the form $Mon\langle X : w = 1\rangle$ (where $w \in X^+$). His proof involves showing that the group of units of such a monoid is a one-relator group and then reducing the word problem for $M$ to the word problem for the group of units of $M$ and appealing to Magnus' result. However it is not known whether all one-relation semigroups $S = Sgp\langle X : u = v\rangle$ (for $u, v \in X^*$) have decidable word problem. Many special cases of this have been solved, but the decidability of the word problem for all one relation semigroups is not known in general. The problem reduces, by a result of Adian and Oganessian[3], to the word problem for all semigroups with a presentation of the form $Sgp\langle X : u = v\rangle$ where $u = asb$ and $v = atc$ are reduced words with $a, b, c \in X$, $b \neq c$ and $s, t \in X^*$.

It seems natural to try to extend Magnus' proof of the decidability of the word problem for one-relator groups to a study of the word problem for one-relator inverse monoids, at least those with a presentation of the form $M = Inv\langle X : w = 1\rangle$. However, even though there is a theory of $HNN$ extensions of inverse monoids[31,83], this approach seems somewhat problematic. As a very first obstacle, Magnus' proof depends on being able to cyclically permute the relator without changing the one-relator group, but this is not the case for one-relator inverse monoids. For example, the inverse monoids $M_1 = Inv\langle a, b : aba = 1\rangle$ and $M_2 = Inv\langle a, b : baa = 1\rangle$ are not isomorphic. Clearly $M_1$ is a group (namely the group of integers $\mathbb{Z}$ since the letter $a$ is a unit and hence $b = a^{-2}$ is also a unit in $M_1$), while $M_2$ is isomorphic to the bicyclic monoid, which is not a group. There does not seem to be a systematic study of the relationship between $Inv\langle X : w = 1\rangle$ and $Inv\langle X : w' = 1\rangle$ where $w'$ is a cyclic conjugate of $w$. In addition, the inverse monoid $M = Inv\langle X : w = 1\rangle$ is not in general isomorphic to the inverse monoid $Inv\langle X : r(w) = 1\rangle$ and this is also not in general isomorphic to the inverse monoid $Inv\langle X : u = 1\rangle$ where $u$ is the cyclically reduced form of $w$.

Of course the word problem for one-relator inverse monoids of the form $M = Inv\langle X : e = 1\rangle$ where $e$ is a Dyck word (i.e. an idempotent in $FIM(X)$), is solvable since this is a special case of the result of Margolis and Meakin mentioned above. In this case the paper by Birget, Margolis and Meakin[8] provides a much more efficient (cubic time) algorithm however. Wang[81] used rewriting techniques for Munn trees to provide a quadratic time algorithm to solve the word problem in this case and also showed that the maximal subgroups of an inverse monoid of the form $Inv\langle X : e = 1\rangle$ (where $e$ is a Dyck word) are finitely generated free groups.

In the paper[30], Ivanov, Margolis and Meakin studied one-relator inverse monoids of the form $M = Inv\langle X : w = 1\rangle$ where $w$ is a reduced word. One of the results in Ref. 30 is the following.

**Theorem 4.1.** *If the word problem is decidable for all inverse monoids of the form $M = Inv\langle X : w = 1\rangle$ for $w$ a reduced word, then the word problem is decidable for all one-relation semigroups $S = Sgp\langle X : u = v\rangle$.*

The proof shows that the one-relation semigroup $S = Sgp\langle X : asb = atc\rangle$ (with $asb, atc$ reduced words with $b \neq c$) embeds in the one-relator inverse monoid $M = Inv\langle X : asbc^{-1}t^{-1}a^{-1} = 1\rangle$. So if the word problem for $M$ is decidable, then the word problem for $S$ must also be decidable. However, the word $asbc^{-1}t^{-1}a^{-1}$ is not *cyclically* reduced. It is not known whether decidability of the word problem for one-relator inverse monoids of the form $Inv\langle X : w = 1\rangle$, where $w$ is a *cyclically* reduced word, implies decidability of the word problem for all one one-relation semigroups. The main theorem in Ref. 30 is the following.

**Theorem 4.2.** *If $w$ is a cyclically reduced word, then the inverse monoid $M = Inv\langle X : w = 1\rangle$ is E-unitary.*

Of course by definition, an inverse monoid $M$ is $E$-unitary if and only if every word that is equal to the identity in the maximum group image $G$ of $M$ is an idempotent of $M$. By van Kampen's Lemma (see Ref. 37), this is equivalent to showing that every cyclically reduced word labeling the boundary of any van Kampen diagram over the presentation is an idempotent in $M$. The proof of Theorem 4.2 involves a combination of arguments involving van Kampen diagrams, Stephen's iterative construction of Schützenberger graphs, and a theorem of Ivanov and Meakin[29] showing that certain two-relator groups are diagrammatically aspherical.

A consequence of Theorem 4.2 is that if $w$ is a cyclically reduced word then $S\Gamma(1)$ (the Schützenberger graph of 1 for the presentation of $M = Inv\langle X : w = 1\rangle$) is naturally embedded in the Cayley graph $\Gamma(G, X)$ of the one-relator group $G = Gp\langle X : w = 1\rangle$. It follows from this that the vertices of $S\Gamma(1)$ may be identified with the elements of $G$ that can be expressed as products of *prefixes* of the cyclically reduced word $w$ in $G$. In fact the following result in Ref. 30 may be deduced from results in Stephen's paper[79].

**Corollary 4.1.** *Let $w$ be a cyclically reduced word, let $M = Inv\langle X : w = 1\rangle$ and let $P_w$ be the prefix monoid of $w$; that is $P_w$ is the submonoid of $G = Gp\langle X : w = 1\rangle$ generated by the prefixes of $w$. Then the word problem for $M$ is decidable if the membership problem for $P_w$ is decidable.*

We remark that the prefix monoid $P_w$ is not invariant under cyclic conjugation of the word $w$. For example, in the inverse monoid $M_1 = Inv\langle a, b : w = 1\rangle$ where $w = aba$, $P_w = \mathbb{Z}$, while in $M_2 = Inv\langle a, b : w' = 1\rangle$ where $w' = baa$, $P_{w'}$ is the subsemigroup of non-positive integers under addition. In general, it is not clear what the relationship between $P_w$ and $P_{w'}$ is when $w'$ is a cyclic conjugate of $w$.

Despite the fact that there is a large literature on one-relator groups, the prefix membership problem discussed above remains open. Several special cases of this problem were solved in the original paper of Ivanov, Margolis and Meakin[30] (including for example, the case where $w$ is a cyclic conjugate of the word $a_1 b_1 a_1^{-1} b_1^{-1} a_2 b_2 a_2^{-1} b_2^{-1} ... a_g b_g a_g^{-1} b_g^{-1}$ corresponding to the fundamental group of an orientable surface of genus $g$, or the case where $w = ab^m a^{-1} b^{-n}$ corresponding to a Baumslag-Solitar group), but the problem remains open in general, as is the word problem for one-relator inverse monoids.

In their paper[42], Margolis, Meakin and Sunić used the notion of *distortion functions* to study the prefix membership problem. Let $M = Inv\langle X : w = 1\rangle$ with $w$ cyclically reduced, let $G = Gp\langle X : w = 1\rangle$ and and let $P = P_w \leq G$ be the corresponding prefix monoid. A non-decreasing function $f : \mathbb{N} \to \mathbb{N}$ is called a *distortion function* for $P$ in $G$ if $|g|_P \leq f(|g|_G)$ for all $g \in P$: here by $|g|_P$ we mean the minimum length of a word $g \in P$ expressible as a product of prefixes of $w$, and by $|g|_G$ we mean the minimum length of $g$ expressible as a product of the generators in $X \cup X^{-1}$ for $G$. The following fact is a special case of a result in Ref. 42.

**Proposition 4.1.** *The membership problem for $P_w$ is decidable if and only if there is a recursive distortion function for $P_w$ in $G$.*

It is easy to find examples of cyclically reduced words $w$ such that $P_w$ is distorted in $G = Gp\langle X : w = 1\rangle$. For example let $w$ be the word $w = aba^n c$ where $a, b$ and $c$ are distinct letters. Then $a^n = b^{-1}a^{-1}c^{-1}$ in $G$ so $|a^n|_G \leq 3$, But $a^n$ is a product of $n$ prefixes of $w$ so $a^n \in P_w$. One can verify, on constructing the Schützenberger graph of $a^n$, that $a^n$ cannot be written as a product of fewer than $n$ prefixes of $w$, so $|a^n|_P = n$. We note that membership in the prefix monoid $P_w$ in $G$ is decidable in this example. This is because the map $\phi$ that sends $a$ and $b$ to 1 and $c$ to $-(n+2)$ extends to a homomorphism from $G$ to $\mathbb{Z}$ that sends every proper prefix of $w$ to a positive integer. Hence all elements of $P_w$ other than 1 map to positive integers and there are only finitely many products of prefixes of $w$ that map to the same positive integer. To decide whether an element $g$ of $G$ is in $P_w$ we compute its value under this map $\phi$ and if $\phi(g)$ is positive we check whether any product of prefixes with the same value as $\phi(g)$ is equal to $g$ in $G$. This is possible since the word problem is decidable in $G$ and there are only finitely many products of prefixes to check. It also follows from results in Ref. 42 that the group of units of the inverse monoid $Inv\langle a, b, c : aba^n c = 1\rangle$ is trivial. This example also shows that although $ST(1)$ is always embedded in $\Gamma(G, X)$, the embedding is not in general an isometry. That is, $ST(1)$ is distorted as a subgraph of $\Gamma(G, X)$.

It follows from the Corollary 1 and Proposition 4 above that the word problem for $M$ is decidable if there is a recursive distortion function for $P_w$ in $G$. It is not known, however, whether decidability of the word problem for $M$ implies decidability of the prefix membership problem for $P_w$. Thus it is not known whether there is a cyclically reduced word $w$ for which $P_w$ has unbounded distortion function, or if there is, whether there is such a word $w$ for which $P_w$ has unbounded distortion in $G$ but the corresponding one-relator inverse monoid $M$ has decidable word problem. Margolis, Meakin and Sunić[42] made use of Proposition 4.1 and related ideas about distortion functions to solve the prefix membership problem for several additional special cases.

One may also try to solve the word problem for inverse monoids of the form $M = Inv\langle X : w = 1\rangle$ by analyzing the language $L(S\Gamma(u))$ directly for all words $u \in (X \cup X^{-1})^*$. This approach was used by Hermiller, Lindblad and Meakin[26] who introduced a combinatorial condition on the cyclically reduced word $w$ that guarantees that successive approximations of the Schützenberger graphs under Stephen's iterative procedure "grow away from the origin". Informally, a word $w$ is defined to be "sparse" if there are "gaps" between segments of the cyclic word $w$ that fold onto occurrences of prefixes or suffixes of $w$ in the iterative construction of the Schützenberger graph $S\Gamma(1)$. The word $w = [a_1, b_1][a_2, b_2]...[a_g, b_g]$ corresponding to a surface group of genus $g$ is an example of a sparse word. See Ref. 26 for a precise definition of this concept. The paper[26] analyzes the structure of $S\Gamma(1)$ for a sparse word $w$ and shows that this graph has a tree-like structure, from which it follows that $L(S\Gamma(1))$ is a deterministic context-free language. This enables us to decide which words represent the identity in $M$.

Of course $S\Gamma(u)$ is a finite union of copies of $S\Gamma(1)$ rooted at the images in $\Gamma(G, X)$ of the vertices in $MT(u)$. This does not immediately appear to imply that the word problem for $M$ is decidable if membership in $L(S\Gamma(1))$ is decidable however, contrary to what was claimed in Ref. 26. Some additional work is needed to decide membership in the language $L(S\Gamma(u))$ for an arbitrary word $u$. It is of interest to know whether it is true in general (not just for sparse words $w$) whether the word problem for $M = Inv\langle X : w = 1\rangle$ is decidable if there is an algorithm to decide which words represent the identity of $M$.

Another natural approach to the word problem for one-relator inverse monoids is to try to modify the ideas developed by Adian[1] in his solution to the word problem for one-relator monoids of the form $S = Mon\langle X : w = 1\rangle$ (where $w$ is a positive word of course). Central to Adian's approach is the calculation of the group of units of such a monoid $S$. It is clear that if there is some word $u$ that is a prefix and a suffix of $w$, (i.e. $w \equiv uv_1 \equiv v_2u$ for some words $u, v_1, v_2$) then $u$ is in the group of units of $S$. So the words $w_1 \equiv v_1u$ and $w_2 \equiv uv_2$ are also equal to 1 in $S$. Then if there is a word $u_1$

that is a prefix of one of the words $w, w_1, w_2$ as well as a suffix of one of these words, then $u_1$ is also in the group of units of $S$, so we may construct new words that are equal to 1 in $S$ as above. This argument may be repeated, but must eventually stop since all the words that are constructed that are equal to 1 in $S$ have the same length as $w$. Adian[1] shows that this process constructs all elements of the group of units of $S$.

However the construction of the group of units of a one-relator inverse monoid $M = Inv\langle X : w = 1\rangle$ is more complicated. For example in Ref. 30, Ivanov, Margolis and Meakin used a combination of arguments involving van Kampen diagrams and Stephen's iterative construction of Schützenberger graphs to show that all cyclic conjugates of the word $abcdacdadabbcdacd$ that begin with the letter $a$ are equal to 1 in the inverse monoid $M_{17} = Inv\langle a, b, c, d : w_{17} = 1\rangle$. Thus all words $abcd, acd, ad$ and $abbcd$ are in the group of units of $M_{17}$. By way of contrast, since no prefix of $w_{17}$ is equal to a suffix of $w_{17}$, the group of units in the monoid $S = Mon\langle a, b, c, d : w_{17} = 1\rangle$ is trivial. For a general cyclically reduced word $w$ it is not known how to decide which cyclic conjugates of $w$ are equal to 1 in the inverse monoid $M = Inv\langle X : w = 1\rangle$, and it is not known how to construct the group of units of $M$ in general.

In his paper[84], Zhang reworks Adian's proof of the decidability of the word problem for one-relator monoids $S = Mon\langle X : w = 1\rangle$ in the language or rewriting systems to provide a short proof of Adian's theorem. It is natural to try to modify Zhang's proof to the case of one-relator inverse monoids of the form $M = Inv\langle X : w = 1\rangle$ for $w$ a cyclically reduced word. However, this approach also seems somewhat problematic, in part because Zhang's proof depends on Adian's construction of the group of units of $S$, and in part because there is not a well developed theory of rewriting systems for Munn trees. Some work in this direction was initiated by Wang[81] but is limited in scope.

One of the difficulties is that one needs to resolve many ambiguities to develop a satisfactory theory of rewriting for Munn trees since there may be many overlaps between two such trees. Another difficulty is that the result of applying a "rewriting rule" $T_1 \to T_2$ to an occurrence of a Munn tree $T_1$ that is a subtree of a given Munn tree $T$ yields different Munn trees, depending on the context in which one reads the copy of $T_1$ inside $T$. As a simple example, consider the bicyclic monoid $B = Inv\langle a, b : ab = 1\rangle$ (Of course this implies that $a = b^{-1}$ and it is very easy to solve the word problem). Try doing this by rewriting $MT(ab) \to MT(1)$. This is problematic. For example, let $u = ba^{-1}a^2b^{-1}b^2a^{-1}a^2$. Then $MT(u)$ contains a path labeled by $ab$, which we would like excise by using the rewriting rule. But what we get depends on how we read over the segment labeled by $ab$. For example, $u = ba^{-1}a(ab)aa^{-2}ab^{-2}b^2a \to u_1 = ba^{-1}a^2a^{-2}ab^{-2}b^2a$ in $FIM(a,b)$ and also $u = ba^{-1}a^2b^{-1}ba^{-1}(ab)a^{-1}a^2 \to u_2 = ba^{-1}a^2b^{-1}ba^{-2}a^2$, but $MT(u_1) \neq MT(u_2)$. Of course $S\Gamma(u_1) = S\Gamma(u_2) = S\Gamma(u)$, but $MT(u)$ rewrites to two distinct Munn trees to which no further rewriting applies.

It seems plausible that additional progress may be made on the word problem for one-relator inverse monoids, either by further developing some of the techniques briefly outlined above, or by placing additional restrictions on the word $w$ or the Schützenberger graphs of the inverse monoid $M = Inv\langle X : w = 1 \rangle$. For example, since each Schützenberger graph is a geodesic metric space it seems natural to study those monoids for which the Schützenberger graphs admit an automatic structure or the structure of a hyperbolic metric space*.

## 5. Adian inverse semigroups

Another interesting class of inverse semigroup presentations arises from some of the work in Adian's paper[1].

Let $R = \{(u_i, v_i) : i = 1, ..., n\}$ where each $u_i, v_i$ is a *positive word* (i.e. $u_i, v_i \in X^+$). Then we may form the semigroup $S(R)$, the group $G(R)$ and the inverse semigroup $I(R)$ with set $X$ of generators and set $R$ of relations. That is,

$$S(R) = Sgp\langle X : R \rangle = Sgp\langle X : u_i = v_i, i = 1, ..., n \rangle,$$

$$G(R) = Gp\langle X : R \rangle = Gp\langle X : u_i = v_i, i = 1, ..., n \rangle, \text{ and}$$

$$I(R) = Inv\langle X : R \rangle = Inv\langle X : u_i = v_i, i = 1, ..., n \rangle.$$

We associate two graphs with a set $R$ of positive relations as follows. The *left graph* $LG(R)$ has set $X$ of vertices and has an undirected edge between $x$ and $y$ corresponding to each relation $u_i = v_i$ in $R$ where $x$ is the first letter of $u_i$ and $y$ is the first letter of $v_i$. The *right graph* $RG(R)$ is formed dually, using the last letters of the two sides of a relation in $R$ instead of the first letters. We say that $R$ is *cycle-free* if both $LG(R)$ and $RG(R)$ have no loops or circuits (i.e. $LG(R)$ and $RG(R)$ are forests). A

---

*Subsequent to submission of this paper, Robert Gray[21] has published a proof that there are one-relator inverse monoids of the form $M = Inv\langle X : w = 1 \rangle$ with undecidable word problem. However it remains unknown whether the word problem for such monoids is decidable if $w$ is a reduced (or a cyclically reduced) word.

semigroup $S$ is called an *Adian semigroup* if $S = S(R)$ for some cycle-free presentation. Similarly $G$ is an *Adian group* and $I$ is an *Adian inverse semigroup* if $G = G(R)$ $[I = I(R)]$ for some cycle-free presentation.

For example, $S_1 = Sgp\langle a, b : ab = ba \rangle$ is obviously an Adian semigroup since both the left graph and the right graph in this case consists of a single edge linking $a$ and $b$. The semigroup $S_2 = Sgp\langle a, b, c : ab = ba, ac = ca, c = b \rangle$ is also an Adian semigroup since it is isomorphic to $S_1$, but the presentation given for $S_2$ is not cycle-free. The presentation $S_3 = Sgp\langle a, b : a^2 = ab \rangle$ is not cycle-free since the left graph has a loop from $a$ to $a$: one can prove easily (see below) that $S_3$ admits *no* cycle-free presentation, so $S_3$ is not an Adian semigroup. On the other hand the group $G = Gp\langle a, b : a^2 = ab \rangle$ is isomorphic to $\mathbb{Z}$, which does admit a cycle-free presentation, so $G$ is an Adian group.

Cycle-free presentations were introduced by Adian[1], who established the following result.

**Theorem 5.1.** *If $R$ is a finite cycle-free set of positive relations, then $S(R)$ is cancellative and $S(R)$ embeds in $G(R)$.*

Remmers[64] gave a geometric proof of Adian's theorem and extended it to not necessarily finite cycle-free presentations by means of *semigroup diagrams*, which are an analogue for semigroup presentations of van Kampen diagrams for group presentations. Note that it follows easily from Adian's theorem that the semigroup $S_3 = Sgp\langle a, b : a^2 = ab \rangle$ discussed above is not an Adian semigroup since it is not cancellative. (The elements $a$ and $b$ are not equal in $S_3$ since no relation applies to either $a$ or $b$.)

In her papers[67,68], Sarkisian claims to prove that if $R$ is a cycle-free set of relations, then the left and right divisibility problems (and hence the word problem) for $S(R)$ are decidable and that this implies that the word problem for $G(R)$ is decidable. Unfortunately there appears to be a gap in the proof and as far as I am aware this has not yet been resolved, so the word problem for Adain groups and Adian semigroups appears to be still open. The word problem for Adian inverse semigroups is also open.

Some structural information about Adian inverse semigroups is obtained in the following theorem of Inam, Meakin and Ruyle[28].

**Theorem 5.2.** *Every Adian inverse semigroup is $E$-unitary.*

The proof makes use of some properties established by Remmers[64] of semigroup diagrams and van Kampen diagrams over cycle-free presentations. Note that by this theorem it easily follows that the inverse semigroup $C_n = Inv\langle a_1, a_2, ..., a_n : a_i a_j = a_j a_i \rangle$ is an Adian inverse semigroup iff $n \leq 2$ since by the observation made earlier in this paper, $C_n$ is $E$-unitary iff $n \leq 2$.

From Theorems 7 and 8 and Proposition 3 above it follows that if $R$ is a cycle-free set of relations and the word problem for the Adian inverse semigroup $I(R)$ is decidable, then the word problem is also decidable for the Adian group $G(R)$ and the Adian semigroup $S(R)$. Some progress on the word problem for Adian inverse semigroups was made by Inam[27], who proved that for cycle-free presentations, if the Schützenberger graphs of all positive words are finite, then the Schützenberger graphs of *all* words are finite. As a consequence, Inam is able to solve the word problem for some special cases of Adian inverse semigroups, for example for Baumslag-Solitar inverse semigroups $Inv\langle a, b : ab^m = b^n a \rangle$. However, the word problem for Adian inverse semigroups in general remains wide open.

We close this section with an undecidability observation.

**Proposition 5.1.** *It is undecidable, on input a finite set $R = \{u_i = v_i, i = 1, ..., n\}$ of relations, whether $S(R)$ is an Adian semigroup, whether $G(R)$ is an Adian group, or whether $I(R)$ is an Adian inverse semigroup.*

**Proof** To see this for $S(R)$, note that being an Adian semigroup is a *Markov property* of semigroups, that is, there is a finitely presented semigroup that is an Adian semigroup, and there is a finitely presented semigroup that cannot be embedded in any Adian semigroup since Adian semigroups are cancellative. Hence by the well-known theorem of Markov[43] it is undecidable, on input $R$, whether $S(R)$ is an Adian semigroup. For $G(R)$, we note that Adian groups are torsion-free, and being torsion-free is a Markov property of groups, so by the Adian-Rabin theorem[2,62] it is undecidable, on input $R$, whether $G(R)$ is an Adian group. For $I(R)$, it suffices to use Theorem 8 and the fact that being $E$-unitary is a Markov property of inverse semigroups, so by a theorem of Yamamura[83] it is undecidable, on input $R$, whether $I(R)$ is an Adian inverse semigroup.

## 6. Inverse semigroups and operator algebras

There is a large and growing literature connecting groupoids, inverse semigroups and operator algebras, particularly $C^*$-algebras. The reader is referred to the monographs by Renault[65], Paterson[58], Raeburn[63], Exel[17], and Resende[66] for some background on this development and for numerous references to this field. The books by Davidson[13] and Fillmore[19] are standard references to the theory of operator algebras. I will make no attempt to survey the extensive literature connecting inverse semigroups and operator algebras here but I will mention a few results where inverse semigroups play a prominent role.

We first recall some standard notation. As is customary, we denote by $\mathcal{B}(\mathcal{H})$ the set of (bounded linear) operators on a Hilbert space $\mathcal{H}$: i.e. the set of linear functions $T : \mathcal{H} \to \mathcal{H}$ such that $\{||Tx|| : x \in \mathcal{H}, ||x|| \leq 1\}$ is bounded. If $\mathcal{H} = \mathbb{C}^n$, then $\mathcal{B}(\mathcal{H}) = M_n(\mathbb{C})$. The *norm* of $T \in \mathcal{B}(\mathcal{H})$ is $||T|| = sup\{||Tx|| : ||x|| \leq 1\}$. The *adjoint* of an operator $T$ is the unique operator $T^*$ such that $\langle Tx, y \rangle = \langle x, T^*y \rangle$ for all $x, y \in \mathcal{H}$. In $M_n$, the adjoint of a matrix $A$ is the conjugate transpose of $A$.

A *projection* is an operator $P \in \mathcal{B}(\mathcal{H})$ such that $P = P^* = P^2$. A projection determines a subspace $M = P(\mathcal{H})$ and $P(M^\perp) = 0$. Conversely every subspace determines a projection onto that subspace. The product $P_1 P_2$ of two projections is a projection if and only if $P_1 P_2 = P_2 P_1$. An operator $U \in \mathcal{B}(\mathcal{H})$ is a *unitary operator* if $UU^* = U^*U = 1$. An *isometry* is an operator $T \in \mathcal{B}(\mathcal{H})$ such that $T^*T = 1$. If $\mathcal{H}$ is finite dimensional, an isometry is necessarily a unitary operator. But on $\ell^2$ the *unilateral shift* $T(z_0, z_1, z_2, \ldots) = (0, z_0, z_1, \ldots)$ is an isometry that is a projection onto a proper subspace.

A *partial isometry* is an operator $T \in \mathcal{B}(\mathcal{H})$ such that $TT^*T = T$ (this implies $T^*TT^* = T^*$). Then $T^*T$ and $TT^*$ are projections: $T$ maps $T^*T(\mathcal{H})$ isometrically onto $TT^*(\mathcal{H})$. The unilateral shift is a partial isometry that is not unitary. A *representation* of an inverse semigroup $S$ is a semigroup homomorphism $\rho : S \to \mathcal{B}(\mathcal{H})$ that sends the inverse operation in $S$ to the adjoint operation in $\mathcal{B}(\mathcal{H})$. Each $T \in \rho(S)$ is a partial isometry in $\mathcal{B}(\mathcal{H})$.

An early result that indicates an important connection between inverse semigroups and the theory of $C^*$-algebras is a result of Duncan and Paterson[16], which is essentially an extension of the Vagner-Preston theorem.

**Theorem 6.1.** *Every inverse semigroup can be faithfully represented as a semigroup of partial isometries of some Hilbert space.*

A *$C^*$-algebra* is a Banach $*$-algebra that satisfies the $C^*$-identity, $||x^*x|| = ||x||^2$. Some key examples of $C^*$-algebras are $\mathcal{B}(\mathcal{H})$ (for a Hilbert space $\mathcal{H}$); the set $C(X)$ of continuous complex-valued functions on a compact space $X$; and the set $C_0(Y)$ of continuous complex valued functions on a locally compact space $Y$ that vanish at infinity. (Here $f(x)^* = \overline{f(x)}$). A standard result in the field, due to Gelfand and Naimark, asserts that every $C^*$-algebra is isomorphic to a norm closed self adjoint subalgebra of some $\mathcal{B}(\mathcal{H})$. A second standard result, again due to Gelfand and Naimark, asserts that up to $*$-isomorphism, every commutative $C^*$-algebra is $C_0(Y)$ for some locally compact space $Y$. See Ref. 19 for proofs and details of the Gelfand-Naimark theorems.

One may construct a $C^*$-algebra $C^*(S)$ associated with every inverse semigroup $S$ as follows. Let $\mathbb{C}(S)$ be the *complex semigroup algebra* of the inverse semigroup $S$ (i.e. the set of all functions $\phi : S \to \mathbb{C}$ for which $\phi(s) = 0$ for all but finitely many $s \in S$). Any representation $\pi : \mathbb{C}(S) \to \mathcal{B}(\mathcal{H})$ is norm decreasing. So define a new norm $||.||_*$ on $\mathbb{C}(S)$ by

$$||x||_* = sup\{||\pi(x)|| : \pi \text{ is a representation of } \mathbb{C}(S)\}.$$

Then the $C^*$-algebra $C^*(S)$ is the completion of $\mathbb{C}(S)$ relative to $||.||_*$. Alternatively, $C^*(S)$ has the universal property that each representation of $S$ lifts to a unique representation of $C^*(S)$. If $S$ has a zero $\theta$ then restrict to representations that send $\theta$ to the zero operator to get the *contracted $C^*$-algebra* $C_0^*(S)$. Then $C^*(S) \cong C_0^*(S) \oplus \mathbb{C}$ and $C^*(S) = \mathbb{C}(S)$ if $S$ is finite. Many important examples of $C^*$-algebras arise from inverse semigroups. We list some of these examples below.

A *finite dimensional* $C^*$-algebra $P$ is a direct sum of matrix algebras over $\mathbb{C}$: $P = \oplus_{i=1}^r M_{m_i}$. If $B_n(G)$ is a *Brandt semigroup*, then $C^*(B_n(G)) = M_n(C^*(G)) \oplus \mathbb{C}$. The *Toeplitz algebra* $\mathcal{T}$ is the $C^*$-subalgebra of $\mathcal{B}(\ell^2)$ generated by the unilateral shift: then $\mathcal{T}$ is the $C^*$-algebra of the bicyclic monoid. The *Cuntz algebra* $\mathcal{O}_n$ is the $C^*$-algebra generated by $n$ isometries $u_1, u_2, \ldots u_n$ subject to the relation $u_1 u_1^* + u_2 u_2^* + \ldots u_n u_n^* = 1$: then $\mathcal{O}_n$ is the contracted $C^*$-algebra $C_0^*(P_n)$ of the polycyclic monoid $P_n$. More generally, the contracted $C^*$-algebra $C_0^*(I(\Gamma))$ of the inverse semigroup of a directed graph $\Gamma$ is the *Toeplitz $C^*$-algebra* of the graph. A presentation for $I(\Gamma)$ is given in the first section of this paper.

In his thesis[22], Haataja provided a description of the $C^*$-algebra of an inverse monoid $S$ with semilattice $E$ as a partial crossed product of $C^*(E)$ by $S$. This decomposition is based on the Munn representation of $S$. Crossed products (and partial crossed products) by groups are somewhat easier to handle than partial crossed products by inverse semigroups. In his thesis[52], Milan provides a construction of $C^*(S)$ as a partial crossed product of $C^*(E)$ by a group if $S$ is $E$-unitary (or more generally strongly $0 - E$-unitary). See also the paper by Milan and Steinberg[53] for additional information.

In some cases, structural information about inverse semigroups provides useful structural information about $C^*$-algebras. One such situation concerns the structure of $C^*$-algebras of *full amalgams* of inverse semigroups. An *amalgam* $[S_1, S_2, U, i_1, i_2]$ of inverse semigroups consists of inverse semigroups $S_1, S_2$ and $U$ together with injective homomorphisms $i_j : U \to S_j$ for $j = 1, 2$. If $S_i = Inv\langle X_i : R_i \rangle$ with $X_1 \cap X_2 = \emptyset$, then the amalgamated free product is $S_1 *_U S_2 = Inv\langle X_1 \cup X_2 : R_1, R_2, i_1(u) = i_2(u), u \in U \rangle$. A theorem of Hall[24] states that inverse semigroups have the *strong amalgamation property*. In terms of the presentation for $S_1 *_U S_2$ given above, this means that the natural maps $\phi_i : S_i \to S_1 *_U S_2$ and $\phi : U \to S_1 *_U S_2$ are injective and $\phi_1(S_1) \cap \phi_2(S_2) = \phi(U)$. This property fails in general in the category of semigroups[12(Vol 2., p. 139)]. The amalgam $[S_1, S_2, U, i_1, i_2]$ is called *full* if $E(S_1) = E(S_2) = E(U)$. For full amalgams, we have the following theorem of Donsig, Haataja and Meakin[15].

**Theorem 6.2.** *Let $[S_1, S_2, U]$ be a full amalgam of inverse semigroups. Then, in the category of complex algebras, $\mathbb{C}(S_1 *_U S_2) \cong \mathbb{C}S_1 *_{\mathbb{C}U} \mathbb{C}S_2$. Also, in the category of $C^*$-algebras, $C^*(S_1 *_U S_2) \cong C^*(S_1) *_{C^*(U)} C^*(S_2)$ and, if $U$ has a zero, then $C_0^*(S_1 *_U S_2) \cong C_0^*(S_1) *_{C_0^*(U)} C_0^*(S_2)$.*

In Ref. 23, Haataja, Margolis and Meakin made use of techniques of Bass-Serre theory (see Serre's book[72]) to determine the structure of the maximal subgroups of $S_1 *_U S_2$ corresponding to a full amalgam of inverse semigroups. Let $[S_1, S_2, U]$ be a full amalgam of inverse semigroups (or regular semigroups). Build a *graph of groups* $(\mathcal{G}, W)$ as follows. The set of *vertices of* $W$ is the disjoint union of the $\mathcal{D}$-classes of $S_1$ and the $\mathcal{D}$-classes of $S_2$. The *edges of* $W$ are described as follows: for each $\mathcal{D}$-class $D$ of $U$ there is a directed edge from the $\mathcal{D}$-class of $S_1$ containing $D$ to the $\mathcal{D}$-class of $S_2$ containing $D$. The *vertex groups and edge groups* are maximal subgroups of the corresponding $\mathcal{D}$-classes, with a natural embedding. The main theorem of Ref. 23 is the following.

**Theorem 6.3.** *For each idempotent $e \in E(U) = E(S_1) = S(S_2)$, the maximal subgroup of $S_1 *_U S_2$ containing $e$ is $\pi_1(\mathcal{G}_e, W_e)$, where $(\mathcal{G}_e, W_e)$ is the restriction of $(\mathcal{G}, W)$ to the connected component of $W$ containing $D_e$.*

When combined with Theorem 6.2, this yields information about the structure of amalgamated free products of $C^*$-algebras in several cases. For example, we can describe the structure of certain amalgams of finite dimensional algebras this way. Let $P = \oplus_{i=1}^r M_{m_i}$ and $Q = \oplus_{i=1}^s M_{n_i}$ be finite-dimensional $C^*$-algebras where $\sum_i m_i = \sum_i n_i = N$. Then

$P \cong C_0^*(S)$ where $S$ is a 0-direct union of combinatorial Brandt semi-groups $B_{m_1}, \dots, B_{m_r}$ and $Q = C_0^*(T)$ where $T$ is the 0-direct union of $B_{n_1}, \dots, B_{n_s}$. Moreover, the *diagonal matrices* form a natural abelian subalgebra of $P$ and $Q$ that can be identified with $\mathbb{C}^N = C_0^*(U)$ for $U = E(S) = E(T)$. Another theorem of Donsig, Haataja and Meakin[15] is the following.

**Theorem 6.4.** $P *_{\mathbb{C}^N} Q = C_0^*(S *_U T)$. *Furthermore this algebra is* $\bigoplus_{i=1}^p M_{k_i}\big(C^*(F_{q_i})\big)$ *where each group $F_{q_i}$ is a free group, $p$ is the number of components of the Bass-Serre graph associated with the amalgam and the $k_i$ (resp. $q_i$) are the number of edges (resp. number of edges not in a spanning tree) of the $i^{th}$ component of this graph.*

Several additional computations of the structure of certain amalgams of Toeplitz algebras or Toeplitz graph algebras are contained in Ref. 15. In general, the structure of amalgamated free products of $C^*$-algebras of inverse semigroups is far from understood.

We close this section of the paper with a brief discussion of some results of Donsig, Fuller and Pitts[14] connecting inverse semigroups and von Neumann algebras. Von Neumann algebras were introduced by Murray and von Neumann[56] in one of a series of seminal papers on rings of operators in the 1930's. We refer to Fillmore[19] for an introduction to the extensive literature on von Neumann algebras. These algebras may be defined in several different ways: we provide an "algebraic" definition below and a basic theorem of von Neumann that links the algebraic definition to a more topological definition.

A *von Neumann algebra* is a self-adjoint subalgebra $A$ of $\mathcal{B}(\mathcal{H})$ which is equal to its double commutant, $A = A''$. (Here by the *commutant $A'$* of a subset $A \subseteq \mathcal{B}(\mathcal{H})$ we mean the set $A' = \{T \in \mathcal{B}(\mathcal{H}) : TS = ST$ for all $S \in A\}$). von Neumann's *double commutant theorem* asserts that a von Neumann algebra may alternatively be defined as a self-adjoint subalgebra $A$ of $\mathcal{B}(\mathcal{H})$ that is closed in the weak operator topology on $\mathcal{B}(\mathcal{H})$ or equivalently as a self-adjoint subalgebra of $\mathcal{B}(\mathcal{H})$ that is closed in the strong operator topology on $\mathcal{B}(\mathcal{H})$.

The maximal abelian subalgebras (MASA's) of a von Neumann algebra play a prominent role in the theory. We now discuss a result of Donsig, Fuller and Pitts[14] that classifies a special class of MASA's via idempotent-separating extensions of inverse monoids.

If $\phi$ is an idempotent-separating morphism from an inverse semigroup $S$ onto an inverse semigroup $T$ then its *kernel* $K = \{a \in S : \phi(a) \in E(T)\}$

is a Clifford inverse semigroup. We write $K \hookrightarrow S \rightarrow T$ and say that $S$ is an *(idempotent-separating) extension* of $T$ with kernel $K$.

Let $A$ be a maximal abelian subalgebra of a von Neumann algebra $V$. We can associate an idempotent-separating extension $K(V, A) \hookrightarrow N(V, A) \rightarrow F(V, A)$ of inverse monoids with the pair $(V, A)$ as follows:

$N(V, A) = \{x \in V : x$ is a partial isometry of $V$ and $xAx^*, x^*Ax \subseteq A\}$,
$K(V, A) = N(V, A) \cap A$, and
$F(V, A)$ is the fundamental image of $N(V, A)$, i.e. $F(V, A) \cong N(V, A)/\mu$
where $\mu$ is the Munn congruence on $N(V, A)$.

Then $K(V, A)$ is the set of all partial isometries in $A$, $E(F(V, A)) \cong E(N(V, A))$
$= E(K(V, A)) = Proj(A)$ (the projections in $A$) and $K(V, A)$ is the kernel of the natural map from $N(V, A)$ onto $F(V, A)$.

While the extension $K(V, A) \hookrightarrow N(V, A) \rightarrow F(V, A)$ can be defined for any pair $(V, A)$, it does not always give useful information. However, in the case that $(V, A)$ is a *Cartan pair*, this extension completely classifies the pair. The concepts of *Cartan pairs* and *Cartan inverse monoids* are briefly defined below: we refer to Ref. 14 for more detail and for the proof of the theorem below, which provides an algebraic (and more general) version of a result due to Feldman and Moore[18].

The pair $(V, A)$ is called a *Cartan pair* if
(a) There exists a faithful, normal conditional expectation $E$ from $V$ onto $A$.
(b) $N(V, A)$ spans a weak - * dense subset of $V$.

An inverse semigroup $S$ with 0 and 1 is called a *Cartan inverse monoid* if
(a) $(E(S), \leq)$ is a Boolean algebra.
(b) $(S, \leq)$ is a meet semilattice.
(c) Every pairwise orthogonal family $\mathcal{S} \subseteq S$, has a join in $S$.
(d) $S$ is fundamental.
(e) The character space $\widehat{E(S)}$ of the complete Boolean lattice $E(S)$ is a hyperstonian topological space.

With this notation we have the following theorem of Donsig, Fuller and Pitts[14].

**Theorem 6.5.** *Let $(V_1, A_1)$ and $(V_2, A_2)$ be Cartan pairs, with associated extensions $K_i \hookrightarrow N_i \rightarrow F_i$. Then each $F_i$ is a Cartan inverse monoid and* the Cartan pairs $(V_1, A_1)$ and $(V_2, A_2)$ are isomorphic iff the associated extensions are equivalent. Conversely, if $K \hookrightarrow N \to F$ is an idempotent-separating extension of inverse monoids with $F$ a Cartan inverse monoid, then there is an associated Cartan pair $(V_F, A_F)$ such that the idempotent-separating extension associated with $(V_F, A_F)$ is equivalent to $K \hookrightarrow N \to F$. There is a one-one correspondence between isomorphism classes of Cartan pairs and equivalence classes of idempotent-separating extensions of Cartan inverse monoids.

## 7. Inverse monoids and immersions of cell complexes

In this section we outline some recent work that shows how inverse monoids may be used to study immersions between cell complexes. Most of the results of this section are taken from the paper[51] of Meakin and Szakács. The results in Ref. 51 extend earlier work along these lines by Margolis and Meakin[38] for graphs (1-dimensional $CW$-complexes) and Meakin and Szakács[50] for 2-dimensional $CW$-complexes.

Recall that under mild restrictions on a topological space $\mathcal{A}$, connected covers of $A$ may be classified via conjugacy classes of the fundamental group $\pi_1(\mathcal{A})$ (see Hatcher[25] for this and basic background on algebraic topology). In order to study *immersions* between spaces, the idea is to replace the fundamental group of the base space by an appropriate inverse monoid and to use closed inverse submonoids of that inverse monoid to classify immersions over the space. We outline how to carry out this program for immersions between finite dimensional $\Delta$-complexes. It is an open problem to determine the extent to which this program may be extended to classify immersions between more general topological spaces.

Recall the following definition[25] of a finite dimensional $CW$-complex $\mathcal{C}$:

(1) Start with a discrete set $\mathcal{C}^0$, the 0-cells of $\mathcal{C}$.
(2) Inductively, form the $n$-skeleton $\mathcal{C}^n$ from $\mathcal{C}^{n-1}$ by attaching $n$-cells $C_\tau^n$ via attaching maps $\varphi_\tau \colon S^{n-1} \to \mathcal{C}^{n-1}$. This means that $\mathcal{C}^n$ is the quotient space of $\mathcal{C}^{n-1} \, \dot{\cup}_\tau \, B_\tau^n$ under the identifications $x \sim \varphi_\tau(x)$ for $x \in \partial B_\tau^n$. The cell $C_\tau^n$ is a homeomorphic image of $B_\tau^n - \partial B_\tau^n$ under the quotient map. (Here $B^n$ is the unit ball in $\mathbb{R}^n$ and $S^{n-1} = \partial B^n$ is its boundary).
(3) Stop the inductive process after a finite number of steps to obtain a finite dimensional $CW$-complex $\mathcal{C}$.

The dimension of the complex is the largest dimension of one of its cells. Note that a 1-dimensional $CW$-complex is just an undirected graph, with the usual topology. We denote the set of $n$-cells of $C$ by $C^{(n)}$. We emphasize that each cell $C_\tau^n$ is open in the topology of the $CW$-complex $C$. A subset $A \subseteq C$ is open iff $A \cap C^n$ is open in $C^n$ for each $n$. Each cell $C_\tau^n$ has a *characteristic map* $\sigma_\tau$, which is defined to be the composition $B_\tau^n \hookrightarrow C^{n-1} \dot{\cup}_\tau B_\tau^n \to C^n \hookrightarrow C$. This is a continuous map whose restriction to the interior of $B_\tau^n$ is a homeomorphism onto $C_\tau^n$ and whose restriction to the boundary of $B_\tau^n$ is the corresponding attaching map $\varphi_\tau$. An alternative way to describe the topology on $C$ is to note that a subset $A \subseteq C$ is open iff $\sigma_\tau^{-1}(A)$ is open in $B_\tau^n$ for each characteristic map $\sigma_\tau$.

Our most general results apply to $\Delta$-complexes, which are $CW$-complexes with an additional restriction on the characteristic maps. The standard $n$-simplex is the set

$$\Delta^n = \{(t_0, ..., t_n) \in \mathbb{R}^{n+1} : \Sigma_i t_i = 1 \text{ and } t_i \geq 0 \text{ for all } i\}.$$

We denote the $n+1$ *vertices* of $\Delta^n$ by $v_i = (0, \ldots, 0, 1, 0, \ldots, 0)$ (1 in $i$th position). We order vertices by $v_i < v_j$ if $i < j$. The *faces* of the simplex are the subsimplices with vertices any non-empty subset of the $v_i$'s. There are $n+1$ faces of dimension $n-1$, namely the faces $\Delta_i^{n-1} = [v_0, ..., v_{i-1}, v_{i+1}, ...v_n]$ for $i = 0, 1, ..., n$ spanned by omitting one vertex. A $\Delta$-*complex* is a quotient space of a collection of disjoint simplices obtained by identifying certain of their faces via the canonical linear homeomorphisms that preserve the ordering of vertices. Equivalently, a $\Delta$-complex is a $CW$-complex $X$ in which each $n$-cell $e_\alpha^n$ has a distinguished characteristic map $\sigma_\alpha : \Delta^n \to X$ such that the restriction of $\sigma_\alpha$ to each $(n-1)$-dimensional face of $\Delta^n$ is the distinguished characteristic map for an $(n-1)$-cell of $X$. We refer the reader to Hatcher's book[25] for more detail and many results about $CW$-complexes and $\Delta$-complexes.

The order on the vertices of the simplex makes it naturally possible to regard each $k$-cell $C_\tau^k$ of a $\Delta$-complex $C$ as a *rooted* cell, with distinguished root the image under the characteristic map $\sigma_\tau^k$ of the minimal 0-cell in the order on 0-cells in $\Delta^k$. We will denote the root of the cell $C$ by $\alpha(C)$. Thus we may regard the 1-skeleton as a digraph with each 1-cell (edge) $e$ directed from its initial vertex (the root of the cell) to its terminal vertex $\omega(e)$ (the image of the maximal 0-cell of $\Delta^1$ under the characteristic map). We will further assume all complexes to be connected.

Let $f$ be a map from the $CW$-complex $\mathcal{D}$ to the $CW$-complex $C$ such that for each $k$-cell $D_\tau^k$ of $\mathcal{D}$, $f(D_\tau^k)$ is a $k$-cell of $C$. Denote the corresponding distinguished characteristic maps of $D_\tau^k$ and $f(D_\tau^k)$ by $\sigma_\tau^k : B^k \to \mathcal{D}$ and

$\gamma_\tau^k : B^k \to C$ respectively. We say that $f$ *commutes with the characteristic maps of* $D$ *and* $C$ if $f \circ \sigma_\tau^k = \gamma_\tau^k$ for all $k$-cells $D_\tau^k$ of $D$. If $f$ commutes with the characteristic maps then it is a continuous map that is a homeomorphism restricted to the (open) cells. An *immersion* from a $CW$-complex $D$ to a $CW$-complex $C$ is a continuous map $f : D \to C$ such that (a) $f$ is a *local homeomorphism onto its image*; that is, for each point $x \in D$ there is an (open) neighborhood $U$ of $x$ such that $f|_U$ is a homeomorphism from $U$ onto $f(U)$ and (b) $f$ commutes with the characteristic maps of $D$ and $C$.

We now describe a way to assign *labels* to all cells of a $\Delta$-complex. Let $B$ be a $\Delta$-complex of dimension $n$ with one 0-cell. Let $\{e_\rho^1 : \rho \in X\}$ be the set of 1-cells and $\{e_\rho^k : \rho \in P_k\}$ the set of $k$-cells of $B$ for $2 \le k \le n$ and let $\beta_\rho^k : \Delta^k \to B$ be the characteristic map of $e_\rho^k$ for $k \ge 1$. Here we assume that the sets $X, P_k$ are all mutually disjoint. We denote this $\Delta$-complex $B$ by $B(X, P_2, ..., P_n, \{\beta_\rho^k\})$, or more briefly by $B(X, P)$ where $P = P_2 \cup ... \cup P_n$. Then $B = B_X$ (the bouquet of $|X|$ circles) if $n = 1$, and $|X|, |P_2|, ..., |P_n|$ are all non-empty sets if $n \ge 2$ by definition of a $\Delta$-complex of dimension $n$. We view $X$ as a set of labels for the 1-cells of $B(X, P)$ and $P_k$ as a set of labels of the $k$-cells of $B(X, P)$ for $2 \le k \le n$. That is, the label on the $k$-cell $e_\rho^k$ is $\ell(e_\rho^k) = \rho$. The 1-skeleton of $B(X, P)$ is $B_X$. We regard this as an $X$-*graph* as usual; i.e. each edge labeled by $x \in X$ is equipped with an inverse edge labeled by $x^{-1}$. The labeling on the 1-cells of $B(X, P)$ extends to a labeling on *paths* in the 1-skeleton of $B(X, P)$ in the obvious way.

We say that a $\Delta$-complex $C$ is *labeled* over a complex $B(X, P)$ if it admits an immersion $f : C \to B(X, P)$. In this case, the labeling on the $k$-cells of $B(X, P)$ induces a labeling on the $k$-cells of $C$ for $k \ge 1$: a $k$-cell $C_\tau^k$ of $C$ has label $\ell(C_\tau^k) = \ell(f(C_\tau^k))$. So $\ell(C_\tau^1) \in X$ and $\ell(C_\tau^k) \in P_k$ if $2 \le k \le n$. Thus cells of $C$ have the same label if and only if they have the same image under $f$. It is not difficult to see that every $\Delta$-complex admits such a labeling. If $C$ is labeled over $B(X, P)$ and $g$ is an immersion $g : D \to C$, then $g$ induces a labeling of $D$ over $B(X, P)$ that respects the labeling. We show how a labeling of a $\Delta$-complex $C$ may be used to construct an inverse monoid given by generators and relations.

If $\Delta^k = [v_0, v_1, ..., v_k]$, then $\Delta^k$ has $(k + 1)$ faces of dimension $k - 1$, namely the $(k-1)$-simplices $\Delta_i^{(k-1)} = [v_0, ..., v_{i-1}, v_{i+1}, ..., v_k]$ for $i = 0, ..., k$. All of these faces except $\Delta_0^{k-1}$ contain the vertex $v_0$. The smallest vertex of $\Delta_0^{k-1}$ under the order on vertices is $v_1$. If $C$ is a $\Delta$-complex of dimension $n$ and $C^k$ is a $k$-cell of $\Delta$, there is a corresponding characteristic map $\sigma^k : \Delta^k \to C$. The restriction of $\sigma^k$ to $\Delta_i^{k-1}$ is a characteristic map $\sigma_i^{k-1}$ of some $(k-1)$-dimensional cell $C_i^{k-1}$ of $C$, by definition of a $\Delta$-

complex. The root of $C^k$ is $\alpha(C^k) = \sigma^k(v_0)$ and the root of $C_i^{k-1}$ is also $\sigma^k(v_0)$ if $i \neq 0$ but the root of $C_0^{k-1}$ is $\sigma^k(v_1)$. Thus the 1-cell $\sigma^k([v_0, v_1])$ is a directed edge in the 1-skeleton of $\mathcal{C}$ from the root of $C^k$ to the root of $C_0^{k-1}$.

For a 2-cell $C^2$ of a $\Delta$-complex $\mathcal{C}$, we denote by $bl(C^2)$ the *boundary label* of $C^2$. This is the label on the image in $\mathcal{C}$ of the path $(v_0, v_1, v_2, v_0)$ in the 1-skeleton of $\Delta^2 = [v_0, v_1, v_2]$ under the corresponding characteristic map from $\Delta^2$ to $\mathcal{C}$. For a $k$-cell $C^k$ of $\mathcal{C}$, denote the image under $\sigma^k$ of the 1-cell $[v_0, v_1]$ of $\Delta^k$ by $e(C^k)$. For $k \geq 3$, denote by $bl(C^k)$ the *label* on the *generalized path* $C_k^{k-1} C_{k-1}^{k-1} ... C_1^{k-1} e(C^k) C_0^{k-1} (e(C^k))^{-1}$ and refer to this as the *boundary label* of the cell $C^k$. It is not difficult to see that if $f$ is an immersion of $\mathcal{C}$ into $B(X, P)$ that defines a labeling of $\mathcal{C}$, then $bl(C^k) = bl(f(C^k))$ for every $k$-cell $C^k$ of $\mathcal{C}$.

It follows that if $f$ is an immersion of $\mathcal{C}$ to $B(X, P)$ that is used to define a labeling of $\mathcal{C}$, then any $k$-cells ($k \geq 2$) of $\mathcal{C}$ that have the same label in $P_k$ have the same boundary label. Thus we may denote the boundary label of a $k$-cell $C_\tau^k$ in $\mathcal{C}$ (or its image in $B(X, P)$) by $bl(\rho)$ where $\rho = \ell(C_\tau^k)$.

We now define the inverse monoid $M(X, P)$ by generators and relations, namely

$$M(X, P) = Inv\langle X \cup P : \rho^2 = \rho \text{ and } \rho = \rho\, bl(\rho) \text{ for all } \rho \in P \rangle$$

We note that if $P = \emptyset$ (i.e. $B(X, P) = B_X$), then $M(X, P) = FIM(X)$, and also that the maximal group image of $M(X, P)$ is $\pi_1(B(X, P))$.

We may define a natural *action* of the inverse monoid $M(X, P)$ by partial one-to-one maps on the set $\mathcal{C}^{(0)}$ of 0-cells of $\mathcal{C}$ as follows. For $x \in X \cup X^{-1}$ and $v \in \mathcal{C}^{(0)}$ define $v.x = w$ if there is an edge labeled by $x$ from $v$ to $w$ in the 1-skeleton of $\mathcal{C}$, and $v.x$ is undefined if there is no such edge. For $\rho \in P_k$ with $k \geq 2$ and $v \in \mathcal{C}^{(0)}$, define $v.\rho = v$ if $v = \alpha(C^k)$ for some $k$-cell $C^k$ with $\alpha(C^k) = v$, and $v.\rho$ is undefined otherwise. The action of the generators $X \cup P$ of $M(X, P)$ on $\mathcal{C}^{(0)}$ extends to a well-defined action of $M(X, P)$ on $\mathcal{C}^{(0)}$. The stabilizer of a vertex in $\mathcal{C}^{(0)}$ under the action by $M(X, P)$ is a closed inverse submonoid of $M(X, P)$, and stabilizers of different vertices in $\mathcal{C}^{(0)}$ are conjugate closed inverse submonoids of $M(X, P)$. Hence the immersion $f : \mathcal{C} \to B(X, P)$ that defines the labeling of $\mathcal{C}$ gives rise to a conjugacy class of closed inverse submonoids of $M(X, P)$.

We may interpret the stabilizer of a vertex $u$ in $\mathcal{C}^{(0)}$ under the action by $M(X, P)$ as the loop monoid $L(\mathcal{C}, u)$ in an appropriate inverse category based at that vertex (see Ref. 51 for details). The greatest group homomorphic image of $L(\mathcal{C}, u)$ is the fundamental group of $\mathcal{C}$. In Ref. 51, the converse is shown: that is, immersions are in one-to-one correspondence with conjugacy classes of closed inverse submonoids of loop monoids.

**Theorem 7.1.** *Let $\mathcal{C}$ be a $\Delta$-complex labeled over some $B(X, P)$, let $u \in \mathcal{C}^{(0)}$, and let $H$ be any closed inverse submonoid of $L(\mathcal{C}, u)$. Then there exists a unique complex $\mathcal{C}_H$ and a unique immersion $f : \mathcal{C}_H \to \mathcal{C}$ with $H = L(\mathcal{C}_H, v)$ for some vertex $v \in \mathcal{C}_H$ with $f(v) = u$.*

The construction extends to the classification of immersions between all finite dimensional $\Delta$-complexes that are labeled over $B(X, P)$. See Ref. 51 for details and a precise statement of the general version of the theorem. The proof follows the general outline of the proof of the corresponding theorem for graph immersions given in Ref. 38, and the extension of that to immersions between 2-dimensional $CW$-complexes given in Ref. 50, but there are significant additional technicalities in the higher dimensional case. The proof is *constructive*. If $X$ and $P$ are finite then $M(X, P)$ has decidable word problem (even though its maximal group image $\pi_1(B(X, P))$ may not) and if $H$ is finitely generated as a closed inverse submonoid of $M(X, P)$, then the associated complex $\mathcal{C}_H$ is finite and effectively constructible.

It would be of interest to provide a "presentation-free" version of the construction of $M(X, P)$. Such a development might provide insight into extensions of this approach to the study of immersions between more general classes of topological spaces. It seems plausible that this approach could at least be extended to a classification of immersions between arbitrary $CW$-complexes, not just $\Delta$-complexes.

## References

1. S.I Adian, "Defining relations and algorithmic problems for groups and semigroups", *Trudy Math. Inst. Steklov* **85**, 1966, (in Russian).
2. S.I. Adian, "Algorithmic insolubility of problems of recognition of certain properties of groups", *Dokl. Akad. Nauk SSSR* **103**, 1955, 533-535, (in Russian).
3. S.I Adian, J. Oganessian, "Problems of equality and divisibility in semigroups with a single defining relation", *Mathem. Zametki* **41**, 1987, 412-421, (in Russian).
4. P. Bennett, "Amalgamated free products of inverse semigroups", *J. Algebra* **198**, 1997, 499-537.
5. P. Bennett, "On the structure of inverse semigroup amalgams", *Inter. J. Algebra and Computation* Vol. **7**, No. 5, 1997, 577-604.
6. P. Bennett, "Normal forms for semigroup amalgams", *Israel J. of Mathematics*, to appear.

7. P. Bennett, "Amalgamation of semilattices of groups", *Inter. J. Algabra and Computation*, to appear.

8. J.-C. Birget, S. Margolis and J. Meakin, "The word problem for inverse monoids presented by one idempotent relator", *Theoretical Computer Science* **123**, 1994, 273-289.

9. A. Cherubini, J. Meakin and B. Piochi, "Amalgams of free inverse semigroups", *Semigroup Forum 54*, 1997, 199-220.

10. A. Cherubini, J. Meakin and B. Piochi, "Amalgams of finite inverse semigroups", *J. Algebra* 285, 2005, 706-725.

11. A. Cherubini and E. Rodaro, "Decidability versus undecidability of the word problem in amalgams of inverse semigroups", *Semigroups, Algebras, Operator Theory* **142**, 2014, 1-25.

12. A. H. Clifford and G. B. Preston, "The Algebraic Theory of Semigroups", Vol. 1 & 2, *Math. Surveys*, No. 7, American Math. Soc., 1961 & 1967.

13. K. R. Davidson, "$C^*$-Algebras by Example", *Amer. Math. Soc.*, Providence, RI, 1996.

14. A. Donsig, A. Fuller and D. Pitts, "von Neumann algebras and extensions of inverse semigroups", *Proc. Edin, Math. Soc.*, **60** (1), 2017, 57-97, arXiv:1409.1624v2.

15. A. Donsig, S. Haataja and J. Meakin, "Amalgams of Inverse Semigroups and C*-algebras", *Indiana University Math. Journal* Vol **60**, No 4, 2011, 1059-1076. http://arxiv.org/abs/1007.1192.

16. J. Duncan, A.L.T. Paterson, "$C^*$-algebras of inverse semigroups", *Proc. Edinburgh Math. Soc. (2)* **28**, no. 1, 1985, 41–58.

17. R. Exel, "Inverse semigroups and combinatorial $C^*$-algebras", *Bull. Braz. Math. Soc. (N.S.)*, **39** (2008), 191-13, arXiv: math.OA/0703182

18. J. Feldman and C.C. Moore, "Ergodic equivalence relations, cohomology and von Neumann algebras I", *Trans. Amer. Math. Soc.* **234**, 1977, No. 2, 289-324.

19. P. A. Fillmore, "A User's Guide to Operator Algebras", *Wiley*, New York, 1996.

20. L.M Gluskin, "Elementary generalized groups", *Mat. Sbornik* **41**, 1957, 23-36, (in Russian).

21. Robert D. Gray, "Undecidability of the word problem for one-relator inverse monoids via right-angled Artin subgroups of one-relator groups, *Invent. Math*, 2019, https://doi.org/10.1007/s00222-019-00920-2.

22. Steven P. Haataja, "Amalgamation of Inverse Semigroups and Operator Algebras", Ph.D. Dissertation, Univ. Nebraska—Lincoln, 2006.

23. S. Haataja, S. Margolis and J. Meakin, "Bass-Serre theory for groupoids and the structure of full regular semigroup amalgams", *J. Algebra* **183**, 1996, 38-54.

24. T. E. Hall, "Free products with amalgamation of inverse semigroups", *J. Algebra* **34**, 1975, 375–385.

25. Allen Hatcher, "Algebraic Topology", *Cambridge University Press*, 2001.

26. S. Hermiller, S. Lindblad and J. Meakin, "Decision problems for inverse monoids presented by a single sparse relator", *Semigroup Forum* **81**, 2010,

128-144. http://arxiv.org/abs/0911.1484.

27. M. Inam, "The word problem for some classes of Adian inverse semigroups", *Semigroup Forum*, to appear.

28. M. Inam, J. Meakin and R. Ruyle, "A structural property of Adian inverse semigroups", *Semigroup Forum* **94**, No. 1, 2017, 93-103.

29. S.V. Ivanov and J.C. Meakin, "On asphericity and the Freiheitssatz for certain finitely presented groups", *J. Pure Appl. Algebra* **159**, 2001, 113-121.

30. S.V. Ivanov, S.W. Margolis and J.C. Meakin, "On one-relator inverse monoids and one-relator groups", *J. Pure Appl. Algebra* **159**, 2001, 83-111.

31. T. Jajcayová, "*HNN* extensions of inverse semigroups", PhD Thesis, University of Nebraska-Lincoln, 1997.

32. T. Jajcayová, "The word problem for *HNN* extensions of free inverse semigroups", *Language and automata theory and applications*, vol. **9618** of *Lecture notes in computer science*, 2016, 506-517.

33. P.R. Jones, S.W. Margolis, J. Meakin and J.B. Stephen, "Free products of inverse semigroups II", *Glasgow Math. J.* **33**, 1991, 373-387.

34. M.V. Lawson, "Inverse Semigroups: the Theory of Partial Symmetries", *World Scientific*, 1998.

35. S. Lindblad, "Inverse monoids presented by a single sparse relator", PhD Thesis, University of Nebraska, 2003.

36. M. Lohrey and N. Ondrusch, "Inverse monoids: decidability and complexity of algebraic questions", *Inf. Comput.*, 205(8), 2007, 1212-1234.

37. R.C. Lyndon and P.E. Schupp, "Combinatorial Group Theory", *Ergebnisse der Mathematik und ihrer Grenzgebiete*, Band 89, Springer-Verlag, Berlin, 1977.

38. S. Margolis and J. Meakin, "Free inverse monoids and graph immersions", *Int. J. Algebra and Computation*, **3**, No. 1, 1993, 79-99.

39. S. Margolis and J. Meakin, "Inverse monoids, trees and context-free languages", *Trans Amer. Math. Soc.* **335** (1), 1993, 259-276.

40. S. Margolis, J. Meakin and M.Sapir, "Algorithmic Problems in Groups, semigroups and Inverse Semigroups", in "Semigroups, Formal Languages and groups," J. Fountain (Ed), *Kluwer*, 1995, 147-214.

41. S. Margolis, J. Meakin and J. Stephen, "Free objects in certain varieties of inverse semigroups", *Canadian J. Math.* **42**, 1990, 1084-1097.

42. S. Margolis, J. Meakin and Z. Šunić, "Distortion functions and the membership problem for submonoids of groups and monoids", *Contemporary Mathematics* Vol. **372**, 2005, 109-129.

43. A. A. Markov, "On the impossibility of certain algorithms in the theory of associative systems", *Dokl. Akad. Nauk SSSR*, **55**, 1947, 587-590, (in Russian); French transl., C. R. (Dokl.) Acad. Sci. URSS II, **55**, 1947, 583-586.

44. D.B. McAlister, "Groups, semilattices and inverse semigroups", *Trans. Amer. Math. Soc.* **192**, 1974, 227-244.

45. D.B. McAlister, "Groups, semilattices and inverse semigroups II", *Trans. Amer. Math. Soc.* **196**, 1974, 251-270.

46. D.B. McAlister and R. McFadden, "The free inverse monoid on two commuting generators", *J. Algebra* **32**, 1974, 215-233.

47. J. Meakin, "On the structure of inverse semigroups", *Semigroup Forum* **12**, 1976, 6-14.

48. J. Meakin, "Groups and semigroups: some connections and contrasts", in "Groups St Andrews 2005, Vol. 2", *London Math Soc Lecture Notes* **340**, 2007, 357-400.

49. J. Meakin and M. Sapir, "The Word Problem in the Variety of Inverse Semigroups with abelian covers", *J. Lond. Math Soc.* (2) **53**, 1996, 79-98.

50. J. Meakin and N. Szakács, "Inverse monoids and immersions of 2-complexes", *Int. J. Algebra and Computation* **25**, Nos. 1-2, 2015, 301-324. http://arxiv.org/abs/1401.2621

51. J. Meakin and N. Szakács, "Inverse monoids and immersions of cell complexes" Preprint , https://arxiv.org/abs/1709.03887

52. D. Milan, "$C^*$-algebras of Inverse Semigroups", Ph.D. Dissertation, Univ. Nebraska—Lincoln, 2008.

53. D. Milan and B. Steinberg, "On inverse semigroup $C^*$-algebras and crossed products", *Groups, Geom. Dynamics*, **8**, 2014, 485-512.

54. W.D. Munn, "Fundamental inverse semigroups", Quarterly J. Math. Oxford (2) **21**, 1970, 157-170.

55. W. D. Munn, "Free inverse semigroups", *Proc. London Math. Soc.* (3) **29**, 1974, 385-404.

56. F.J. Murray and J. von Neumann, "On rings of operators", *Ann. Math.* **37**, 1936, 116-229.

57. M. Nivat and J-F. Perrot, "Une généralisation du monoïde bicyclique", *Comptes Rendus de l'Academie des Sciences de Paris* **271**, 1970, 824-827.

58. Alan L.T. Paterson, "Groupoids, Inverse Semigroups, and their Operator Algebras", *Birkhäuser*, 1998.

59. G. B. Preston, "Inverse semi-groups", *J. London Math. Soc.*, **29**, 1954, 396-403.

60. G.B. Preston, "Free inverse semigroups", *J. Australian math. Soc.* **16**, 1973, 443-453.

61. M.O. Rabin, "Decidability of Second Order Theories and Automata on Infinite Trees", *Trans. Amer. Math. Soc.* **141**, 1969, 1-35.

62. M.O. Rabin, "Recursive unsolvability of group theoretic problems", *Annals of Mathematics*, **67**, 1958, 172-194.

63. I. Raeburn, "Graph algebras", *CBMS Regional Conference Series in Mathematics*, **103**, American Mathematical Society, Providence, RI, 2005. vi+113 pp.

64. J.H. Remmers, "On the geometry of semigroup presentations", *Adv. in Math.* **36** (3), 1980, 283-296.

65. J. Renault, "A groupoid approach to C*-algebras", *Lecture Notes in Mathematics* vol. **793**, Springer, 1980.

66. P. Resende, "Lectures on étale groupoids, inverse semigroups, and quantales", *Lecture Notes for the GAMAP IP Meeting*, Antwep, 2006, 115pp.

67. O. Sarkisian, "Some relations bewtween the word and divisibility problems for groups and semigroups", *Izv. Akad. Nauk SSSR Ser. Mat*, in Russian). **43**, 1979, 909-921; English transl. in *Math, USSR Izv.* **15**, 1980.

68. O. Sarkisian, "On the word and divisibility problems in semigroups and groups without cycles", *Izv. Akad. Nauk SSSR, Ser. Mat.* **45** (6), 1981, 1424-1441, (in Russian); English transl. in *Math USSR Izv.* **19**, 1982.

69. H.E. Scheiblich, "Free inverse semigroups", *Proc. Amer. Math. Soc.* **38**, 1973, 1-7.

70. B.M. Schein, "Representations of generalized groups", *Izv. Vyss. Ucebn. Zav. Mat.* **3**, 1962, 164-176, (in Russian).

71. B.M. Schein, "Free inverse semigroups are not finitely presentable", *Acta Math. Acad. Sci. Hungaricae*, **26**, 1975, 41-52.

72. J.P. Serre, "Trees", *Springer-Verlag*, New York, 1980.

73. P. V. Silva, "Word Equations and Inverse Monoid Presentations", in *Semigroups and Applications, Including Semigroup Rings*, ed. S. Kublanovsky, A. Mikhalev, P. Higgins, J. Ponizovskii, "Severny Ochag", St. Petersburg, 1999.

74. J. Stallings, "Adian groups and pregroups", in "Essays in group theory", vol 8 of *Math. Sci. Res. Inst. Publ.*, pp. 321-342, *Springer*, New York, 1987.

75. B. Steinberg, "Fundamental groups, inverse Schützenberger automata, and monoid presentations", *Comm. Algebra* **28**, 2000, 5235-5253.

76. B. Steinberg, "A topological approach to inverse and regular semigroups", *Pacific J. Math.*, **208**(2), 2003, 367-396.

77. J.B.Stephen, "Applications of automata theory to presentations of monoids and inverse monoids" Ph.D Thesis, Univ. of Nebraska, 1987.

78. J.B. Stephen, "Presentations of inverse monoids", *J. Pure and Appl. Algebra* **63**, 1990, 81-112.

79. J.B. Stephen, "Inverse monoids and rational subsets of related groups", *Semigroup Forum* **46** (1), 1993, 98-108.

80. V.V. Wagner, "Generalized groups", *Doklady Acad. Nauk. SSSR*, **84**, 1952, 1119-1122 (in Russian).

81. Kaicheng Wang, "Rewriting reduction and pruning reduction on Munn trees", Ph.D Thesis, Univ. of Nebraska, 1996.

82. A. Yamamura, "*HNN* extensions of inverse semigroups", Ph.D Thesis, Univ.of Nebraska, 1996.

83. A. Yamamura, "*HNN* extensions of inverse semigroups and applications", *Inter. J. Algebra and Computation* Vol. **7**, No 5, 1997, 605-624.

84. L. Zhang, "A short proof of a theorem of Adian", *Proc. Amer. Math. Soc.* Vol **116**, No. 1, 1992, 1-3.

# Replicators, Manin white product of binary operads and average operators

Jun Pei

*School of Mathematics and Statistics, Southwest University,
Chongqing 400715, China*

*peitsun@swu.edu.cn*

Chengming Bai

*Chern Institute of Mathematics & LPMC, Nankai University,
Tianjin 300071, China*

*baicm@nankai.edu.cn*

Li Guo

*Department of Mathematics and Computer Science, Rutgers University,
Newark, NJ 07102*

*liguo@rutgers.edu*

Xiang Ni

*Department of Mathematics, Caltech,
Pasadena, CA 91125, USA*

*nixiang85@gmail.com*

We consider the notions of the replicators, including the duplicator and triplicator, of a binary operad. As in the previous notions of di-Var-algebra and tri-Var-algebra in Ref. 14, they provide a general operadic definition for the recent constructions of replicating the operations of algebraic structures. We show that taking replicators is in Koszul dual to taking successors in Ref. 3 for binary quadratic operads and is equivalent to taking the white product with certain operads such as Perm. We also relate the replicators to the actions of average operators.

## 1. Introduction

Motivated by the study of the periodicity in algebraic $K$-theory, J.-L. Loday[24] introduced the concept of a Leibniz algebra twenty years ago as a non-skew-symmetric generalization of the Lie algebra. He then defined the diassociative algebra[25] as the enveloping algebra of the Leibniz algebra in analogue to the associative algebra as the enveloping algebra of the Lie algebra. The dendriform algebra was introduced as the Koszul dual of the diassociative algebra. These structures were studied systematically in the next few years in connection with operads[29], homology[11,12], Hopf algebras[2,18,30,35], arithmetic[26], combinatorics[10,31], quantum field theory[10] and Rota-Baxter algebras[1].

The diassociative and dendriform algebras extend the associative algebra in two directions. While the diassociative algebra "doubles" the associative algebra in the sense that it has two associative operations with certain compatible conditions, the dendriform algebra "splits" the associative algebra in the sense that it has two binary operations with relations between them so that the sum of the two operations is associative.

Into this century, more algebraic structures with multiple binary operations emerged, beginning with the triassociative algebra that "triples" the associative algebra and the tridendriform algebra that gives a "three way splitting" of the associative algebra[30]. Since then, quite a few more "multiway splitting" structures, such as the quadri-algebra[2], the ennea-algebra, the NS-algebra, the dendriform-Nijenhuis algebra, the octo-algebra[20-22] and eventually a whole class of algebras[8,29], were introduced. These structures have a common property of "splitting" the associativity into multiple pieces. Furthermore, analogues of the dendriform algebra, quadri-algebra and octo-algebra for the Lie algebra, commutative algebra, Jordan algebra, alternative algebra and Poisson algebra have been obtained[1,4,16,23,28,33], such as the pre-Lie and Zinbiel algebras. More recently, these constructions can be put into the framework of operad products (Manin black square product)[7,27,38].

In Ref. 3, the notion of "successors" was introduced to give the precise meaning of two way and three way splitting of a binary operad and thus put the previous constructions in a uniform framework. This notion is also related to the Manin black products that had only been dealt with in special cases before, as indicated above. It is also shown to be related to the action of the Rota-Baxter operator, completing a long series of studies starting from the beginning of the century[1]. In Ref. 34, we introduced the notion of a configuration to give a uniform treatment of different splitting patterns which include the bisuccessor and trisuccessor in Ref. 3 as two special cases.

In this paper, we take a similar approach to another class of structures starting from the diassociative (resp. triassociative) algebra. That is, we seek to understand the phenomena of "replicating" the operations in an operad. After the completion of the paper, we learned that the closely related notions di-Var-algebra and tri-Var-algebra had been introduced in Ref. 14 (see also Ref. 15,19) by P. S. Kolesnikov and his coauthors. In fact their notions also apply to not necessarily binary operads. We thank Kolesnikov for informing us to their studies. In this regards, the current paper provides an alternative treatment of these notations for binary operads. Indeed this notion is equivalent to the notion of di-Var-algebra in Ref. 14 for binary operads with nontrivial relations. See the remark before Section 5.1.

In Section 2 we set up a general framework to make precise the notion of "replicating" any binary algebraic operad including the previously well-known di-type (resp. tri-type) algebras, such as the Leibniz algebra for the Lie algebra and the permutative algebra for the commutative algebra, as well as the recently defined pre-Lie dialgebra[9]. In general, it gives a general rule to replicate algebraic structures associated to any binary operad.

In Section 3, examples of duplicators and triplicators of binary operads are provided for various algebras. In Section 4, we study the relationship among a binary operad, its duplicator and triplicator.

We show in Section 5 that taking the replicator of a binary quadratic operad is in Koszul dual with taking the successor of the dual operad. A direct application of this duality (Theorems 5.1 and 5.2) is to explicitly compute the Koszul dual of the operads of existing algebras, for example the Koszul dual of the commutative tridendriform algebra of Loday[28]. We also relate replicating to the Manin white product in the case of binary quadratic operads. In fact taking the duplicator (resp. triplicator) of such an operad with nontrivial relations is isomorphic to taking the white product of the operad Perm (resp. ComTriass) with this operad, as in the case of taking di-Var-algebras and tri-Var-algebras[14]. Thus the notations of duplicators and triplicators are equivalent to those of di-Var-algebras and tri-Var-algebras.

Finally, in Section 6, we relate the replicating process to the action of average operators on binary quadratic operads.

Thus there are relationships among the three operations applied to a binary operad $\mathcal{P}$: taking its duplicator (resp. triplicator), taking its Manin white product with *Perm* (resp. *ComTria*), when the operad is quadratic, and applying a di-average operator (resp. tri-average operator) to it, as summarized in the following diagram.

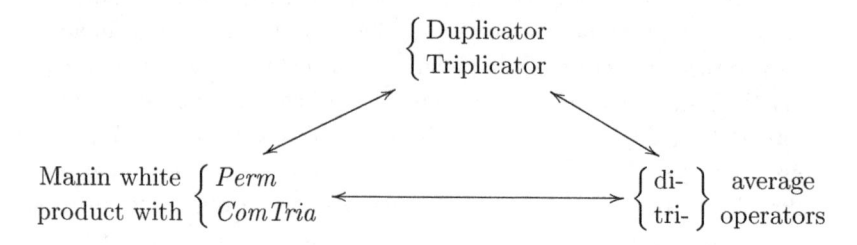

Combining the replicators with the successors introduced in Ref. 3 allows us to put the splitting and replicating processes together, as exemplified in the following diagram, shown on the level of categories. The arrows should be reversed for operads.

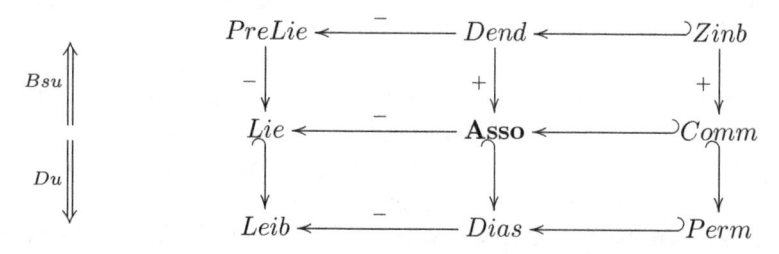

Here the vertical arrows in the upper half of the diagram are addition of the two operations given in Ref. 3 (Proposition 2.31.(a)), while those in the lower half of the diagram are given in Proposition 4.1 (a). The horizontal arrows in the left half of the diagram are anti-symmetrization of the binary operations while those in the right half of the diagram are induced by the identity maps on the binary operations. In the diagram, the Koszul dual of an operad is the reflection across the center. A similar commutative diagram holds for the trisuccessors and triplicators.

## 2. The replicators of a binary operad

Since binary operads can be represented by labeled planar binary trees, we first introduce the concepts of the replicators, namely the duplicator and triplicator, of a labeled planar binary tree which are then applied to define replicators for (symmetric) operads and nonsymmetric operads. The concepts of the replicators are parallel to those of successors[3] and bear in close resemblance with those of the di-Var-algebra and tri-Var-algebra in Ref. 14.

## 2.1. *Replicators of planar binary trees*

### 2.1.1. *Labeled trees*

We recall some basic notions on trees and operads. For more details see Ref. 32.

**Definition 2.1.**

(1) Let $\mathscr{T}$ denote the set of planar binary reduced rooted trees together with the trivial tree $|$. If $t \in \mathscr{T}$ has $n$ leaves, we call $t$ an $n$-**tree**.

The trivial tree $|$ has one leaf.

(2) Let $\Omega$ be a set and $t$ be an $n$-tree. By a **decoration of $t$ by $\Omega$** we mean a decoration of the vertices of $t$ by elements of $\Omega$ and the leafs of $t$ by distinct variables $x_1, x_2, \cdots, x_n$ from left to right. An $n$-tree $t$ together with one of its decorations is called a **labeled $n$-tree**. Let $t(\Omega)$ denote the set of decorations of $t$ and denote

$$\mathscr{T}(\Omega) := \coprod_{t \in \mathscr{T}} t(\Omega).$$

(3) For $\tau \in \mathscr{T}(\Omega)$, we let $Vin(\tau)$ (resp. $Lin(\tau)$) denote the (multi-)set (resp. ordered set) of labels of the vertices (resp. leaves) of $\tau$.

(4) The **grafting** of trees $t_1, t_2$, denoted by $t = t_1 \vee t_2$, is the tree obtained by joining the roots of the trees $t_1, t_2$ (ordered from left to right) to a new root. The trees $t_1, t_2$ are called the **branches** of $t$.

(5) Let $\tau \in \mathscr{T}(\Omega)$ with $|Lin(\tau)| > 1$ be a labeled tree from $t \in \mathscr{T}$. There exists a unique pair of planar binary reduced trees $(t_\ell, t_r)$ such that $t = t_\ell \vee t_r$. Correspondingly, let $\tau = \tau_\ell \vee_w \tau_r$ denote the unique decomposition of $\tau$ as a grafting of $\tau_\ell$ and $\tau_r$ in $\mathscr{T}(\Omega)$ along $w \in \Omega$.

Let $V$ be a vector space, regarded as an arity graded vector space concentrated in arity 2: $V = V_2$. Recall Ref. 32 (Section 5.8.5) that the free nonsymmetric operad $\mathcal{T}_{ns}(V)$ on $V$ is given by the vector space

$$\mathcal{T}_{ns}(V) := \bigoplus_{t \in \mathscr{T}} t[V] \, ,$$

where $t[V]$ is the treewise tensor module associated to $t$, explicitly given by

$$t[V] := \bigotimes_{v \in \mathrm{V}in(t)} V_{|\mathrm{I}n(v)|} \, .$$

Here $|\mathrm{I}n(v)|$ denotes the number of incoming edges of $v$. A basis $\mathcal{V}$ of $V$ induces a basis $t(\mathcal{V})$ of $t[V]$ and a basis $\mathscr{T}(\mathcal{V})$ of $\mathcal{T}_{hs}(V)$. Consequently any element of $t[V]$ can be represented as a linear combination of elements in $t(\mathcal{V})$.

### 2.1.2. *Duplicators*

**Definition 2.2.** Let $V$ be a vector space with a basis $\mathcal{V}$. Define a vector space

$$\mathrm{Du}(V) = V \otimes (\mathbf{k} \dashv \oplus \mathbf{k} \vdash) \, , \tag{1}$$

where we denote $(\omega \otimes \dashv)$ (resp. $(\omega \otimes \vdash)$) by $\left(\begin{smallmatrix}\omega\\\dashv\end{smallmatrix}\right)$ $\left(\text{resp. } \left(\begin{smallmatrix}\omega\\\vdash\end{smallmatrix}\right)\right)$ for $\omega \in \mathcal{V}$. The set $\displaystyle\bigcup_{\omega \in \mathcal{V}} \left\{ \left(\begin{smallmatrix}\omega\\\dashv\end{smallmatrix}\right), \left(\begin{smallmatrix}\omega\\\vdash\end{smallmatrix}\right) \right\}$ is a basis of $\mathrm{Du}(V)$.

**Definition 2.3.** For $\tau \in \mathcal{T}(V)$, the **duplicator** $\mathrm{Du}_x(\tau)$ of $\tau$ with respect to a leaf $x \in \mathrm{L}in(\tau)$ is the subset of $\mathcal{T}_{hs}(\mathrm{Du}(\mathcal{V}))$ defined by relabeling each vertex $\omega$ of $\mathrm{V}in(\tau)$ by

$$\begin{cases} \left(\begin{smallmatrix}\omega\\\dashv\end{smallmatrix}\right), \text{ the path from the root of } \tau \text{ to } x \text{ turns left at } \omega; \\ \left(\begin{smallmatrix}\omega\\\vdash\end{smallmatrix}\right), \text{ the path from the root of } \tau \text{ to } x \text{ turns right at } \omega; \\ \left(\begin{smallmatrix}\omega\\\dagger\end{smallmatrix}\right), \omega \text{ is not on the path from the root of } \tau \text{ to } x, \end{cases}$$

where $\left(\begin{smallmatrix}\omega\\\dagger\end{smallmatrix}\right) := \left\{ \left(\begin{smallmatrix}\omega\\\dashv\end{smallmatrix}\right), \left(\begin{smallmatrix}\omega\\\vdash\end{smallmatrix}\right) \right\}$. For labeled $n$-trees $\tau_i$ with the same set of leaf decorations and $c_i \in \mathbf{k}, 1 \leq i \leq r$, define

$$\mathrm{Du}_x \left( \sum_{i=1}^{r} c_i \tau_i \right) := \sum_{i=1}^{r} c_i \mathrm{Du}_x(\tau_i). \tag{2}$$

Here and in the rest of the paper we use the notation

$$\sum_{i=1}^{r} c_i W_i := \left\{ \sum_{i=1}^{r} c_i w_i \,\middle|\, w_i \in W_i, 1 \leq i \leq r \right\}, \tag{3}$$

for nonempty subsets $W_i, 1 \leq i \leq r$, of a $\mathbf{k}$-module.

**Example 2.1.** $\mathrm{Du}_{x_2}$

### 2.1.3. *Triplicators*

**Definition 2.4.** Let $V$ be a vector space with a basis $\mathcal{V}$. Define a vector space

$$\mathrm{Tri}(V) = V \otimes (\mathbf{k}\dashv \oplus \mathbf{k}\vdash \oplus \mathbf{k}\perp)\,, \tag{4}$$

where we denote $(\omega \otimes \dashv)$ (resp. $(\omega \otimes \vdash)$, resp. $(\omega \otimes \perp)$) by $\binom{\omega}{\dashv}$ $\Big($resp. $\binom{\omega}{\vdash}$, resp. $\binom{\omega}{\perp}\Big)$ for $\omega \in \mathcal{V}$. The set $\bigcup_{\omega \in \mathcal{V}} \left\{\binom{\omega}{\dashv}, \binom{\omega}{\vdash}, \binom{\omega}{\perp}\right\}$ is a basis of $\mathrm{Tri}(V)$.

**Proposition 2.1.** *For $\tau \in \mathscr{T}(V)$, the **triplicator** $\mathrm{Tri}_J(\tau)$ of $\tau$ with respect to a nonempty subset $J \subseteq \mathrm{Lin}(\tau)$ is the subset of $\mathcal{T}_{hs}(\mathrm{Tri}(V))$ defined by relabeling each vertex $\omega$ of $\mathrm{Vin}(\tau)$ by*

*(1) Suppose $\omega$ is on the paths from the root of $\tau$ to some (possibly multiple) $x$ in $J$. Then*
   *(i) replace $\omega$ by $\binom{\omega}{\dashv}$ if all of such paths turn left at $\omega$;*
   *(ii) replace $\omega$ by $\binom{\omega}{\vdash}$ if all of such paths turn right at $\omega$;*
   *(iii) replace $\omega$ by $\binom{\omega}{\perp}$ if some of such paths turn left at $\omega$ and some of such paths turn right at $\omega$.*

*(2) Suppose $\omega$ is not on the path from the root of $\tau$ to any $x \in J$. Then replace $\omega$ by $\binom{\omega}{\dagger} := \left\{\binom{\omega}{\dashv}, \binom{\omega}{\vdash}, \binom{\omega}{\perp}\right\}$.*

**Example 2.2.** $\mathrm{Tri}_{\{1,2\}}$ (tree diagram) $=$ (tree diagram)

$$= \left\{ \text{(tree diagram)} \;,\; \text{(tree diagram)} \;,\; \text{(tree diagram)} \right\}$$

## 2.2. Replicators of binary operads

Let $V = V(2)$ be an $\mathbb{S}$-module concentrated in arity two with a linear basis $\mathcal{V}$. For any finite set $\mathcal{X}$ of cardinal $n$, define the coinvariant space

$$V(\mathcal{X}) := \left( \bigoplus_{f:\underline{n} \to \mathcal{X}} V(n) \right)_{\mathbb{S}_n},$$

where the sum is over all the bijections from $\underline{n} := \{1, \ldots, n\}$ to $\mathcal{X}$ and where the symmetric group acts diagonally.

Let $\mathbb{T}$ denote the set of isomorphism classes of reduced binary trees[32](Appendix C). For $\mathsf{t} \in \mathbb{T}$, define the treewise tensor $\mathbb{S}$-module associated to $\mathsf{t}$, explicitly given by

$$\mathsf{t}[V] := \bigotimes_{v \in Vin(\mathsf{t})} V(In(v)),$$

see Ref. 32 (Section 5.5.1). Then the free operad $\mathcal{T}(V)$ on an $\mathbb{S}$-module $V = V(2)$ is given by the $\mathbb{S}$-module

$$\mathcal{T}(V) := \bigoplus_{\mathsf{t} \in \mathbb{T}} \mathsf{t}[V].$$

Each tree $\mathsf{t}$ in $\mathbb{T}$ can be represented by a planar tree $t$ in $\mathscr{T}$ by choosing a total order on the set of inputs of each vertex of $\mathsf{t}$. Further, $t[V] \cong \mathsf{t}[V]$[17(Section 2.8)]. Fixing such a choice $t$ for each $\mathsf{t} \in \mathbb{T}$ gives a subset $\mathfrak{R} \subseteq \mathscr{T}$ with a bijection $\mathbb{T} \cong \mathfrak{R}$. Then we have

$$\mathcal{T}(V) \cong \bigoplus_{t \in \mathfrak{R}} t[V].$$

**Definition 2.5.** Let $V$ be a vector space with a basis $\mathcal{V}$.

(1) An element

$$r := \sum_{i=1}^{r} c_i \tau_i, \quad c_i \in \mathbf{k}, \tau_i \in \mathscr{T}(\mathcal{V}),$$

in $\mathcal{T}_{hs}(V)$ is called **homogeneous** if $Lin(\tau_i)$ are the same for $1 \leq i \leq r$. Then denote $Lin(r) = Lin(\tau_i)$ for any $1 \leq i \leq r$.

(2) A collection of elements

$$r_s := \sum_{i=1}^{r} c_{s,i}\tau_{s,i}, \quad c_{s,i} \in \mathbf{k}, \tau_{s,i} \in \mathcal{T}(V), 1 \leq s \leq k, k \geq 1,$$

in $\mathcal{T}_{hs}(V)$ is called **locally homogenous** if each element $r_s$, $1 \leq s \leq k$, is homogeneous.

**Definition 2.6.** Let $\mathcal{P} = \mathcal{T}(V)/(R)$ be a binary operad where the $\mathbb{S}$-module $V$ is concentrated in arity 2: $V = V(2)$ with an $\mathbb{S}_2$-basis $\mathcal{V}$ and the space of relations is generated, as an $\mathbb{S}$-module, by a set $R$ of locally homogeneous elements

$$r_s := \sum_{i} c_{s,i}\tau_{s,i}, \quad c_{s,i} \in \mathbf{k}, \tau_{s,i} \in \bigcup_{t \in \mathfrak{R}} t(V), \ 1 \leq s \leq k. \tag{5}$$

(1) The **duplicator** of $\mathcal{P}$ is defined to be the binary operad

$$\mathrm{Du}(\mathcal{P}) = \mathcal{T}(\mathrm{Du}(V))/(\mathrm{Du}(R))$$

where the $\mathbb{S}_2$-action on $\mathrm{Du}(V) = V \otimes (\mathbf{k} \dashv \oplus \mathbf{k} \vdash)$ is given by

$$\left(\begin{smallmatrix} \omega \\ \dashv \end{smallmatrix}\right)^{(12)} := \left(\begin{smallmatrix} \omega^{(12)} \\ \vdash \end{smallmatrix}\right), \quad \left(\begin{smallmatrix} \omega \\ \vdash \end{smallmatrix}\right)^{(12)} := \left(\begin{smallmatrix} \omega^{(12)} \\ \dashv \end{smallmatrix}\right), \quad \omega \in V,$$

and the space of relations is generated, as an $\mathbb{S}$-module, by

$$\mathrm{Du}(R) := \bigcup_{s=1}^{k} \left( \bigcup_{x \in Lin(r_s)} \mathrm{Du}_x(r_s) \right) \tag{6}$$

with $\mathrm{Du}_x(r_s) := \sum_{i} c_{s,i}\mathrm{Du}_x(\tau_{s,i})$.

(2) The **triplicator** of $\mathcal{P}$ is defined to be the binary operad

$$\mathrm{Tri}(\mathcal{P}) = \mathcal{T}(\mathrm{Tri}(V))/(\mathrm{Tri}(R))$$

where the $\mathbb{S}_2$-action on $\mathrm{Tri}(V) = V \otimes (\mathbf{k} \dashv \oplus \mathbf{k} \vdash \oplus \mathbf{k} \perp)$ is given by

$$\left(\begin{smallmatrix} \omega \\ \dashv \end{smallmatrix}\right)^{(12)} := \left(\begin{smallmatrix} \omega^{(12)} \\ \vdash \end{smallmatrix}\right), \quad \left(\begin{smallmatrix} \omega \\ \vdash \end{smallmatrix}\right)^{(12)} := \left(\begin{smallmatrix} \omega^{(12)} \\ \dashv \end{smallmatrix}\right), \quad \left(\begin{smallmatrix} \omega \\ \perp \end{smallmatrix}\right)^{(12)} := \left(\begin{smallmatrix} \omega^{(12)} \\ \perp \end{smallmatrix}\right), \quad \omega \in V,$$

and the space of relations is generated, as an $\mathbb{S}$-module, by

$$\mathrm{Tri}(R) := \bigcup_{s=1}^{k} \left( \bigcup_{\varnothing \neq J \subseteq Lin(r_s)} \mathrm{Tri}_J(r_s) \right)$$

with $\mathrm{Tri}_J(r_s) := \sum_{i} c_{s,i}\mathrm{Tri}_J(\tau_{s,i})$.

The above definitions are also valid for non-symmetric operads.

See Ref. 14 for the closely related notions of the di-Var-algebra and tri-Var-algebra, and Ref. 15 for these notions for not necessarily binary operads. The latter can also be achieved by applying the approach of this paper to the context of Ref. 34 in terms of configurations.

We conclude this section by showing that taking replicator of an operad is independent of the presentation of the operad. Since the case of duplicators is parallel to the case of triplicators. We will only consider the latter case for which we will prove in full detail, in Corollary 2.1.

Let $\eta : \mathcal{P} \longrightarrow \mathcal{Q}$ be a binary operad morphism. We take the generating vector space $V$ (resp. $W$) of $\mathcal{P} = \mathcal{T}(V)/(R_{\mathcal{P}})$ (resp. $\mathcal{Q} = \mathcal{T}(W)/(R_{\mathcal{Q}})$) to be $\mathcal{P}(2)$ (resp. $\mathcal{Q}(2)$). Thus for the binary operad $\mathcal{P}$ (and similarly for $\mathcal{Q}$), we take the free binary operad $\mathcal{T}(V)$ where $V = \mathcal{P}(2)$. We then define the operad morphism $p_V : \mathcal{T}(V) \to \mathcal{P}$ induced by the identity map $V \to \mathcal{P}$ of $\mathbb{S}$-modules and take $(R_{\mathcal{P}})$ to be the kernel.

The operad morphism $\eta$ defines an $\mathbb{S}_2$-equivalent maps $\eta_2 : V = \mathcal{P}(2) \longrightarrow W = \mathcal{Q}(2)$. Define a map $\theta : \mathrm{Tri}(V) \longrightarrow \mathrm{Tri}(W)$ by

$$\theta : \mathrm{Tri}(V) \longrightarrow \mathrm{Tri}(W), \quad \binom{\omega}{\clubsuit} \longmapsto \binom{\eta_2(\omega)}{\clubsuit}, \quad \omega \in V, \; \clubsuit \in \{\dashv, \vdash, \perp\}.$$

Then $\theta$ induces a binary operad morphism $\bar{\theta} : \mathcal{T}(\mathrm{Tri}(V)) \longrightarrow \mathcal{T}(\mathrm{Tri}(W))$. Let $p_{\mathrm{Tri}(V)} : \mathcal{T}(\mathrm{Tri}(V)) \longrightarrow \mathrm{Tri}(\mathcal{P})$ be the projection. We have the following result.

**Theorem 2.1.** *Any operad morphism $\eta : \mathcal{P} \longrightarrow \mathcal{Q}$ induces a unique operad morphism $\mathrm{Tri}(\eta) : \mathrm{Tri}(\mathcal{P}) \longrightarrow \mathrm{Tri}(\mathcal{Q})$ such that*

$$p_{\mathrm{Tri}(W)} \circ \bar{\theta} = \mathrm{Tri}(\eta) \circ p_{\mathrm{Tri}(V)}. \tag{7}$$

**Proof.** We use the following diagram to keep track of the maps that we will use below.

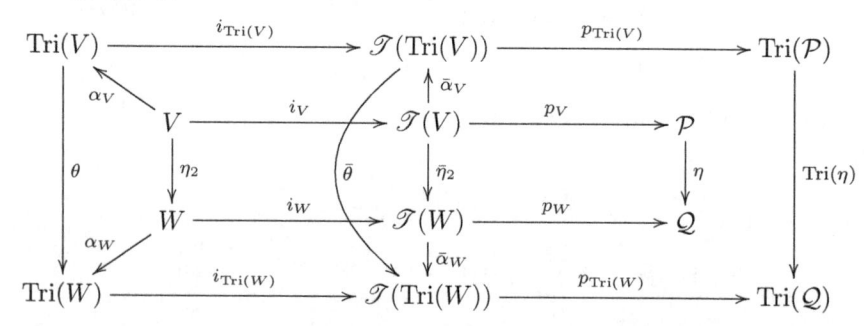

It follows from the fact that $\eta$ is a binary operad morphism and the universal property of $\mathscr{T}(V)$ that $(\bar{\eta}_2(R_\mathcal{P})) \subseteq (R_\mathcal{Q})$. On the other hand, for any tree $\tau \in \mathscr{T}(V)$ and nonempty subset $J \subseteq Lin(\tau)$, we have $\mathrm{Tri}_J(\bar{\eta}(\tau)) = \bar{\theta}(\mathrm{Tri}_J(\tau))$. Therefore, $(\bar{\theta}(\mathrm{Tri}(R_\mathcal{P}))) = (\mathrm{Tri}(\bar{\eta}(R_\mathcal{P}))) \subseteq (\mathrm{Tri}(R_\mathcal{Q}))$. So there exists a unique morphism $\mathrm{Tri}(\eta) : \mathrm{Tri}(\mathcal{P}) \longrightarrow \mathrm{Tri}(\mathcal{Q})$ such that $\mathrm{Tri}(\eta) \circ p_{\mathrm{Tri}(V)} = p_{\mathrm{Tri}(W)} \circ \bar{\theta}$. $\qquad\square$

**Corollary 2.1.** *Let* $\mathscr{T}(V)/(R_\mathcal{P}') \cong \mathscr{T}(W)/(R_\mathcal{P}'')$ *be two presentations of an operad* $\mathcal{P}$. *Then*

$$\mathscr{T}(\mathrm{Tri}(V))/(\mathrm{Tri}(R_\mathcal{P}')) \cong \mathscr{T}(\mathrm{Tri}(W))/(\mathrm{Tri}(R_\mathcal{P}'')).$$

*The same result holds for a nonsymmetric operad.*

**Proof.** Let

$$f : \mathscr{T}(V)/(R_\mathcal{P}') \longrightarrow \mathscr{T}(W)/(R_\mathcal{P}''),$$
$$g : \mathscr{T}(W)/(R_\mathcal{P}'') \longrightarrow \mathscr{T}(V)/(R_\mathcal{P}'),$$

be such that $g \circ f = id_{\mathscr{T}(V)/(R_\mathcal{P}')}, f \circ g = id_{\mathscr{T}(W)/(R_\mathcal{P}'')}$. By applying Theorem 2.1 three times: for $\mathcal{P} = \mathscr{T}(V)/(R_\mathcal{P}'), \mathcal{Q} = \mathscr{T}(W)/(R_\mathcal{P}''), \eta = f$; for $\mathcal{P} = \mathscr{T}(W)/(R_\mathcal{P}''), \mathcal{Q} = \mathscr{T}(V)/(R_\mathcal{P}'), \eta = g$ and for $\mathcal{P} = \mathscr{T}(V)/(R_\mathcal{P}'), \mathcal{Q} = \mathscr{T}(V)/(R_\mathcal{P}'), \eta = id_{\mathscr{T}(V)/(R_\mathcal{P}')}$, we obtain

$$\mathrm{Tri}(g) \circ \mathrm{Tri}(f) = \mathrm{Tri}(id_{\mathscr{T}(V)/(R_\mathcal{P}')}) = id_{\mathscr{T}(\mathrm{Tri}(V))/(\mathrm{Tri}(R_\mathcal{P}'))}.$$

Similarly, we have $\mathrm{Tri}(f) \circ \mathrm{Tri}(g) = id_{\mathscr{T}(\mathrm{Tri}(W))/(\mathrm{Tri}(R_\mathcal{P}''))}$, which completes the proof. $\qquad\square$

## 3. Examples of duplicators and triplicators

In this section, a list of examples of duplicators and triplicators is provided for a nonsymmetric operad and a (symmetric) operad.

### 3.1. *The nonsymmetric case*

Let $As$ be the nonsymmetric operad of the associative algebra with product $\cdot$. Using the abbreviations $\dashv := \left(\begin{smallmatrix} \cdot \\ \dashv \end{smallmatrix}\right)$ and $\vdash := \left(\begin{smallmatrix} \cdot \\ \vdash \end{smallmatrix}\right)$, we have

$$\mathrm{Du}_y((x \cdot y) \cdot z - x \cdot (y \cdot z)) = \{(x \vdash y) \dashv z - x \vdash (y \dashv z)\},$$
$$\mathrm{Du}_x((x \cdot y) \cdot z - x \cdot (y \cdot z)) = \{(x \dashv y) \dashv z - x \dashv (y \dashv z),$$
$$(x \dashv y) \dashv z - x \dashv (y \vdash z)\},$$
$$\mathrm{Du}_z((x \cdot y) \cdot z - x \cdot (y \cdot z)) = \{(x \dashv y) \vdash z - x \vdash (y \vdash z),$$

$$(x \vdash y) \vdash z - x \vdash (y \vdash z)\},$$

giving the five relations of the **diassociative algebra** of Loday[25]. Therefore the duplicator of $As$ is $Dias$.

A similar computation shows that the triplicator of $As$ is the operad *Trias* of the **triassociative algebra** of Loday and Ronco[30]. For example,

$$\mathrm{Du}_{\{x\}}((xy)z - x(yz)) = \{(x \dashv y) \dashv z - x \dashv (y \dashv z),$$
$$(x \dashv y) \dashv z - x \dashv (y \vdash z),$$
$$(x \dashv y) \dashv z - x \dashv (y \perp z)\},$$
$$\mathrm{Du}_{\{x,y\}}((xy)z - x(yz)) = \{(x \perp y) \dashv z - x \perp (y \dashv z)\},$$
$$\mathrm{Du}_{\{x,y,z\}}((xy)z - x(yz)) = \{(x \perp y) \perp z - x \perp (y \perp z)\}.$$

### 3.2. *The symmetric case*

Let $V$ be an $\mathbb{S}$-module concentrated in arity two. Then we have

$$\mathcal{T}(V)(3) = (V \otimes_{\mathbb{S}_2} (V \otimes \mathbf{k} \oplus \mathbf{k} \otimes V)) \otimes_{\mathbb{S}_2} \mathbf{k}[\mathbb{S}_3],$$

which can be identify with 3 copies of $V \otimes V$, denoted by $V \circ_{\mathrm{I}} V, V \circ_{\mathrm{II}} V$ and $V \circ_{\mathrm{III}} V$, following the convention in Ref. 38. Then, as an abelian group, $\mathcal{T}(V)(3)$ is generated by elements of the form

$$\omega \circ_{\mathrm{I}} \nu \, (\leftrightarrow (x\,\nu\,y)\,\omega\,z), \quad \omega \circ_{\mathrm{II}} \nu \, (\leftrightarrow (y\,\nu\,z)\,\omega\,x),$$
$$\omega \circ_{\mathrm{III}} \nu \, (\leftrightarrow (z\,\nu\,x)\,\omega\,y), \, \forall \omega, \, \nu \in V. \tag{8}$$

For an operad where the space of generators $V$ is equal to $\mathbf{k}[\mathbb{S}_2] = \mu.\mathbf{k} \oplus \mu'.\mathbf{k}$ with $\mu.(12) = \mu'$, we will adopt the convention in Ref. 38 (p. 129) and denote the 12 elements of $\mathcal{T}(V)(3)$ by $v_i, 1 \leq i \leq 12$, in the following table.

| $v_1$ | $\mu \circ_{\mathrm{I}} \mu \leftrightarrow (xy)z$ | $v_5$ | $\mu \circ_{\mathrm{III}} \mu \leftrightarrow (zx)y$ | $v_9$ | $\mu \circ_{\mathrm{II}} \mu \leftrightarrow (yz)x$ |
|---|---|---|---|---|---|
| $v_2$ | $\mu' \circ_{\mathrm{II}} \mu \leftrightarrow x(yz)$ | $v_6$ | $\mu' \circ_{\mathrm{I}} \mu \leftrightarrow z(xy)$ | $v_{10}$ | $\mu' \circ_{\mathrm{III}} \mu \leftrightarrow y(zx)$ |
| $v_3$ | $\mu' \circ_{\mathrm{II}} \mu' \leftrightarrow x(zy)$ | $v_7$ | $\mu' \circ_{\mathrm{I}} \mu' \leftrightarrow z(yx)$ | $v_{11}$ | $\mu' \circ_{\mathrm{III}} \mu' \leftrightarrow y(xz)$ |
| $v_4$ | $\mu \circ_{\mathrm{III}} \mu' \leftrightarrow (xz)y$ | $v_8$ | $\mu \circ_{\mathrm{II}} \mu' \leftrightarrow (zy)x$ | $v_{12}$ | $\mu \circ_{\mathrm{I}} \mu' \leftrightarrow (yx)z$ |

### 3.2.1. *Examples of duplicators*

Recall that a **(left) Leibniz algebra**[25] is defined by a bilinear operation $\{,\}$ and a relation

$$\{x, \{y, z\}\} = \{\{x, y\}, z\} + \{y, \{x, z\}\}.$$

**Proposition 3.1.** *The operad Leib of the Leibniz algebra is the duplicator of Lie, the operad of the Lie algebra.*

**Proof.** Let $\mu$ denote the operation of the operad *Lie*. The space of relations of *Lie* is generated as an $\mathbb{S}_3$-module by

$$v_1 + v_5 + v_9 = \mu \circ_{\mathrm{I}} \mu + \mu \circ_{\mathrm{II}} \mu + \mu \circ_{\mathrm{III}} \mu = (x\mu y)\mu z + (z\mu x)\mu y + (y\mu z)\mu x. \quad (9)$$

Use the abbreviations $\dashv := \left(\begin{smallmatrix}\mu \\ \dashv\end{smallmatrix}\right)$ and $\vdash := \left(\begin{smallmatrix}\mu \\ \vdash\end{smallmatrix}\right)$. Then from $\left(\begin{smallmatrix}\mu \\ \dashv\end{smallmatrix}\right)^{(12)} = \left(\begin{smallmatrix}\mu^{(12)} \\ \vdash\end{smallmatrix}\right) = -\left(\begin{smallmatrix}\mu \\ \vdash\end{smallmatrix}\right)$, we have $\dashv^{(12)} = -\vdash$. Then we have

$$\begin{aligned}
\mathrm{Du}_z(v_1 + v_5 + v_9) &= \{(x \vdash y) \vdash z + (y \vdash z) \dashv x + (z \dashv x) \dashv y, \\
&\quad (x \dashv y) \vdash z + (y \vdash z) \dashv x + (z \dashv x) \dashv y\} \\
&= \{(x \vdash y) \vdash z - x \vdash (y \vdash z) + y \vdash (x \vdash z), \\
&\quad y \vdash (x \vdash z) - (y \vdash x) \vdash z - x \vdash (y \vdash z)\},
\end{aligned}$$

with similar computations for $\mathrm{Du}_x$ and $\mathrm{Du}_y$. Replacing the operation $\vdash$ by $\{,\}$, we see that the underlined relation is precisely the relation of the Leibniz algebra while the other relations are obtained from this relation by a permutation of the variables. Therefore $\mathrm{Du}(Lie) = Leib$. $\square$

Recall that a **(left) permutative algebra**[6] (also called commutative diassociative algebra) is defined by one bilinear operation $\cdot$ and the relations

$$x \cdot (y \cdot z) = (x \cdot y) \cdot z = (y \cdot x) \cdot z.$$

**Proposition 3.2.** *The operad Perm of the permutative algebra is the duplicator of Comm, the operad of the commutative associative algebra.*

**Proof.** Let $\omega$ denote the operation of the operad *Comm*. Setting $\dashv := \left(\begin{smallmatrix}\omega \\ \dashv\end{smallmatrix}\right)$ and $\vdash := \left(\begin{smallmatrix}\omega \\ \vdash\end{smallmatrix}\right)$, then from $\left(\begin{smallmatrix}\omega \\ \vdash\end{smallmatrix}\right)^{(12)} = \left(\begin{smallmatrix}\omega^{(12)} \\ \vdash\end{smallmatrix}\right) = \left(\begin{smallmatrix}\omega \\ \vdash\end{smallmatrix}\right)$ we have $\dashv^{(12)} = \vdash$. The space of relations of *Comm* is generated as an $\mathbb{S}_3$-module by

$$v_1 - v_9 = \omega \circ_{\mathrm{I}} \omega - \omega \circ_{\mathrm{II}} \omega = (x\omega y)\omega z - (y\omega z)\omega x.$$

Then we have

$$\begin{aligned}
\mathrm{Du}_z(v_1 - v_9) &= \{(x \dashv y) \vdash z - (y \vdash z) \dashv x, (x \vdash y) \vdash z - (y \vdash z) \dashv x\} \\
&= \{(y \vdash x) \vdash z - x \vdash (y \vdash z), (x \vdash y) \vdash z - x \vdash (y \vdash z)\},
\end{aligned}$$

with similar computations for $\mathrm{Du}_x$ and $\mathrm{Du}_y$. Replacing the operation $\vdash$ by $\cdot$ and following the same proof as in Proposition 3.1, we get $\mathrm{Du}(Comm) = Perm$. $\square$

A **(left) Poisson algebra** is defined to be a k-vector space with two bilinear operations $\{,\}$ and $\circ$ such that $\{,\}$ is the Lie bracket, $\circ$ is the product of commutative associative algebra, and the two operations are compatible in the sense that

$$\{x, y \circ z\} = \{x, y\} \circ z + y \circ \{x, z\}.$$

A **dual (left) pre-Poisson algebra**[1] is defined to be a k-vector space with two bilinear operations $\{,\}$ and $\circ$ such that $\{,\}$ is a Leibniz bracket and $\circ$ is a permutative algebra product, and they are compatible in the sense that

$$\{x, y \circ z\} = \{x, y\} \circ z + y \circ \{x, z\}, \quad \{x \circ y, z\} = x \circ \{y, z\} + y \circ \{x, z\},$$
$$\{x, y\} \circ z = -\{y, x\} \circ z.$$

By a similar argument as in Proposition 3.1, we obtain

**Proposition 3.3.** *The duplicator of Pois, the operad of the Poisson algebra, is DualPrePois, the operad of the dual pre-Poisson algebra.*

We next consider the duplicator of the the operad *preLie* of **(left) pre-Lie** algebra (also called left-symmetric algebra). A pre-Lie algebra is defined by a bilinear operation $\{\,,\,\}$ that satisfies

$$R_{preLie} := \{\{x, y\}, z\} - \{x, \{y, z\}\} - \{\{y, x\}, z\} + \{y, \{x, z\}\} = 0.$$

By Definition 2.6 and the abbreviations $\dashv := \left(\begin{smallmatrix}\omega\\\dashv\end{smallmatrix}\right), \vdash := \left(\begin{smallmatrix}\omega\\\vdash\end{smallmatrix}\right)$, we have

$$
\begin{aligned}
\mathrm{Du}(R_{preLie}) = \Big\{ & x \dashv (y \dashv z) - x \dashv (y \vdash z), \ y \dashv (x \dashv z) - y \dashv (x \vdash z),\\
& (x \vdash y) \vdash z - (x \dashv y) \vdash z, \ (y \vdash x) \vdash z - (y \dashv x) \vdash z,\\
& \underline{x \dashv (y \dashv z) - (x \dashv y) \dashv z - y \vdash (x \dashv z) + (y \vdash x) \dashv z},\\
& \underline{x \vdash (y \dashv z) - (x \vdash y) \dashv z - y \dashv (x \dashv z) + (y \dashv x) \dashv z},\\
& \underline{x \vdash (y \vdash z) - (x \vdash y) \vdash z - y \vdash (x \vdash z) + (y \vdash x) \vdash z} \Big\}
\end{aligned}
$$

These underlined relations coincide with the axioms of pre-Lie dialgebra (left-symmetric dialgebra) defined in Ref. 9, and the other relations are obtained from these relations by a permutation of the variables. Then we have

**Proposition 3.4.** *The duplicator of preLie, the operad of the pre-Lie algebra, is DipreLie, the operad of the pre-Lie dialgebra.*

### 3.2.2. *Examples of triplicators*

We similarly have the following examples of triplicators of operads.

A **commutative trialgebra**[30] is a vector space $A$ equipped with a product $\star$ and a commutative product $\bullet$ satisfying the following equations:

$$(x \star y) \star z = \star(y \star z), \quad x \star (y \star z) = x \star (y \bullet z),$$
$$x \bullet (y \star z) = (x \bullet y) \star z, \quad (x \bullet y) \bullet z = x \bullet (y \bullet z).$$

**Proposition 3.5.** *The operad ComTria of the commutative trialgebra is the triplicator of Comm.*

**Proof.** Continuing with the notations of the operad *Comm* in the proof of Proposition 3.2, we have, for example,

$$\begin{aligned}
\mathrm{Tri}_x(v_1 - v_9) &= \{(x \dashv y) \dashv z - (y \dashv z) \vdash x, (x \dashv y) \dashv z - (y \vdash z) \vdash x, \\
&\quad (x \dashv y) \dashv z - (y \perp z) \vdash x\} \\
&= \{(x \dashv y) \dashv z - x \dashv (y \dashv z), (x \dashv y) \dashv z - x \dashv (z \dashv y), \\
&\quad \underline{(x \dashv y) \dashv z - x \dashv (y \perp z)}\}; \\
\mathrm{Tri}_{\{x,y\}}(v_1 - v_9) &= \{(x \perp y) \dashv z - (y \dashv z) \perp x\}; \\
\mathrm{Tri}_{\{x,y,z\}}(v_1 - v_9) &= \{(x \perp y) \perp z - (y \perp z) \perp x\} \\
&= \{\underline{(x \perp y) \perp z - x \perp (y \perp z)}\}.
\end{aligned}$$

Replacing the operation $\dashv$ by $\star$ and $\perp$ by $\bullet$, we see that the underlined relations are equivalent to the relations of the commutative trialgebra. The other relations can be obtained from these relations by a permutation of the variables and the commutativity of $\perp$. Thus we get $\mathrm{Tri}(Comm) = ComTria$. $\square$

We next consider the triplicator of *Lie*. Let $\mu$ be the operation of the operad *Lie*. Set $\dashv := \left(\begin{smallmatrix}\mu\\\dashv\end{smallmatrix}\right)$, $\vdash := \left(\begin{smallmatrix}\mu\\\vdash\end{smallmatrix}\right)$ and $\perp := \left(\begin{smallmatrix}\mu\\\perp\end{smallmatrix}\right)$. Since $\left(\begin{smallmatrix}\mu\\\dashv\end{smallmatrix}\right)^{(12)} = \left(\begin{smallmatrix}\mu^{(12)}\\\vdash\end{smallmatrix}\right) = -\left(\begin{smallmatrix}\mu\\\vdash\end{smallmatrix}\right)$ and $\left(\begin{smallmatrix}\mu\\\perp\end{smallmatrix}\right)^{(12)} = \left(\begin{smallmatrix}\mu^{(12)}\\\perp\end{smallmatrix}\right) = -\left(\begin{smallmatrix}\mu\\\perp\end{smallmatrix}\right)$, we have $\dashv^{(12)} = -\vdash$ and $\perp^{(12)} = -\perp$. The space of relations of *Lie* is generated as an $\mathbb{S}_3$-module by

$$v_1 + v_5 + v_9 = \mu \circ_\mathrm{I} \mu + \mu \circ_\mathrm{II} \mu + \mu \circ_\mathrm{III} \mu = (x\mu y)\mu z + (z\mu x)\mu y + (y\mu z)\mu x.$$

Then we compute

$$\begin{aligned}
\mathrm{Tri}_{\{x\}}(v_1 + v_5 + v_9) &= \{(x \dashv y) \dashv z + (z \vdash x) \dashv y + (y \dashv z) \vdash x, \\
&\quad (x \dashv y) \dashv z + (z \vdash x) \dashv y + (y \vdash z) \vdash x, \\
&\quad (x \dashv y) \dashv z + (z \vdash x) \dashv y + (y \perp z) \vdash x\}
\end{aligned}$$

$$= \{(x \dashv y) \dashv z - (x \dashv z) \dashv y - x \dashv (y \dashv z),$$
$$(x \dashv y) \dashv z - (x \dashv z) \dashv y + x \dashv (z \dashv y),$$
$$\underline{(x \dashv y) \dashv z - (x \dashv z) \dashv y - x \dashv (y \perp z)}\};$$
$$\mathrm{Tri}_{\{x,y\}}(v_1 + v_5 + v_9) = \{(x \perp y) \dashv z + (z \vdash x) \perp y + (y \dashv z) \perp x\};$$
$$\mathrm{Tri}_{\{x,y,z\}}(v_1 + v_5 + v_9) = \{(x \perp y) \perp z + (z \perp x) \perp y + (y \perp z) \perp x\},$$

and other computations yield the same relations up to permutations.

Replacing the operation $\dashv$ by $\diamond$ and $\perp$ by $[,]$, then $[\,,\,]$ is skew-symmetric and the underlined relations are

$$[x, [y, z]] + [y, [z, x]] + [z, [x, y]] = 0,$$
$$x \diamond [y, z] = x \diamond (y \diamond z), \quad [x, y] \diamond z = [x \diamond z, y] + [x, y \diamond z], \qquad (10)$$
$$(x \diamond y) \diamond z = x \diamond (y \diamond z) + (x \diamond z) \diamond y.$$

Then in particular $(A, \diamond)$ is a right Leibniz algebra. Since the duplicator of *Lie* is *Leib*, the operad of the Leibniz algebra, we tentatively call the new algebra **Leibniz trialgebra**. In summary, we obtain

**Proposition 3.6.** *The triplicator of Lie is TriLeib, the operad of the Leibniz trialgebra.*

As we will see in Section 5.1, *TriLeib* is precisely the Koszul dual of the operad $CTD = ComTriDend$ of the commutative tridendriform algebra, namely the **Dual CTD algebra** in Ref. 39.

## 4. Operads, their duplicators and triplicators

In this section, we study the relationship among a binary operad, its duplicator and triplicator.

### 4.1. *Relationship between an operad and its duplicator and triplicator*

For a given $\mathbb{S}$-module $V$ concentrated in ariry 2: $V = V(2)$. Let $i_V : V \to \mathcal{T}(V)$ denote the natural embedding to the free operad $\mathcal{T}(V)$. Let $\mathcal{P} := \mathcal{T}(V)/(R)$ be a binary operad and let $j_V : V \to \mathcal{P}$ be $p_V \circ i_V$, where $p_V : \mathcal{T}(V) \to \mathcal{P}$ is the operad projection. Similarly define the maps $i_{\mathrm{Du}(V)} : \mathrm{Du}(V) \to \mathcal{T}(\mathrm{Du}(V))$ and operad morphism $p_{\mathrm{Du}(V)} : \mathcal{T}(\mathrm{Du}(V)) \to \mathrm{Du}(\mathcal{P})$ and $j_{\mathrm{Du}(V)} := p_{\mathrm{Du}(V)} \circ i_{\mathrm{Du}(V)}$, as well as the corresponding map and operad morphisms for $\mathrm{Tri}(V)$.

**Proposition 4.1.** *Let $\mathcal{P} = \mathcal{T}(V)/(R)$ be a binary operad.*

*(1) The linear map*

$$\eta : \mathrm{Du}(V) \to V, \quad \left(\begin{smallmatrix} \omega \\ u \end{smallmatrix}\right) \longmapsto \omega \ \text{for all} \ \left(\begin{smallmatrix} \omega \\ u \end{smallmatrix}\right) \in \mathrm{Du}(V), u \in \{\dashv, \vdash\} \tag{11}$$

*induces a unique operad morphism*

$$\tilde{\eta} : \mathcal{T}(\mathrm{Du}(V)) \to \mathcal{P}$$

*such that* $\tilde{\eta} \circ j_{\mathrm{Du}(V)} = j_V \circ \eta$.

*(2) The linear map*

$$\zeta : \mathrm{Tri}(V) \to V,$$
$$\left(\begin{smallmatrix} \omega \\ u \end{smallmatrix}\right) \longmapsto \omega \ \text{for all} \ \left(\begin{smallmatrix} \omega \\ u \end{smallmatrix}\right) \in \mathrm{Tri}(V), u \in \{\dashv, \vdash, \bot\} \tag{12}$$

*induces a unique operad morphism*

$$\tilde{\zeta} : \mathcal{T}(\mathrm{Tri}(V)) \to \mathcal{P}$$

*such that* $\tilde{\zeta} \circ j_{\mathrm{Tri}(V)} = j_V \circ \zeta$.

*(3) There is a morphism $\rho$ of operads from $\mathrm{Tri}(\mathcal{P})$ to $\mathcal{P}$ which extends the linear map from $\mathrm{Tri}(V)$ to $V$ defined by*

$$\left(\begin{smallmatrix} \omega \\ \bot \end{smallmatrix}\right) \longmapsto \omega, \quad \left(\begin{smallmatrix} \omega \\ u \end{smallmatrix}\right) \longmapsto 0, \quad \text{where} \ u \in \{\dashv, \vdash\}. \tag{13}$$

**Proof.** Let $R$ be the set of locally homogeneous elements

$$r_s := \sum_i c_{s,i} \tau_{s,i}, \ c_{s,i} \in \mathbf{k}, \tau_{s,i} \in \bigcup_{t \in \mathfrak{R}} t(V), \ 1 \le s \le k,$$

as given in Eq.(5).

(1) By the universal property of the free operad $\mathcal{T}(\mathrm{Du}(V))$ on the $\mathbb{S}$-module $\mathrm{Du}(V)$, the $\mathbb{S}$-module morphism $i_V \circ \eta : \mathrm{Du}(V) \to \mathcal{T}(V)$ induces a unique operad morphism $\bar{\eta} : \mathcal{T}(\mathrm{Du}(V)) \to \mathcal{T}(V)$ such that $i_{\mathrm{Du}(V)} \circ \bar{\eta} = i_V \circ \eta$.

For any $x \in \mathrm{Lin}(r_s)$ and $1 \le s \le k$, by the definition of $\mathrm{Du}_x(\tau_{s,i}))$ in Definition 2.3 and $\eta$ in Eq. (11), the element $\eta(\mathrm{Du}(\tau_{s,i}))$ is obtained by replacing each decoration $\left(\begin{smallmatrix} \omega \\ u \end{smallmatrix}\right)$ of the vertices of $\mathrm{Du}(\tau_{s,i})$ by $\omega$, where $\omega \in V$ and $u \in \{\dashv, \vdash\}$. Thus $\bar{\eta}(\mathrm{Du}(\tau_{s,i})) = \tau_{s,i}$. Then we have

$$\bar{f}\left(\sum_i c_{s,i} \mathrm{Du}_x(\tau_{s,i})\right) = \sum_i c_{s,i} \tau_{s,i} \equiv 0 \mod (R).$$

By Eq. (6), we see that $(\mathrm{Du}(R)) \subseteq \ker(\eta)$. Thus there is a unique operad morphism $\tilde{\eta} : \mathrm{Du}(\mathcal{P}) := \mathcal{T}(\mathrm{Du}(V))/(\mathrm{Du}(R)) \to \mathcal{P} := \mathcal{T}(V)/(R)$ such that

$\tilde{\eta} \circ p_{\mathrm{Du}(V)} = p_V \circ \bar{\eta}$. We then have $\tilde{\eta} \circ j_{\mathrm{Du}(V)} = j_V \circ \eta$. In summary, we have the following diagram in which each square commutes.

$$
\begin{array}{ccccc}
\mathrm{Du}(V) & \xrightarrow{i_{\mathrm{Du}(V)}} & \mathcal{T}(\mathrm{Du}(V)) & \xrightarrow{p_{\mathrm{Du}(V)}} & \mathrm{Du}(\mathcal{P}) \\
\downarrow{\scriptstyle\eta} & & \downarrow{\scriptstyle\bar{\eta}} & & \downarrow{\scriptstyle\tilde{\eta}} \\
V & \xrightarrow{i_V} & \mathcal{T}(V) & \xrightarrow{p_V} & \mathcal{P}
\end{array}
$$

Suppose $\tilde{\eta}' : \mathrm{Du}(\mathcal{P}) \to \mathcal{P}$ is another operad morphism such that $\tilde{\eta}' \circ j_{\mathrm{Du}(V)} = j_V \circ \eta$. Then we have $\tilde{\eta}' \circ j_{\mathrm{Du}(V)} = \tilde{\eta}' \circ p_{\mathrm{Du}(V)} \circ i_{\mathrm{Du}(V)}$ and $j_V \circ \eta = p_V \circ j_V \circ \eta = p_V \circ \bar{\eta} \circ i_{\mathrm{Du}(V)}$. By the universal property of the free operad $\mathcal{T}(\mathrm{Du}(V))$, we obtain $\tilde{\eta}' \circ p_{\mathrm{Du}(V)} = p_V \circ \bar{\eta} = \tilde{\eta} \circ p_{\mathrm{Du}(V)}$. Since $p_{\mathrm{Du}(V)}$ is surjective, we obtain $\tilde{\eta}' = \tilde{\eta}$. This proves the uniqueness of $\tilde{\eta}$.

(2) The proof is similar to the proof of Item (1).

(3) By the description of $\mathrm{Tri}_{\{x\}}(\tau_{s,i})$ in Proposition 2.1, $\rho(\mathrm{Tri}_{\{x\}}(\tau_{s,i}))$ is obtained by replacing $\left(\begin{smallmatrix}\omega\\u\end{smallmatrix}\right)$ by $\rho(\left(\begin{smallmatrix}\omega\\u\end{smallmatrix}\right))$. Since $\rho(\left(\begin{smallmatrix}\omega\\\dashv\end{smallmatrix}\right)) = 0$, $\rho(\left(\begin{smallmatrix}\omega\\\vdash\end{smallmatrix}\right)) = 0$ and $\rho(\left(\begin{smallmatrix}\omega\\\perp\end{smallmatrix}\right)) = \omega$, it is easy to see that if $J \neq Lin(\tau)$, then $\rho(\sum_i c_{s,i} \mathrm{Tri}_J(\tau_{s,i})) = \sum_i c_{s,i} \tau_{s,i} = 0$, and, if $J = Lin(\tau)$, then $\rho(\sum_i c_{s,i} \mathrm{Tri}_{Lin(\tau)}(\tau_{s,i})) = \sum_i c_{s,i} \tau_{s,i} \equiv 0 \mod (R)$. Thus $\rho(\mathrm{Tri}(R)) \subseteq R$ and $\rho$ induces the desired operad morphism. □

## 4.2. *Relationship between the duplicator and the triplicator of a binary operad*

The following results relates the duplicator and the triplicator of a binary algebraic operad.

**Proposition 4.2.** *Let* $\mathcal{P} = \mathcal{T}(V)/(R)$ *be a binary algebraic operad. There is a morphism of operads from* $\mathrm{Tri}(\mathcal{P})$ *to* $\mathrm{Du}(\mathcal{P})$ *that extends the linear map defined by*

$$
\left(\begin{smallmatrix}\omega\\\dashv\end{smallmatrix}\right) \to \left(\begin{smallmatrix}\omega\\\dashv\end{smallmatrix}\right), \quad \left(\begin{smallmatrix}\omega\\\vdash\end{smallmatrix}\right) \to \left(\begin{smallmatrix}\omega\\\vdash\end{smallmatrix}\right), \quad \left(\begin{smallmatrix}\omega\\\perp\end{smallmatrix}\right) \to 0 \text{ for all } \omega \in V. \tag{14}
$$

**Proof.** The linear map $\phi : \mathrm{Tri}(V) \to \mathrm{Du}(V)$ defined by Eq.(14) is $\mathbb{S}_2$-equivariant. Hence it induces a morphism of the free operads $\phi : \mathcal{T}(\mathrm{Tri}(V)) \to \mathcal{T}(\mathrm{Du}(V))$ which, by composing with the quotient map, induces the morphism of operads

$$
\phi : \mathcal{T}(\mathrm{Tri}(V)) \to \mathrm{Du}(\mathcal{P}) = \mathcal{T}(\mathrm{Du}(V))/(\mathrm{Du}(R)).
$$

Let $\mathrm{Tri}_J(r) \in \mathrm{Tri}(R)$ be one of the generators of $(\mathrm{Tri}(R))$ with $r = \sum_i c_i \tau_i \in R$ in Eq. (5) and $\varnothing \neq J \subseteq \mathrm{Lin}(r)$. If $J$ is the singleton $\{x\}$ for some $x \in \mathrm{Lin}(r)$, then by the description of $\mathrm{Tri}_{\{x\}}(\tau_i)$ in Proposition 2.1, $\phi(\mathrm{Tri}_{\{x\}}(\tau))$ is obtained by keeping all the $\left(\begin{smallmatrix} \omega \\ \dashv \end{smallmatrix}\right)$ and $\left(\begin{smallmatrix} \omega \\ \vdash \end{smallmatrix}\right)$, and by replacing all $\left(\begin{smallmatrix} \omega \\ \perp \end{smallmatrix}\right)$, $\omega \in V$ by zero. Thus in Case (b) of Proposition 2.1 we have $\phi(\mathrm{Tri}_{\{x\}}(\tau_i)) = \mathrm{Du}_x(\tau_i)$. Also Case (a)(iii) cannot occur for the singleton $\{x\}$. Thus in Case (a) of Proposition 2.1, we also have $\phi(\mathrm{Tri}_{\{x\}}(\tau_i)) = \mathrm{Du}_x(\tau_i)$. Thus $\phi(\mathrm{Tri}_{\{x\}}(r)) = \mathrm{Du}_x(r)$ and hence is in $\mathrm{Du}(R)$.

If $J$ contains more than one element, then at least one of the vertices of $\mathrm{Tri}_J(\tau_i)$ is $\left(\begin{smallmatrix} \omega \\ \perp \end{smallmatrix}\right)$ and hence the corresponding vertex of $\phi(\mathrm{TSu}_J(\tau_i))$ is zero. Thus we have $\phi(\mathrm{Tri}_J(\tau_i)) = 0$, $\phi(\mathrm{Tri}_J(r)) = 0$ and hence $\phi(\mathrm{Tri}_J(R)) = 0$. Thus, for any $J \neq \varnothing$ and $r \in R$, we have $\phi(\mathrm{Tri}_J(r)) \in \mathrm{Du}(R)$ and hence $\phi(\mathrm{Tri}(R))$ is a subset of $\mathrm{Du}(R)$.

In summary, we have $\phi((\mathrm{Tri}(R)) \subseteq \mathrm{Du}(R)$. Thus the morphism $\phi : \mathcal{T}(\mathrm{Tri}(V)) \to \mathrm{Du}(\mathcal{P})$ induces a unique morphism $\phi : \mathrm{Tri}(\mathcal{P}) \to \mathrm{Du}(\mathcal{P})$. $\qquad\square$

If we take $\mathcal{P}$ to be the operad of the associative algebra, then we obtain the following result of Loday and Ronco[30]:

**Corollary 4.1.** *Let* $(A, \dashv, \vdash)$ *be an associative dialgebra. Then* $(A, \dashv, \vdash, 0)$ *is an associative trialgebra, where 0 denotes the trivial product.*

## 5. Duality of replicators with successors and Manin products

The similarity between the definitions of the replicators and successors[3] (see below) suggests that there is a close relationship between the two constructions. We show that this is indeed the case. More precisely, taking the replicator of a binary quadratic operad is in Koszul dual with taking the successor of the dual operad. This, in particular, allows us to identify the duplicator (resp. triplicator) of a binary quadratic operad $\mathcal{P}$ with the Manin white product of *Perm* (resp. *ComTria*) with $\mathcal{P}$, providing an easy way to compute these white products.

Since it is shown in Ref. 14 that taking di-Var and tri-Var are also isomorphic to taking these Manin products, taking duplicator (resp. triplicator) is isomorphic to taking di-Var (resp. tri-Var) other than the case of free operads.

### 5.1. *The duality of replicators with successors*

For later reference, we recall the definition of successors[3].

**Definition 5.1.** The **bisuccessor**[3] of a binary operad $\mathcal{P} = \mathcal{T}(V)/(R)$ is defined to be the binary operad $\mathrm{Su}(\mathcal{P}) = \mathcal{T}(\widetilde{V})/(\mathrm{Su}(R))$ where the $\mathbb{S}_2$-action on $\widetilde{V}$ is given by

$$\left(\begin{smallmatrix}\omega\\\prec\end{smallmatrix}\right)^{(12)} := \left(\begin{smallmatrix}\omega\,(12)\\\succ\end{smallmatrix}\right), \quad \left(\begin{smallmatrix}\omega\\\succ\end{smallmatrix}\right)^{(12)} := \left(\begin{smallmatrix}\omega\,(12)\\\prec\end{smallmatrix}\right), \quad \omega \in V,$$

and the space of relations is generated, as an $\mathbb{S}$-module, by

$$\mathrm{Su}(R) := \left\{ \mathrm{Su}_x(r_s) := \sum_i c_{s,i}\mathrm{Su}_x(t_{s,i}) \,\Big|\, x \in Lin(r_s),\, 1 \le s \le k \right\}. \quad (15)$$

Here for $\tau \in \mathcal{T}(V)$, $x \in Lin(\tau)$, $\mathrm{Su}_x(\tau)$ is defined by relabeling each vertex $\omega$ of $Vin(\tau)$ by

$$\begin{cases} \left(\begin{smallmatrix}\omega\\\prec\end{smallmatrix}\right), \text{ the path from the root of } \tau \text{ to } x \text{ turns left at } \omega; \\ \left(\begin{smallmatrix}\omega\\\succ\end{smallmatrix}\right), \text{ the path from the root of } \tau \text{ to } x \text{ turns right at } \omega; \\ \left(\begin{smallmatrix}\omega\\\star\end{smallmatrix}\right), \omega \text{ is not on the path from the root of } \tau \text{ to } x, \end{cases}$$

where $\left(\begin{smallmatrix}\omega\\\star\end{smallmatrix}\right) := \left\{ \left(\begin{smallmatrix}\omega\\\prec\end{smallmatrix}\right) + \left(\begin{smallmatrix}\omega\\\succ\end{smallmatrix}\right) \right\}$. Similarly, there is a notion of the trisuccessor of a binary operad[3].

Let $\mathcal{P} = \mathcal{T}(V)/(R)$ be a binary quadratic operad. Then with the notations in Section 3.2, we have $\mathcal{T}(V)(3) = 3V \otimes V = \bigoplus_{u \in \{\mathrm{I},\mathrm{II},\mathrm{III}\}} V \circ_u V$.

**Proposition 5.1.** *Let $\mathbf{k}$ be an infinite field. Let $W$ be a non-zero $\mathbb{S}$-submodule of $3V \otimes V$. Then there is a basis $\{e_1, \cdots, e_n\}$ of $V$ such that the restriction to $W$ of the coordinate projections*

$$p_{i,j,u} : 3V \otimes V = \bigoplus_{1 \le k,\ell \le n, v \in \{\mathrm{I},\mathrm{II},\mathrm{III}\}} \mathbf{k}\, e_k \circ_v e_\ell \to \mathbf{k}\, e_i \circ_u e_j,$$

*are non-zero and hence surjective for all $1 \le i,j \le n$ and $u \in \{\mathrm{I},\mathrm{II},\mathrm{III}\}$.*

**Proof.** Fix a $0 \ne w \in W$ and write $w = w_{\mathrm{I}} + w_{\mathrm{II}} + w_{\mathrm{III}}$ with $w_u \in V \circ_u V, u \in \{\mathrm{I},\mathrm{II},\mathrm{III}\}$. Then at least one of the three terms is non-zero. Since $W$ is an $\mathbb{S}$-module and $(w_u)^{(123)} = w_{u+\mathrm{I}}$ (where $\mathrm{III} + \mathrm{I}$ is taken to be $\mathrm{I}$), we might assume that $w \in W$ is chosen so that $w_{\mathrm{I}} \ne 0$. Fix a basis $\{v_1, \cdots, v_n\}$ of $V$. Then there are $c_{ij} \in \mathbf{k}, 1 \le i,j \le n$, that are not all zero such that $w_{\mathrm{I}} = \sum_{1 \le i,j \le n} c_{ij} v_i \circ_{\mathrm{I}} v_j$.

Consider the set of polynomials

$$f_{k\ell}(x_{rs}) := f_{k\ell}(\{x_{rs}\}) := \sum_{1 \le i,j \le n} c_{ij} x_{ik} x_{j\ell} \in \mathbf{k}[x_{rs} \,|\, 1 \le r,s \le n],$$

for $1 \leq k,\, \ell \leq n$. Then the polynomial $\prod_{1\leq k,\ell\leq n} f_{k\ell}(x_{rs})$ is non-zero since at least one of $c_{ij}$ is non-zero, giving a monomial $\prod_{1\leq r,s\leq n} c_{ij}x_{ir}x_{js}$ in the product with non-zero coefficient. Hence the product

$$f(x_{rs}) := \det(x_{rs}) \prod_{1\leq k,\ell\leq n} f_{k\ell}(x_{rs})$$

is non-zero since $\det(x_{rs}) := \prod_{\sigma\in\mathbb{S}_n} x_{1\sigma(1)} \cdots x_{n\sigma(n)}$ is also a non-zero polynomial. Thus, by our assumption that $\mathbf{k}$ is an infinite field, there are $d_{rs} \in \mathbf{k},\, 1 \leq r, s \leq n$, such that $f(d_{rs}) \neq 0$. Thus $D := (d_{rs}) \in M_{n\times n}(\mathbf{k})$ is invertible and $f_{k\ell}(d_{rs}) \neq 0,\, 1 \leq k, \ell \leq n$.

Fix such a matrix $D = (d_{rs})$ and define

$$(e_1, \cdots, e_n)^T := D^{-1}(v_1, \cdots, v_n)^T.$$

Then $\{e_1, \cdots, e_n\}$ is a basis of $V$ and $v_i = \sum_{k=1}^n d_{ik}e_k$. Further

$$w_{\mathrm{I}} = \sum_{1\leq i,j\leq n} c_{ij}v_i \circ_{\mathrm{I}} v_j = \sum_{1\leq i,j\leq n} c_{ij}\left( \sum_{1\leq k,\ell n} d_{ik}d_{j\ell}e_k \circ_{\mathrm{I}} e_\ell \right)$$

$$= \sum_{1\leq k,\ell\leq n} \left( \sum_{1\leq i,j\leq n} c_{ij}d_{ik}d_{j\ell} \right) e_k \circ_{\mathrm{I}} e_\ell.$$

The coefficients are $f_{k\ell}(d_{rs})$ and are non-zero by the choice of $D$. Thus $p_{i,j,\mathrm{I}}(w) = p_{i,j,\mathrm{I}}(w_{\mathrm{I}})$ is non-zero and hence $p_{i,j,\mathrm{I}}(W)$ is onto for all $1 \leq i, j \leq n$.

Since $W$ is an $\mathbb{S}$-module, we have $w^{(123)} \in W$ and $(w^{(123)})_{\mathrm{II}} = (w_{\mathrm{I}})^{(123)}$. Thus $p_{i,j,\mathrm{II}}(w^{(123)}) = p_{i,j,\mathrm{II}}((w_{\mathrm{I}})^{(123)})$ is non-zero and hence $p_{i,j,\mathrm{II}}(W)$ is onto for all $1 \leq i, j \leq n$. By the same argument, $p_{i,j,\mathrm{III}}(W)$ is onto for all $1 \leq i, j \leq n$, completing the proof. $\qquad\square$

**Lemma 5.1.** *Let $W$ be a non-zero $\mathbb{S}$-submodule of $3V \otimes V$ and let $\{e_1, \cdots, e_n\}$ be a basis as chosen in Proposition 5.1. Let $\{r_1, \cdots, r_m\}$ be a basis of $U$ and write*

$$r_k = \sum_{1\leq i,j\leq n} c^k_{iju}e_i \circ_u e_j, \quad c^k_{iju} \in \mathbf{k},\, 1 \leq i, j \leq n,\, u \in \{\mathrm{I}, \mathrm{II}, \mathrm{III}\},\, 1 \leq k \leq m.$$

*Then for each $1 \leq i, j \leq n$ and $u \in \{\mathrm{I}, \mathrm{II}, \mathrm{III}\}$, there is $1 \leq k \leq m$, such that $c^k_{iju}$ is not zero.*

**Proof.** Suppose there is $1 \leq i, j \leq n$ and $u \in \{\mathrm{I}, \mathrm{II}, \mathrm{III}\}$ such that $c^k_{iju} = 0$ for all $1 \leq k \leq m$. Then $p_{iju}(r_k) = 0$ and hence $p_{iju}(W) = 0$. This contradicts Proposition 5.1. $\qquad\square$

Let $\mathcal{P} = \mathcal{T}(V)/(R)$ be a binary quadratic operad. Fix a **k**-basis $\{e_1, e_2, \cdots, e_n\}$ for $V$. The space $\mathcal{T}(V)(3)$ is spanned by the basis $\{e_i \circ_u e_j \mid 1 \leq i, j \leq n, u \in \{\mathrm{I}, \mathrm{II}, \mathrm{III}\}\}$. Thus if $f \in \mathcal{T}(V)(3)$, we have

$$f = \sum_{i,j} a_{i,j} e_i \circ_{\mathrm{I}} e_j + \sum_{i,j} b_{i,j} e_i \circ_{\mathrm{II}} e_j + \sum_{i,j} c_{i,j} e_i \circ_{\mathrm{III}} e_j.$$

Then we can take the relation space $(R) \subset \mathcal{T}(V)(3)$ to be generated by $m$ linearly independent relations

$$R = \left\{ f_k = \sum_{i,j} a_{i,j}^k e_i \circ_{\mathrm{I}} e_j + \sum_{i,j} b_{i,j}^k e_i \circ_{\mathrm{II}} e_j + \sum_{i,j} c_{i,j}^k e_i \circ_{\mathrm{III}} e_j \ \middle| \ 1 \leq k \leq m \right\}.$$
(16)

We state the following easy fact for later applications.

**Lemma 5.2.** *Let $f_i, 1 \leq i \leq m$, be a basis of $(R)$. Then $\{\mathrm{BSu}_x(f_i) \mid x \in \mathrm{Lin}(f_i), 1 \leq i \leq m\}$ is a linear spanning set of $(\mathrm{BSu}(R))$ and $\{\mathrm{Du}_x(f_i) \mid x \in \mathrm{Lin}(f_i), 1 \leq i \leq m\}$ is a linear spanning set of $(\mathrm{Du}(R))$.*

**Proof.** Let $L$ be the linear span of $\{\mathrm{BSu}_x(f_i) \mid x \in \mathrm{Lin}(f_i), 1 \leq i \leq m\}$. Then from $\mathrm{BSu}_x(f_i) \in (\mathrm{BSu}(R))$ we obtain $L \subseteq (\mathrm{BSu}(R))$. On the other hand, by Ref. 3 (Lemma 2.6), $L$ is already an $\mathbb{S}$-submodule. Thus from $\mathrm{BSu}(R) \subseteq L$ we obtain $(\mathrm{BSu}(R))_{\mathbb{S}} \subseteq L$. The proof for $(\mathrm{Du}(R))$ is the same. $\qquad\square$

For the finite dimensional $\mathbb{S}_2$-module $V$, we define its Czech dual $V^\vee = V^* \otimes sgn_2$. There is a natural pairing with respect to this duality given by:

$$\langle , \rangle : \mathcal{T}(V^\vee)(3) \otimes \mathcal{T}(V)(3) \longrightarrow \mathbf{k}, \quad \langle e_i^\vee \circ_u e_j^\vee, e_k \circ_v e_\ell \rangle := \delta_{(i,k)} \delta_{(j,\ell)} \delta_{(u,v)}.$$

We denote by $R^\perp$ the annihilator of $R$ with respect to this pairing. Given relations as in Eq. (16), we can express a basis of $(R^\perp)$ as

$$R^\perp = \left\{ g_\ell = \sum_{i,j} \alpha_{i,j}^\ell e_i^\vee \circ_{\mathrm{I}} e_j^\vee + \sum_{i,j} \beta_{i,j}^\ell e_i^\vee \circ_{\mathrm{II}} e_j^\vee \right.$$
$$\left. + \sum_{i,j} \gamma_{i,j}^\ell e_i^\vee \circ_{\mathrm{III}} e_j^\vee \ \middle| \ 1 \leq \ell \leq 3n^2 - m \right\}, \quad (17)$$

where, for all $k$ and $\ell$, we have

$$\sum_{i,j} a_{i,j}^k \alpha_{i,j}^\ell + \sum_{i,j} b_{i,j}^k \beta_{i,j}^\ell + \sum_{i,j} c_{i,j}^k \gamma_{i,j}^\ell = 0. \quad (18)$$

Further for any $(x_{i,j}, y_{i,j}, z_{i,j}) \in \mathbf{k}^3, 1 \leq i, j \leq n$, if

$$\sum_{i,j} a_{i,j}^k x_{i,j} + \sum_{i,j} b_{i,j}^k y_{i,j} + \sum_{i,j} c_{i,j}^k z_{i,j} = 0 \ \text{ for all } 1 \leq k \leq m,$$

then $\sum_{i,j} x_{i,j} e_i^\vee \circ_I e_j^\vee + \sum_{i,j} y_{i,j} e_i^\vee \circ_{II} e_j^\vee + \sum_{i,j} z_{i,j} e_i^\vee \circ_{III} e_j^\vee$ is in $R^\perp$ and hence is of the form $\sum_{\ell=1}^{3n^2 - m} d_\ell g_\ell$ for some $d_\ell \in \mathbf{k}$. Thus

$$(x_{i,j}, y_{i,j}, z_{i,j}) = \sum_{\ell=1}^{3n^2 - m} d_\ell \left( \alpha_{i,j}^\ell, \beta_{i,j}^\ell, \gamma_{i,j}^\ell \right). \tag{19}$$

By definition, we have

$$\mathrm{Du}_x(e_i \circ_I e_j) = \left\{ \left( {}^{e_i}_{\dashv} \right) \circ_I \left( {}^{e_j}_{\dashv} \right) \right\}, \mathrm{Du}_x(e_i \circ_{II} e_j) = \left\{ \left( {}^{e_i}_{\vdash} \right) \circ_{II} \left( {}^{e_j}_{\vdash} \right) \right\},$$

$$\mathrm{Du}_x(e_i \circ_{III} e_j) = \left\{ \left( {}^{e_i}_{\dashv} \right) \circ_{III} \left( {}^{e_j}_{\vdash} \right) \right\},$$

$$\mathrm{Du}_y(e_i \circ_I e_j) = \left\{ \left( {}^{e_i}_{\dashv} \right) \circ_I \left( {}^{e_j}_{\vdash} \right) \right\}, \mathrm{Du}_y(e_i \circ_{II} e_j) = \left\{ \left( {}^{e_i}_{\dashv} \right) \circ_{II} \left( {}^{e_j}_{\dashv} \right) \right\},$$

$$\mathrm{Du}_y(e_i \circ_{III} e_j) = \left\{ \left( {}^{e_i}_{\vdash} \right) \circ_{III} \left( {}^{e_j}_{\vdash} \right) \right\}, \tag{20}$$

$$\mathrm{Du}_z(e_i \circ_I e_j) = \left\{ \left( {}^{e_i}_{\vdash} \right) \circ_I \left( {}^{e_j}_{\vdash} \right) \right\}, \mathrm{Du}_z(e_i \circ_{II} e_j) = \left\{ \left( {}^{e_i}_{\dashv} \right) \circ_{II} \left( {}^{e_j}_{\vdash} \right) \right\},$$

$$\mathrm{Du}_z(e_i \circ_{III} e_j) = \left\{ \left( {}^{e_i}_{\dashv} \right) \circ_{III} \left( {}^{e_j}_{\dashv} \right) \right\},$$

where $\left( {}^{e_i}_{\vdash} \right) \circ_u \left( {}^{e_j}_{\dashv} \right) := \left\{ \left( {}^{e_i}_{\vdash} \right) \circ_u \left( {}^{e_j}_{\vdash} \right), \left( {}^{e_i}_{\vdash} \right) \circ_u \left( {}^{e_j}_{\dashv} \right) \right\}, u \in \{\mathrm{I}, \mathrm{II}, \mathrm{III}\}$.

Let $\mathrm{BSu}(\mathcal{P}^!)$ be the bisuccessor of the dual operad $\mathcal{P}^!$ recalled in Definition 5.1. Then we also have

$$\mathrm{BSu}_x(e_i^\vee \circ_I e_j^\vee) = \left\{ \left( {}^{e_i^\vee}_{\prec} \right) \circ_I \left( {}^{e_j^\vee}_{\prec} \right) \right\}, \mathrm{BSu}_x(e_i^\vee \circ_{II} e_j^\vee) = \left\{ \left( {}^{e_i^\vee}_{\succ} \right) \circ_{II} \left( {}^{e_j^\vee}_{\star} \right) \right\},$$

$$\mathrm{BSu}_x(e_i^\vee \circ_{III} e_j^\vee) = \left\{ \left( {}^{e_i^\vee}_{\prec} \right) \circ_{III} \left( {}^{e_j^\vee}_{\succ} \right) \right\},$$

$$\mathrm{BSu}_y(e_i^\vee \circ_I e_j^\vee) = \left\{ \left( {}^{e_i^\vee}_{\prec} \right) \circ_I \left( {}^{e_j^\vee}_{\succ} \right) \right\}, \mathrm{BSu}_y(e_i^\vee \circ_{II} e_j^\vee) = \left\{ \left( {}^{e_i^\vee}_{\prec} \right) \circ_{II} \left( {}^{e_j^\vee}_{\prec} \right) \right\},$$

$$\mathrm{BSu}_y(e_i^\vee \circ_{III} e_j^\vee) = \left\{ \left( {}^{e_i^\vee}_{\succ} \right) \circ_{III} \left( {}^{e_j^\vee}_{\star} \right) \right\}, \tag{21}$$

$$\mathrm{BSu}_z(e_i^\vee \circ_I e_j^\vee) = \left\{ \left( {}^{e_i^\vee}_{\succ} \right) \circ_I \left( {}^{e_j^\vee}_{\star} \right) \right\}, \mathrm{BSu}_z(e_i^\vee \circ_{II} e_j^\vee) = \left\{ \left( {}^{e_i^\vee}_{\prec} \right) \circ_{II} \left( {}^{e_j^\vee}_{\succ} \right) \right\},$$

$$\mathrm{BSu}_z(e_i^\vee \circ_{III} e_j^\vee) = \left\{ \left( {}^{e_i^\vee}_{\prec} \right) \circ_{III} \left( {}^{e_j^\vee}_{\prec} \right) \right\},$$

where $\star = \prec + \succ$.

**Theorem 5.1.** *Let $\mathbf{k}$ be an infinite field. Let $\mathcal{P} = \mathcal{T}(V)/(R)$ be a binary quadratic operad. Then*

$$\mathrm{Du}(\mathcal{P})^! = \mathrm{BSu}(\mathcal{P}^!)$$

*if and only if $R \neq 0$.*

**Proof.** For the "if" part, let $\mathcal{P} = \mathcal{T}(V)/(R)$ be a binary quadratic operad with $R \neq 0$. Take $W = (R)$ in Proposition 5.1 and fix a **k**-basis $\{e_1, e_2, \cdots, e_n\}$ of $V$ as in the proposition. Let $f_k, 1 \leq k \leq m$, be the basis of $(R)$ as defined in Eq. (16).

By Eq. (20), we have

$$\mathrm{Du}_x(f_k) = \left\{ \sum_{i,j} a^k_{i,j} \left(\tfrac{e_i}{\cdot}\right) \circ_{\mathrm{I}} \left(\tfrac{e_j}{\cdot}\right) + \sum_{i,j} b^k_{i,j} \left(\tfrac{e_i}{\cdot}\right) \circ_{\mathrm{II}} \left(\tfrac{e_j}{\cdot}\right) + \sum_{i,j} c^k_{i,j} \left(\tfrac{e_i}{\cdot}\right) \circ_{\mathrm{III}} \left(\tfrac{e_j}{\cdot}\right) \right\},$$

$$\mathrm{Du}_y(f_k) = \left\{ \sum_{i,j} a^k_{i,j} \left(\tfrac{e_i}{\cdot}\right) \circ_{\mathrm{I}} \left(\tfrac{e_j}{\cdot}\right) + \sum_{i,j} b^k_{i,j} \left(\tfrac{e_i}{\cdot}\right) \circ_{\mathrm{II}} \left(\tfrac{e_j}{\cdot}\right) + \sum_{i,j} c^k_{i,j} \left(\tfrac{e_i}{\cdot}\right) \circ_{\mathrm{III}} \left(\tfrac{e_j}{\cdot}\right) \right\},$$

$$\mathrm{Du}_z(f_k) = \left\{ \sum_{i,j} a^k_{i,j} \left(\tfrac{e_i}{\cdot}\right) \circ_{\mathrm{I}} \left(\tfrac{e_j}{\cdot}\right) + \sum_{i,j} b^k_{i,j} \left(\tfrac{e_i}{\cdot}\right) \circ_{\mathrm{II}} \left(\tfrac{e_j}{\cdot}\right) + \sum_{i,j} c^k_{i,j} \left(\tfrac{e_i}{\cdot}\right) \circ_{\mathrm{III}} \left(\tfrac{e_j}{\cdot}\right) \right\}.$$

From Eq. (21), we similarly obtain

$$\mathrm{BSu}_x(g_\ell) = \left\{ \sum_{i,j} \alpha^\ell_{i,j} \left(\tfrac{e_i}{\vee}\right) \circ_{\mathrm{I}} \left(\tfrac{e_j}{\vee}\right) + \sum_{i,j} \beta^\ell_{i,j} \left(\tfrac{e_i}{\vee}\right) \circ_{\mathrm{II}} \left(\tfrac{e_j}{\vee}\right) + \sum_{i,j} \gamma^\ell_{i,j} \left(\tfrac{e_i}{\vee}\right) \circ_{\mathrm{III}} \left(\tfrac{e_j}{\vee}\right) \right\},$$

$$\mathrm{BSu}_y(g_\ell) = \left\{ \sum_{i,j} \alpha^\ell_{i,j} \left(\tfrac{e_i}{\vee}\right) \circ_{\mathrm{I}} \left(\tfrac{e_j}{\vee}\right) + \sum_{i,j} \beta^\ell_{i,j} \left(\tfrac{e_i}{\vee}\right) \circ_{\mathrm{II}} \left(\tfrac{e_j}{\vee}\right) + \sum_{i,j} \gamma^\ell_{i,j} \left(\tfrac{e_i}{\vee}\right) \circ_{\mathrm{III}} \left(\tfrac{e_j}{\vee}\right) \right\},$$

$$\mathrm{BSu}_z(g_\ell) = \left\{ \sum_{i,j} \alpha^\ell_{i,j} \left(\tfrac{e_i}{\vee}\right) \circ_{\mathrm{I}} \left(\tfrac{e_j}{\vee}\right) + \sum_{i,j} \beta^\ell_{i,j} \left(\tfrac{e_i}{\vee}\right) \circ_{\mathrm{II}} \left(\tfrac{e_j}{\vee}\right) + \sum_{i,j} \gamma^\ell_{i,j} \left(\tfrac{e_i}{\vee}\right) \circ_{\mathrm{III}} \left(\tfrac{e_j}{\vee}\right) \right\}.$$

By Lemma 5.2, we have

$$(\mathrm{Du}(R)) = \sum_{k=1}^{m} \mathbf{k}\mathrm{Du}(f_k) = \sum_{k} \left( \mathbf{k}\mathrm{Du}_x(f_k) + \mathbf{k}\mathrm{Du}_y(f_k) + \mathbf{k}\mathrm{Du}_z(f_k) \right),$$

$$\mathrm{BSu}(R^\perp) = \sum_{\ell=1}^{3n^2-m} \mathbf{k}\mathrm{BSu}(g_\ell) = \sum_{\ell=1}^{3n^2-m} \left( \mathbf{k}(\mathrm{BSu}_x(g_\ell) + \mathbf{k}\mathrm{BSu}_y(g_\ell) + \mathbf{k}\mathrm{BSu}_z(g_\ell)) \right).$$

To reach our conclusion, it suffices to show the equality $(\mathrm{Du}(R)^\perp) = (\mathrm{BSu}(R^\perp))$ of $\mathbb{S}$-modules under the condition $R \neq 0$. For all $1 \leq k \leq m$ and $1 \leq \ell \leq 3n^2 - m$, by Eq. (18), we have

$$\langle \mathrm{BSu}_p(g_\ell), \mathrm{Du}_q(f_k) \rangle = 0, \quad \text{where } p, q \in \{x, y, z\}.$$

Thus $\langle \mathrm{BSu}(g_\ell), \mathrm{Du}(f_k)\rangle = 0$ and hence $\mathrm{BSu}(R^\perp) \subset \mathrm{Du}(R)^\perp$, implying that $(\mathrm{BSu}(R^\perp)) \subseteq (\mathrm{Du}(R)^\perp)$. On the other hand, if

$$h = \sum_{i,j,u,v} x_{i,j,u,v}\left(\overset{\vee}{\underset{u}{e_i}}\right)\circ_{\mathrm{I}}\left(\overset{\vee}{\underset{v}{e_j}}\right) + \sum_{i,j,u,v} y_{i,j,u,v}\left(\overset{\vee}{\underset{u}{e_i}}\right)\circ_{\mathrm{II}}\left(\overset{\vee}{\underset{v}{e_j}}\right) + \sum_{i,j,u,v} z_{i,j,u,v}\left(\overset{\vee}{\underset{u}{e_i}}\right)\circ_{\mathrm{III}}\left(\overset{\vee}{\underset{v}{e_j}}\right)$$

is in $\mathrm{Du}(R)^\perp$, where $u, v \in \{\prec, \succ\}$. Then for all $1 \le k \le m$, we have

$$\langle h, \mathrm{Du}_x(f_k)\rangle = 0, \langle h, \mathrm{Du}_y(f_k)\rangle = 0, \langle h, \mathrm{Du}_z(f_k)\rangle = 0.$$

Since $R \ne 0$, by Proposition 5.1, for any fixed $i_0, j_0 \in \{1, 2, \cdots, n\}$, there exists $1 \le k_0 \le m$, such that $b^{k_0}_{i_0,j_0} \ne 0$. Then, for any $k$, by the definition of $\mathrm{Du}_x$ we see that the relations

$$F_1 := \sum_{i,j} a^k_{i,j}\left(\overset{e_i}{\underset{\prec}{}}\right)\circ_{\mathrm{I}}\left(\overset{e_j}{\underset{\prec}{}}\right) + b^k_{i_0,j_0}\left(\overset{e_i}{\underset{\prec}{}}\right)\circ_{\mathrm{II}}\left(\overset{e_j}{\underset{\vdash}{}}\right) +$$
$$\sum_{i\ne i_0,j\ne j_0} b^k_{i,j}\left(\overset{e_i}{\underset{\vdash}{}}\right)\circ_{\mathrm{II}}\left(\overset{e_j}{\underset{\prec}{}}\right) + \sum_{i,j} c^k_{i,j}\left(\overset{e_i}{\underset{\prec}{}}\right)\circ_{\mathrm{III}}\left(\overset{e_j}{\underset{\vdash}{}}\right),$$

$$F_2 := \sum_{i,j} a^k_{i,j}\left(\overset{e_i}{\underset{\prec}{}}\right)\circ_{\mathrm{I}}\left(\overset{e_j}{\underset{\prec}{}}\right) + b^k_{i_0,j_0}\left(\overset{e_i}{\underset{\vdash}{}}\right)\circ_{\mathrm{II}}\left(\overset{e_j}{\underset{\vdash}{}}\right) +$$
$$\sum_{i\ne i_0,j\ne j_0} b^k_{i,j}\left(\overset{e_i}{\underset{\vdash}{}}\right)\circ_{\mathrm{II}}\left(\overset{e_j}{\underset{\prec}{}}\right) + \sum_{i,j} c^k_{i,j}\left(\overset{e_i}{\underset{\prec}{}}\right)\circ_{\mathrm{III}}\left(\overset{e_j}{\underset{\vdash}{}}\right)$$

are in $\mathrm{Du}_x(f_k)$. Thus, for $1 \le k \le m$, we obtain

$$\sum_{i,j} a^k_{i,j} x_{i,j,\prec,\prec} + b^k_{i_0,j_0} y_{i_0,j_0,\succ,\prec} + \sum_{i\ne i_0,j\ne j_0} b^k_{i,j} y_{i,j,\succ,\prec}$$
$$+ \sum_{i,j} c^k_{i,j} z_{i,j,\prec,\succ} = \langle h, F_1\rangle = 0,$$

$$\sum_{i,j} a^k_{i,j} x_{i,j,\prec,\prec} + b^k_{i_0,j_0} y_{i_0,j_0,\succ,\succ} + \sum_{i\ne i_0,j\ne j_0} b^k_{i,j} y_{i,j,\succ,\prec}$$
$$+ \sum_{i,j} c^k_{i,j} z_{i,j,\prec,\succ} = \langle h, F_2\rangle = 0.$$

Comparing the two equations and applying $b^{k_0}_{i_0,j_0} \ne 0$, we obtain $y_{i_0,j_0,\succ,\succ} = y_{i_0,j_0,\succ,\prec}$ for all $1 \le i_0, j_0 \le n$. From the second equation and Eq. (19), we also have

$$(x_{i,j,\prec,\prec}, y_{i,j,\succ,\prec}, z_{i,j,\prec,\succ}) = \sum_{\ell=1}^{3n^2-m} d_\ell(\alpha^\ell_{i,j}, \beta^\ell_{i,j}, \gamma^\ell_{i,j}),$$

for some $d_\ell \in \mathbf{k}$. Thus we obtain

$$h_x := \sum_{i,j} x_{i,j,\prec,\prec}\left(\overset{\vee}{\underset{\prec}{e_i}}\right)\circ_{\mathrm{I}}\left(\overset{\vee}{\underset{\prec}{e_j}}\right) + \sum_{i,j} y_{i,j,\succ,\prec}\left(\overset{\vee}{\underset{\succ}{e_i}}\right)\circ_{\mathrm{II}}\left(\overset{\vee}{\underset{v}{e_j}}\right)$$

$$+ \sum_{i,j} y_{i,j,\succ,\succ} \left( {e_i^\vee \atop \succ} \right) \circ_{\mathrm{II}} \left( {e_j^\vee \atop v} \right) + \sum_{i,j} z_{i,j} \left( {e_i^\vee \atop \prec} \right) \circ_{\mathrm{III}} \left( {e_j^\vee \atop \succ} \right)$$

$$= \sum_{i,j} x_{i,j,\prec,\prec} \left( {e_i^\vee \atop \prec} \right) \circ_{\mathrm{I}} \left( {e_j^\vee \atop \prec} \right) + \sum_{i,j} y_{i,j,\succ,\prec} \left( \left( {e_i^\vee \atop \succ} \right) \circ_{\mathrm{II}} \left( {e_j^\vee \atop v} \right) + \left( {e_i^\vee \atop \succ} \right) \circ_{\mathrm{II}} \left( {e_j^\vee \atop v} \right) \right)$$

$$+ \sum_{i,j} z_{i,j} \left( {e_i^\vee \atop \prec} \right) \circ_{\mathrm{III}} \left( {e_j^\vee \atop \succ} \right)$$

$$= \sum_{\ell=1}^{3n^2-m} d_\ell \left( \sum_{i,j} \alpha_{i,j}^\ell \left( {e_i^\vee \atop \prec} \right) \circ_{\mathrm{I}} \left( {e_j^\vee \atop \prec} \right) + \sum_{i,j} \beta_{i,j}^\ell \left( \left( {e_i^\vee \atop \succ} \right) \circ_{\mathrm{II}} \left( {e_j^\vee \atop v} \right) + \left( {e_i^\vee \atop \succ} \right) \circ_{\mathrm{II}} \left( {e_j^\vee \atop \succ} \right) \right) \right.$$

$$\left. + \sum_{i,j} \gamma_{i,j}^\ell \left( {e_i^\vee \atop \prec} \right) \circ_{\mathrm{III}} \left( {e_j^\vee \atop \succ} \right) \right).$$

This is in $\sum_{\ell=1}^{3n^2-m} \mathbf{kBSu}_x(g_\ell)$. By the same argument, we find that

$$h_y := \sum_{i,j} x_{i,j,\prec,\succ} \left( {e_i^\vee \atop \prec} \right) \circ_{\mathrm{I}} \left( {e_j^\vee \atop \prec} \right) + \sum_{i,j} y_{i,j,\prec,\prec} \left( {e_i^\vee \atop \succ} \right) \circ_{\mathrm{II}} \left( {e_j^\vee \atop \prec} \right) +$$

$$\sum_{i,j,v} z_{i,j,\succ,\prec} \left( {e_i^\vee \atop \prec} \right) \circ_{\mathrm{III}} \left( {e_j^\vee \atop \prec} \right) + \sum_{i,j,\succ,\succ} z_{i,j,\succ,\succ} \left( {e_i^\vee \atop \succ} \right) \circ_{\mathrm{III}} \left( {e_j^\vee \atop \succ} \right)$$

is in $\sum_{\ell=1}^{3n^2-m} \mathbf{kBSu}_y(g_\ell)$ and

$$h_z := \sum_{i,j,\succ,\prec} x_{i,j,\succ,v} \left( {e_i^\vee \atop \succ} \right) \circ_{\mathrm{I}} \left( {e_j^\vee \atop \succ} \right) + \sum_{i,j,\succ,\succ} x_{i,j,\succ,\succ} \left( {e_i^\vee \atop \succ} \right) \circ_{\mathrm{I}} \left( {e_j^\vee \atop \succ} \right) +$$

$$\sum_{i,j} y_{i,j,\prec,\succ} \left( {e_i^\vee \atop \prec} \right) \circ_{\mathrm{II}} \left( {e_j^\vee \atop \succ} \right) + \sum_{i,j} z_{i,j,\prec,\prec} \left( {e_i^\vee \atop \prec} \right) \circ_{\mathrm{III}} \left( {e_j^\vee \atop \prec} \right)$$

is in $\sum_{\ell=1}^{3n^2-m} \mathbf{kBSu}_z(g_\ell)$. Note that $h = h_x + h_y + h_z$. Thus in summary, we find that $h$ is in

$$\sum_\ell \mathbf{kBSu}_x(g_\ell) + \sum_\ell \mathbf{kBSu}_y(g_\ell) + \sum_\ell \mathbf{kBSu}_z(g_\ell)$$

and hence is in the $\mathbb{S}$-module generated by $\mathrm{BSu}(R^\perp)$. Thus we have the equality $(\mathrm{Du}(R)^\perp) = (\mathrm{BSu}(R^\perp))$ of $\mathbb{S}$-modules. Therefore

$$\mathrm{Du}(\mathcal{P})^! = \mathrm{BSu}(\mathcal{P}^!) \text{ and } \mathrm{Du}(\mathcal{P}) = \mathrm{BSu}(\mathcal{P}^!)^!.$$

To prove the "only if" part, suppose $R = 0$. Then we have $\mathrm{Du}(R) = 0 \subseteq \mathcal{T}(\mathrm{Du}(V))$ and hence $\mathrm{Du}(R)^\perp = \mathcal{T}(\mathrm{BSu}(V^\vee))(3)$. On the other hand, $R^\perp = \mathcal{T}(V^\vee)(3)$ which has a basis $e_i^\vee \circ_u e_j^\vee, 1 \le i, j \le n, u \in \{\mathrm{I}, \mathrm{II}, \mathrm{III}\}$.

Then a linear spanning set of $\mathrm{BSu}(\mathcal{T}(V^\vee)(3))$ is given by $\mathrm{BSu}_v(e_i^\vee \circ_u e_j^\vee), 1 \leq i, j \leq n, v \in \{x, y, z\}, u \in \{\mathrm{I}, \mathrm{II}, \mathrm{III}\}$ in Eq. (21). Thus the dimension of $\mathrm{BSu}(\mathcal{T}(V^\vee)(3))$ is at most $9n^2$, while the dimension of $\mathcal{T}(\mathrm{BSu}(V^\vee))(3)$ is

$$3 \dim(\mathrm{BSu}(V^\vee)^{\otimes 2}) = 3(2n)^2 = 12n^2.$$

Hence $\mathrm{BSu}(\mathcal{T}(V^\vee)(3))$ is a proper subspace of $\mathcal{T}(\mathrm{BSu}(V^\vee))(3)$ and thus $\mathrm{Du}(\mathcal{P})^! \neq \mathrm{BSu}(\mathcal{P}^!)$. $\square$

**Theorem 5.2.** *Let* $\mathbf{k}$ *be an infinite field. Let* $\mathcal{P} = \mathcal{T}(V)/(R)$ *be a binary quadratic operad. Then*

$$\mathrm{Tri}(\mathcal{P})^! = \mathrm{TSu}(\mathcal{P}^!)$$

*if and only if* $R \neq 0$.

**Proof.** The proof is similar to Theorem 5.1. $\square$

Taking $\mathcal{P}$ to be the operad of the associative algebra in Theorem 5.2, we get the result of Loday and Ronco [30 (Theorem 3.1)] that the triassociative algebra and the tridendriform algebra are in Koszul dual to each other.

More generally, Theorem 5.1 and Theorem 5.2 make it straightforward to compute the generating and relation spaces of the Koszul duals of the operads of some existing algebras. We give the following examples as illustrations.

(1) The operad $DualCTD$ [39] is defined to be the Koszul dual of the operad $CTD$ of the commutative tridendriform algebra. Since the latter operad is $\mathrm{TSu}(Comm)^3$, we have

$$CTD^! = \mathrm{TSu}(Comm)^! = \mathrm{Tri}(Comm^!) = \mathrm{Tri}(Lie),$$

which is precise $TriLeib$, the operad of the tri-Leibniz algebra in Proposition 3.6. Thus we easily obtain the relations of $DualCTD$, see Eq. (10)

(2) The operad of the commutative quadri-algebra is the Kozul dual of $\mathrm{BSu}(Zinb)$ and hence is $\mathrm{Du}(Leib)$. Thus its relations can be easily computed.

(3) The Kozul dual of $\mathrm{BSu}(PreLie)$, the operad of the L-dendriform algebra, is $\mathrm{Du}(Perm)$ and hence can be easily computed.

(4) The operad $L\text{-}quad$ [4] of the L-quadri-algebra is shown to be $\mathrm{BSu}(L\text{-}dend) = \mathrm{BSu}(\mathrm{BSu}(Lie))$ in Ref. 3. Thus the dual of $L\text{-}quad$ is $\mathrm{Du}(\mathrm{Du}(Perm))$ and can be easily computed.

## 5.2. *Replicators and Manin white products*

As a preparation for later discussions, we recall concepts and notations on Manin white product, most following[38].

Ginzburg and Kapranov defined in Ref. 13 a morphism of operads $\Phi : \mathcal{T}(V \otimes W) \rightarrowtail \mathcal{T}(V) \otimes \mathcal{T}(W)$. Let $\mathcal{P} = \mathcal{T}(V)/(R)$ and $\mathcal{Q} = \mathcal{T}(W)/(S)$ be two binary quadratic operads with finite-dimensional generating spaces. Consider the composition of morphisms of operads

$$\mathcal{T}(V \otimes W) \xrightarrow{\Phi} \mathcal{T}(V) \otimes \mathcal{T}(W) \xrightarrow{\pi_{\mathcal{P}} \otimes \pi_{\mathcal{Q}}} \mathcal{P} \otimes \mathcal{Q},$$

where $\pi_{\mathcal{P}} : \mathcal{T}(V) \to \mathcal{P}$ and $\pi_{\mathcal{Q}} : \mathcal{T}(W) \to \mathcal{Q}$ are the natural projections. Its kernel is $(\Phi^{-1}(R \otimes \mathcal{T}(W) + \mathcal{T}(V) \otimes S))$, the ideal generated by $\Phi^{-1}(R \otimes \mathcal{T}(W) + \mathcal{T}(V) \otimes S)$.

**Definition 5.2.** [13,38] Let $\mathcal{P} = \mathcal{T}(V)/(R)$ and $\mathcal{Q} = \mathcal{T}(W)/(S)$ be two binary quadratic operads with finite-dimensional generating spaces. The **Manin white product** of $\mathcal{P}$ and $\mathcal{Q}$ is defined by

$$\mathcal{P} \bigcirc \mathcal{Q} := \mathcal{T}(V \otimes W)/(\Phi^{-1}(R \otimes \mathcal{T}(W) + \mathcal{T}(V) \otimes S)).$$

In general, it is difficult to compute the white Manin product explicitly when the operads are given in terms of generators and relations. Theorem 5.3 provides a convenient way to compute the white Manin product of a binary quadratic operad with the operad *Perm* or *ComTria* by relating them to the duplicator and triplicator.

**Theorem 5.3.** *Let $\mathcal{P} = \mathcal{T}(V)/(R)$ be a binary quadratic operad with $R \neq 0$. We have the isomorphism of operads*

$$\mathrm{Du}(\mathcal{P}) \cong Perm \bigcirc \mathcal{P}, \quad \mathrm{Tri}(\mathcal{P}) \cong ComTria \bigcirc \mathcal{P}.$$

**Proof.** By Ref. 3, we have the isomorphisms of operads

$$\mathrm{BSu}(\mathcal{P}^!) \cong PreLie \bullet \mathcal{P}^!, \quad \mathrm{TSu}(\mathcal{P}^!) \cong PostLie \bullet \mathcal{P}^!.$$

Since $PreLie^! \cong Perm$, $PostLie^! = ComTria$ and $(\mathcal{P} \bullet \mathcal{Q})^! \cong \mathcal{P}^! \bigcirc \mathcal{Q}^!$, we obtain

$$\mathrm{Du}(\mathcal{P}) \cong (\mathrm{BSu}(\mathcal{P}^!))^! \cong (PreLie \bullet \mathcal{P}^!)^! \cong Perm \bigcirc \mathcal{P}.$$

Similarly $\mathrm{Tri}(\mathcal{P}) \cong ComTria \bigcirc \mathcal{P}$. □

By taking replicators of suitable operads $\mathcal{P}$, we immediately get

**Corollary 5.1.**

(1) *(Ref. 38) Perm $\bigcirc$ Lie $=$ Leib and Perm $\bigcirc$ Ass $=$ Dias.*
(2) *(Ref. 37) Perm $\bigcirc$ Pois $=$ DualPrePois.*
(3) *ComTria $\bigcirc$ Ass $=$ Triass.*

By an argument similar to the one for Theorem 5.3 we obtain

**Proposition 5.2.** *Let $\mathcal{P} = \mathcal{T}_{ns}(V)/(R)$ be a binary quadratic nonsymmetric operad with $R \neq 0$. There is an isomorphism of nonsymmetric operads*

$$\mathrm{Du}(\mathcal{P}) \cong Dias \,\square\, \mathcal{P} \;, \quad \mathrm{Tri}(\mathcal{P}) \cong Triass \,\square\, \mathcal{P} \;.$$

*where $\square$ denotes the white square product[38].*

## 6. Replicators and Average operators on operads

In this section we establish the relationship between the duplicator and triplicator of an operad on one hand and the actions of the di-average and tri-average operators on the operad on the other hand. We will work with symmetric operads, but all the results also hold for nonsymmetric operads.

### 6.1. *Duplicators and di-average operators*

Average operators have been studied for associative algebras by Rota since 1960 and for other algebraic structures more recently[1,5,36,37].

**Definition 6.1.** Let $(A, \cdot)$ be a **k**-module $A$ with a binary operation $\cdot$.

(1) A **di-average operator** on $A$ is a **k**-linear map $P : A \longrightarrow A$ such that

$$P(x \cdot P(y)) = P(x) \cdot P(y) = P(P(x) \cdot y), \quad \text{for all } x, y \in A. \quad (22)$$

(2) Let $\lambda \in \mathbf{k}$. A tri-average operator of weight $\lambda$ on $A$ is a **k**-linear map $P : A \longrightarrow A$ such that Eq. (22) holds and

$$P(x) \cdot P(y) = \lambda P(xy), \quad \text{for all } x, y \in A. \quad (23)$$

We note that a tri-average operator of weight zero is not a di-average operator. So we cannot give a uniform definition of the average operators as in the case of Rota-Baxter algebras of weight $\lambda$.

We next consider the operation of average operators at the level of operads.

**Definition 6.2.** Let $V = V(2)$ be an $\mathbb{S}$-module concentrated in arity 2.

(1) Let $V_P$ denote the $\mathbb{S}$-module concentrated in arity 1 and arity 2 with $V_P(2) = V$ and $V_P(1) = \mathbf{k}\,P$, where $P$ is a symbol. Let $\mathcal{T}(V_P)$ be the free operad generated by binary operations $V$ and an unary operation $P \neq \mathrm{id}$.

(2) Define $\mathrm{Du}(V) = V \otimes (\mathbf{k} \dashv \oplus \mathbf{k} \vdash)$ as in Eq. (1), regarded as an $\mathbb{S}$-module concentrated in arity 2. Define a linear map of graded vector spaces from $\mathrm{Du}(V)$ to $V_P$ by the following correspondence:

$$\xi: \quad \begin{pmatrix} \omega \\ \dashv \end{pmatrix} \mapsto \omega \circ (\mathrm{id} \otimes P), \quad \begin{pmatrix} \omega \\ \vdash \end{pmatrix} \mapsto \omega \circ (P \otimes \mathrm{id}) \text{ for all } \omega \in V,$$

where $\circ$ is the operadic composition. By the universality of the free operad, $\xi$ induces a homomorphism of operads that we still denote by $\xi$:

$$\xi : \mathcal{T}(\mathrm{Du}(V)) \to \mathcal{T}(V_P).$$

(3) Let $\mathcal{P} = \mathcal{T}(V)/(R_{\mathcal{P}})$ be a binary operad defined by generating operations $V$ and relations $R_{\mathcal{P}}$. Let

$$\mathrm{DA}_{\mathcal{P}} := \{ \omega \circ (P \otimes P) - P \circ \omega \circ (P \otimes \mathrm{id}),$$
$$\omega \circ (P \otimes P) - P \circ \omega \circ (\mathrm{id} \otimes P) \mid \omega \in V \}.$$

Define the **operad of di-average $\mathcal{P}$-algebras** by

$$\mathrm{DA}(\mathcal{P}) := \mathcal{T}(V_P)/(R_{\mathcal{P}}, \mathrm{DA}_{\mathcal{P}}).$$

Let $p_1 : \mathcal{T}(V_P) \to \mathrm{DA}(\mathcal{P})$ denote the operadic projection.

**Theorem 6.1.**

*(1) Let $\mathcal{P}$ be a binary operad. There is a morphism of operads*

$$\mathrm{Du}(\mathcal{P}) \longrightarrow \mathrm{DA}(\mathcal{P}),$$

*which extends the map $\xi$ given in Definition 6.2.*

*(2) Let $A$ be a $\mathcal{P}$-algebra. Let $P : A \to A$ be a di-average operator. Then the following operations make $A$ into a $\mathrm{Du}(\mathcal{P})$-algebra:*

$$x \dashv_j y := x \circ_j P(y), \quad x \vdash_j y := P(x) \circ_j y, \quad \forall \circ_j \in \mathcal{P}(2),$$

*for all $x, y \in A$.*

The proof is parallel to the case of triplicators in Theorem 6.2 for which we will prove in full detail.

When we take $\mathcal{P}$ to be the operad of the associative algebra, Lie algebra or Poisson algebra, we obtain the following results of Aguiar[1].

**Corollary 6.1.**

(1) *Let $(A, \cdot)$ be an associative algebra and $P : A \longrightarrow A$ be a di-averaging operator. Define two new operations on $A$ by $x \vdash y = P(x) \cdot y$ and $x \dashv y = x \cdot P(y)$. Then $(A, \vdash, \dashv)$ is an associative dialgebra.*

(2) *Let $(A, [,])$ be a Lie algebra and $P : A \longrightarrow A$ be a di-averaging operator. Define a new operation on $A$ by $\{x, y\} = [P(x), y]$. Then $(A, \{, \})$ is a left Leibniz algebra.*

(3) *Let $(A, \cdot, [,])$ be a Poisson algebra and let $P : A \rightarrow A$ be a di-averaging operator. Define two new products on $A$ by $x \circ y := P(x) \cdot y$, and $\{x, y\} := [P(x), y]$. Then $(A, \circ, \{, \})$ is a dual left prePoisson algebra.*

Combining Theorem 6.1 with Theorem 5.3, we obtain the following relation between the Manin white product and the action of the di-average operator. It can be regarded as the interpretation of Ref. 37 (Theorem 3.2) at the level of operads.

**Proposition 6.1.** *For any binary quadratic operad $\mathcal{P} = \mathcal{T}(V)/(R)$, there is a morphism of operads*

$$Perm \bigcirc \mathcal{P} \longrightarrow DA(\mathcal{P}),$$

*defined by the following map*

$$Perm(2) \bigcirc \mathcal{P}(2) \longrightarrow DA(\mathcal{P}),$$
$$\mu \otimes \omega \longmapsto \omega \circ (id \otimes P),$$
$$\mu' \otimes \omega \longmapsto \omega \circ (P \otimes id), \quad \omega \in \mathcal{P}(2),$$

*where $\mu$ denotes the generating operation of the operad $Perm$.*

### 6.2. *Triplicators and tri-average operators*

In this section, we establish the relationship between the triplicator of an operad and the action of the tri-average operator with a non-zero weight on the operad. For simplicity, we assume that the weight of the tri-average operator is one.

**Definition 6.3.** Let $V = V(2), V_P$ and $\mathcal{T}(V_P)$ as defined in Definition 6.2.

(1) Let $\mathrm{Tri}(V) = V \otimes (\mathbf{k} \dashv \oplus \mathbf{k} \vdash \oplus \mathbf{k} \perp)$ in Eq. (4), viewed as an $\mathbb{S}$-module concentrated in arity 2. Define a linear map of graded vector spaces from $\mathrm{Tri}(V)$ to $V_P$ by the correspondence

$$\eta: \quad \left(\begin{smallmatrix} \omega \\ \dashv \end{smallmatrix}\right) \mapsto \omega \circ (\mathrm{id} \otimes P), \quad \left(\begin{smallmatrix} \omega \\ \vdash \end{smallmatrix}\right) \mapsto \omega \circ (P \otimes \mathrm{id}), \quad \left(\begin{smallmatrix} \omega \\ \perp \end{smallmatrix}\right) \mapsto \omega,$$

where $\circ$ is the operadic composition. By the universality of the free operad, $\eta$ induces a homomorphism of operads:

$$\eta: \mathcal{T}(\mathrm{Tri}(V)) \to \mathcal{T}(V_P).$$

(2) Let $\mathcal{P} = \mathcal{T}(V)/(R_\mathcal{P})$ be a binary operad defined by generating operations $V$ and relations $R_\mathcal{P}$. Let

$$\begin{aligned} \mathrm{TA}_\mathcal{P} := \{ &\omega \circ (P \otimes P) - P \circ \omega \circ (P \otimes \mathrm{id}), \\ &\omega \circ (P \otimes P) - P \circ \omega \circ (\mathrm{id} \otimes P), \\ &\omega \circ (P \otimes P) - P \circ \omega \mid \omega \in V \}. \end{aligned}$$

Define the **operad of tri-average $\mathcal{P}$-algebras of weight one** by

$$\mathrm{TA}(\mathcal{P}) := \mathcal{T}(V_P)/(R_\mathcal{P}, \mathrm{TA}_\mathcal{P}).$$

Let $p_1 : \mathcal{T}(V_P) \to \mathrm{TA}(\mathcal{P})$ denote the operadic projection.

We first prove a lemma relating triplicators and tri-average operators.

**Lemma 6.1.** *Let $\mathcal{P} = \mathcal{T}(V)/(R_\mathcal{P})$ be a binary operad and let $\tau \in \mathscr{T}(V)$ with $\mathrm{Lin}(\tau) = n$.*

*(1) For each $\bar{\tau} \in \mathrm{Tri}(\tau)$, we have*

$$P \circ \eta(\bar{\tau}) \equiv \tau \circ P^{\otimes n} \quad \mathrm{mod} \ (R_\mathcal{P}, \mathrm{TA}_\mathcal{P}). \tag{24}$$

*(2) For $\varnothing \neq J \subseteq \mathrm{Lin}(\tau)$, let $P^{\otimes n, J}$ denote the n-th tensor power of $P$ but with the component from $J$ replaced by the identity map. So, for example, for the two inputs $x_1$ and $x_2$ of $P^{\otimes 2}$, we have $P^{\otimes 2, \{x_1\}} = P \otimes \mathrm{id}$ and $P^{\otimes 2, \{x_1, x_2\}} = \mathrm{id} \otimes \mathrm{id}$. Then for each $\bar{\tau}_J \in \mathrm{Tri}_J(\tau)$, we have*

$$\eta(\bar{\tau}_J) \equiv \tau \circ (P^{\otimes n, J}) \quad \mathrm{mod} \ (R_\mathcal{P}, \mathrm{TA}_\mathcal{P}). \tag{25}$$

**Proof.** (1). We prove the statement by induction on $|Lin(\tau)| \geq 1$. When $|Lin(\tau)| = 1$, $\tau$ is the tree with one leaf standing for the identity map. Then we have $\eta(\,\mathrm{Tri}(\tau)\,) = \tau$, $P \circ \eta(\,\mathrm{Tri}(\tau)\,) = P = \tau \circ P$. Assume that the statement has been proved for $\tau$ with $|Lin(\tau)| = k$ and consider a $\tau$ with $|Lin(\tau)| = k + 1$. Then from the decomposition $\tau = \tau_\ell \vee_w \tau_r$, we have $\mathrm{Tri}(\tau) = \mathrm{Tri}(\tau_\ell) \vee_{\binom{w}{\dagger}} \mathrm{Tri}(\tau_r)$. Recall that $\mathrm{Tri}(\tau)$ is a set of labeled trees. For each $\bar{\tau} \in \mathrm{Tri}(\tau)$, there exist $\bar{\tau}_\ell \in \mathrm{Tri}(\tau)_\ell$ and $\bar{\tau}_r \in \mathrm{Tri}(\tau_r)$ such that

$$\bar{\tau} \in \left\{ \bar{\tau}_\ell \vee_{\binom{w}{\vdash}} \bar{\tau}_r, \bar{\tau}_\ell \vee_{\binom{w}{\dashv}} \bar{\tau}_r, \bar{\tau}_\ell \vee_{\binom{w}{\cdot}} \bar{\tau}_r \right\}.$$

If $\bar{\tau} = \bar{\tau}_\ell \vee_{\binom{w}{\vdash}} \bar{\tau}_r$, then we have

$$
\begin{aligned}
P \circ \eta(\bar{\tau}) &= P \circ \eta(\bar{\tau}_\ell \vee_{\binom{w}{\vdash}} \bar{\tau}_r) \\
&= P \circ w \circ ((P \circ \eta(\bar{\tau}_\ell)) \otimes \eta(\bar{\tau}_r)) \\
&\equiv w \circ ((P \circ \eta(\bar{\tau}_\ell)) \otimes (P \circ \eta(\bar{\tau}_r))) \quad \mathrm{mod}\ (\mathrm{R}_\mathcal{P}, \mathrm{TA}_\mathcal{P}) \\
&= w \circ ((\bar{\tau}_\ell \circ P^{\otimes |Lin(\tau_\ell)|}) \otimes (\bar{\tau}_r \circ P^{\otimes |Lin(\tau_r)|})) \quad \text{(by induction} \\
&= w \circ (\bar{\tau}_\ell \otimes \bar{\tau}_r) \circ P^{\otimes(k+1)} \qquad\qquad\qquad \text{hypothesis)} \\
&= (\bar{\tau}_\ell \vee_{\binom{w}{\vdash}} \bar{\tau}_r) \circ P^{\otimes(k+1)} \\
&= \bar{\tau} \circ P^{\otimes(k+1)}.
\end{aligned}
$$

Similarly, we have

$$P \circ \eta(\,\bar{\tau}_\ell \vee_{\binom{w}{\dashv}} \bar{\tau}_r\,) \equiv \bar{\tau} \circ P^{\otimes(k+1)} \quad \mathrm{mod}\ (\mathrm{R}_\mathcal{P}, \mathrm{TA}_\mathcal{P}),$$

$$P \circ \eta(\,\bar{\tau}_\ell \vee_{\binom{w}{\cdot}} \bar{\tau}_r\,) \equiv \bar{\tau} \circ P^{\otimes(k+1)} \quad \mathrm{mod}\ (\mathrm{R}_\mathcal{P}, \mathrm{TA}_\mathcal{P}).$$

(2). We again prove the statement by induction on $|Lin(\tau)|$. When $|Lin(\tau)| = 1$, then $x$ is the only leaf label of $\tau$ and $|\mathrm{Tri}_x(\tau)| = 1$. Thus we have

$$\eta(\bar{\tau}_x) = \eta(x) = x = \tau \circ (P^{\otimes 1, x}).$$

Assume that the statement has been proved for all $\tau$ with $|Lin(\tau)| = k$ and consider $\tau$ with $|Lin(\tau)| = k + 1$. Write $\tau = \tau_\ell \vee_w \tau_r$. Let $J$ be a non-empty subset of $Lin(\tau)$. If $J \subseteq Lin(\tau_\ell)$, then by the definition of $\mathrm{Tri}_J(\tau)$, for each $\bar{\tau}_J \in \mathrm{Tri}_J(\tau)$, there exist $\bar{\tau}_{J,\ell} \in \mathrm{Tri}_J \tau_\ell$ and $\bar{\tau}_{J,r} \in \mathrm{Tri}_\varnothing \tau_r$ such that $\bar{\tau}_J = \bar{\tau}_{J,\ell} \vee_{\binom{w}{\dashv}} \bar{\tau}_{J,r}$. Then we have

$$\eta(\bar{\tau}_J) = \eta(\bar{\tau}_{J,\ell} \vee_{\binom{w}{\dashv}} \bar{\tau}_{J,r}) = w \circ (\eta(\bar{\tau}_{J,\ell} \otimes P \circ \eta(\bar{\tau}_{J,r}))$$

$$\equiv \omega \circ \left( (\tau_\ell \circ P^{\otimes |Lin(\tau_\ell)|, J}) \otimes (\tau_r \circ P^{\otimes |Lin(\tau_r)|}) \right) \quad \mathrm{mod} \ (\mathrm{R}_\mathcal{P}, \mathrm{TA}_\mathcal{P}),$$

$$\text{(by induction hypothesis and Item (1))}$$

$$= \tau \circ P^{\otimes (k+1), J}.$$

When $J \subseteq Lin(\tau_r)$, the proof is the same. When $J \not\subseteq Lin(\tau_\ell)$ and $J \not\subseteq Lin(\tau_r)$, for each $\bar{\tau}_J \in \mathrm{Tri}_J(\tau)$, there exist $\bar{\tau}_{J,\ell} \in \mathrm{Tri}_{J \cap Lin(\tau_\ell)} \tau_\ell$ and $\bar{\tau}_{J,r} \in \mathrm{Tri}_{J \cap Lin(\tau_r)} \tau_r$ such that $\bar{\tau}_J = \bar{\tau}_{J,\ell} \vee_{\left(\overset{\omega}{\cdot}\right)} \bar{\tau}_{J,r}$. Then by the same argument we have

$$\eta(\bar{\tau}_J) \equiv \omega \circ \left( (\tau_\ell \circ P^{\otimes |Lin(\tau_\ell)|, J \cap Lin(\tau_\ell)}) \otimes (\tau_r \circ P^{\otimes |Lin(\tau_r)|, J \cap Lin(\tau_r)}) \right)$$

$$\mathrm{mod} \ (\mathrm{R}_\mathcal{P}, \mathrm{TA}_\mathcal{P}),$$

$$= \tau \circ P^{\otimes (k+1), J}.$$

This completes the induction.

$\square$

**Theorem 6.2.** *Let* $\mathcal{P}$ *be a binary operad.*

(1) *There is a morphism of operads*

$$\mathrm{Tri}(\mathcal{P}) \longrightarrow \mathrm{TA}(\mathcal{P}),$$

*which extends the map* $\eta$ *given in Definition 6.3.*

(2) *Let* $A$ *be a* $\mathcal{P}$-*algebra. Let* $P : A \longrightarrow A$ *be a tri-average operator of weight one. Then the following operations make* $A$ *into a* $\mathrm{Tri}(\mathcal{P})$-*algebra:*

$$x \dashv_j y = x \circ_j P(y), \quad x \vdash_j y = P(x) \circ_j y,$$

$$x \cdot_j y = x \circ_j y, \quad \text{for all } \circ_j \in \mathcal{P}(2).$$

**Proof.** The second statement is just the interpretation of the first statement on the level of algebras. So we just need to prove the first statement. Let $R_{\mathrm{Tri}(\mathcal{P})}$ be the relation space of $\mathrm{Tri}(\mathcal{P})$. By definition, the relations of $\mathrm{Tri}(\mathcal{P})$ are generated by $\mathrm{Tri}_J(r)$ for locally homogeneous $r = \sum_i c_i \tau_i \in R_\mathcal{P}$, where $\varnothing \neq J \subseteq Lin(\tau_i)$. By Eqs. (24) and (25), we have

$$\eta \left( \sum_i c_i (\bar{\tau}_i)_J \right) = \sum_i c_i \eta((\bar{\tau}_i)_J) \equiv \sum_i c_i \tau_i \circ P^{\otimes n, J}$$

$$\equiv \left( \sum_i c_i \tau_i \right) \circ P^{\otimes n, J} \quad \mathrm{mod} \ (\mathrm{R}_\mathcal{P}, \mathrm{TA}_\mathcal{P}).$$

Hence $\eta(R_{\mathrm{Tri}(\mathcal{P})}) \subseteq (\mathrm{R}_{\mathcal{P}}, \mathrm{T}A_{\mathcal{P}})$ and $\eta$ induces a morphism of operads

$$\bar{\eta} : \mathrm{Tri}(\mathcal{P}) \longrightarrow \mathrm{T}A(\mathcal{P}).$$

This proves the first statement. $\qquad\square$

**Corollary 6.2.**

*(1) Let $A$ be an associative algebra and let $P : A \longrightarrow A$ be a tri-average operator on $A$. Then the new operations defined in Theorem 6.2 (2) makes it into an associative trialgebra.*

*(2) Let $L$ be a Lie algebra and let $P : L \longrightarrow L$ be a tri-average operator on $L$, then the operations defined in Theorem 6.2 (2) make it into a Leibniz trialgebra.*

**Acknowledgments**

This work is supported by NNSF of China (Grant Nos. 11271202, 11221091, 11771190, 11425104 and 11501466) and SRFDP (20120031110022). The authors thank P. Kolesnikov for informing them the related papers[14,15,19].

**References**

1. M. Aguiar, Pre-Poisson algebras, *Lett. Math. Phys.* **54** (2000) 263-277.
2. M. Aguiar and J.-L. Loday, Quadri-algebras, *J. Pure Appl. Algebra* **191** (2004) 205-221.
3. C. Bai, O. Bellier, L. Guo and X. Ni, Spliting of operations, Manin products and Rota-Baxter operators, *IMRN* **2013** (2013), 485-524.
4. C. Bai, L. Liu and X. Ni, Some results on L-dendriform algebras, *J. Geom. Phys.* **60** (2010) 940-950.
5. W. Cao, An algebraic study of averaging operators, Ph.D. Thesis, Rutgers University at Newark, 2000.
6. F. Chapoton, Un endofoncteur de la catégorie des opérades, in: Dialgebras and related operads, *Lect. Notes Math.* **1763**, Springer-Verlag, 2001, 105-110.
7. K. Ebrahimi-Fard and L. Guo, On products and duality of binary, quadratic, regular operads, *J. Pure. Appl. Algebra*, **200**, (2005), 293-317.
8. K. Ebrahimi-Fard and L. Guo, Unit actions on operads and Hopf algebras, *Theory and Applications of Categories* **18** (2007), 348-371.
9. R. Felipe, A brief foundation of the left-symmetric dialgebras, *Comunicación del CIMAT*, NO I, (2011).
10. L. Foissy, Les algebres de Hopf des arbes enracines decores II, *Bull. Sci. Math.* **126** (2002), 249-288.

11. A. Frabetti, Dialgebra homology of associative algebras, *C. R. Acad. Sci. Paris* **325** (1997), 135-140.
12. A. Frabetti, Leibniz homology of dialgebras of matrices, *J. Pure Appl. Alg.* **129** (1998), 121-141.
13. V. Ginzburg and M. M. Kapranov, Koszul duality for operads, *Duke Math. J.* **76** (1995) 203-272.
14. V. Yu. Gubarev and P. S. Kolesnikov, On embedding of dendriform algebras into Rota-Baxter algebras, *Cent. Eur. Jour. Math.* **11** (2013), 226-245.
15. V. Yu. Gubarev and P. S. Kolesnikov, Operads of decorated trees and their duals, *Comment. Math. Univ. Carolin.* **55** (2014), 421-445.
16. D. Hou, X. Ni and C. Bai, Pre-Jordan algebras, *Math. Scand.* **112** (2013), 19-48.
17. E. Hoffbeck, A Poincaré-Birkhoff-Witt criterion for Koszul operads, *Manuscripta Math.* **131** (2010) 87-110.
18. R. Holtkamp, Comparision of Hopf algebras on trees, *Archiv der Mathematik* **80** (2003), 368-383.
19. P. Kolesnikov, Varieties of dialgebras and conformal algebras (Russian), *Sib. Mat. Zh.* **49** (2008), 323-340.
20. P. Leroux, Construction of Nijenhuis operators and dendriform trialgebras, *Int. J. Math. Sci.* (2004) 2595-2615.
21. P. Leroux, On some remarkable operads constructed from Baxter operators, arXiv: math. QA/0311214.
22. P. Leroux, Ennea-algebras, *J. Algebra*, **281**, (2004), 287-302.
23. L. Liu, X. Ni and C. Bai, L-quadri-algebras (in Chinese), *Sci. Sin. Math.* **42** (2011) 105-124.
24. J.-L. Loday, Une version non commutative des algèbres de Lie: les algèbres de Leibniz, *Enseign. Math.* **39** (1993) 269-293.
25. J.-L. Loday, Dialgebras, in Dialgebras and related operads, *Lecture Notes in Math.* **1763** (2002), 7-66.
26. J.-L. Loday, Arithmetree, *J. Algebra* **258** (2002), 275-309.
27. J.-L. Loday, Completing the operadic butterfly, *Georgian Math. J.* **13** (2006)741-749.
28. J.-L. Loday, On the algebra of quasi-shuffles, *Manuscripta Math.* **123** (2007) 79-93.
29. J.-L. Loday, Scindement d'associativté et algébres de Hopf. in the proceedings of conference in honor of Jean Leray, Nantes (2002), *Séminaire et Congrés (SMF)* **9** (2004), 155-172.
30. J.-L. Loday and M. Ronco, Trialgebras and families of polytopes, in "Homotopy Theory: Relations with Algebraic Geometry, Group Cohomology, and Algebraic K-theory", *Comtep. Math.* **346** (2004) 369-398.
31. J.-L. Loday and M. Ronco, Algébre de Hopf colibres, *C. R. Acad. Sci. Paris* **337** (2003), 153-158.
32. J.-L. Loday and B. Vallette, Algebraic Operads, *Grundlehren Math. Wiss.* **346**, Springer, Heidelberg, 2012.
33. X. Ni and C. Bai, Prealternative algebras and prealternative bialgebras, *Pacific J. Math.*, **248** (2010) 355-391.

34. J. Pei, C. Bai and L. Guo, Splitting of Operads and Rota-Baxter Operators on Operads, *Appl. Categor. Struct.* **25** (2017), 505-538.

35. M. R. Ronco, Eulerian idempotents and Milnor-Moore theorem for certain non-commutative Hopf algebras, *J. Algebra* **254** (2002), 152-172.

36. G.-C. Rota, Baxter operators, an intoduction, in Gian-Carlo Rota on Combinatics: Introductory papers and commentaries (Joseph P.S. Kung, Ed), Birkhäuser, Boston (1995).

37. K. Uchino, Derived bracket construction and Manin products, *Lett. Math. Phys.* **93** (2010) 37-53.

38. B. Vallette, Manin products, Koszul duality, Loday algebras and Deligne conjecture, *J. Reine Angew. Math.* **620** (2008) 105-164.

39. G.W. Zinbiel, Encyclopedia of types of algebras 2010, in: "Operads and Universal Algebra", *Nankai Series in Pure, Applied Mathematics and Theoretical Physics* **9**, 217-296, World Scientific, Singapore, 2011.

354

# Kac-Moody Groups and Their Representations*

Dmitriy Rumynin

*Department of Mathematics, University of Warwick,*
*Coventry, CV4 7AL, UK*
*Associated member of Laboratory of Algebraic Geometry,*
*National Research University Higher School of Economics, Russia*

*D.Rumynin@warwick.ac.uk*

To Leonid Arkadievich Bokut with admiration

In this expository paper we review some recent results about representations of Kac-Moody groups. We sketch the construction of these groups. If practical, we present the ideas behind the proofs of theorems. At the end we pose open questions.

*2010 Mathematics subject classification:* 20G44, 20G05.

*Keywords*: Kac-Moody group, Kac-Moody algebra, adjoint representation, smooth representations, completion, Davis realisation

## 1. Introduction

Kac-Moody Lie algebras are well-known generalisations of simple finite-dimensional Lie algebras, subject of 1533 research papers on MathSciNet and at least 3 beautiful monographs [11,16,30] Kac-Moody groups are less well-known cousins, subject of only 214 research papers on MathSciNet. One issue with them is that there are several different notions of a Kac-Moody group:

- a group valued functor on commutative rings defined by Tits, a generalisation of $R \mapsto \mathrm{SL}_n(R[z, z^{-1}])$,
- a locally compact totally disconnected group, a generalisation of $\mathrm{SL}_n(\mathbb{F}_q((z)))$,

---

*The research was partially supported by the Russian Academic Excellence Project '5–100' and by Leverhulme Foundation. The author would like to thank Inna Capdeboscq and Timothée Marquis for valuable discussions.

- an ind-algebraic group, a generalisation of $SL_n(\mathbb{C}((z)))$,
- a more complicated topological group, e.g., $SL_n(\mathbb{Q}_p((z)))$.

In this survey, we review some new results about the first two types of Kac-Moody groups and their representations. We give examples and sketch proofs whenever it is practical. The only completely new results are in Section 2.4 where full proofs are given.

There are instructional sources about their Group Theory and Geometry[7,18,22] but not about their Representation Theory. A reader interested in ind-algebraic Kac-Moody groups can consult a monograph[17] but someone who wants to learn about more complicated Kac-Moody groups will need to look at scholarly sources[23,24]. We start without further ado.

## 2. Representations of uncompleted group

### 2.1. *Kac-Moody Lie algebra*

Let $\mathcal{A} = (A_{i,j})_{n \times n}$ be a square matrix with coefficients in a commutative ring $\mathbb{K}$. A *realisation of* $\mathcal{A}$ is a collection $\mathcal{R} = (\mathfrak{h}, h_1, \ldots h_n, \alpha_1, \ldots \alpha_n)$ where $\mathfrak{h}$ is a finitely generated free $\mathbb{K}$-module, $h_i$ are $\mathbb{K}$-linearly independent elements of $\mathfrak{h}$, $\alpha_j$ are $\mathbb{K}$-linearly independent elements of $\mathfrak{h}^*$, and $\alpha_j(h_i) = A_{ij}$ for all $i$ and $j$.

A realisation gives several interesting Lie $\mathbb{A}$-algebras for any commutative $\mathbb{K}$-algebra $\mathbb{A}$. The first Lie $\mathbb{A}$-algebra is $\widetilde{L}_{\mathcal{R}}(\mathbb{A})$: it is generated by $\mathfrak{h}_{\mathbb{A}} = \mathfrak{h} \otimes_{\mathbb{K}} \mathbb{A}$ and elements $e_1, \ldots e_n, f_1 \ldots f_n$ subject to the relations

$$[h, e_i] = \alpha_i(h)e_i, \ [h, f_j] = -\alpha_j(h)f_j, \ [h', h] = 0,$$
$$[e_i, f_i] = h_i, \ [e_i, f_j] = 0 \text{ if } i \neq j$$

for all $h', h \in \mathfrak{h}_{\mathbb{A}}$. The Lie algebra $\widetilde{L}_{\mathcal{R}}(\mathbb{A})$ is graded by the root group $X(\mathcal{R})$, the free abelian group generated by elements $\alpha_i$. The grading is given by

$$\deg(h) = 0, \ \deg(e_i) = \alpha_i, \ \deg(f_i) = -\alpha_i .$$

Let $I_{\mathbb{A}}$ be the sum of all ideals of $\widetilde{L}_{\mathcal{R}}(\mathbb{A})$, contained in the non-zero graded part $\oplus_{\gamma \neq 0} \widetilde{L}_{\mathcal{R}}(\mathbb{A})_{\gamma}$. The second Lie algebra is $\widehat{L}_{\mathcal{R}}(\mathbb{A}) := \widetilde{L}_{\mathcal{R}}(\mathbb{A})/I_{\mathbb{A}}$ and the third Lie algebra is $L_{\mathcal{R}}(\mathbb{A}) := \widehat{L}_{\mathcal{R}}(\mathbb{K}) \otimes_{\mathbb{K}} \mathbb{A}$. Although there is some literature on $L_{\mathcal{R}}(\mathbb{A})$ for a general $\mathcal{A}$[26], these algebras merit further investigation (cf. 4.1 and 4.2).

If $\mathcal{A}$ is a generalised Cartan matrix, we set $\mathbb{K} = \mathbb{Z}$, call a realisation (over $\mathbb{Z}$) *a root datum* and denote it $\mathcal{D}$. While both $\widehat{L}_{\mathcal{D}}(\mathbb{A})$ and $L_{\mathcal{D}}(\mathbb{A})$ deserve to be called *Kac-Moody algebras*, the actual definition of a Kac-Moody algebra is different. Let $\mathcal{U}_{\mathbb{Z}}$ be the divided powers integral form of the universal enveloping algebra $U(L_{\mathcal{D}}(\mathbb{C}))$. Then *a Kac-Moody algebra* is defined as

$$\mathfrak{g}_{\mathbb{Z}} := L_{\mathcal{D}}(\mathbb{C}) \cap \mathcal{U}_{\mathbb{Z}} \,, \quad \mathfrak{g}_{\mathbb{A}} := \mathfrak{g}_{\mathbb{Z}} \otimes_{\mathbb{Z}} \mathbb{A} \,.$$

It inherits a triangular decomposition $\mathfrak{g}_{\mathbb{A}} = (\mathfrak{n}_- \otimes \mathbb{A}) \oplus (\mathfrak{h} \otimes \mathbb{A}) \oplus (\mathfrak{n}_+ \otimes \mathbb{A})$ from $\mathfrak{g}_{\mathbb{Z}} = \mathfrak{n}_- \oplus \mathfrak{h} \oplus \mathfrak{n}_+$ where $\mathfrak{n}_{-\mathbb{C}}$ is the Lie subalgebra of $\mathfrak{g}_{\mathbb{C}}$ generated by all $f_i$, $\mathcal{U}_-$ is the divided powers $\mathbb{Z}$-form of $U(\mathfrak{n}_{-\mathbb{C}})$ and $\mathfrak{n}_- := \mathfrak{n}_{-\mathbb{C}} \cap \mathcal{U}_-$ (ditto for $\mathfrak{n}_+$ using $e_i$-s and $\mathcal{U}_+$). If $\mathbb{F}$ is a field of characteristic $p$, the Lie algebra $\mathfrak{g}_{\mathbb{F}}$ is restricted with the $p$-operation

$$(h \otimes 1)^{[p]} = h \otimes 1, \quad (x \otimes 1)^{[p]} = x^p \otimes 1 \quad \text{where} \quad h \in \mathfrak{h}, \ x \in \mathfrak{n}_{\pm}$$

where $x^p$ is calculated inside the associative $\mathbb{Z}$-algebra $\mathcal{U}_{\pm} \leq \mathcal{U}^{18(\text{Th. } 4.39)}$.

If $p > \max_{i \neq j}(-A_{i,j})$, then all the three Kac-Moody coincide: $\widehat{L}_{\mathcal{D}}(\mathbb{F}) = L_{\mathcal{D}}(\mathbb{F}) = \mathfrak{g}_{\mathbb{F}}$[18,24] but it is probably no longer true for small primes.

## 2.2. *Kac-Moody group*

The algebras $\mathcal{U}_{\mathbb{Z}}$ and $\mathfrak{g}_{\mathbb{Z}}$ inherit the grading by $X(\mathbb{R})$. It is also known as the root decomposition

$$\mathfrak{g}_{\mathbb{C}} = \bigoplus_{\alpha \in \Phi \subseteq X(\mathcal{D})} \mathfrak{g}_{\mathbb{C}\,\alpha} \,.$$

The set of roots splits into two disjoint parts: real roots $\Phi^{re} := W\{\alpha_1, \ldots \alpha_n\}$ (where $W$ is the Weyl group) and imaginary roots $\Phi^{im} := \Phi \setminus \Phi^{re}$.

*The Kac-Moody group* is a functor $G_{\mathcal{D}}$ from commutative rings to groups. Its value on a field $\mathbb{F}$ can be described as

$$G_{\mathcal{D}}(\mathbb{F}) = T * *_{\alpha \in \Phi^{re}} U_{\alpha} / \langle \text{ Tits' relations} \rangle, \quad T = \mathfrak{h} \otimes_{\mathbb{Z}} \mathbb{F}^{\times}, \quad U_{\alpha} \cong \mathbb{F}^+,$$

where $T$ is a torus and

$U_{\alpha} = \{X_{\alpha}(\mathbf{t})\}$, $X_{\alpha}(\mathbf{t})X_{\alpha}(\mathbf{s}) = X_{\alpha}(\mathbf{t} + \mathbf{s})$ is a root subgroup. There are different ways to write Tits' relations: the reader should consult classical papers[10,29] for succinct presentations. Note that Tits' Relations have infinitely many generators and relations unless $\mathcal{A}$ is of finite type.

However, if the field $\mathbb{F} = \mathbb{F}_q$, $q = p^m$ is finite and under mild assumptions on $\mathcal{A}$, the groups $G_{\mathcal{D}}(\mathbb{F}_q)$ are finitely presented[2] (cf. Ref. 5 for concrete finite presentations of affine groups) and simple[8]. Thus, the groups $G_{\mathcal{D}}(\mathbb{F}_q)$ form a good source of finitely-presented simple (non-linear) groups.

It is important for us that they have a BN-pair with $B = T \ltimes U_+$ where $U_+$ is the subgroup generated by all $U_{\alpha}$ for positive real roots $\alpha$.

## 2.3. *Adjoint representation*

The group $G_{\mathcal{D}}(\mathbb{F})$ acts the Lie algebra $\mathfrak{g}_{\mathbb{F}}$ via adjoint action [18,22]. The torus action comes from the $X(\mathcal{D})$-grading

$$\mathrm{Ad}(h \otimes \mathbf{t})(a) = \mathbf{t}^{\alpha(h)} a \quad \text{where} \quad a \in (\mathfrak{g}_{\mathbb{F}})_{\alpha}$$

and the action of $U_{\alpha}$ is exponential:

$$\mathrm{Ad}(X_{\alpha}(\mathbf{t}))(a) = \mathbf{e}^{\mathrm{ad}(\mathbf{t}e_{\alpha})}(a) = \sum_{n=0}^{\infty} \mathbf{t}^n \, \mathrm{ad}(e_{\alpha}^{(n)})(a)$$

where $e_{\alpha}$ (rather than $e_{\alpha} \otimes 1$) is a non-zero element of $\mathfrak{g}_{\mathbb{F}\,\alpha}$ and $e_{\alpha}^{(n)} \in \mathcal{U} \otimes \mathbb{F}$ is its divided power. Notice that $\mathfrak{g}_{\mathbb{F}\,\alpha}$ is one-dimensional for a real root $\alpha$.

We denote the image of Ad by $G_{\mathcal{D}}^{ad}(\mathbb{F})$.

## 2.4. *Over-restricted representations*

Let $\mathbb{F}$ be a field of positive characteristic $p$ in this section. A representation $(V, \rho)$ of the Lie algebra $\mathfrak{g}_{\mathbb{F}}$ is called *restricted* if $\rho(x)^p = \rho(x^{[p]})$ for all $x \in \mathfrak{g}_{\mathbb{F}}$. Each real root $\alpha$ yields an additive family of linear operators on a restricted representation

$$Y_{\alpha}(\mathbf{t}) := \mathbf{e}^{\rho(e_{\alpha})} = \sum_{k=0}^{p-1} \frac{1}{k!} \rho(e_{\alpha})^k.$$

These operators do not define an action of $G_{\mathcal{D}}(\mathbb{F})$ in general. The concept of *an over-restricted* representation, proposed recently to integrate representations from Lie algebras to algebraic groups [27], proves beneficial here as well. We say that a restricted representation $(V, \rho)$ of $\mathfrak{g}_{\mathbb{F}}$ is *over-restricted* if $\rho(e_{\alpha})^{\lfloor (p+1)/2 \rfloor} = 0$ for any real root $\alpha$.

**Proposition 2.1.** *(cf. Ref. 27) Suppose that $(V, \rho)$ is an over-restricted representation $\mathfrak{g}_{\mathbb{F}}$. If $\mathrm{ad}(e_{\alpha}^{(p)})(x) = 0$ for some $x \in \mathfrak{g}_{\mathbb{F}}$, then*

$$\rho\big( \mathrm{Ad}(X_{\alpha}(\mathbf{t}))(x) \big) = Y_{\alpha}(\mathbf{t}) \rho(x) Y_{\alpha}(-\mathbf{t}) . \tag{1}$$

**Proof.** Observe by induction that for each $k = 1, 2, \ldots p-1$

$$\rho\big(\frac{1}{k!} \mathrm{ad}(e_{\alpha})^k(x)\big) = \sum_{j=0}^{k} \frac{(-1)^j}{(k-j)! j!} \rho(e_{\alpha})^{k-j} \rho(x) \rho(e_{\alpha})^j . \tag{2}$$

The condition $\mathrm{ad}(e_{\alpha}^{(p)})(x) = 0$ implies that $\mathrm{ad}(e_{\alpha}^{(n)})(x) = 0$ for all $n \geq p$ and $\rho\big( \mathrm{Ad}(X_{\alpha}(\mathbf{t}))(x) \big) = \sum_{k=0}^{p-1} \rho(\frac{1}{k!} \mathrm{ad}(\mathbf{t}e_{\alpha})^k(x))$ stops at degree $p-1$. Using

Formula (2), this is equal to

$$\sum_{i+j=0}^{p-1} \frac{(-1)^j}{i!j!}\rho(te_\alpha)^i\rho(x)\rho(te_\alpha)^j \overset{\spadesuit}{=} \sum_{i,j=0}^{p-1} \frac{(-1)^j}{i!j!}\rho(te_\alpha)^i\rho(x)\rho(te_\alpha)^j$$
$$=\mathbf{e}^{\rho(te_\alpha)}\rho(x)\mathbf{e}^{-\rho(te_\alpha)},$$

exactly the right hand side. Notice that Equality $\spadesuit$ holds because $(V,\rho)$ is over-restricted: terms on the right, missing from the left, are all zero.    □

Consider an $X(\mathcal{D})$-graded restricted representation $(V,\rho)$ of $\mathfrak{g}$. Grading gives an action of $T$ on $V$ by $\widehat{\rho}(h \otimes \mathbf{t})(v_\alpha) = \mathbf{t}^{\alpha(h)}v_\alpha$. An analogue of Proposition 2.1 holds for $T$:

$$\rho\big(\operatorname{Ad}(h \otimes \mathbf{t})(x)\big) = \widehat{\rho}(h \otimes \mathbf{t})\rho(x)\widehat{\rho}(h \otimes \mathbf{t}^{-1}) . \tag{3}$$

Let $G_V$ be the subgroup of $\operatorname{GL}(V)$ generated by $\widehat{\rho}(T)$ and all $Y_\alpha(\mathbf{t})$.

**Theorem 2.1.** *Suppose that $p > \max_{i\neq j}(-A_{i,j})$. If $(V,\rho)$ is an $X(\mathcal{D})$-graded over-restricted representations of $\mathfrak{g}_{\mathbb{F}}$, faithful on both $T$ and $\mathfrak{g}_{\mathbb{F}}$, then*

$$\phi : G_V \to G_\mathcal{D}^{ad}(\mathbb{F}), \quad \phi(Y_\alpha(\mathbf{s})) = X_\alpha(\mathbf{s}), \quad \phi(\widehat{\rho}(t)) = t \quad for \ \ t \in T$$

*is a surjective homomorphism of groups whose kernel is central and consists of $\mathfrak{g}_{\mathbb{F}}$-automorphisms of $V$.*

**Proof.** Let $H$ be the free product of $T$ and all additive groups $U_\alpha, \alpha \in \Phi^{re}$. Both $G_V$ and $G_\mathcal{D}^{ad}(\mathbb{F})$ are naturally quotients of $H$. If $x_1 * \ldots * x_n \in \ker(H \to G_V)$ where all $x_i$ are from the constituent groups then

$$\phi(x_1)\phi(x_2)\ldots\phi(x_n) = I_V .$$

Formulas (1) and (3) imply that

$$\rho\big([\operatorname{Ad}(x_1)\operatorname{Ad}(x_2)\ldots\operatorname{Ad}(x_n)](e_\beta)\big) = \rho(e_\beta),$$
$$\rho\big([\operatorname{Ad}(x_1)\operatorname{Ad}(x_2)\ldots\operatorname{Ad}(x_n)](h)\big) = \rho(h)$$

for all $h \in \mathfrak{h}_{\mathbb{F}}$ and real roots $\beta$. Our restriction on $p$ imply that $\mathfrak{g}_{\mathbb{F}}$ is generated by $\mathfrak{h}_{\mathbb{F}}$ and all $e_\beta$[24]. Consequently,

$$\rho\big([\operatorname{Ad}(x_1)\operatorname{Ad}(x_2)\ldots\operatorname{Ad}(x_n)](y)\big) = \rho(y)$$

for all $y \in \mathfrak{g}_{\mathbb{F}}$. Since $\rho$ is injective it follows that

$$[\operatorname{Ad}(x_1)\operatorname{Ad}(x_2)\ldots\operatorname{Ad}(x_n)](y) = I_{\mathfrak{g}}$$

and $x_1 * \ldots * x_n \in \ker(H \to G_\mathcal{D}^{ad}(\mathbb{F}))$. Hence, $\phi$ is well-defined.

It remains to determine the kernel of $\phi$. Suppose $y = x_1 x_2 \ldots x_n \in \ker(\phi)$ with all $x_i$ are either $Y_\alpha(\mathbf{s})$, or in $T$. Arguing as above, $\rho(z) = \rho(\phi(y)(z)) = y\rho(z)y^{-1}$ for all $z \in \mathfrak{g}$. So $y \in \operatorname{End}(V, \rho)$: it commutes with all $\rho(e_\alpha)$, hence with all $Y_\alpha(\mathbf{s})$. Since $T$ acts faithfully, $y$ commutes with $T$ as well. Commuting with all generators of $G_V$, $y$ is inevitably central. $\quad\square$

As soon as there are few endomorphisms, the map $\phi$ in Theorem 2.1 can be "reversed" to define a projective representation of the Kac-Moody group.

**Corollary 2.1.** *Suppose that in the conditions of Theorem 2.1 the representation $(V, \rho)$ is a brick, i.e., $\operatorname{End}(V, \rho) = \mathbb{F}$. Then*

$$\theta : G_{\mathcal{D}}(\mathbb{F}) \to \operatorname{GL}(V), \ \theta(X_\alpha(\mathbf{s})) = Y_\alpha(\mathbf{s}), \ \theta(t) = \widehat{\rho}(t) \ \ for \ t \in T$$

*extends to a projective representation of $G_{\mathcal{D}}(\mathbb{F})$.*

If the root datum is simply-connected, i.e., $\alpha_i$ form a basis of $\mathfrak{h}$, then the group $G_{\mathcal{D}}(\mathbb{F})$ is generated by $U_\alpha$-s[10]. Hence, no grading is needed to define a representation of $G_{\mathcal{D}}(\mathbb{F})$, with all the proofs going through as before:

**Corollary 2.2.** *Suppose that $\mathcal{D}$ is simply-connected and $p > \max_{i \neq j}(-A_{i,j})$. If $(V, \rho)$ is a faithful, over-restricted brick for $\mathfrak{g}_{\mathbb{F}}$, then*

$$\theta : G_{\mathcal{D}}(\mathbb{F}) \to \operatorname{GL}(V), \ \theta(X_\alpha(\mathbf{s})) = Y_\alpha(\mathbf{s})$$

*extends to a projective representation of $G_{\mathcal{D}}(\mathbb{F})$.*

## 3. Representations of completed group

### 3.1. *Completion*

The group $G_{\mathcal{D}}(\mathbb{F})$ is also known as the "minimal" Kac-Moody group, while some of its various completions $\widehat{G_{\mathcal{D}}(\mathbb{F})}$ go under the name a "maximal" Kac-Moody group.

Let us consider a group $G$ with a BN-pair $(B, N)$. Let $\widehat{G}$ be a completion of $G$ with respect to some topology. Is $(\overline{B}, N)$ (where $\overline{B}$ is the closure of $B$ in $\widehat{G}$) a BN-pair on $\widehat{G}$? It depends on circumstances. For example, consider a simple split group scheme $\mathcal{G}$, $G = \mathcal{G}(\mathbb{F}_q[z, z^{-1}])$, the group of monomial matrices $N \le G$ and positive and negative Iwahori subgroups $I_\pm = [\mathcal{G}(\mathbb{F}_q[z^{\pm 1}]) \xrightarrow{z^{\pm 1} \mapsto 0} \mathcal{G}(\mathbb{F}_q)]^{-1}(B)$. Both pairs $(I_\pm, N)$ are BN-pairs on $G$ but only $(\overline{I_+}, N)$ is a BN-pair on the positive completion $\widehat{G} = \mathcal{G}(\mathbb{F}_q((z)))$: the countable groups $\overline{I_-} = I_-$ and $N$ cannot generate uncountable $\widehat{G}$. The following theorem pinpoints the completion process for groups with a BN-pair under some conditions:

**Theorem 3.1.** [6](Th. 1.2) *Let $G$ be a group with a BN-pair $(B, N)$ with Weyl group $(W, S)$ where $S$ is finite. Suppose further that a topology $\mathcal{T}$ on $B$ is given such that the four conditions (1)–(4) hold.*

*(1) $(B, \mathcal{T})$ is a topological group.*

*(2) The completion $\widehat{B}$ is a group.*

*(3) $\mathcal{T}_1 := \{A \in \mathcal{T} \mid 1 \in A\}$ is a basis at 1 of topology on each minimal parabolic $P_s$, $s \in S$ that defines a structure of topological group on $P_s$.*

*(4) The index $|P_s : B|$ is finite for each $s \in S$.*

*Under these conditions the following statements hold:*

*(a) $\mathcal{T}_1$ is a basis at 1 of topology on $G$ that defines a structure of topological group on $G$.*

*(b) The completion $\widehat{G}$ is a group. The completion $\widehat{B}$ is equal to the closure $\overline{B}$.*

*(c) The completion $\widehat{G}$ is isomorphic to the amalgam $\underset{\mathfrak{B}}{*} H$ where*
$$\mathfrak{B} = \{\overline{B}, N, \overline{P_s}; s \in S\}.$$

*(d) The pair $(\overline{B}, N)$ is a BN-pair on the completed group $\widehat{G}$.*

**Proof.** It is a well-known theorem of Tits that a group with a BN-pair is an amalgam of $N$ and its minimal parabolics [17](Prop. 5.1.7). Later on Tits has shown how to back-engineer this group from such amalgam [28] (cf. Ref. 17(Th. 5.1.8)).

This is the heart of the proof: we pinpoint the completed group in part (c) but need to check numerous technical conditions of the Tits theorem. See Ref. 6 for full details.    □

Theorem 3.1 gives us *a locally pro-p-complete* Kac-Moody group $G^{lpp} := \widehat{G_{\mathcal{D}}(\mathbb{F}_q)}$, $q = p^m$ by choosing the pro-p-topology on $B$: its basis at 1 is $\{A \leq B \mid |B : A| = p^a \text{ for some } a \in \mathbb{N}\}$. The Borel and the minimal parabolic subgroups of $G_{\mathcal{D}}(\mathbb{F}_p)$ are split [7](6.2) :

$$B = T \ltimes U_+, \quad P_s = L_s \ltimes U_s$$

where $L_s = \langle U_{\alpha_i} \cup U_{-\alpha_i} \rangle T$, $s = s_i$, $U_s := U_+ \cap sU_+s^{-1}$. In particular, $|P_i : B|$ is finite for all $i \in I$ so that, by Theorem 3.1, we can complete

$G_{\mathcal{D}}(\mathbb{F}_q)$ with respect to the pro-$p$-topology on $B$ (or, in fact, any "$M_s$-equivariant" topology). The group $G^{lpp}$ has a BN-pair $(\widehat{B}, N)$ where $\widehat{B} = H \ltimes \widehat{U}_+$ and $\widehat{U}_+$ is the full pro-$p$ completion of $U$.

The *congruence subgroup* $C(G^{lpp}) = \cap_{g \in G^{lpp}} \widehat{U}_+^{\,g}$ is of crucial interest. Suppose that $A$ is irreducible and the root datum $\mathcal{D}$ is simply connected. Let $Z'(G^{lpp}) := Z((G_{\mathcal{D}}(\mathbb{F}_q)) \times C(G^{lpp})$ (note that the intersection is trivial).

**Theorem 3.2.** [6,9,19] *Under these conditions $G^{lpp}/Z'(G^{lpp})$ is a topologically simple group. Moreover, if $A$ is 2-spherical, then $G^{lpp}/Z'(G^{lpp})$ is an abstractly simple group.*

## 3.2. *Comparison to other completions*

It is instructive to compare $G^{lpp}$ with other completions of $G_{\mathcal{D}}(\mathbb{F}_q)$, a.k.a topological Kac-Moody groups (cf. Ref. 6,9,18,21,24). Let us list them:

- the *Caprace-Rémy-Ronan group* $G^{crr}$, a completion in the topology of the action on Bruhat-Tits building,
- the *Carbone-Garland group* $G^{c\lambda}$, a completion in the topology of the action on the integrable simple module with a highest weight $\lambda$,
- the *Mathieu-Rousseau group* $G^{ma+}$, an analogue of the ind-algebraic completion, also obtained as an amalgam $\underset{\mathfrak{B}}{*} H$ where $\mathfrak{B} = \{\widetilde{B}, N, \widetilde{P_s}\}$ with specially constructed groups $\widetilde{P_s}$.
- another *Mathieu-Rousseau group* $G^+$, the closure of $G$ in $G^{ma+}$,
- the *Belyaev group* $G^b$, the "largest" completion with compact totally disconnected $\overline{U_+}$.
- the *Schlichting group* $G^s$, the "smallest" completion with compact totally disconnected $\overline{U_+}$.

If $p > \max_{i \neq j}(-A_{ij})$, then $G^+ = G^{ma+}$ but they could be different, in general [24(6.11)]. The precise meaning of the "largest" and the "smallest" of the last two groups is a certain universal property (consult Ref. 21 for precise statement). The action on the Bruhat-Tits building ensures that $G^s = G^{crr}$. Theorem 3.1 gives $G^b$ by considering the profinite topology on $B$ instead of the pro-$p$-topology: its basis at 1 is $\{A \leq B \mid |B : A| < \infty\}$. The following theorem compares the known completions:

**Theorem 3.3.** [6,21,24] *There are open continuous surjective group homomorphisms:*

$$G^b \twoheadrightarrow G^{lpp} \twoheadrightarrow G^+ \twoheadrightarrow G^{c\lambda} \twoheadrightarrow G^{crr} \overset{\cong}{\to} G^s.$$

### 3.3. *Davis Building*

Let $G^\star$ be one of the locally compact, totally disconnected groups from Section 3.2. It admits a BN-pair $(B^\star, N)$ with the same Weyl group $(W, S)$ as the Kac-Moody algebra $\mathfrak{g}_{\mathbb{C}}$. Consequently, $G^\star$ acts on two simplicial complexes: the Bruhat-Tits building $\mathfrak{B}$ and *the Davis building* $\mathfrak{D}$ (also known as the Davis realisation [1 (Section 12.4)] or the geometric realisation [12 (Section 18.2)]). Notice that there are variations in this definition: the original Davis' definition produces a cell complex, while $\mathfrak{D}$ (defined below) is a simplicial complex, a subdivision of this cell complex.

While the building $\mathfrak{B}$ is well-known, it is still instructive to recall its definition. Let $\mathcal{P}(G^\star)$ be the set of all proper parabolic subgroups of $G^\star$. A parabolic $P \in \mathcal{P}(G^\star)$ is conjugate to precisely one of the standard parabolics $P_J := \langle B^\star, \dot{s} \rangle_{s \in J}$ where $J \subset S$, $\dot{s} \in G^\star$ is a lift of the element $s \in W = N(T)/T$. Thus, we can define *the type* and *the rank* of each parabolic by

$$t(P) = J, \; r(P) = |J| \;\; \text{whenever} \;\; P \sim P_J.$$

The building $\mathfrak{B}$ is an $n$-dimensional simplicial complex ($n = |S|$) whose set of $k$-dimensional simplices $\mathfrak{B}_k$ is equal to $t^{-1}(n - k) = \{P \mid r(P) = n - k\}$. A simplex $P'$ is a face of $P$ if and only if $P \subseteq P'$. The group $G^\star$ acts on $\mathfrak{B}$ in the obvious way: $^g P = gPg^{-1}$. Since parabolic subgroups are self-normalising, the stabiliser of $P$ is $P$ itself. One drawback of this action is that stabilisers of simplices are not necessarily compact. This drawback is fixed in the Davis building.

A subset $J \subset S$ is called *spherical* if the Coxeter subgroup $\langle J \rangle$ is finite. If $\mathcal{A}$ has no components of finite type, $\mathcal{P}^{sp}(G^\star)$ is the subset of $\mathcal{P}(G^\star)$ that consists of parabolics of spherical type. In general, $G^\star = G_1^\star \times \ldots \times G_k^\star$ with $G_j^\star$ corresponding to connected components of $\mathcal{A}$. The elements of $\mathcal{P}^{sp}(G^\star)$ are *marked parabolics* $H \le P$ where $P = P_1 \times \ldots \times P_k$ is a finite type parabolic, $P_j \le G_j$, $H = H_1 \times \ldots \times H_k$ and each $H_j$ is either $\{1\}$ (if $P_j \ne G_j$), or a Borel subgroup (if $P_j = G_j$).

The set $\mathcal{P}^{sp}(G^\star)$ is partially ordered: $(H' \le P') \preceq (H \le P)$ if and only if $H \subseteq P' \subseteq P$. The Davis building $\mathfrak{D}$ is the geometric realisation of the poset $\mathcal{P}^{sp}(G^\star)$, i.e., its set $\mathfrak{D}_k$ of $k$-dimensional simplices consists of $(k + 1)$-long chains of marked spherical parabolics

$$P_0 \prec P_1 \prec \ldots \prec P_{k-1} \prec P_k \,.$$

Faces of a simplex are its subchains. The group $G^\star$ acts on $\mathfrak{D}$ in the same obvious way: $g(H_i \le P_i) = (gH_ig^{-1} \le gP_ig^{-1})$. The stabiliser of a chain

$(P_i)$ is an open subgroup $P_0'$ of $P_0$. Thus, all stabilisers are compact because spherical parabolic subgroups are necessarily compact.

One interesting example is *a generic* Kac-Moody group. Suppose $A_{i,j} A_{j,i} \geq 4$ for all $i$, $j$. Then the only spherical subsets of $S$ are the empty set and one element subsets. Consequently, any chain in $\mathcal{P}^{sp}(G^\star)$ is of length at most 1 and $\mathfrak{D}$ is a tree.

If $\mathcal{A}$ has no irreducible components of finite type, both buildings $\mathfrak{B}$ and $\mathfrak{D}$ are contractible. If $\mathcal{A}$ has an irreducible component of finite type, then $\mathfrak{D}$ is still contractible, while $\mathfrak{B}$ is not (see Ref. 12 for this as well as detailed study of $\mathfrak{D}$). We finish this section by stating Davis' Theorem:

**Theorem 3.4.** [13] $\mathfrak{D}$ *admits a locally Euclidean, $G^\star$-invariant metric that turns $\mathfrak{D}$ into complete, CAT(0) geodesic space.*

### 3.4. *Projective Dimension of Smooth Representations*

We study representations of $G^\star$ over a field $\mathbb{K}$ of characteristic zero. A representation $(V, \rho)$ is called *a smooth representation* if for all $v \in V$ there exists a compact open subgroup $K_v \leq G^\star$ such that $\rho(g)v = v$ for all $g \in K_v$. Equivalently, the action $G \times V \to V$ is required to be continuous with respect to the discrete topology on $V$ (and standard topologies in $G^\star$ and the product).

The category $\mathcal{M}(H)$ of smooth representations of a locally compact totally disconnected topological group $H$ is abelian with enough projectives [3,4]. In case of the group $G^\star$ we can say more by examining its action on $\mathfrak{D}$:

**Theorem 3.5.** [15] *Let $d$ be the dimension of $\mathfrak{D}$. Then*

$$\operatorname{coh.dim}(\mathcal{M}(G^\star)) \leq d.$$

**Proof.** Let $C_i = C_i(\mathfrak{D}, \mathbb{K})$ be the group of $\mathbb{K}$-linear chains on $\mathfrak{D}$. The chain complex $\mathscr{C} = (C_d \xrightarrow{\partial} C_{d-1} \cdots \xrightarrow{\partial} C_0)$ is acyclic since $\mathfrak{D}$ is contractible, i.e., all homology groups are trivial except for $H_0(\mathscr{C}) = \mathbb{K}$. This gives an exact sequence

$$0 \to C_d \xrightarrow{\partial} C_{d-1} \xrightarrow{\partial} \cdots \xrightarrow{\partial} C_0 \to \mathbb{K} \to 0$$

of smooth representations of $G^\star$ where $\mathbb{K}$ is the trivial representation.

Let $\sigma = ((P_i), \tau)$ be an oriented simplex in $\mathfrak{D}$. Its stabiliser $\operatorname{Stab}_{G^\star}(\sigma)$ is open: it is either $P_0'$, or its subgroup of index 2, depending on whether an element of $G^\star$ can reverse the orientation $\tau$ or not. The one-dimensional space $\mathbb{K}[\sigma]$ is a smooth representation of $P_0'$. Since $P_0'$ is compact, $\mathbb{K}[\sigma]$ is projective in $\mathcal{M}(P_0')$. Since $P_0'$ is open, the algebraic induction $\mathbb{K}G^\star \otimes_{\mathbb{K}P_0'}$ is left adjoint to the restriction functor $\mathcal{M}(G^\star) \to \mathcal{M}(P_0')$. Hence, $\mathbb{K}G^\star \otimes_{\mathbb{K}P_0'} \mathbb{K}[\sigma]$ is a projective module in $\mathcal{M}(G^\star)$. Observe that

$$C_m \cong \oplus_{(P_i)} \mathbb{K}G^\star \otimes_{\mathbb{K}P_0'} \mathbb{K}[((P_i), \tau)]$$

where the sum is taken over representatives of $G^\star$-orbits on $\mathfrak{D}_m$. It follows that $\mathscr{C}$ is a projective resolution of the trivial representation $\mathbb{K}$.

Let $V$ be an object in $\mathcal{M}(G^\star)$. Tensor product of representations $\otimes V$ is an exact functor $\mathcal{M}(G^\star) \to \mathcal{M}(G^\star)$ so that

$$0 \to C_d \otimes V \to C_{d-1} \otimes V \to \cdots \to C_0 \otimes V \to V \to 0$$

is an exact sequence. We claim that it is a projective resolution of $V$. Indeed, the functor $\mathscr{F} = \hom(C_m, \underline{\ \ })$ is exact since $C_m$ is projective. The functor of all linear maps $\mathscr{E} = \hom_{\mathbb{K}}(V, \underline{\ \ })$ is also exact. The composition of two exact functors is exact, so $\mathscr{F}\mathscr{E} = \hom(C_m \otimes V, \underline{\ \ })$ is exact and $C_m \otimes V$ are projective objects. $\qquad\square$

### 3.5. *Localisation*

One should put Theorem 3.5 into a broader perspective of Schneider-Stuhler Localisation[15,25]. By localisation we understand an equivalence of two categories: a representation theoretic category ($\mathcal{M}(G^\star)$ for us) is equivalent to ("localised to") a geometric category. The key geometric category is the category $\mathrm{Csh}_{G^\star}(\mathfrak{D})$ of $G^\star$-equivariant cosheaves on $\mathfrak{D}$.

A *$G^\star$-equivariant cosheaf*, a.k.a. a coefficient system for homology, is a datum $\mathcal{C} = (\mathcal{C}_F, r_{F'}^F, g_F)$ where $\mathcal{C}_F$ is a $\mathbb{K}$-vector space for each face $F$ of $\mathfrak{D}$, $r_{F'}^F : \mathcal{C}_F \to \mathcal{C}_{F'}$ is a linear map for each pair of faces $F' \subseteq F$, $g_F : \mathcal{C}_F \to \mathcal{C}_{gF}$ is a linear map for all $g \in G^\star$ and a face $F$ that are subject to the following axioms:

(i) $r_F^F = \mathrm{id}_F$ for every face $F$,
(ii) $r_{F''}^{F'} \circ r_{F'}^F = r_{F''}^F$ for faces $F'' \subseteq F' \subseteq F$,
(iii) $g_{hF} \circ h_F = (gh)_F$ for all $g, h$ and $F$,
(iv) $\mathcal{C}_F$ is a smooth representation of the stabiliser $G_F^\star$ for all $F$,

(v) The square $\begin{array}{ccc} \mathcal{C}_F & \xrightarrow{\ g_F\ } & \mathcal{C}_{gF} \\ {\scriptstyle r_{F'}^F}\downarrow & & \downarrow{\scriptstyle r_{gF'}^{gF}} \\ \mathcal{C}_{F'} & \xrightarrow{\ g_{F'}\ } & \mathcal{C}_{gF'} \end{array}$ is commutative for all $g$ and $F' \subseteq F$.

A morphism of equivariant cosheaves $\psi : \mathcal{C} \to \mathcal{E}$ is a system of linear maps $\psi_F : \mathcal{C}_F \to \mathcal{E}_F$, commuting with actions and restrictions, i.e, the squares

$$
\begin{array}{ccc}
\mathcal{C}_F & \xrightarrow{\psi_F} & \mathcal{E}_F \\
\downarrow{r^F_{F'}} & & \downarrow{r^F_{F'}} \\
\mathcal{C}_{F'} & \xrightarrow{\psi_{F'}} & \mathcal{E}_{F'}
\end{array}
\qquad \text{and} \qquad
\begin{array}{ccc}
\mathcal{C}_F & \xrightarrow{\psi_F} & \mathcal{E}_F \\
\downarrow{g_F} & & \downarrow{g_F} \\
\mathcal{C}_{gF} & \xrightarrow{\psi_{gF}} & \mathcal{E}_{gF}
\end{array}
$$

are commutative for all $g$ and $F' \subseteq F$.

The category of equivariant cosheaves $\mathrm{Csh}_{G^\star}(\mathfrak{D})$ is an abelian category[25]: kernels and cokernels can be computed simplexwise. There are several functors connecting the key categories $\mathcal{M}(G^\star)$ and $\mathrm{Csh}_{G^\star}(\mathfrak{D})$. For instance, *the trivial cosheaf functor* $\mathscr{L}$ associates a cosheaf $\underline{V} \in \mathrm{Csh}_{G^\star}(\mathfrak{D})$ to $(V, \rho) \in \mathcal{M}(G^\star)$:

$$
\underline{V}_F = V, \quad r^F_{F'} = \mathrm{id}_V, \quad g_F = \rho(g).
$$

In the opposite direction, if $\mathcal{C}$ is a $G^\star$-equivariant cosheaf, the group $G^\star$ acts on the vector space of oriented $i$-chains (with finite support) $C_i(\mathfrak{D}, \mathcal{C})$ with coefficients in $\mathcal{C}$. In fact, $G^\star$ acts on the space of more general chains as well but the finite support ensures that $C_i(\mathfrak{D}, \mathcal{C}) \in \mathcal{M}(G^\star)$, a functor in the opposite direction! Furthermore, the chain complex

$$
\mathscr{C}(\mathcal{C}) : 0 \to C_d(\mathfrak{D}, \mathcal{C}) \xrightarrow{\partial} C_{d-1}(\mathfrak{D}, \mathcal{C}) \xrightarrow{\partial} \cdots \xrightarrow{\partial} C_0(\mathfrak{D}, \mathcal{C}) \to 0
$$

is a chain complex in $\mathcal{M}(G^\star)$. The functor $\mathscr{C}$ allows us to paraphrase Theorem 3.5:

**Corollary 3.1.** *Given a smooth representation $V$, the complex $\mathscr{C}(\underline{V})$ is a projective resolution of $V$.*

Resolutions of the form $\mathscr{C}(\mathcal{C})$ are quite useful. The category $\mathrm{Csh}_{G^\star}(\mathfrak{D})$ is *Noetherian*: a subobject of a finitely-generated object is finitely-generated. Hence, a finitely-generated infinite-dimensional object $V$ admits a finitely-generated projective resolution, yet $\mathscr{C}(\underline{V})$ is not finitely generated. We call a finitely-generated projective resolution of the form $\mathscr{C}(\mathcal{C})$ a *Schneider-Stuhler resolution*. Do they exist (cf. Section 4.8)?

It may be possible to construct them using systems of subgroups or system of idempotents[15,20]. In fact, Ref. 15 contains a positive answer to existence of Schneider-Stuhler resolutions modulo a (yet open) conjecture on homology of a CAT(0)-complex. To satisfy the reader's curiosity we state this conjecture in full in Section 4.9.

Let us turn our attention to the localisation. We have the trivial cosheaf and the 0-th homology functors going between $\mathcal{M}(G^\star)$ and $\mathrm{Csh}_{G^\star}(\mathfrak{D})$:

$$\mathscr{L}((V,\rho)) = \underline{V}, \quad \mathscr{H}(\mathscr{C}) = H_0(\mathfrak{D},\mathscr{C}).$$

Let $\Sigma \subset \mathrm{Mor}(\mathrm{Csh}_{G^\star}(\mathfrak{D}))$ be the class of morphisms $\psi$ such that $\mathscr{H}(\psi)$ is an isomorphism. Consider the category of left fractions $\mathrm{Csh}_{G^\star}(\mathfrak{D})[\Sigma^{-1}]$ and the fraction functor $\mathscr{Q}_\Sigma : \mathrm{Csh}_{G^\star}(\mathfrak{D}) \to \mathrm{Csh}_{G^\star}(\mathfrak{D})[\Sigma^{-1}]$. Note that while these fractions always exist, $\Sigma$ needs to satisfy *the left Ore condition* (a.k.a. admit a calculus of left fractions) for these objects to be malleable[14]. The 0-th cohomology functor extends to a functor from the category of left fractions $\mathscr{H}[\Sigma^{-1}] : \mathrm{Csh}_{G^\star}(\mathfrak{D})[\Sigma^{-1}] \to \mathcal{M}(G^\star)$. We are ready for the localisation theorem, a generalisation of Schneider-Stuhler Localisation[25]:

**Theorem 3.6.** [15] *Under the notations established above, the following statements hold:*

(i) *The class $\Sigma$ satisfies the left Ore condition.*

(ii) *The functor $\mathscr{H}[\Sigma^{-1}] : \mathrm{Csh}_{G^\star}(\mathfrak{D})[\Sigma^{-1}] \to \mathcal{M}(G^\star)$ is an equivalence of categories.*

(iii) *$\mathscr{Q}_\Sigma \circ \mathscr{L}$ is a quasi-inverse of $\mathscr{H}[\Sigma^{-1}]$.*

## 4. Questions

### 4.1. *Isomorphism Problem*

Find necessary and sufficient conditions on realisations $\mathcal{R}$ and $\mathcal{S}$ for $\widehat{L}_\mathcal{R}$ and $\widehat{L}_\mathcal{S}$ to be equivalent as functors to graded Lie algebras.

### 4.2. *Existence of Restricted Structure*

Suppose $\mathbb{F}$ is a field of positive characteristic. Find necessary and sufficient conditions on realisation for $\widehat{L}_\mathcal{R}(\mathbb{F})$ to admit a structure of restricted Lie algebra. In particular, if $\mathcal{A}$ is a generalised Cartan matrix of general type and $p \leq \max_{i \neq j}(-A_{i,j})$, could $\widehat{L}_\mathcal{D}(\mathbb{F})$ be restricted?

### 4.3. *Humphreys-Verma Conjecture*

Consider "natural" restricted $\mathfrak{g}_\mathbb{F}$-modules, e.g., irreducible, projective, injective. Do they admit an action of $G_\mathcal{D}(\mathbb{F})$ such that for each real root $\alpha$ the differential of the $U_\alpha$-action is the $\mathfrak{g}_{\mathbb{F}\,\alpha}$-action?

### 4.4. *Theory of Over-restricted Representations*

Investigate algebraic properties of the over-restricted enveloping algebra $U(\mathfrak{g}_{\mathbb{F}})/(x^p - x^{[p]}, e_\alpha^{\lfloor (p+1)/2 \rfloor})$ and its representations.

### 4.5. *Congruence Kernel*

Develop techniques for computing $C(G^{lpp})$. Find necessary and sufficient conditions for $C(G^{lpp})$ to be trivial (central, finitely pro-$p$-generated, etc.).

### 4.6. *Lattices in Locally Pro-p-complete Kac-Moody Groups*

Find minimal covolume of lattices (uniform and overall) in $G^{lpp}$.

### 4.7. *Completions*

Investigate the completions. Find necessary and sufficient conditions the following completions to be equal $G^b \overset{?}{=} G^{lpp}$, $G^+ \overset{?}{=} G^{ma+}$, $G^{c\lambda} \overset{?}{=} G^{crr}$.

### 4.8. *Schneider-Stuhler Resolution*

Does a Schneider-Stuhler resolution exist for any finitely-generated object $V \in \mathcal{M}(G^\star)$? What about irreducible objects? More precisely, does there exist a family of functors $\mathscr{T}_k : \mathcal{M}(G^\star) \to \mathrm{Csh}_{G^\star}(\mathfrak{D})$, indexed by natural numbers, such that for each irreducible $L \in \mathcal{M}(G^\star)$ there exists $N \in \mathbb{N}$ such that $\mathscr{C}(\mathscr{T}_k(L))$ is a Schneider-Stuhler resolution of $L$ for all $k > N$.

### 4.9. *Homology of CAT(0)-Complex*

Let $\mathfrak{X}$ be a CAT(0)-simplicial complex, $A$ an abelian group. Suppose we have an idempotent operator $\Lambda_x : A \to A$ for each vertex $x$ of $\mathfrak{X}$. We call this system of idempotents *geodesic* if the following conditions hold:

(i) $\Lambda_x \Lambda_y = \Lambda_y \Lambda_x$ if $x$ and $y$ are adjacent,
(ii) $\Lambda_x \Lambda_z \Lambda_y = \Lambda_x \Lambda_y$ and $\Lambda_x \Lambda_z = \Lambda_z \Lambda_x$ if $z$ is any vertex of the first simplex along the geodesic $[x, y]$ for all vertices $x$ and $y$.

Such geodesic system gives a cosheaf $\underline{\underline{A_\Lambda}}$ where $\underline{\underline{A_\Lambda}}_F$ is the image of the product $\prod_x \Lambda_x$ taken over all faces of $F$ and $r^F_{F'}$ are natural inclusions.

Is it true that $H_m(\mathfrak{X}, \underline{\underline{A_\Lambda}}) = 0$ for all $m > 0$?

A positive answer to this question for Bruhat-Tits buildings can be obtained by the methods of Meyer and Solleveld [20].

## References

1. P. Abramenko, K. Brown, *Buildings: Theory and applications*, Springer, New York, 2008.
2. P. Abramenko, B. Muhlherr, *Presentations de certaines BN-paires jumeles comme sommes amalgames*, C. R. Acad. Sci. Paris, Ser. I **325** (7) (1997) 701–706.
3. J. Bernstein, *Representations of p-adic groups*, http://math.harvard.edu/~gaitsgde/Jerusalem_2010/GradStudentSeminar/p-adic.pdf, Harvard University, 1992.
4. C. Bushnell, G. Henniart, *The Local Langlands Conjecture for GL(2)*, Springer, 2006.
5. I. Capdeboscq, K. Kirkina, D. Rumynin, *Presentations of affine Kac-Moody groups*, Forum Math. Sigma **6** (2018), e21, 35 pp.
6. I. Capdeboscq, D. Rumynin, *Kac-Moody groups and completions*, arXiv:1706.08374
7. P.-E. Caprace, B. Rémy, *Groups with a root group datum*, Innov. Incidence Geom. **9** (2009), 5–77.
8. P.-E. Caprace, B. Rémy, *Simplicity and superrigidity of twin building lattices*, Invent. Math. **176** (2009), 169–221.
9. L. Carbone, M. Ershov, G. Ritter, *Abstract simplicity of complete Kac-Moody groups over finite fields*, J. Pure Appl. Algebra **212** (2008), 2147–2162.
10. R. Carter, Y. Chen, *Automorphisms of Affine Kac-Moody Groups and Related Chevalley Groups over Rings*, Journal of Algebra **155** (1993), 44–54.
11. R. Carter, *Lie algebras of finite and affine type*, Cambridge University Press, Cambridge, 2005.
12. M. Davis, *The Geometry and Topology of Coxeter groups*, LMS Monographs, London, 2008.
13. M. Davis, *Buildings are CAT(0)*, Geometry and Cohomology in Group Theory, *1994*, Durham, 108 – 123.
14. P. Gabriel, M. Zisman, *Calculus of Fractions and Homotopy Theory*, Berlin-Heidelberg-New York, Springer, 1967.
15. K. Hristova, D. Rumynin, *Kac-Moody groups and cosheaves on Davis building*, J. Algebra **515** (2018), 202–235.
16. V. Kac, *Infinite-dimensional Lie algebras*, Cambridge University Press, Cambridge, 1985.
17. S. Kumar, *Kac-Moody groups, their flag varieties and representation theory*, Birkhäuser, Boston, 2002
18. T. Marquis, *Topological KacMoody groups and their subgroups*, Ph.D. Thesis, Universite Catholique de Louvain, 2013.
19. T. Marquis, *Abstract simplicity of locally compact Kac-Moody groups*, Compos. Math. **150** (2014), 713–728.
20. R. Meyer, M. Solleveld, *Resolutions for representations of reductive p-adic groups via their buildings*, J. Reine Angew. Math. **647** (2010), 115–150.
21. C. Reid, P. Wesolek, *Homomorphisms into totally disconnected, locally compact groups with dense image*, arXiv:1509.00156.

22. B. Rémy, *Groupes de Kac-Moody déployés et presque déployés*, Astérisque No. 277, 2002.

23. G. Rousseau, *Groupes de Kac-Moody déployés sur un corps local, immeubles microaffines*, Compos. Math. **142** (2006), 501–528.

24. G. Rousseau, *Groupes de Kac-Moody déployés sur un corps local II. Masures ordonnées*, Bull. Soc. Math. France **144** (2016), 613–692.

25. P. Schneider, U. Stuhler, *Representation theory and sheaves on the Bruhat-Tits building*, Publications Mathmatiques de l'IHS, Springer, 1997.

26. B. Veisfeiler, V. Kac, *Exponentials in Lie algebras of characteristic p*, Izv. Akad. Nauk SSSR Ser. Mat. **35** (1971), 762–788.

27. D. Rumynin, M. Westaway, *Integration of modules - II: exponentials*, arXiv:1807.08698.

28. J. Tits, *Définition par générateurs et relations de groups avec BN-paires*, C. R. Acad. Sci. Paris Sér. I Math. **293** (1981), 317–322.

29. J. Tits, *Uniqueness and presentation of Kac-Moody groups over fields*, J. Algebra **105** (1987), 542–573.

30. Z. X. Wan, *Introduction to Kac-Moody algebra*, World Scientific, 1991.

# On the heredity of $V$-modules over Noetherian nonsingular rings*

Jarunee Soontharanon

*Dept. of Mathematics, Faculty of Science,
Mahidol University, Bangkok 10400, Thailand*

*ja_soon@hotmail.com*

Dinh Van Huynh

*Dept. of Mathematics, Ohio University,Athens, OH 45700*

*huynhohio@gmail.com*

Nguyen van Sanh

*Department of Mathematics, Faculty of Science, Mahidol University
and Center of Excellence in Mathematics, CHE Bangkok 10400, Thailand*

*nguyen.san@mahidol.ac.th*

A right module $M_R$ over a ring $R$ is called a V-module if every simple right $R$-module is $M$-injective. A module $N_R$ is defined to be hereditary if every submodule of $N$ is projective. With these concepts, the main result of Ref. 6 can be restated in the form: a V-module $P_R$ over a prime right Noetherian ring $R$ is hereditary if and only if $R_R$ is hereditary and $P$ is $R$-projective.

In this paper we consider a general case for a right Noetherian right nonsingular ring $R$ having a nonsingular V-module $P_R$, and show that all uniform right ideal $U \subseteq R$ which are embedded in $P_R$ generate a nonzero two-sided ideal $\mathcal{U}_R(P) \subseteq R$ such that $\mathcal{U}_R(P)_R$ is a V-module. Further, we show that, if $\mathcal{U}_R(P)_R$ is hereditary, then $P_R$ is hereditary if and only if $P$ is $\mathcal{U}_R(P)_P$-projective (Theorem 3.1). If $R$ is a right Noetherian right hereditary ring, then all $R$-projective V-module $P_R$ are hereditary (Corollary 3.1).

If $R$ is semisimple right Noetherian ring, then a V-module $P_R$ is hereditary if and only if $\mathcal{U}_R(P)_R$ is hereditary and $P_R$ is $\mathcal{U}_R(P)_R$-projective. In this case, $\mathcal{U}_R(P)$ is a ring-direct summand of $R$ (Theorem 2.1). There are known examples of prime Noetherian V-rings which are not hereditary. Over these rings, there are no nonzero hereditary modules.

---

*Funded by The Thailand Research Fund (Contract No. PH.D/0208/2552).

## 1. Introduction

A ring $R$ is called a right V-ring if every simple right $R$-module is injective. V-rings are named after Villamayor who studied them and showed that these rings are characterized by the property that every right module has zero Jacobson radical or, equivalently, that every right ideal is an intersection of maximal right ideals(see, e.g., Ref. 9(7.32A). Due to these and other interesting properties, V-rings have drawn much attention from researchers since their inception.

A natural generalization of V-rings is the concept of V-modules. A right module $M$ over a ring $R$ is called a V-module if every simple right $R$-module $S$ is $M$-injective. This means, for any submodule $A \subseteq M$, any homomorphism $f \colon A \to S$ can be extended to a homomorphism $f^* \colon M \to S$. It is obvious that subfactors of a V-module are also V-modules. In many cases, studying V-modules has an advantage that V-condition is assumed only on the module under consideration, while other modules do not need to be V. If we assume the ring $R$ to be a right V-ring, then all right $R$-modules are V-modules.

For more information on V-rings and V-modules we refer to Ref. 17 and the recent work[7] and references listed therein.

We consider associative rings with identity. All modules are unitary modules. Dual to the injectivity, the projectivity of modules is defined as follows.

Let $M, N$ be right $R$- modules. The module $M$ is called $N$-projective if for each exact sequence

$$0 \longrightarrow H \longrightarrow N \overset{g}{\longrightarrow} K \longrightarrow 0$$

in Mod-$R$, and any homomorphism $f \colon M \to K$, there is a homomorphism $f' \colon M \to N$ such that $gf' = f$.

A right $R$-module $P$ is defined to be a projective module if $P$ is $N$-projective for any $N \in$ Mod-$R$. A module $H_R$ is said to be hereditary if all submodules of $H_R$ are projective. A ring $R$ is called right (left) hereditary if $R_R$ ($_RR$) is hereditary.

For basic properties of projective modules as well as concepts of modules and rings not defined here, we refer to the texts[1,3,9,11,14,16,18].

Unlike the injectivity of modules, an $R$- projective module may not be projective, in general. As an example for that, we consider the ring $\mathbb{Q}$ of rational numbers as a module over the ring $\mathbb{Z}$ of integers. The module $\mathbb{Q}_{\mathbb{Z}}$ is $\mathbb{Z}$-projective. But, obviously, $\mathbb{Q}_{\mathbb{Z}}$ is not projective. Note that $\mathbb{Z}$ is a Noetherian hereditary domain, it is even a commutative PID.

Motivated by this, in Ref. 6, the projectivity of modules over prime right hereditary right Noetherian right V-ring was studied, and it was shown that over such rings $R$, a module $P_R$ is projective if and only if $P$ is $R$-projective. In this paper we show that a right Noetherian right nonsingular ring $R$ with a nonsingular V-module $P_R$ contains a uniquely determined ideal $\mathcal{U}_R(P)$ such that $\mathcal{U}_R(P)_R$ is the smallest V-submodule of $R_R$ which contains all those uniform V-submodules of $R_R$ which are embedded in $P_R$ (Lemma 1). Further, we show that if $R$ is a semiprime right Noetherian ring such that $\mathcal{U}_R(P)_R$ is hereditary, then $\mathcal{U}_R(P)$ is a ring-direct summand of $R$. Moreover, we also show that $P_R$ is hereditary if and only if $\mathcal{U}_R(P)_R$ is hereditary and $P$ is $\mathcal{U}_R(P)$-projective (Theorem 2.1).

The general case of $\mathcal{U}_R(P)_R$ for right Noetherian right nonsingular rings is discussed in Section 3. In Theorem 3.1, we show that, if $\mathcal{U}_R(P)_R$ is hereditary, then $P_R$ is hereditary if and only if $P_R$ is $\mathcal{U}_R(P)$-projective.

From Theorem 3.1, we obtain a result which states that, over a right Noetherian right hereditary ring $R$, a V-module $P_R$ is hereditary if and only if $P_R$ is $R$- projective. (Corollary 3.1). This includes the main result of Ref. 6.

A submodule $E$ of a module $M$ is called an essential submodule if for any nonzero submodule $A \subseteq M, E \cap A \neq 0$. A non zero module $M$ is called uniform if every nonzero submodule of $M$ is essential in $M$.

A right $R$-module $N$ is called nonsingular if for any nonzero element $x \in N$, the right annihilator $ann_R(x)$ of $x$ in $R$ is not an essential right ideal of $R$. A right $R$-module $S$ is called a singular module if the annihilator in $R$ of each element of $S$ is an essential right ideal of $R$. Every $R$- module $M$ has the largest singular submodule $Z(M)$, which contains all singular submodules of $M$. This is a fully invariant submodule of $M$ and it is called the singular submodule of $M$. Clearly, $M$ is nonsingular if and only if $Z(M) = 0$. For a ring $R$, if $Z(R_R) = 0$ (resp., $Z(_R R) = 0$), then $R$ is called right (left) nonsingular. (See, e.g., Ref. 10(p. 5)). To indicate that $M$ is a right (left) module over $R$ we write $M_R$ (resp., $_R M$). If $A$ is a set then $M^{(A)}$ denotes the direct sum of $|A|$ copies of $M$. If $A$ is finite, say $|A| = n$ where $n$ is a positive integer, then instead of $M^{(A)}$ we simply write $M^n$. If a module $M$

has finite uniform dimension, we denote its dimension by $u$-$\dim(M)$, and this module is called a finite dimensional module.

## 2. Semiprime Noetherian Rings with nonsingular V-modules

A simple right module over a ring is always a V-module. This implies that every semisimple module is a V-module. Hence every rings has nonzero V-modules. But we are interested only on nonsingular V-modules over Noetherian nonsingular rings.

**Lemma 2.1.** *Let $R$ be a right Noetherian right nonsingular ring which has a nonsingular V-module $P_R$. Then, the sum $A$ of all those uniform right ideals of $R$ which are embedded in $P_R$ is a nonzero two-sided ideal of $R$. Moreover, $A_R$ is a V-module. This ideal $A$ is uniquely determined by $P$, and we denote it by $\mathcal{U}_R(P)$.*

**Proof.** Let $P_R$ be a nonsingular V-module. For $0 \neq x \in P$, $xR \cong R/ann_R(x)$. As $xR$ is nonsingular, $ann_R(x)$ is not essential in $R_R$. This implies that $xR$ contains a copy of a uniform right ideal $U \subseteq R$. In particular, $U_R$ is a V-module.

Now, let $\{Y_\alpha\}_{\alpha \in \Omega}$ be the system of all uniform right ideals of $R$ such that each $(Y_\alpha)_R$ is embedded in $P_R$. In particular, each $(Y_\alpha)_R$ is a V-module. For any $r \in R$, and $Y_\alpha \in \{Y_\alpha\}_{\alpha \in \Omega}$, $R \supseteq rY_\alpha \cong Y_\alpha/ann_{Y_\alpha}(r)$. But, as $R_R$ is nonsingular and $Y_\alpha$ is a uniform module, $rY_\alpha$ is either zero or $rY_\alpha \in \{Y_\alpha\}_{\alpha \in \Omega}$. This particularly shows that the right ideal

$$A = \sum_{\alpha \in \Omega} Y_\alpha$$

is a nonzero two-sided ideal of $R$.

Moreover, let $Y = \bigoplus_{\alpha \in \Omega} Y_\alpha$. For any simple right $R$-module $S$, $S$ is $Y_\alpha$-injective. Thus by Ref. 1(16.13(2)), $S$ is $Y$-injective. This means $Y_R$ is a V-module . Let

$$\psi \colon Y_R \longrightarrow A_R$$

be a map which is defined by

$$\psi \colon (x_{\alpha_1} r_1, \cdots, x_{\alpha_k} r_k) \longmapsto x_{\alpha_1} r_1 + \cdots + x_{\alpha_k} r_k, \forall x_{\alpha_i} \in Y_{\alpha_i}, \forall r_i \in R$$

where $k$ is any positive integer. Then $\psi$ is an epimorphism. Thus, as an epimorphic image of the V-module $Y_R$, $A_R$ is a V-module. Since $A$

is uniquely determined by the nonsingular V-module $P$, we denote $A$ by $\mathcal{U}_R(P)$. □

**Lemma 2.2.** [6(Corollary 4)]. *If $R$ is a prime right Noetherian right hereditary right V-ring, then a module $P_R$ is hereditary if and only if $P_R$ is $R$-projective..*

The following lemma is a specal case of Ref. 18(39.7) where hereditary modules are from Mod-$R$ :

**Lemma 2.3.** *Let $\{P_\lambda\}_{\lambda \in \Lambda}$ be a family of hereditary modules. Then $P = \bigoplus_{\lambda \in \Lambda} P_\lambda$ is hereditary, and any submodule of $P$ is isomorphic to $\bigoplus_{\lambda \in \Lambda} Q_\lambda$ where each $Q_\lambda$ is a submodule of $P_\lambda$.*

Lemma 2.3 generalizes a result of Kaplansky [12] from hereditary rings to hereditary modules.

Now we are ready to prove the main result of this section.

**Theorem 2.1.** *Let $R$ be a semiprime right Noetherian ring having a nonsingular V-module $P_R$ which determines the ideal $\mathcal{U}_R(P) \subseteq R$ in the sense of Lemma 2.1.*
*(I) If $\mathcal{U}_R(P)_R$ is hereditary, then $R = \mathcal{U}_R(P) \oplus B$ for some ideal $B \subset R$ such that $PB = 0$.*
*(II) The following conditions are equivalent:*
  *(a) $P_R$ is hereditary;*
  *(b) $\mathcal{U}_R(P)$ is hereditary and $P_R$ is $\mathcal{U}_R(P)$-projective.*

**Proof.** Note that a semiprime right Noetherian ring is right nonsingular. Hence $\mathcal{U}_R(P)_R$ is well-defined by Lemma 2.1.

(I) Assume that $\mathcal{U}_R(P)_R$ is hereditary. By the construction of $\mathcal{U}_R(P)$, $\mathcal{U}_R(P) = \sum_{\alpha \in \Gamma} U_\alpha$ where each $U_\alpha$ is a uniform right ideal of $R$ which is embedded in $P_R$. In particular, each $U_\alpha$ is hereditary. Since $R$ is a right Noetherian, $\mathcal{U}_R(P)$ is a finitely generated submodule of $R_R$. Hence $\mathcal{U}_R(P) = \sum_{i=1}^{m} U_{\alpha_i}$ for some positive integer $m$. First we show that every submodule $V$ of $\mathcal{U}_R(P)_R$ is isomorphic to a direct sum of uniform submodules of $\mathcal{U}_R(P)_R$ that are embedded in $P_R$. Since $R$ is semiprime, $\mathcal{U}_R(P)V \neq 0$. There is a $U_{\alpha_i}$ with $U_{\alpha_i} V \neq 0$. Therefore there is $0 \neq x_i \in U_{\alpha_i}$ with $x_i V \neq 0$. As $x_i V$ is projective, the epimorphism $V \to x_i V (v \mapsto x_i v, \forall v \in V)$ splits. Hence $V \cong x_i V \oplus ann_V(x_i)$. Since $x_i V \subseteq U_{\alpha_i}$, $x_i V$ is embedded in

$P_R$. If $V_1 := ann_V(x_i)$ is nonzero, there is an $U_{\alpha_i}$ with $U_{\alpha_j} V_1 \neq 0$. By a similar argument, we get $V_1 \cong x_j V_1 \oplus ann_{V_1}(x_j)$ where $x_j V_1 \subseteq U_{\alpha_j}$. Hence $V \cong x_i V \oplus x_j V_1 \oplus ann_{V_1}(x_j)$ where $x_i V$ and $x_j V_1$ are embedded in $P_R$. Since $u$-dim$(V_R)$ is finite, continuing this way we finally obtain the fact that $V$ is isomorphic to a direct sum of uniform submodules of $R_R$ that are embedded in $P_R$. From this, we particularly conclude that

$(I_1)$ Every uniform submodule of $\mathcal{U}_R(P)_R$ is embedded some $U_{\alpha_i}$ and hence in $P_R$,

$(I_2)$ For a right ideal $U \subseteq R$, if $U_R$ is isomorphic to a submodule of $\mathcal{U}_R(P)_R$, then $U \subseteq \mathcal{U}_R(P)$.

Let $T$ be a complement of $\mathcal{U}_R(P)$ in $R_R$. Then $\mathcal{U}_R(P) \oplus T$ is essential in $R_R$. Hence there is a regular element $c$ of $R$ with $c \in \mathcal{U}_R(P) \oplus T$ (cf.,e.g., Ref. 3(1.10)). Write $c = d+t, d \in \mathcal{U}_R(P), t \in T$. Since $T\mathcal{U}_R(P) \subseteq T \cap \mathcal{U}_R(P) = 0$, we have $ann_{\mathcal{U}_R(P)}(d) = 0$, for otherwise $c$ would not be a regular element of $R$. The map $\varphi \colon R \to dR \subseteq \mathcal{U}_R(P)$ $(r \mapsto dr, \forall r \in R)$ is a homomorphism with $Ker(\varphi) \cap \mathcal{U}_R(P) = 0$. Since $\mathcal{U}_R(P)_R$ is hereditary, $\varphi$ splits, i.e.,

$$R = D \oplus Ker(\varphi)$$

for some right ideal $D$ with $D \cong dR$. By $(I_2)$, $D \subseteq \mathcal{U}_R(P)$, and so $\mathcal{U}_R(P) = D \oplus (Ker(\varphi) \cap \mathcal{U}_R(P))$. This implies $D = \mathcal{U}_R(P)$. Whence

$$R = \mathcal{U}_R(P) \oplus B \tag{1}$$

where $B = Ker(\varphi)$. For $0 \neq x \in \mathcal{U}_R(P)$, if $xB \neq 0$, then, as $xB$ is projective, $B \cong xB \oplus ann_B(x)$. Then by $(I_2)$, $\mathcal{U}_R(P) \cap B \neq 0$, a contradiction. Thus $\mathcal{U}_R(P)B = 0$. This means $B$ is an ideal of $R$, and so (1) is a ring-direct sum. If there is an element $0 \neq y \in P$ such that $yB \neq 0$, then, as $yB \subseteq P$ is nonsingular, $ann_B(y)$ is not essential in $B_R$. This implies the fact that $yB$ has a uniform submodule which is isomorphic to a submodule of $B_R$, a contradiction to the fact that $\mathcal{U}_R(P) \cap B = 0$. Thus $PB = 0$, proving (I).

(II) (a)$\Rightarrow$(b). By (a), as a hereditary module, $P_R$ is $\mathcal{U}_R(P)$-projective. Again we have $\mathcal{U}_R(P) = \sum_{i=1}^{m} U_{\alpha_i}$ for some positive integer $m$ where each $U_{\alpha_i}$ is uniform and can be embedded in $P_R$. Hence each $U_{\alpha_i}$ is hereditary. In the first part of the proof of (I), our argument is depending only on the hereditary of each $U_{\alpha_i}$. Therefore, we can apply it here to see that $(I_1)$ holds, and also that $\mathcal{U}_R(P)_R$ is isomorphic to a direct sum of uniform submodules which are hereditary because of $(I_1)$ and the heredity of $P_R$. By Lemma 2.3, $\mathcal{U}_R(P)_R$ is hereditary.

(b)⇒(a). By (I), $R = \mathcal{U}_R(P) \oplus B$ for some ideal $B \subseteq R$ such that $PB = 0$. In particular, $\mathcal{U}_R(P)$ is a semiprime right Noetherian right V-ring, and $P_R$ is a $\mathcal{U}_R(P)$-projective right $\mathcal{U}_R(P)$-module.

By Ref. 9(7.36A), $\mathcal{U}_R(P) = \bigoplus_{i=1}^{t} A_i$ (a ring-direct sum) where each $A_i$ is a prime ring with identity $e_i$. Then $e_i = \sum_{i=1}^{t} e_i$ is the identity of $\mathcal{U}_R(P)$ and we have $P = Pe = \bigoplus_{i=1}^{t} Pe_i$. In particular, each $Pe_i$ is a V-module over the prime right Noetherian right hereditary right V-ring $A_i$, and $Pe_i$ is $A_i$-projective. Hence by Lemma 2.2, each $Pe_i$ is projective. Thus $P_R$ is projective, and therefore $P_R$ is isomorphic to a direct summand of the free module $\mathcal{U}_R(P)^{(\Omega)}$ for some set $\Omega$. Hence by a theorem of Kaplansky[12] (or Lemma 2.3), $P_R$ and each submodule of it are projective, proving that $P_R$ is hereditary. □

## 3. For Non-semiprime Case, When Is A $\mathcal{U}_R(P)$-projective V-Module Hereditary?

In this section we will use $\mathcal{U}_R(P)$ to investigate the question of when a V-module $P_R$ is hereditary. For this purpose we need one more concept:

Let $M$ and $N$ be modules. We say that $M$ is $N$-generated if there is an epimorphism $N^{(\Delta)} \to M$ for some set $\Delta$. In this case, every submodule of $M$ is called an $N$-subgenerated module.

**Lemma 3.1.** *Let $R$ be a right Noetherian right nonsingular ring with a nonsingular V-module $P_R$ such that $\mathcal{U}_R(P)_R$ is hereditary and $P$ is $\mathcal{U}_R(P)$-projective. Then*
*(a) Every finite dimensional submodule of $P_R$ is embedded in some finite direct sum of copies of $\mathcal{U}_R(P)$;.*
*(b) $P_R$ is $\mathcal{U}_R(P)$-subgenerated.*

**Proof.** By our assumptions, $\mathcal{U}_R(P) \neq 0$.

(a) Let $M \subseteq P$ be a submodule with $u\text{-dim}(M_R) = k$, where $k$ is a positive integer. Then there are $k$ cyclic uniform independent submodules $X_i \subseteq M$ such that $X_1 \oplus \cdots \oplus X_k$ is essential in $M$. Let $X_i = x_iR$ for some $x_i \in X_i$. Since $x_iR \cong R/ann_R(x_i)$ and $ann_R(x_i)$ is not essential in $R_R$, $x_iR$ contains a cyclic uniform submodule $Y_i$ which can be embedded in $R_R$. In particular, by the assumption of $\mathcal{U}_R(P)$ we see that $Y_i$ is hereditary. Moreover, as $Y_i$ is isomorphic to a submodule of $\mathcal{U}_R(P)_R$, $P_R$ is $(Y_1 \oplus \cdots \oplus$

$Y_k$)-projective (see Ref. 1(16.12)). Note that $Y_1 \oplus \cdots \oplus Y_k$ is a hereditary right $R$-module.

From here we can use a similar argument of Ref. 6(Lemma 2(b)) to see the existence of a homomorphism $\varphi \colon M_R \longrightarrow Y_1 \oplus \cdots \oplus Y_k$ such that

$$M \cong \varphi(M) \oplus K, \text{ and } \varphi(M) \neq 0 \qquad (1)$$

where $K = Ker(\varphi)$.

Now we prove the claim by induction on $k$. For $k = 1$, $M$ is uniform, and hence $K = 0$ and so $M$ is embedded in $\mathcal{U}_R(P)$. Now assume that the claim is true for any submodule with uniform dimension $< k$. If in (1), $K \neq 0$, then both $\varphi(M)$ and $K$ have uniform dimension $< k$, and hence they are embedded in a finite direct sum of copies of $\mathcal{U}_R(P)$. Therefore, by (1), $M$ is embedded in a direct sum of finitely many copies of $\mathcal{U}_R(P)$. If $K = 0$, then $M \cong \varphi(M)$, and so $M$ is embedded in $\bigoplus_{i=1}^{k} Y_i$ which is isomorphic to a submodule of $\bigoplus_{i=1}^{k} \mathcal{U}_R(P)$.

(b) Write $P_R = \sum_{\alpha \in \Omega} x_\alpha R$ for some index set $\Omega$. By (a), as a Noetherian submodule of $P_R$, $x_\alpha R$ is embedded in $\mathcal{U}_R(P)^{n_\alpha}$ where $n_\alpha$ is a positive integer. Set $T_\alpha = \mathcal{U}_R(P)^{n_\alpha}$. Then we can consider $\bigoplus_{\alpha \in \Omega} x_\alpha R$ as a submodule of $\bigoplus_{\alpha \in \Omega} T_\alpha$. There is an epimorphism $\varphi \colon \bigoplus_{\alpha \in \Omega} x_\alpha R \longrightarrow \sum_{\alpha \in \Omega} x_\alpha R = P$ (see the proof of Lemma 2.1). This shows that $P_R$ is isomorphic to a submodule of $(\bigoplus_{\alpha \in \Omega} T_\alpha)/Ker(\varphi)$. Hence $P_R$ is subgenerated by $\mathcal{U}_R(P)$. $\qquad \square$

Now we are ready to prove the main result of this section.

**Theorem 3.1.** *Let $P_R$ be a nonsingular V-module over a right Noetherian right nonsingular ring $R$. If $\mathcal{U}_R(P)_R$ is hereditary, then the following conditions are equivalent:*

*(i) $P$ is $\mathcal{U}_R(P)$-projective;*

*(ii) $P_R$ is hereditary.*

**Proof.** The ideal $\mathcal{U}_R(P)$ ($\subseteq R$) is determined by $P_R$ as in Lemma 2.1.

(ii)$\Rightarrow$(i) is obvious because (i) is a special case of (ii) (see, e.g., Ref. 1(16.12(1))).

(i)$\Rightarrow$(ii): As $P_R$ is nonsingular, $P \neq 0$. Write $P = \sum_{\alpha \in \Delta} x_\alpha R$. Let $\leq$ be a well ordering of $\Delta$. Then $\Delta$ can be viewed as the set $\{1, 2, 3, \ldots, \omega, \omega+1, \ldots\}$ of ordinal numbers. Let $\gamma = ord(\Delta)$. Hence $P = \sum\limits_{\alpha \leq \gamma} x_\alpha R$. For any $\alpha \in \Delta$, set $P(\alpha) = \sum\limits_{\beta \leq \alpha} x_\beta R$. We use a transfinite induction proof to show that $P(\alpha)$ is embedded in some direct sum $\mathcal{U}_R(P)^{(\Lambda_\alpha)}$ for each $\alpha \in \Delta$. By Lemma 3.1(a), our claim is true for any finite $\alpha \in \Delta$ because in this case, $P(\alpha)$ is Noetherian and hence finite dimensional.

Now, for a fixed $\alpha < \gamma$, we assume that $P(\beta)$ is embedded in $\mathcal{U}_R(P)^{(\Lambda_\beta)}$ for all $\beta < \alpha$. By Lemma 2.3, as a submodule of $\mathcal{U}_R(P)^{(\Lambda_\beta)}$, we have

$$P(\beta) = \bigoplus_{i \in K_\beta} P_{\beta_i}$$

where each $P_{\beta_i}$ is isomorphic to a submodule of $\mathcal{U}_R(P)$.

If $\alpha = \beta + 1$, then $P(\alpha) = P(\beta) + x_\alpha R$. As $x_\alpha R$ is Noetherian, there is a finite subset $F_1 \subseteq K_\beta$, such that $P(\beta) \cap x_\alpha R \subseteq (\bigoplus_{i \in F_1} P_{\beta_i})$. Set $Q = \bigoplus_{i \in K_\beta \setminus F_1} P_{\beta_i}$. Again, since $Q \cap [(\bigoplus_{i \in F_1} P_{\beta_i}) + x_\alpha R]$ is Noetherian, there is a finite subset $F_2 \subseteq K_\beta \setminus F_1$ such that $Q \cap [(\bigoplus_{i \in F_1} P_{\beta_i}) + x_\alpha R] \subseteq \bigoplus_{i \in F_2} P_{\beta_i}$. Set $F = F_1 \cup F_2$. Then $(\bigoplus_{i \in K_\beta \setminus F} P_{\beta_i}) \cap [(\bigoplus_{i \in F} P_{\beta_i}) + x_\alpha R] = 0$. Thus

$$P(\alpha) = (\bigoplus_{i \in K_\beta \setminus F} P_{\beta_i}) \oplus [(\bigoplus_{i \in F} P_{\beta_i}) + x_\alpha R]. \tag{2}$$

Now, as $(\bigoplus_{i \in F} P_{\beta_i}) + x_\alpha R$ is finitely generated, $(\bigoplus_{i \in F} P_{\beta_i}) + x_\alpha R$ is Noetherian, by Lemma 3.1(a), it is embedded in a direct sum of finitely many copies of $\mathcal{U}_R(P)$. By Lemma 2.3, $(\bigoplus_{i \in F} P_{\beta_i}) + x_\alpha R$ is a direct sum of submodules of $\mathcal{U}_R(P)$. Putting this back into (2), we see that $P(\alpha)$ is embedded in $\mathcal{U}_R(P)^{\Lambda_\alpha}$ for some suitable set $\Lambda_\alpha$.

If $\alpha$ is a limit ordinal, we have $P(\alpha) = \bigcup\limits_{\beta < \alpha} P(\beta) \subseteq P$. By Lemma 3.1(b), $P$ is subgenerated by $\mathcal{U}_R(P)$, hence so is $P(\alpha)$. This means there is an epimorphism $\varphi \colon \mathcal{U}_R(P)^{(A)} \longrightarrow Q_R$ such that $P(\alpha)$ is a submodule of $Q_R$ where $A$ is a suitable index set. Let $V = \varphi^{-1}(P(\alpha)) \subseteq \mathcal{U}_R(P)^{(A)}$. By Lemma 2.3, $V = \bigoplus\limits_{\lambda \in A} V_\lambda$ where each $V_\lambda$ is isomorphic to some submodule of $\mathcal{U}_R(P)$. Set $\psi = \varphi|_V$. Note that $\{P(\beta)\}_{\beta < \alpha}$ is an ascending chain of projective submodules of $P$.

Let $T_\beta = \{x \in V \mid \psi(x) \in P(\beta)\}$, i.e., $T_\beta = \psi^{-1}(P(\beta)) \subseteq V$. It is clear that $Ker(\psi) \subseteq T_\beta$. Hence $T_\beta = W_\beta \oplus Ker(\psi)$ for some submodule $W_\beta \subseteq T_\beta$ such that $\psi(W_\beta) = P(\beta)$. Observe that for any $\beta < \beta' \lneqq \alpha$, $P(\beta) \subseteq P(\beta')$ if and only if $W_\beta \subseteq W_{\beta'}$. Hence the system $\{W_\beta\}_{\beta < \alpha}$ forms an ascending chain in $V$. Set $W = \bigcup_{\beta < \alpha} W_\beta$. We see that $W$ is a submodule of $V$. It is obvious that $\psi(W) \subseteq P(\alpha)$. Now let $y \in P(\alpha)$. Then, there is an ordinal $\beta_i < \alpha$ such that $y \in P(\beta_i)$. By the construction of $T_{\beta_i}$, there is an $x \in W_{\beta_i}$ such that $\psi(x) = y$. Hence $\psi(W) \supseteq P(\alpha)$, and so $\psi(W) = P(\alpha)$. Let $0 \neq z \in W$. There is a $W_{\beta_j}$, $\beta_j < \alpha$ such that $z \in W_{\beta_j}$. As $\psi|_{W_{\beta_j}} : W_{\beta_j} \to P(\beta_j)$ is an isomorphism, we have $\psi(z) \neq 0$. Thus $\varphi|_V$ is an isomorphism, or equivalently, $W \cong P(\alpha)$. This shows that $P(\alpha)$ is embedded in $\mathcal{U}_R(P)^{(\Lambda_\alpha)}$ for some suitable index set $\Lambda_\alpha$.

Thus our induction proof is complete, and so we conclude that $P_R$ is embedded in $\mathcal{U}_R(P)^{(\Lambda)}$ for some set $\Lambda$. By Lemma 2.3, we further conclude that $P_R$ is hereditary. $\qquad\square$

Unlike Theorem 2.1, in Theorem 3.1 we are not able to get the hereditary of $\mathcal{U}_R(P)_R$ from that of $P_R$. We also do not know if, for the ring $R$ in Theorem 3.1, $\mathcal{U}_R(P)$ is a direct summand of $R$ or not. In a right V-ring , any right ideal $I$ of it is idempotent, i.e., $I^2 = I$. We do not know if the same holds for the V-module $\mathcal{U}_R(P)$. If yes, then $\mathcal{U}_R(P) \cap N = 0$ where $N$ is the prime radical of $R$.

**Corollary 3.1.** *Let $R$ be a right hereditary right Noetherian ring. Then a V-module $P_R$ is hereditary if and only if $P$ is $R$-projective..*

**Proof.** One direction is clear. Assume that $P$ is $R$-projective. First we show that $P_R$ is nonsingular (i.e., $Z(P) = 0$.) As $R$ is a right hereditary, $R$ is right nonsingular. Assume on the contrary that $Z(P) \neq 0$. This means that there is a nonzero element $a \in Z(P)$ with $aR \in Z(P) \subset P$. Setting $T = aR$ is a nonzero singular cyclic submodule $T$ which is contained in $P$. If $P_R$ is a V-module, then so is $T$. Since $T$ is finitely generated, $T$ contains a maximal submodule $X$. As $T/X$ is simple, $T/X$ is injective by definition of V-module.

Consider a homomorphism $\varphi : T/X \overset{\iota}{\longrightarrow} P/X$, where $T \subseteq P$. Since $T/X$ is injective, $Im(\varphi)$ is a direct summand of $P/X$ and then we can write $P/X = T/X \oplus U/X$ for some submodule $U$ of $P$ with $X \subset U$. Thus $P/U \cong (P/X)/(U/X) \cong T/X$.

On the other hand, there exists a maximal right ideal $B$ of $R$ such that $R/B \cong T/X$ as a right $R$-module. This means there is a homomorphism $f : P \to R/B$ with $Ker(f) = U$. By the definition of $R$-projectivity, there exists a homomorphism $f' : P \to R_R$ such that $gf' = f$, where $g$ is the canonical epimorphism $R \to R/B$. This is impossible because if $R$ is a right nonsingular, then $Ker(f')$ must contain the singular submodule $T$ which implies $Ker(gf') \neq Ker(f)$. Therefore $P$ does not contain a nonzero singular submodule, proving that $P_R$ is nonsingular. Hence $\mathcal{U}_R(P) \neq 0$ by Lemma 1. Obviously, $P_R$ is $\mathcal{U}_R(P)$-projective. Moreover, as an ideal of $R$, $\mathcal{U}_R(P)_R$ is hereditary. Thus $P_R$ is hereditary by Theorem 3.1.    $\square$

Corollary 3.1 includes the main result of Ref. 6. Since the module $\mathbb{Q}_{\mathbb{Z}}$ is $\mathbb{Z}$-projective however $\mathbb{Q}_{\mathbb{Z}}$ is not projective in Mod-$\mathbb{Z}$, the V-condition of $P_R$ in Theorem 2.1 and 3.1, in Corollary 3.1 cannot be removed. Also, since, for some prime number $p$, the $\mathbb{Z}$-module $C(p^\infty)$ is $\mathbb{Z}$-projective while $C(p^\infty)$ is not projective in Mod-$\mathbb{Z}$, the non-singularity assumption of $P_R$ in our results above cannot be removed.

Finally, we remark that there are prime two-sided Noetherian V-rings that are not hereditary (see Ref. 5 and the discussion in Ref. 2, Ref. 13; see also Ref. 8 for a related conjecture by A. Boyle on prime Noetherian V-rings). An example of a prime Noetherian hereditary V-ring which is not a division ring was given in Ref. 4.

## References

1. F.W. Anderson and K.R. Fuller, *Rings and Categories of Modules*, Springer Verlag, Berlin-Heidelberg-New York, 2nd Edition, 1992.
2. A.K. Boyle and K.R. Goodearl, *Rings over which certain modules are injective*, Pacific J. Math. **58** (1975), 43–53.
3. A.W. Chatters and C.R. Hajarnavis, *Rings with Chain Conditions*, Pitman, London, 1980
4. J.H. Cozzens, *Homlogical properties of the ring of differential polynomials*, Bull. Amer. Math. Soc., **76** (1970), 75–79.
5. J.H. Cozzens and J. Johnson, *Some applications of differential algebra to ring theory*, Proc. Amer. Math. Soc., **72** (1972), 354–356.
6. H.Q. Dinh, C.J. Holston and D.V. Huynh, *Quasi-projective modules over prime hereditary Noetherian V-ring s are projective or injective*, J. Algebra, **360** (2012), 87–91.
7. H.Q. Dinh, C.J. Holston and D.V. Huynh, *Some results on V-rings and WV-rings*, J. Pure Appl. Algebra, (2012), to appear.
8. C. Faith, *On hereditary rings and Boyle's Conjecture*, Arch. Math., **27** (1976), 113–119.

9. C. Faith, "Algebra I: Rings, Modules and Categories", Springer Verlag, Berlin-Heideberg-New York, 1973.

10. K.R. Goodearl, "Singular Torsion and the splitting Properties", Memoirs. Amer. Math. Soc., **124**, Providence, 1972.

11. K.R. Goodearl and R.B. Warfield, Jr., "An Introduction to Noncommutative Noetherian Rings", London Math. Soc. Texts **72**, Cambridge Univ. Press, 1989.

12. I. Kaplansky, *Projective modules*, Ann. Math., **68** (1958), 372-377.

13. K.A. Kosler, *On hereditary rings and Noetherian V-rings*, Pacific J. Math., **103** (1982), 467-473.

14. T.Y. Lam, *Lectures on Modules and Rings*, Springer Verlag Berlin-Heideberg-New York, GTM 190, 1998.

15. G.O. Michler and O.E. Villamayor, *On rings whose simple modules are injective*, J. Algebra, **25** (1973), 185–201.

16. L.H. Rowen, *Rings Theory, Vol. I*, Acadamic Press, New York, 1988.

17. A. Tuganbaev, *Max Rings and V-rings*, Handbook of Algebra, Holland, Amsterdam, **3** (2003), 565–584.

18. R. Wisbauer, *Foundations of Module and Ring Theory*, Reading, London, 1991.

# Quasigroups, hyperquasigroups, and vector spaces over fields with one or more elements

Jonathan D. H. Smith

*Department of Mathematics*
*Iowa State University*
*Ames, Iowa 50011, U.S.A.*

*jdhsmith@iastate.edu*
*URL: http://orion.math.iastate.edu/jdhsmith/*

Dedicated to L.A. Bokut' on the occasion of his eightieth birthday.

The paper presents an elementary and unified approach to vector spaces over fields of order greater than or equal to one (the latter reducing to sets), based on three key principles. Firstly, use of quasigroups enables the field concept to be redefined in a way that admits a field of order one. Secondly, use of hyperquasigroups provides a recursive definition of linear combination that applies equally well to vector spaces over fields of order greater than or equal to one. Thirdly, it is recognized that relations rather than functions provide the correct morphisms for a category of sets to behave like categories of vector spaces over fields of order greater than one.

*2010 Mathematics subject classification:* 12-01, 15-01, 05A30, 20N05.

*Keywords:* field, linear combination, quasigroup, hyperquasigroup, Hopf algebra, Gaussian binomial coefficient

## 1. Introduction

For various reasons, mathematicians have pondered the existence of a field with one element (compare Ref. 2–4,12,13,16,23,24 and Ref. 6(0.4.24.2) for example). The most ambitious motivation for such considerations is the desire to transfer Weil's proof of the Riemann hypothesis for curves over finite fields to a proof of the classical Riemann hypothesis, interpreting $\mathbb{Z}$ as a curve over a field of order one.

The goal of the current paper is much more modest and elementary. The primary motivation is the observation that in algebra, combinatorial structures on sets and linear algebraic structures on vector spaces often appear in parallel, so that it would be desirable to have a fully unified approach that embraces both. One example of such parallel appearance concerns groups over sets and Hopf algebras over vector spaces. A second example appears in the work of Bokut, Chen and Mo[1], juxtaposing Evans' proof that every countably generated semigroup can be embedded into a two-generated semigroup[8] alongside Malcev's proof that every countably generated associative (linear) algebra can be embedded into a two-generated associative (linear) algebra[17].

We begin by making a small change to the classical definition of a field so that it also embraces a field $\mathsf{GF}(1)$ of order one (Definition 2.2), but admits no other extraneous fields (Theorem 2.1). The new definition involves some rudimentary quasigroup theory, covered in §§2.1–2.3.

Once the field of order one is admitted, the next step is to construe sets as vector spaces, sustaining linear combinations. This is achieved using a new recursive definition of linear combinations (Definition 3.4) which is based on the concept of a linear hyperquasigroup. Hypergroups, which provide a more symmetrical version of quasigroup theory (enlarging left/right duality to triality) are introduced in §3.1. Linear hyperquasigroups are covered in §3.2. On sets, as vector spaces over $\mathsf{GF}(1)$, the recursive step in the new definition of a linear combination is never called. In vector spaces over fields with nonzero elements, on the other hand, the recursive construction of general linear combinations is modeled by parsing trees in linear hyperquasigroups, as illustrated by an example in Figure 2. Theorem 3.3 shows that the new definition of a linear combination agrees with the classical definition over fields with more than one element. Section 3.4 covers spans and subspaces over fields with one or more elements, while §3.5 deals with linear transformations.

One of the major obstacles to developing a theory of sets as vector spaces over a singleton field has been the fact that the category of sets and functions is quite unlike categories of vector spaces. In particular, the latter have biproducts, while coproducts of sets, namely disjoint unions, do not function as products. In the final chapter of the paper, we overcome this obstacle by turning to the category **Rel** of relations between sets, whose properties are summarized in §4.1. A unified treatment of linear algebra, working with the usual categories of finite-dimensional vector spaces in parallel with the category of relations between finite sets, is sketched in §4.2. Finally, in §4.3, unified interpretations of $q$-numbers, $q$-factorials and $q$-binomial coefficients are provided for enumeration questions in vector spaces over $\mathsf{GF}(q)$ for $q \geq 1$.

Readers are generally referred to Ref. 22 for those notational conventions and definitions that are not explicitly stated in the paper. While our general preference is for algebraic notation (first the argument, then the function, reading from left to right), the opposite Eulerian notation is used for the discussion of linear transformations in §3.5.

## 2. Quasigroups, and fields with one or more elements

### 2.1. *Combinatorial quasigroups*

Let $(Q, \cdot)$ be a magma, a (possibly empty) set $Q$ equipped with a binary operation $x \cdot y$ or $xy$ of *multiplication*. For each element $q$ of $Q$, define the *left multiplication*

$$L(q) \colon Q \to Q; \ x \mapsto q \cdot x \qquad (1)$$

and *right multiplication*

$$R(q) \colon Q \to Q; \ x \mapsto x \cdot q. \qquad (2)$$

The algebra $(Q, \cdot)$ is said to be a (*combinatorial*) *quasigroup* if all the left and right multiplications are permutations of $Q$.

**Example 2.1.** (a) A group $(Q, \cdot)$ is a non-empty quasigroup satisfying the associative law $xy \cdot z = x \cdot yz$. The associative law is expressed here with the governing convention that multiplications denoted by juxtaposition bind more strongly than explicitly written multiplications.

(b) The empty quasigroup $(\varnothing, \cdot)$ vacuously satisfies the associative law.

### 2.2. *Equational quasigroups*

The combinatorial specification of a quasigroup does not admit the use of universal algebraic techniques. For example, a combinatorial quasigroup $(Q, \cdot)$ may be the domain of a magma homomorphism $f \colon (Q, \cdot) \to (P, \cdot)$ whose image is not a combinatorial quasigroup, in violation of the First Isomorphism Theorem[22(Ch. I, Ex. 2.2.1)]. In 1949, Evans[7] redefined quasigroups in the form of *equational quasigroups*, namely universal algebras $(Q, \cdot, /, \backslash)$ equipped with three binary operations of multiplication, *right division* $/$ and *left division* $\backslash$ satisfying the identities

| | | | | |
|---|---|---|---|---|
| (IL) | $v\backslash(v \cdot w) = w$ | | $w = (w \cdot v)/v$ | (IR) |
| (SL) | $v \cdot (v\backslash w) = w$ | | $w = (w/v) \cdot v$ | (SR) |

The identities (IL), (IR) serve to yield the injectivity of the left and right multiplications, while (SL), (SR) give their surjectivity.

It is important to observe the symmetry of the equational quasigroup identities about the vertical line separating left from right. In other words, the theory of equational quasigroups possesses a left/right or chiral duality symmetry.

An equational quasigroup $(Q, \cdot, /, \backslash)$ yields a combinatorial quasigroup $(Q, \cdot)$. Conversely, a combinatorial quasigroup $(Q, \cdot)$ yields an equational quasigroup $(Q, \cdot, /, \backslash)$ with divisions $x/y = xR(y)^{-1}$ and $x \backslash y = yL(x)^{-1}$.

**Example 2.2.** Let $G$ be a group generated by a subset $\{R, L\}$ with at most two elements. Let $M$ be a right $G$-module. Then a quasigroup structure is defined on $M$ by $x \cdot y = xR + yL$. Quasigroups $(M, \cdot)$ of this type are described as being *linear*. The quasigroup structure, in combination with identification of 0 in $M$, serves to specify the module together with the $G$-action. Note that $xR = x \cdot 0$ and $yL = 0 \cdot y$ for $x, y \in M$. Then $x + y = (x/0) \cdot (0 \backslash y)$ and $-y = 0/[0 \backslash (y \cdot 0)]$.

Equational quasigroups form a variety in the sense of universal algebra, so the images of equational quasigroups, under homomorphisms of equational quasigroups, are themselves equational quasigroups[22 (p. 314)]. Evans' reformulation of the quasigroup concept opened up combinatorial questions about quasigroups and Latin squares to analysis with algebraic techniques[9]. In particular, we may note the following.

**Lemma 2.1.** *Suppose that $\theta: (Q, \cdot) \to (Q', \cdot)$ is a magma homomorphism between combinatorial quasigroups. Then $\theta: (Q, \cdot, /, \backslash) \to (Q', \cdot, /, \backslash)$ is a homomorphism of equational quasigroups.*

**Proof.** By (SR), the relation $x = (x/y) \cdot y$ holds for all $x, y \in Q$. Thus the relation $x^\theta = (x/y)^\theta \cdot y^\theta$ holds in $(Q', \cdot)$, so that

$$x^\theta / y^\theta = [(x/y)^\theta \cdot y^\theta]/y^\theta = (x/y)^\theta$$

by (SR) in $Q'$, and $\theta$ preserves right division. The proof that $\theta$ preserves left division follows by chiral duality. □

### 2.3. *Cayley's Theorem*

Example 2.1(a) noted that groups are associative quasigroups. As shown by Example 2.1(b), the converse statement is false. Nevertheless, it is almost true.

**Proposition 2.1 (Cayley's Theorem).** *A nonempty associative quasi-group is isomorphic to a permutation group.*

**Proof.** Let $(Q, \cdot, /, \backslash)$ be a nonempty associative quasigroup. Consider the right multiplication function

$$R \colon Q \to Q!; \ y \mapsto R(y) \tag{3}$$

from $Q$ to the group $Q!$ of all permutations of the set $Q$. Then the function is injective, since $R(y) = R(z)$ for $y, z \in Q$ implies

$$y = x \backslash (xy) = x \backslash [xR(y)] = x \backslash [xR(z)] = x \backslash (xz) = z$$

by (IL), using an arbitrary element $x$ of the nonempty set $Q$.

The associative law $xy \cdot z = x \cdot yz$ in $Q$ may be formulated as

$$\forall \, x, y, z \in Q, \ xR(y)R(z) = xR(yz)$$

or

$$\forall \, y, z \in Q, \ R(y)R(z) = R(yz),$$

so the right multiplication map (3) is a magma homomorphism. Then by Lemma 2.1, it is a homomorphism of equational quasigroups. The First Isomorphism Theorem for equational quasigroups shows that the domain $Q$ of the injective right multiplication homomorphism (3) is isomorphic to its image, a subgroup of $Q!$.  □

**Corollary 2.1.** *A nonempty associative quasigroup is a group.*

**Remark 2.1.** Suppose that $Q$ is the empty (associative) quasigroup. Then the right multiplication map (3) is the insertion $\emptyset \hookrightarrow \{1_\emptyset\}$. As such, it is an injective quasigroup homomorphism, whose image is the empty subquasigroup of the singleton group $\emptyset!$.

### 2.4. *Fields with one or more elements*

The classical definition of a field may be encapsulated as follows.

**Definition 2.1.** A *field* is a commutative ring, in which the set of nonzero elements forms a group under the ring multiplication.

Here, "ring" may denote a nonunital or unital ring, respectively meaning without or with the requirement for a multiplicative identity in the ring. Indeed, the multiplicative identity for the ring is imposed (along with the equation $0 \cdot 1 = 0$) directly by the group requirement in Definition 2.1. The "classical" Definition 2.1 will now be replaced by the following.

**Definition 2.2.** A *field* is a commutative ring, in which the set of nonzero elements forms a quasigroup under the ring multiplication.

**Proposition 2.2.** *Under Definition 2.2, the one-element commutative ring* $\mathsf{GF}(1)$ *is a field.*

**Proof.** Since $\mathsf{GF}(1) = \{0\}$, the set of nonzero elements of $\mathsf{GF}(1)$ is empty. As such, it forms a quasigroup. □

**Remark 2.2.** While some authors have previously insisted on a distinction $0 \neq 1$ between the additive and multiplicative identities of a unital ring, it should be noted that such inequalities are incompatible with the algebraic nature of the unital ring concept.

We now observe that the new Definition 2.2 does not make any change to the specification of fields with more than one element.

**Theorem 2.1.** *Let $F$ be a set of cardinality greater than one. Then $F$ is a field in the sense of Definition 2.2 if and only if it is a field in the sense of Definition 2.1.*

**Proof.** If $F$ is a field in the sense of Definition 2.1, Example 2.1(a) shows that it is a field in the sense of Definition 2.2. Conversely, suppose that $F$ is a field in the sense of Definition 2.2. Consider the set $Q$ of nonzero elements of $F$. Since $|F| > |\{0\}|$, the set $Q$ is nonempty. By Definition 2.2, $Q$ is a nonempty, commutative, associative quasigroup. By Corollary 2.1, $Q$ is a group. Then $F$ is a field in the traditional sense of Definition 2.1. □

## 3. Hyperquasigroups and linear combinations

### 3.1. *Hyperquasigroups*

As observed in §2.2, the equational theory of quasigroups is endowed with the two-fold symmetry of left/right duality. This symmetry was exploited in Evans' solution of the word problem for quasigroups[7]. Nevertheless, his solution still left many separate cases to consider, at least in principle. Subsequently, the solution of the word problem was drastically simplified by the explicit use of a stronger triality symmetry or $S_3$-action* that interchanges

---

*This triality symmetry was identified in Ref. 19 as *syntactic triality*. That paper also discussed *semantic triality*, which is more closely related to the triality symmetry of Moufang loops and the Coxeter-Dynkin diagram of type $D_4$ (compare Ref. 5).

all three equational quasigroup operations and their opposites[19]. While this triality symmetry is already implicit in the theory of equational quasi-groups, and its presence had long been recognized, the choice of specific operations in the equational theory was an impediment to its use in practice. Indeed, implementation of the symmetry entailed the introduction of a new approach to quasigroups, by means of the concept of a hyperquasi-group[18,20]. Hyperquasigroups may be considered as a further step beyond the progression from combinatorial quasigroups to equational quasigroups. They involve the auxiliary concept of a reflexion-inversion space:

**Definition 3.1.** A *reflexion-inversion space* $(\Omega, \sigma, \tau)$ is a set $\Omega$ equipped with two involutive actions, a *reflexion*

$$\sigma : \Omega \to \Omega; \ \omega \mapsto \sigma\omega \tag{4}$$

and an *inversion*

$$\tau : \Omega \to \Omega; \ \omega \mapsto \tau\omega . \tag{5}$$

**Example 3.1.** The most basic reflexion-inversion space is the symmetric group $S_3 = \{1, 2, 3\}!$, with reflexion and inversion implemented as the left multiplications by the respective transpositions (1 2) and (2 3). In this context, it is convenient to identify each element of $S_3$ as the image of the identity permutation under the left action of a series of reflexions and inversions.

**Definition 3.2.** A *hyperquasigroup* $(Q, \Omega)$ is a pair consisting of a set $Q$ and a reflexion-inversion space $\Omega$, together with a binary action

$$Q^2 \times \Omega \to Q; \ (x, y, \omega) \mapsto xy\,\underline{\omega} \tag{6}$$

of $\Omega$ on $Q$, such that the *hypercommutative law*

$$xy\,\underline{\sigma\omega} = yx\,\underline{\omega} \tag{7}$$

and the *hypercancellation law*

$$x\,(xy\,\underline{\omega})\,\underline{\tau\omega} = y \tag{8}$$

are satisfied for all $x$, $y$ in $Q$ and $\omega$ in $\Omega$.

Definition 3.2 may be expressed in graphical form. Suppose that $(Q, \Omega)$ is a hyperquasigroup. For elements $x$ of $Q$ and $\omega$ of $\Omega$, define the (*left*) *translation*

$$L_\omega(x) : Q \to Q; \ y \mapsto xy\underline{\omega} \tag{9}$$

and *right translation*

$$R_\omega(x) : Q \to Q; \ y \mapsto yx\underline{\omega} \tag{10}$$

by analogy with (1) and (2). Note that

$$R_\omega(x) = L_{\sigma\omega}(x) \tag{11}$$

by hypercommutativity, and

$$L_\omega(x)^{-1} = L_{\tau\omega}(x) \tag{12}$$

by hypercancellativity. The relation (12) serves to justify the use of the term "inversion" for $\tau$ in Definition 3.1.

Definition 3.2 is then summarized by the diagram

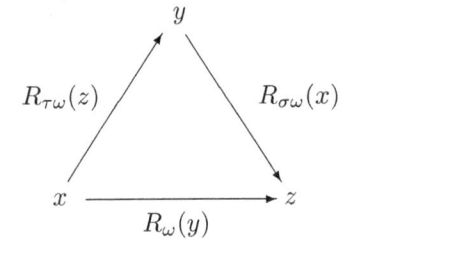

$$\tag{13}$$

in terms of the right translations, or by the diagram

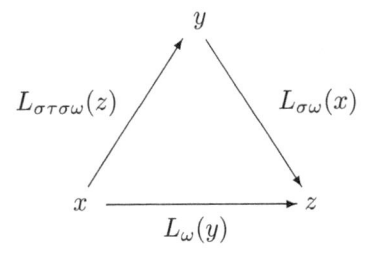

$$\tag{14}$$

using left translations, with $\omega$ in $\Omega$ and $x$, $y$, $z$ in $Q$. For example, the bottom line of (13) gives $z = xy\underline{\omega}$. The right leg then yields the hypercommutativity (7), while the left leg yields the hypercancellativity (8). The equivalence of (13) with (14) follows by replacement of $\omega$ with $\sigma\omega$, and use of (11).

The theory of quasigroups is embedded in the theory of hyperquasigroups, as described by the following two results which pass back and forth between quasigroups and hyperquasigroups.

**Proposition 3.1.** $^{18\,(Prop.\ 5.2)}$ *Let $(Q, \cdot, /, \backslash)$ be an equational quasigroup. Let $\Omega$ be the symmetric group $S_3$, interpreted as a reflexion-inversion space according to Example 3.1. Setting*

$$xy\,\underline{(1)} = x \cdot y, \quad xy\,\underline{(13)} = x/y, \quad xy\,\underline{(23)} = x\backslash y,$$
$$xy\,\underline{(12)} = y \cdot x, \quad xy\,\underline{(123)} = y/x, \quad xy\,\underline{(132)} = y\backslash x,$$

*the pair $(Q, \Omega)$ becomes a hyperquasigroup.*

**Theorem 3.1.** $^{18\,(Th.\ 6.1,\ Cor.\ 6.2)}$ *Let $(Q, \Omega)$ be a hyperquasigroup. Then for each element $\omega$ of the reflexion-inversion space $\Omega$, there is an equational quasigroup $(Q, \underline{\sigma\omega}, \underline{\sigma\tau\omega}, \underline{\tau\sigma\omega})$. In particular, for each element $\omega$ of the reflexion-inversion space $\Omega$, there is a combinatorial quasigroup $(Q, \underline{\omega})$.*

Theorem 3.1 exhibits a typical phenomenon whereby a reflexion-inversion space is a moduli space for a collection of quasigroup structures.

### 3.2. *Linear hyperquasigroups*

In Example 2.2, it was shown how linear representations of two-generated groups may be captured by a quasigroup structure. We will now discuss how certain hyperquasigroups are able to capture linear representations of arbitrary groups.

Let $A$ be an arbitrary group of automorphisms of an abelian group (or right $A$-module) $M$. Define sets

$$\begin{cases} \Omega_A^{+,+} = A \times A, \\ \Omega_A^{-,+} = (-A) \times A, \\ \Omega_A^{+,-} = A \times (-A), \end{cases} \tag{15}$$

known respectively as the *positive cone* or *first* or $2^0$-*th orthant*, the *second* or $2^1$-*st orthant*, and the the *fourth* or $2^2$-*nd orthant*. The orthant notation

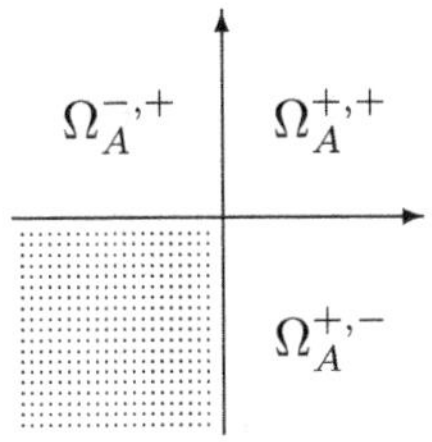


Fig. 1. Orthant structure

is motivated by the case where $M$ is the real line, and $A$ is the group of positive scalars (see Figure 1.) Define

$$\Omega_A = \Omega_A^{+,+} \cup \Omega_A^{-,+} \cup \Omega_A^{+,-} . \tag{16}$$

Define a reflexion

$$\sigma : \Omega_A \to \Omega_A;\ (r, s) \mapsto (s, r) \tag{17}$$

and an inversion

$$\tau : \Omega_A \to \Omega_A;\ (r, s) \mapsto (-rs^{-1}, s^{-1}) \tag{18}$$

to make $\Omega_A$ a reflexion-inversion space. The actions of the reflexion and inversion on the orthants are given by the following Cayley diagram:

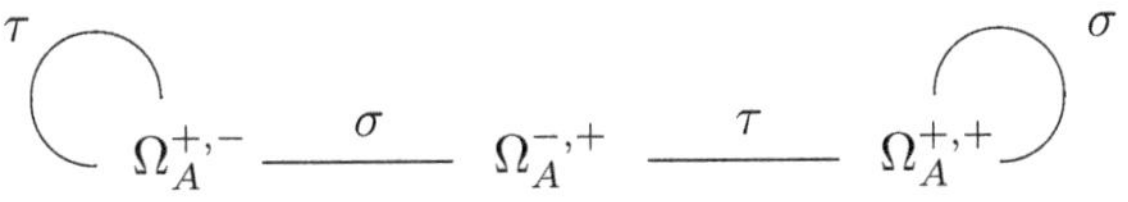


(19)

The inherent triality symmetry is given explicitly here by the elements $\sigma$ and $\tau$ generating $S_3$. At the elementary level, the Cayley diagram appears as follows:

$$\begin{array}{ccccc}
(r, s) & \overset{\tau}{\text{———}} & (-rs^{-1}, s^{-1}) & \overset{\sigma}{\text{———}} & (s^{-1}, -rs^{-1}) \\
\sigma \Big| & & & & \Big| \tau \\
(s, r) & \underset{\tau}{\text{———}} & (-sr^{-1}, r^{-1}) & \underset{\sigma}{\text{———}} & (r^{-1}, -sr^{-1})
\end{array}$$

For $(r, s)$ in $\Omega_A$, define a binary action on $M$ by

$$xy\,\underline{(r, s)} = xr + ys \tag{20}$$

for $x$, $y$ in $M$.

**Definition 3.3.** A hyperquasigroup is said to be *linear* if it has the form $(M, \Omega_A)$, its structure being given by (17)–(20), for a group $A$ of automorphisms of an abelian group $M$. It is *pointed* if $\Omega_A$ is pointed by $(1, 1)$.

Pointed linear hyperquasigroups turn out to be equivalent to faithful group representations (as automorphisms of an abelian group). Indeed, the abelian group $M$, automorphism group $A$, and action of $A$ on $M$ are all recovered from the pointed linear hyperquasigroup structure $(M, \Omega_A)$.

**Theorem 3.2.** [21 (Th. 11.5)] *Consider a pointed linear hyperquasigroup* $(M, \Omega_A)$. *Then:*

(a) *The addition and subtraction in the abelian group $M$ are given by*

$$x + y = xy\,\underline{(1, 1)}$$

*and*

$$x - y = xy\,\underline{(1, -1)} = xy\,\sigma\tau(1, 1)$$

*for $x$, $y$ in $M$, using the pointed element $(1, 1)$ of $\Omega_A$;*
(b) *The zero element of $M$ is given as*

$$0 = xy\,\underline{(1, -1)} = xx\,\sigma\tau(1, 1) \tag{21}$$

*for any element $x$ of $M$, using the pointed element $(1, 1)$ of $\Omega_A$;*
(c) *The set*

$$P = \{L_{(r,s)}(x) \mid (r, s) \in \Omega_A^{+,+},\ x \in M\}$$

*of left translations from the positive cone forms a group;*
(d) *The group $A$ is the stabilizer $P_0$ of $0$ in the action of $P$ on $M$;*
(e) *For elements $m$ of $M$ and $a$ of $A$, the equation*

$$ma = 0m\,\underline{(1, a)}$$

*gives the action of $a$ on $m$.*

### 3.3. *Linear combinations*

Vector spaces are naturally understood as sets that are equipped with an algebraic structure given by linear combinations. In particular, vector spaces over GF(1) will be sets, and one is left with the question of the appropriate linear combinations.[†] Now there are various ways to define linear combinations. A recursive definition is chosen here, based on the use of an appropriate linear hyperquaisgroup.

**Definition 3.4.** Let $V$ be a vector space over a field $F$, with associative, commutative quasigroup $F^*$ of nonzero elements. If $|F| > 1$, consider the linear hyperquasigroup $(V, \Omega_{F^*})$ obtained from the multiplication action of nonzero scalars. Then an $F$-*linear combination* of vectors from $V$ is defined recursively as follows:

(a) Each vector $v \in V$ is an $F$-linear combination of vectors from $V$;
(b) If $x, y$ are $F$-linear combinations of vectors from $V$, and $\alpha, \beta$ are nonzero elements of $F$, then $x\alpha + y\beta = xy\,\underline{(\alpha, \beta)}$ is an $F$-linear combination of vectors from $V$.

Since there are no nonzero elements of the field GF(1), the recursive step of Definition 3.4(b) is never called when $F = \mathsf{GF}(1)$. Thus each set is a vector space over GF(1). To investigate the import of the $F$-linear combinations of Definition 3.4 for traditional fields, some additional concepts are needed.

**Definition 3.5.** Let $V$ be a vector space over a field $F$. Then the *support, scalar list,* and *argument list* of an $F$-linear combination of vectors from $V$ are defined recursively as follows:

(a) For each vector $v \in V$, the support of the $F$-linear combination $v$ is $\{v\}$, its scalar list is $(1)$, and its argument list is $(v)$;
(b) Suppose that $x$ and $y$ are $F$-linear combinations of vectors from $V$ with respective supports $X, Y$; scalar lists $(\alpha_1, \ldots \alpha_r), (\beta_1, \ldots \beta_s)$; and argument lists $(x_1, \ldots, x_r), (y_1, \ldots, y_s)$. For nonzero elements $\alpha, \beta$ of $F$, the support of the $F$-linear combination $x\alpha + y\beta$ is $X \cup Y$. Its scalar list is $(\alpha_1\alpha, \ldots, \alpha_r\alpha, \beta_1\beta, \ldots, \beta_s\beta)$, and its argument list is $(x_1, \ldots, x_r, y_1, \ldots, y_s)$.

---

[†]Compare Cohn's observation: "I know of no … way to make sense of vector spaces over $\mathbb{F}_1$" [3](p.489).

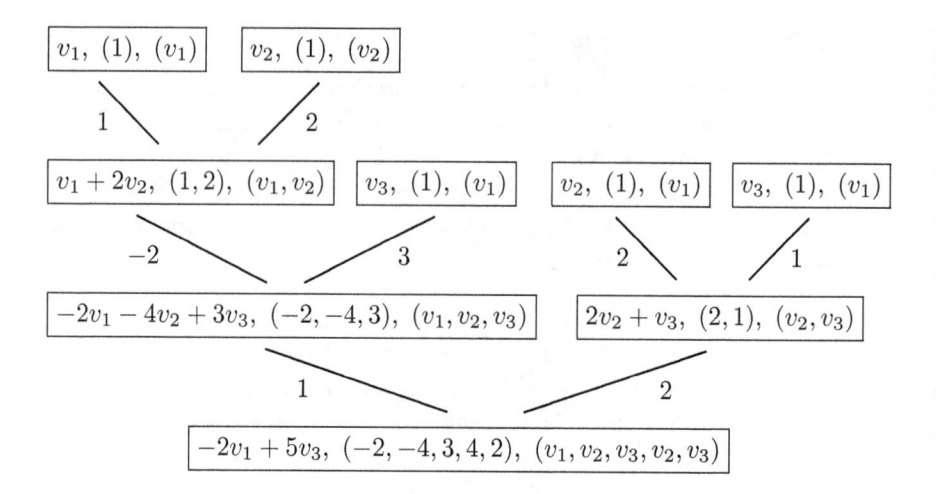

Fig. 2.    Recursive construction of $F$-linear combinations of vectors $v_1, v_2, \ldots$ over a traditional field $F$ with $6 \neq 0$. The nodes of the tree show respective linear combinations (written with left scalar action to conserve space), scalar lists, and argument lists. The edges of the tree are labeled with nonzero scalars from the field. Over $\mathsf{GF}(1)$, the only possible "trees" are isolated leaves of the form $\boxed{v, \ (1), \ (v)}$ .

Figure 2 illustrates the recursive construction of $F$-linear combinations, along with their scalar and argument lists, in a (traditional) field $F$ where $6 \neq 0$. If the underlying vector space is $V$, then the tree structure of Figure 2 reflects a parsing tree in the linear hyperquasigroup $(V, \Omega_{F^*})$ obtained from the multiplication action of nonzero scalars.

The proof of the following result (by induction on the positive integer $r$) is left to the reader.

**Lemma 3.1.** *Let $v$ be an $F$-linear combination of vectors from $V$, with support $X$, scalar list $(\alpha_1, \ldots, \alpha_r)$, and argument list $(v_1, \ldots, v_r)$.*

    (a) *The support is nonempty, while the scalar and argument lists always have positive length $r$.*

    (b) *The support $X$ is the set $\{v_1, \ldots, v_r\}$ of vectors appearing (possibly repeatedly) in the argument list.*

**Definition 3.6.** Suppose that $X$ is a nonempty subset of a vector space $V$ over a field $F$. Then a vector $v$ is an $F$-*linear combination of $X$-vectors* if it appears as an $F$-linear combination whose support is a nonempty subset of $X$.

**Lemma 3.2.** *Suppose that $Y$ is a nonempty subset of a subset $X$ of a vector space $V$ over a field $F$. If a vector $v$ is an $F$-linear combination of $Y$-vectors, then it is an $F$-linear combination of $X$-vectors.*

For fields with more that one element, the definitions given here may be compared with the classical definition:

**Definition 3.7.** [10][(§6)] Let $V$ be a vector space over a field $F$ with $|F| > 1$. Let $r$ be a natural number. Suppose that

$$v = \sum_{i=1}^{r} v_i \alpha_i \qquad (22)$$

for a vector $v$ in $V$, a subset $\{v_1, \ldots, v_r\}$ of $V$, and a subset $\{\alpha_1, \ldots, \alpha_r\}$ of $F$. Then $v$ is said to be a *classical linear combination* of the set $\{v_1, \ldots, v_r\}$ of vectors.

**Remark 3.1.** Consider the situation of Definition 3.7. If $r$ is the natural number 0, then the vector $v$ of (22) is the zero vector in $V$[10][(p.9)].

The following theorem describes the relationship between the classical and new definitions of a linear combination.

**Theorem 3.3.** *Let $V$ be a vector space over a field $F$ with $|F| > 1$.*

   (a) *The zero vector $0$ of $V$ is an $F$-linear combination of $\{0\}$-vectors.*
   (b) *In the situation of Definition 3.7, with $r$ positive, the vector $v$ is an $F$-linear combination of $\{v_1, \ldots, v_r\}$-vectors.*
   (c) *Each $F$-linear combination of vectors from $V$ may be written as a classical linear combination (22) of its support.*

**Proof.** (a) The statement follows by Definition 3.5(a).

(b) The proof is by induction on $r$. For the induction basis, suppose that $r = 1$. If $\alpha_1 = 1$, then $v$ is simply the $F$-linear combination $v = v_1$ with support $\{v_1\}$. If $\alpha_1 \neq 1$, then $v$ may be written as the $F$-linear combination $v = v_1(1) + v_1(\alpha_1 - 1) = v_1 v_1 (1, \alpha_1 - 1)$ with support $\{v_1\}$. For the induction step, suppose that $r \geq 2$, and that the result is true for expressions (22) with a positive number of summands that is less than $r$. Then

$$v = \sum_{i=1}^{r} \alpha_i v_i = \sum_{i=1}^{r-1} \alpha_i v_i + \alpha_r v_r \,.$$

Now $\sum_{i=1}^{r-1} \alpha_i v_i$ is an $F$-linear combination whose support is a nonempty subset $Z$ of $\{v_1, \ldots, v_{r-1}\}$, by the induction assumption. If $\alpha_r = 0$, the induction step is then complete. Otherwise, for $\alpha_r \neq 0$, the vector $v_r$ is itself an $F$-linear combination with support $\{v_r\}$, and then

$$v = \sum_{i=1}^{r-1} v_i \alpha_i(1) + v_r(\alpha_r) = \left( \sum_{i=1}^{r-1} v_i \alpha_i \right) v_r \, (1, \alpha_r)$$

is an $F$-linear combination whose support is the nonempty subset $Z \cup \{v_r\}$ of $\{v_1, \ldots, v_r\}$.

(c) Suppose that $v$ is an $F$-linear combination of vectors with scalar and argument lists of length $r$. The result will be proved by induction on $r$. If $r = 1$, then $\alpha_1 = 1$, and (22) holds with $v_1 = v$. Now suppose that $r > 1$, and that the result is true for all $F$-linear combinations with scalar and argument lists of lengths less than $r$. Then $v = x\alpha + y\beta$ for nonzero scalars $\alpha, \beta$ and $F$-linear combinations $x, y$ with respective scalar lists $(\alpha'_1, \ldots, \alpha'_s)$, $(\beta'_1, \ldots, \beta'_t)$ and argument lists $(v_1, \ldots, v_s)$, $(w_1, \ldots, w_t)$ of positive lengths $s, t$ summing to $r$. By the induction hypothesis, there are classical linear combinations

$$x = \sum_{i=1}^{s} v_i \alpha'_i \quad \text{and} \quad y = \sum_{j=1}^{t} w_j \beta'_j$$

for the subsets $\{v_1, \ldots, v_s\}$, $\{w_1, \ldots, w_t\}$ of $V$ and $\{\alpha'_1, \ldots, \alpha'_s\}$, $\{\beta'_1, \ldots, \beta'_t\}$ of $F$. Then the computation

$$v = x\alpha + y\beta = \sum_{i=1}^{s} v_i \alpha'_i \alpha + \sum_{j=1}^{t} w_j \beta'_j \beta = \sum_{i=1}^{s} v_i(\alpha'_i \alpha) + \sum_{j=1}^{t} w_j(\beta'_j \beta)$$

expresses the $F$-linear combination $v$ with support $\{v_1, \ldots, v_s\} \cup \{w_1, \ldots, w_t\}$ as a classical linear combination of $\{v_1, \ldots, v_s\} \cup \{w_1, \ldots, w_t\}$. $\qquad \square$

### 3.4. *Spans and subspaces*

**Definition 3.8.** Let $V$ be a vector space over a field $F$.

(a) If $|F| > 1$, the *span* $\operatorname{Span} \emptyset$ of the empty set is $\{0\}$.

(b) If $|F| = 1$, the *span* $\operatorname{Span} \emptyset$ of the empty set is $\emptyset$.

(c) Let $X$ be a nonempty subset of $V$. Then the *span* $\operatorname{Span} X$ of $X$ is the set of all $F$-linear combinations of $X$-vectors.

**Lemma 3.3.** *Let $X$ be a subset of a vector space $V$ over a field $F$. Then $X \subseteq \operatorname{Span} X$.*

**Lemma 3.4.** *Let $V$ be a vector space over a field $F$ with $|F| > 1$. Then the zero vector $0$ of $V$ lies in the span of every nonempty subset of $V$.*

**Proof.** Consider an element $v$ of a nonempty subset $X$ of $V$. Then using the pointed element $(1, 1)$ of $\Omega_{F^*}$, the relation

$$0 = v(1) + v(-1) = vv\,(1, -1) = vv\,\sigma\tau(1, 1)$$

corresponding to (21) serves to express $0$ as an $F$-linear combination whose support is the nonempty subset $\{v\}$ of $X$. $\qquad\square$

**Definition 3.9.** Let $V$ be a vector space over a field $F$.

   (a) A subset $S$ of $V$ is a (*vector*) *subspace* if $S = \operatorname{Span} S$.
   (b) If $|F| > 1$, a subset $S$ of $V$ is a *classical subspace* if and only if it contains each classical linear combination of each finite subset of $S$.

**Theorem 3.4.** *Let $V$ be a vector space over a field $F$.*

   (a) *If $|F| = 1$, then each subset of $V$ is a subspace.*
   (b) *If $|F| > 1$, then a subset of $V$ is a subspace if and only if it is a classical subspace.*

**Proof.** (a) Let $S$ be a subset of $V$. If $S = \varnothing$, then Definition 3.8(b) implies that $\varnothing = \operatorname{Span} \varnothing$, so $S$ is a subspace. If $S$ is nonempty, the recursive step Definition 3.4(b) is never called when $F$-linear combinations are formed, so the only $F$-linear combinations of $S$-vectors are the elements of $S$ itself, and $S = \operatorname{Span} S$.

(b) Suppose that $S = \operatorname{Span} S$ for a subset $S$ of $V$. By Definition 3.8(a), $S$ is nonempty, since $\varnothing \neq \operatorname{Span} \varnothing$. Then by Lemma 3.4, the zero vector, namely the unique classical linear combination of the empty subset of $S$, lies in $S$. Now consider a classical linear combination (22) of a nonempty, finite subset $\{v_1, \ldots, v_r\}$ of $S$. By Theorem 3.3(b), it follows that this classical linear combination $v$ may be expressed as an $F$-linear combination of $\{v_1, \ldots, v_r\}$-vectors. By Lemma 3.2, it follows that $v$ is an $F$-linear combination of $S$-vectors, and so lies in $S$. Thus $S$ is a classical subspace.

    Conversely, suppose that $S$ is a classical subspace. Let $v$ be an element of $\operatorname{Span} S$, namely an $F$-linear combination with support $\{v_1, \ldots, v_r\} \subseteq S$. By Theorem 3.3(c), $v$ is a classical linear combination of $\{v_1, \ldots, v_r\}$. Then since $S$ is a classical subspace, $v$ lies in $S$. Conversely, $S \subseteq \operatorname{Span} S$ by Lemma 3.3. Thus $S = \operatorname{Span} S$, and $S$ is a subspace. $\qquad\square$

### 3.5. *Linear transformations*

In this section, linear transformations are defined for fields with one or more elements. It is convenient to use Eulerian rather than algebraic notation for these linear transformations here, so that the preservation of scalar multiplications appears as the mixed associative law in the second equation of (23).

**Definition 3.10.** Let $U$ and $V$ be vector spaces over a field $F$.

    (a) A function $f\colon U \to V$ is said to be a *linear transformation* if, whenever an element $u$ of $U$ is an $F$-linear combination with scalar list $(\alpha_1, \ldots, \alpha_r)$, with argument list $(u_1, \ldots, u_r)$, and with support $\{u_1, \ldots, u_r\} \subseteq U$, then $f(u)$ is an $F$-linear combination with scalar list $(\alpha_1, \ldots, \alpha_r)$ and argument list $\bigl(f(u_1), \ldots, f(u_r)\bigr)$.

    (b) The spaces $U$ and $V$ are *linearly isomorphic* if there is a bijective linear transformation $f\colon U \to V$.

**Theorem 3.5.** *Let $f\colon U \to V$ be a function between vector spaces $U, V$ over a field $F$.*

    (a) *If $|F| = 1$, then $f$ is a linear transformation.*

    (b) *If $|F| > 1$, then $f$ is a linear transformation if and only if*

$$f(u_1 + u_2) = f(u_1) + f(u_2) \quad and \quad f(u\alpha) = f(u)\alpha \qquad (23)$$

*for all $u, u_1, u_2$ in $U$ and $\alpha$ in $F$.*

**Proof.** (a) If $U = \varnothing$, then the linear transformation condition is satisfied vacuously. Now suppose that $U$ is nonempty. Then an element $u$ of the vector space $U$ appears as an $F$-linear combination only with scalar list $(1)$ and argument list $(u)$. Since $f(u)$ appears as an $F$-linear combination with scalar list $(1)$ and argument list $\bigl(f(u)\bigr)$, it follows that $f\colon U \to V$ is a linear transformation.

(b) First, suppose that $f\colon U \to V$ is a linear transformation. For vectors $u_1, u_2$ in $U$, the vector $u' = u_1 + u_2$ in $U$ is an $F$-linear combination with scalar list $(1, 1)$ and argument list $(u_1, u_2)$, so the vector $f(u')$ in $V$ is an $F$-linear combination with scalar list $(1, 1)$ and argument list $\bigl(f(u_1), f(u_2)\bigr)$. It follows that $f(u_1 + u_2) = f(u_1) + f(u_2)$.

Now consider a vector $u$ in $U$ and a scalar $\alpha$ in $F$. For $\alpha = 1$, one has $f(u\alpha) = f(u) = f(u)\alpha$. For $\alpha \neq 1$, the vector $u\alpha$ is an $F$-linear combination with scalar list $(1, \alpha - 1)$ and argument list $(u, u)$. Then $f(u\alpha)$ is an $F$-linear combination with scalar list $(1, \alpha - 1)$ and argument list $(f(u), f(u))$. It follows that $f(u\alpha) = f(u) + f(u)(\alpha - 1) = f(u)\alpha$, so the equations (23) are satisfied.

Conversely, suppose that the equations (23) are satisfied. The linear transformation conditions of Definition 3.10(a) are shown to be satisfied by induction on the length $r$ of the scalar and argument lists in an $F$-linear combination $u$. The induction basis $r = 1$ is trivial. For the induction step, suppose that an element $u$ of $U$ is an $F$-linear combination with scalar list $(\alpha_1\alpha, \ldots, \alpha_s\alpha, \beta_1\beta, \ldots, \beta_t\beta)$, with argument list $(u_1, \ldots, u_s, v_1, \ldots, v_t)$, and support $\{u_1, \ldots, u_s, v_1, \ldots, v_t\} \subseteq U$. Consider the element $x$ of $U$ with scalar list $(\alpha_1, \ldots, \alpha_s)$ and argument list $(u_1, \ldots, u_s)$, along with the element $y$ of $U$ with scalar list $(\beta_1, \ldots, \beta_t)$ and argument list $(v_1, \ldots, v_t)$. Then $u = x\alpha + y\beta$. By the induction hypothesis, it follows that $f(x)$ is an $F$-linear combination with scalar list $(\alpha_1, \ldots, \alpha_s)$ and argument list $(f(u_1), \ldots, f(u_s))$, while $f(y)$ is an $F$-linear combination with scalar list $(\beta_1, \ldots, \beta_t)$ and argument list $(f(v_1), \ldots, f(v_t))$. Then by the equations (23), the image $f(u) = f(x\alpha + y\beta) = f(x)\alpha + f(y)\beta$ is an $F$-linear combination with scalar list $(\alpha_1\alpha, \ldots, \alpha_s\alpha, \beta_1\beta, \ldots, \beta_t\beta)$ and argument list $(f(u_1), \ldots, f(u_s), f(v_1), \ldots, f(v_t))$. □

## 4. Categories of relations

### 4.1. *Relations between sets*

Over a field with more than one element, the category of linear transformations between finite-dimensional vector spaces is self-dual. In particular, each singleton vector space $\{0\}$ is both initial and terminal. On the other hand, even restricting to finite sets, the category of functions between sets is not self-dual. For example, the initial object is the empty set, while the terminal objects are the singletons. In order to construe sets as objects of a category of vector spaces over the one-element field $\mathsf{GF}(1)$, the appropriate category is the category **Rel** of relations between sets, not the category **Set** of functions between sets.

Thus the object class of **Rel** is the class of sets. For sets $X$ and $Y$, the morphism set $\mathbf{Rel}(X, Y)$ is the set of relations $\rho$ from $X$ to $Y$, i.e. the set of subsets $\rho$ of the Cartesian product $X \times Y$ in **Set**. The composition is given by

$$\mathbf{Rel}(X,Y) \times \mathbf{Rel}(Y,Z) \to \mathbf{Rel}(X,Z); \ (\rho,\sigma) \mapsto \rho \circ \sigma$$

with the *relation product*

$$\rho \circ \sigma = \{(x,z) \in X \times Z \mid \exists\, y \in Y : (x,y) \in \rho \text{ and } (y,z) \in \sigma\}.$$

A function $f \colon X \to Y$ may be identified with its *graph*

$$\{(x,y) \in X \times Y \mid xf = y\}$$

so that the relational product restricts to the composition of functions. Then **Set** is included as a subcategory of **Rel**. In particular, the identity at an object $X$ of **Rel** is the equality relation on $X$, the graph $1_X$ of the identity function $1_X \colon X \to X$.

**Lemma 4.1.** *The disjoint union $X \oplus Y$ of two sets $X$ and $Y$ serves as a biproduct in* **Rel**, *both the product and coproduct of $X$ and $Y$.*

**Proof.** The insertions into the biproduct are given by the (graphs of the) usual insertions $\iota_X \colon X \to X \oplus Y$ and $\iota_Y \colon X \to X \oplus Y$ into the disjoint union. Recall the right distributive law $(X \oplus Y) \times Z = (X \times Z) \oplus (Y \times Z)$ in **Set**. Then for relations $\rho \colon X \to Z$ and $\sigma \colon Y \to Z$, the sum $\rho + \sigma \colon X \oplus Y \to Z$ is the disjoint union $\rho \oplus \sigma$ as a subset of $(X \times Z) \oplus (Y \times Z)$. Thus

$$\iota_X \circ (\rho + \sigma) = \rho \quad \text{and} \quad \iota_Y \circ (\rho + \sigma) = \sigma$$

as required.

The respective projections from the biproduct are the relations

$$\pi_X = \{(x\iota_X, x) \in (X \oplus Y) \times X \mid x \in X\}$$

and

$$\pi_Y = \{(y\iota_X, y) \in (X \oplus Y) \times Y \mid y \in Y\}.$$

Recall the left distributive law $Z \times (X \oplus Y) = (Z \times X) \oplus (Z \times Y)$ in **Set**. Then for relations $\rho \colon Z \to X$ and $\sigma \colon Z \to Y$, the product $\rho \times \sigma \colon Z \to X \oplus Y$ is the disjoint union $\rho \oplus \sigma$ as a subset of $(Z \times X) \oplus (Z \times Y)$. Thus

$$(\rho \times \sigma) \circ \pi_X = \rho \quad \text{and} \quad (\rho \times \sigma) \circ \pi_Y = \sigma$$

as required. $\qquad\qquad\square$

**Lemma 4.2.** *The category* **Rel** *has an internal hom functor* **rel** *given by*

$$\mathbf{rel}(Y,Z) = Y \times Z$$

*for sets $Y, Z$.*

**Proof.** Observe that

$$\mathbf{Rel}\big(\mathsf{GF}(1), Y \times Z\big) \cong 2^{Y \times Z} = \mathbf{Rel}(Y, Z)$$

via the natural isomorphism sending a relation $\rho \subseteq \mathsf{GF}(1) \times (Y \times Z)$ to the subset $\{(y, z) \mid \big(0, (y, z)\big) \in \rho\}$ of $Y \times Z$. $\qquad\Box$

**Remark 4.1.** Lemma 4.2 may also be justified on the basis that **Rel** is the Kleisli category for the covariant power set endofunctor of **Set**, noting the nature of the underlying set functor for the Kleisli category [14 (Th. VI.5.1)].

**Lemma 4.3.** *The Cartesian product $X \otimes Y$ of two sets $X$ and $Y$ serves as a tensor product in* **Rel**, *by virtue of the adjunction*

$$\mathbf{Rel}(X \otimes Y, Z) \cong \mathbf{Rel}(X, \mathbf{rel}(Y, Z)) \qquad (24)$$

*between $Y \mapsto X \otimes Y$ and $Y \mapsto \mathbf{rel}(Y, Z)$.*

**Proof.** Each side of (24) is the power set $2^{X \times Y \times Z}$. $\qquad\Box$

We may summarize as follows.

**Theorem 4.1.** *The category* **Rel** *supports symmetric monoidal category structures $\big(\mathbf{Rel}, \oplus, \varnothing\big)$ and $\big(\mathbf{Rel}, \otimes, \mathsf{GF}(1)\big)$, with the disjoint union $X \oplus Y$ as the biproduct and the Cartesian product $X \otimes Y$ as the tensor product of sets $X$ and $Y$.*

For a field $F$ with more that one element, write $\underline{\underline{F}}^{<\omega}$ for the category of linear transformations between finite-dimensional vector spaces over $F$. Theorem 4.1 may then be viewed against the following.

**Theorem 4.2.** *The category $\underline{\underline{F}}^{<\omega}$ supports symmetric monoidal category structures $\big(\underline{\underline{F}}^{<\omega}, \oplus, \{0\}\big)$ and $\big(\underline{\underline{F}}^{<\omega}, \otimes, F\big)$.*

In order to unify the notation for fields with one or more elements, we introduce the following definition.

**Definition 4.1.** The category $\underline{\underline{\mathsf{GF}(1)}}^{<\omega}$ is the full subcategory of **Rel** whose object class is the class of finite sets.

With this notation established, Theorem 4.2 applies equally well for fields $F$ with one or more elements.

## 4.2. *Linear algebra*

The categories $\underline{F}^{<\omega}$ specified in the preceding section provide support for a unified treatment of linear algebra covering fields $F$ with one or more elements. We offer some illustrative fragments. In particular, the following definition provides an answer to Ref. 3(Puzzle 1), which asked "In what way is an $n$-element set like $\mathbb{F}_1^n$?"

**Definition 4.2.** Let $F$ be a field with one or more elements. For a natural number $d$, define the reference space

$$\overbrace{F^d = F \oplus F \oplus \ldots \oplus F}^{d \text{ summands}} \qquad (25)$$

as the $d$-th direct power of $F$ in the category $\underline{F}^{<\omega}$.

With $d = 0$, Definition 4.2 returns the zero object of $\underline{F}^{<\omega}$, namely $\{0\}$ for $|F| > 1$ or $\varnothing$ for $|F| = 1$.

**Definition 4.3.** Let $V$ be a vector space over a field $F$. Suppose that $d$ is a natural number. Then $V$ is said to have (*finite*) *dimension* $d$ if it is linearly isomorphic to the reference space $F^d$.

**Proposition 4.1.** *Let $V$ be a vector space over a field $F$. Let $d$ be a natural number.*

  (a) *If $|F| = 1$, then $V$ has dimension $d$ if and only if $|V| = d$.*
  (b) *If $|F| > 1$, then $V$ has dimension $d$ in the sense of Definition 4.3 if and only if it has dimension $d$ in the classical sense*[10](§8).

**Proof.** (a) By Theorem 3.5(a), the set $V$ has dimension $d$ if and only if, as a set, it is isomorphic to $F^d$. But by Definition 4.2, $|F^d| = d$.

(b) By Theorem 3.5(b), the space $V$ has dimension $d$ if and only if it is classically isomorphic to the vector space $F^d$. $\qquad\qquad\square$

The following definition is introduced to show how the concepts of linear independence and basis may be worked in to the present setting.

**Definition 4.4.** Let $V$ be a vector space over a field $F$. Consider a (possibly empty) ordered list $(v_1, \ldots, v_d)$ of vectors from $V$.

  (a) The ordered list is *linearly independent* if the span of $\{v_1, \ldots, v_d\}$ is linearly isomorphic to the reference space $F^d$.

(b) The ordered list is an *ordered basis* for $V$ if it is linearly independent and spans $V$.

Now consider a linear transformation $f: V \to W$ from a vector space $V$ of dimension $m$ to a vector space $W$ of dimension $n$. Suppose that $V$ has an ordered basis $(v_1, \ldots, v_m)$ and $W$ has an ordered basis $(w_1, \ldots, w_n)$. We present a unified way to specify the linear transformation $f: V \to W$ by its matrix $[f_{ij}]_{m \times n}$ with respect to the ordered bases $(v_1, \ldots, v_m)$ and $(w_1, \ldots, w_n)$. Let the ordered bases correspond to linear isomorphisms $v: V \to F^m$ and $\omega: W \to F^n$. Then for $1 \le i \le m$ and $1 \le j \le n$, the diagram

$$
\begin{array}{ccccccc}
F^m & \xrightarrow{\ v^{-1}\ } & V & \xrightarrow{\ f\ } & W & \xrightarrow{\ \omega\ } & F^n \\
\| & & & & & & \| \\
\bigoplus_{k=1}^{m} F & \xleftarrow{\ \iota_i\ } & F & \xrightarrow[f_{ij}]{} & F & \xleftarrow{\ \pi_j\ } & \bigoplus_{k=1}^{n} F
\end{array}
$$

in $\underline{F}^{<\omega}$ specifies the entries of the matrix as morphisms $f_{ij}: F \to F$. When $|F| > 1$, these morphisms are scalar multiplications, and then the matrix entries are usually considered as the corresponding scalars. If $|F| = 1$, the morphisms are relations from $F$ to $F$. There are only two such relations, namely $\emptyset$ and $1_F$. Thus the specification

$$
f_{ij} = \begin{cases} 1_F & \text{if } v_i f = w_j; \\ \emptyset & \text{otherwise} \end{cases}
$$

describes the matrix directly in this case. In combinatorial terms, we have the incidence matrix of the subset $f$ of $V \times W$. Incidence matrices are usually written with entries 0 and 1, imagined as elements of some nontrivial unital commutative ring or semiring, but in our approach, the entries are just elements of the endomorphism monoid $\{\emptyset, 1_F\}$ of $F$ in $\underline{F}^{<\omega}$ or **Rel**.

## 4.3. *Counting*

For certain counting problems in linear algebra over finite fields $\mathsf{GF}(q)$, the number $q$ is viewed as a parameter. The answers to such counting problems are then formulated in terms of so-called "$q$-analogs" of ordinary numbers and functions [3(§2), 11, 15(p.6)]. Our goal is to show that some well-known counting results of this type carry over seamlessly in our approach to the case $q = 1$.

**Definition 4.5.** Consider a parameter $q$.

(a) For a natural number $n$, the quantity

$$[n]_q = q^{n-1} + q^{n-2} + \ldots + q + 1$$

is called a *q-number*. In particular, $[0]_q$ is the empty sum 0.

(b) For a natural number $n$, the product

$$[n]_q^! = [n]_q [n-1]_q \ldots [2]_q [1]_q$$

is called a *q-factorial*. In particular, $[0]_q^!$ is the empty product 1.

(c) For natural numbers $k \leq n$, the quotient

$$\binom{n}{k}_q = \frac{[n]_q^!}{[n-k]_q^! [k]_q^!}$$

is called a *q-binomial coefficient* or a *Gaussian binomial coefficient*.

The names and notation of Definition 4.5 may be justified as follows.

**Lemma 4.4.** *Consider a parameter $q$.*

(a) *For a natural number $n$, one has $[n]_q = n$ for $q = 1$.*
(b) *For a natural number $n$, one has $[n]_q^! = n!$ for $q = 1$.*
(c) *For natural numbers $k \leq n$, one has $\binom{n}{k}_q = \frac{n!}{(n-k)!k!} = \binom{n}{k}$ for $q = 1$.*

Our approach to the field with one element thus allows us to formulate the following extensions of well-known results for $q > 1$.

**Theorem 4.3.** *Within an $n$-dimensional vector space over a field* $\mathrm{GF}(q)$, *where $q \geq 1$, the q-binomial coefficient*

$$\binom{n}{k}_q$$

*enumerates the $k$-dimensional vector subspaces. In particular, the number of 1-dimensional subspaces of an $n$-dimensional space is given by the q-number $[n]_q$.*

Now recall that in any vector space, a *flag* is a chain of proper inclusions of subspaces:

$$S_0 \subset S_1 \subset \ldots \subset S_r . \tag{26}$$

The flag (26) is said to have *length $r$*. Note that the properness of the inclusions implies that for $1 \leq i < j \leq r$, the dimension of $S_i$ is strictly less than the dimension of $S_j$. In a vector space of dimension $r$, a flag (26) of length $r$ is described as *maximal*.

**Theorem 4.4.** *Let $n$ be a natural number. Then in a vector space $V$ of dimension $n$ over the field $\mathsf{GF}(q)$, there are $[n]_q^!$ maximal flags.*

**Proof.** A maximal flag has the form (26) with $r = n$. Note that $S_0$ is just the unique zero-dimensional subspace. It will be shown, by induction on $i$, that the number of choices for the initial segment $S_0 \subset S_1 \subset \ldots \subset S_i$ of a maximal flag (26) of length $n$ is

$$[n]_q \cdot [n-1]_q \cdot \ldots \cdot [n-(i-1)]_q \tag{27}$$

for $1 \leq i \leq n$. The desired result follows on noting that (27) takes the form $[n]_q^!$ when $i = n$.

For the induction basis $i = 1$ with $q > 1$, there are $q^n - 1$ nonzero vectors that may be chosen to span $S_1$, and any $q - 1$ of these span the same space $S_1$. Thus the number of choices for $S_1$ is $(q^n - 1)/(q - 1) = [n]_q$. Similarly, for $q = 1$, any one of the $n = [n]_1$ elements of $V$ may be chosen as the unique element of $S_1$. This completes the induction basis.

Now suppose that (27) is a correct count for the initial segments of length $i$, where $1 \leq i < n$. Consider the problem of determining $S_{i+1}$.

(a) If $q = 1$, any one of the $n - i = [n-i]_q$ elements $x$ of $V \smallsetminus S_i$ may be chosen to build $S_{i+1}$ as $\mathrm{Span}(S_i \cup \{x\}) = S_i \cup \{x\}$, so there are $[n]_q \cdot \ldots \cdot [n-i]_q$ possible initial segments of length $i + 1$.

(b) If $q > 1$, any one of the $q^n - q^i$ elements $x$ of $V \smallsetminus S_i$ may be chosen to build $S_{i+1}$ as $\mathrm{Span}(S_i \cup \{x\})$. In this case, any of the $q^{i+1} - q^i$ elements of $S_{i+1} \smallsetminus S_i$ will work along with $S_i$ to span the same space $S_{i+1}$. Thus the number of choices for a space $S_{i+1}$ extending the given initial segment is

$$\frac{q^n - q^i}{q^{i+1} - q^i} = \frac{q^{n-i} - 1}{q - 1} = [n-i]_q\,,$$

and again there are $[n]_q \cdot \ldots \cdot [n-i]_q$ possible initial segments of length $i + 1$.

This completes the induction step. $\qquad\qquad\qquad\square$

## 5. Conclusion

By use of quasigroup features, we have augmented the definition of a field to include the field of order one. By use of linear hyperquasigroups, we have broadened the concept of a linear combination to construe sets as vector spaces over the field with one element. We have then shown that the category of relations is the appropriate category for extending linear algebra to cover these vector spaces over the field with one element. The application of linear-algebraic counting formulae involving $q$-numbers, $q$-factorials, and $q$-binomial coefficients is then extended to the case where $q = 1$.

# References

1. L.A. Bokut, Y. Chen and Q. Mo, Gröbner-Shirshov bases and embeddings of algebras, *Internat. J. Algebra Comput.* **20** (2010), 875–900.
2. J.M. Borger, $\Lambda$-*rings and the field with one element*, arxiv.org/abs/0906.3146, 2009.
3. H. Cohn, Projective geometry over $\mathbf{F}_1$ and the Gaussian binomial coefficients, *Amer. Math. Monthly* **111** (2004), 487–495.
4. A. Connes and C. Consani, On the notion of geometry over $\mathbb{F}_1$, *J. Algebraic Geom.* **20** (2011), 525-557.
5. S. Doro, Simple Moufang loops, *Math. Proc. Cambridge Philos. Soc.* **83** (1978), 377–392.
6. N. Durov, *New Approach to Arakelov Geometry*, arxiv.org/abs/0704.2030v1, 2007.
7. T. Evans, Homomorphisms of non-associative systems, *J. London Math. Soc.* **24** (1949), 254–260.
8. T. Evans, Embedding theorems for multiplicative systems and projective geometries, *Proc. Amer. Math. Soc.* **3** (1952), 614–620.
9. T. Evans, Varieties of loops and quasigroups, pp. 1–26 in *Quasigroups and Loops: Theory and Applications* (O. Chein, H.O. Pflugfelder and J. D. H. Smith, eds.), Heldermann, Berlin, 1990.
10. P.R. Halmos, *Finite-Dimensional Vector Spaces* (2nd. edition), Van Nostrand, Princeton, NJ, 1958.
11. V. Kac and P. Cheung, *Quantum Calculus*, Springer, New York, NY, 2002.
12. M. Kapranov and A. Smirnov, *Cohomology determinants and reciprocity laws: number field case*, Preprint series, Institut für experimentelle Mathematik, Essen, 1995. http://www.neverendingbooks.org/DATA/KapranovSmirnov.pdf
13. O. Lorscheid, *Algebraic groups over the field with one element*, arxiv.org/abs/0907.3824v1, 2009.
14. S. Mac Lane, *Categories for the Working Mathematician*, Springer, New York, NY, 1971.
15. S. Majid, *A Quantum Groups Primer*, Cambridge University Press, Cambridge, 2002.
16. Yu.I. Manin, (1995), Lectures on zeta functions and motives (according to Deninger and Kurokawa), *Astérisque* **228** 121-163.

17. A.I. Malcev, On a representation of nonassociative rings (Russian), *Uspekhi Mat. Nauk N.S.* **7** (1952), 181–185.
18. J. D. H. Smith, Axiomatization of quasigroups, *Discuss. Math. Gen. Alg. and Appl.* **27** (2007), 21–33.
19. J. D. H. Smith, Evans' normal form theorem revisited, *Internat. J. Algebra Comput.* **17** (2007), 1577–1592.
20. J. D. H. Smith, Ternary quasigroups and the modular group, *Comment. Math. Univ. Carol.* **49** (2008), 309–317.
21. J. D. H. Smith, Groups, triality, and hyperquasigroups, *J. Pure Appl. Algebra* **216** (2012), 811-825.
22. J. D. H. Smith and A.B. Romanowska, *Post-Modern Algebra*, Wiley, New York, NY, 1999.
23. C. Soulé, Les varietés sur le corps à un élément, *Mosc. Math. J.* **4** (2004), 217–244.
24. J. Tits, Sur les analogues algébriques des groupes semi-simples complexes, pp.261-289 in *Colloque d'Algèbre Supérieure*, Ceuterick, Louvain, 1957.

# Gröbner-Shirshov bases for associative conformal modules*

Yuqun Chen

*School of Mathematical Sciences, South China Normal University
Guangzhou 510631, P. R. China*

*yqchen@scnu.edu.cn*

Lili Ni[†]

*School of Mathematics and Statistics, Taishan University,
Taian 271000, P. R. China*

*nilili2009@163.com*

We construct free modules over an associative conformal algebra. We establish Composition-Diamond lemma for associative conformal modules. As applications, Gröbner-Shirshov bases of the Virasoro conformal module and module over the semidirect product of Virasoro conformal algebra and current algebra are given respectively.

*2010 Mathematics subject classification:* 17B69, 16S15, 13P10.

*Keywords:* Gröbner-Shirshov basis, conformal algebra, free associative conformal module

## 1. Introduction

The subject of conformal algebras is closely related to vertex algebras (see, V. Kac[27]). Implicitly, vertex algebras were introduced by Belavin, Polyakov, and Zamolodchikov in 1984[1]. Explicitly, the definition of vertex algebras was given by R. Borcherds in 1986[3], which led to his solution of the Conway-Norton conjecture in the theory of finite simple groups[4,24]. As pointed out by Kac[27,28], conformal and vertex algebras provide a rigorous mathematical study of the "locality axiom" which came from Wightman's axioms of quantum field theory[38]. M. Roitman studied free (Lie and associative) conformal and vertex algebras in Ref. 36. Free vertex algebras

---

*Supported by the NNSF of China (no. 11571121) and the Science and Technology Program of Guangzhou (no. 201707010137).

†Supported by the talent fund of Taishan University (no. Y-01-2017001).

were mentioned in the original paper of Borcherds[3]. Since conformal and vertex algebras are not varieties in the sense of universal algebra (see, P.M. Cohn[22]), the existence of free conformal and free vertex algebras is not guaranteed by the general theory and should be proved. It was done by Roitman[36]. The free associative conformal algebra generated by a set $B$ with a locality function $N(-, -) : B \times B \to \mathbb{Z}_{\geq 0}$ is constructed by Bokut, Fong, and Ke in 1997[13].

Conformal module is a basic tool for the construction of free field realization of infinite dimensional Lie (super)algebras in conformal field theory. Finite irreducible conformal modules over the Virasoro conformal algebra were determined in Ref. 20. The Lie conformal algebra of a Block type was introduced and free intermediate series modules were classified in Ref. 25.

Gröbner bases and Gröbner-Shirshov bases were invented independently by A.I. Shirshov for ideals of free (commutative, anti-commutative) non-associative algebras[37,39], free Lie algebras[39] and implicitly free associative algebras[39] (see also Ref. 2,5), by H. Hironaka[26] for ideals of the power series algebras (both formal and convergent), and by B. Buchberger[17] for ideals of the polynomial algebras.

Gröbner bases and Gröbner-Shirshov bases theories have been proved to be very useful in different branches of mathematics, including commutative algebra and combinatorial algebra. It is a powerful tool to solve the following classical problems: normal form; word problem; conjugacy problem; rewriting system; automaton; embedding theorem; PBW theorem; extension; homology; growth function; Dehn function; complexity; etc.

Up to now, different versions of Composition-Diamond lemma are known for the following classes of algebras apart those mentioned above: (color) Lie super-algebras[30–32], tensor product of a free algebra and a polynomial algebra[33], tensor product of two free algebras[6], Lie $p$-algebras[31], associative conformal algebras[15,34], shuffle operads[23], modules[21,29] (see also Ref. 18), right-symmetric algebras[10], dialgebras[9], associative algebras with multiple operators[12], Rota-Baxter algebras[8], Lie algebras over a polynomial algebra[7], metabelian Lie algebras[19], semirings[11], and so on.

A Composition-Diamond lemma for associative conformal algebras is firstly established by Bokut, Fong, and Ke in 2004[15] which claims that if (i) $S$ is a Gröbner-Shirshov basis in $C(B, N)$, then (ii) the set of $S$-irreducible words is a linear basis of the quotient conformal algebra $C(B, N|S)$, but not conversely. By introducing some new definitions of normal $S$-words, compositions and compositions to be trivial, a new Composition-Diamond lemma for associative conformal algebras is established by Ni and Chen in

$2016^{34}$ which makes the conditions (i) and (ii) equivalent.

In this article we establish Gröbner-Shirshov bases method for modules over an associative conformal algebra and give some applications. The article is organized as follows. In section 2, we introduce the concepts of conformal algebra, associative (Lie) conformal algebra, and modules over an associative (Lie) conformal algebra. In section 3, we construct free modules over a free associative conformal algebra and free modules over an associative conformal algebra. In section 4, we establish Composition-Diamond lemma for associative conformal modules. In section 5, we give some applications of the Composition-Diamond lemma for associative conformal modules: Gröbner-Shirshov bases of the Virasoro conformal module and module over the semidirect product of Virasoro conformal algebra and current algebra are given and then linear bases of them are obtained respectively.

## 2. Preliminaries

In this section, we introduce some related concepts.

### 2.1. *Associative (Lie) conformal algebras*

We begin with the formal definition of a conformal algebra.

**Definition 2.1.** [13,14,16,27,36] A conformal algebra $C = (C, (n), n \in \mathbb{Z}_{\geq 0}, D)$ is a linear space over a field **k** of characteristic 0, equipped with bilinear multiplications $a_{(n)}b$, $n \in \mathbb{Z}_{\geq 0} = \{0, 1, 2, \cdots\}$, and a linear map $D$, such that the following axioms are valid:

(C1) (locality) For any $a, b \in C$, there exists a nonnegative integer $N(a, b)$ such that $a_{(n)}b = 0$ for $n \geq N(a, b)$ ($N(a, b)$ is called the order of locality of $a$ and $b$);

(C2) $D(a_{(n)}b) = Da_{(n)}b + a_{(n)}Db$ for any $a, b \in C$ and $n \in \mathbb{Z}_{\geq 0}$;

(C3) $Da_{(n)}b = -na_{(n-1)}b$ for any $a, b \in C$ and $n \in \mathbb{Z}_{\geq 0}$, and $Da_{(0)}b = 0$.

A conformal algebra $C$ is called *associative* if the following identity holds for all $a, b, c \in C$, $m, n \in \mathbb{Z}_{\geq 0}$,

$$(a_{(n)}b)_{(m)}c = \sum_{t \geq 0}(-1)^t \binom{n}{t} a_{(n-t)}(b_{(m+t)}c).$$

A conformal algebra $L = \langle L, [n], n \in \mathbb{Z}_{\geq 0}, D \rangle$ is called a Lie conformal algebra if $L$ satisfies the following two axioms:

- (Anti-commutativity) $a_{[n]}b = -\{b_{[n]}a\}$, where

$$\{b_{[n]}a\} = \sum_{k \geq 0} (-1)^{n+k} \frac{1}{k!} D^k(b_{[n+k]}a).$$

- (Jacobi identity)

$$(a_{[n]}b)_{[m]}c = \sum_{k \geq 0} (-1)^k \binom{n}{k} (a_{[n-k]}(b_{[m+k]}c) - b_{[m+k]}(a_{[n-k]}c)).$$

### 2.2. *Modules over a Lie or associative conformal algebra*

**Definition 2.2.**[20,27,35] Let $C$ be an associative (Lie, resp.) conformal algebra. An associative (Lie, resp.) conformal module $_C M$ is a $\mathbf{k}[D]$-module $M$ endowed with a series of operations $(n) : C \times M \to M$, $n \in \mathbb{Z}_{\geq 0}$, such that for any $a, b \in C$ and $v \in M$,

(i) (locality) there exists a nonnegative integer $N(a, v)$ such that $a_{(n)}v = 0$ for $n \geq N(a, v)$ ($N(a, v)$ is called the order of locality of $a$ and $v$);

(ii) $D(a_{(n)}v) = Da_{(n)}v + a_{(n)}Dv$ for $n \in \mathbb{Z}_{\geq 0}$;

(iii) $Da_{(n)}v = -na_{(n-1)}v$ for $n \in \mathbb{Z}_{\geq 0}$, and $Da_{(0)}v = 0$;

(iv) $(a_{(n)}b)_{(m)}v = \sum_{t \geq 0} (-1)^t \binom{n}{t} a_{(n-t)}(b_{(m+t)}v)$ if $C$ is associative;

$(a_{[n]}b)_{(m)}v = \sum_{t \geq 0} (-1)^t \binom{n}{t} (a_{(n-t)}(b_{(m+t)}v) - b_{(m+t)}(a_{(n-t)}v))$ if $C$ is Lie.

A *submodule* $M_1$ of a conformal module $_C M$ is a $\mathbf{k}[D]$-submodule such that for all $n \in \mathbb{Z}_{\geq 0}$, $C_{(n)}M_1 \subseteq M_1$. If $S \subseteq_C M$, $subm(S)$ means the submodule of $_C M$ generated by $S$.

## 3. Free associative conformal modules

**Definition 3.1.** An associative conformal module $mod_C(Y)$ over an associative conformal algebra $C$ is called the free associative conformal module generated by a set $Y$ with the locality function $N : C \times Y \longrightarrow \mathbb{Z}_{\geq 0}$, if for any associative conformal module $_C M$ and any mapping $\varepsilon : Y \longrightarrow M$ with $c_{(n)}\varepsilon(y) = 0$ for all $c \in C$, $y \in Y$ and $n \geq N(c, y)$, there exists a unique $C$-module homomorphism $\varphi : mod_C(Y) \to M$ such that the following diagram is commutative:

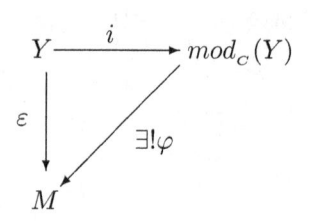

where $i$ is the inclusion map.

In this section, we construct the free associative conformal module generated by a set $Y$.

### 3.1. Double free associative conformal modules

Let $C(B, N)$ be the free associative conformal algebra generated by $B$ with the locality function $N(-, -) : B \times B \longrightarrow \mathbb{Z}_{\geq 0}$ over a field $\mathbf{k}$ of characteristic 0. Let $Y$ be a set. We construct the free module over $C(B, N)$ generated by $Y$ which is called double free associative conformal module. We extend the mapping $N(-, -)$ to $B \times (B \cup Y) \to \mathbb{Z}_{\geq 0}$. Let

$$
\begin{aligned}
U = \{ & b_{1(n_1)} b_{2(n_2)} \cdots b_{k(n_k)} D^i y \mid b_j, b_k \in B, \ 0 \leq n_j < N(b_j, b_{j+1}), \\
& 1 \leq j \leq k-1, 0 \leq n_k < N(b_k, y), \ i \geq 0, \ k \geq 0, \ y \in Y \}, \\
T = \{ & b_{1(n_1)} \cdots b_{k(n_k)} D^i b_{k+1} \mid b_j, b_{k+1} \in B, 0 \leq n_j < N(b_j, b_{j+1}), \\
& 1 \leq j \leq k, i, k \geq 0 \}
\end{aligned}
$$

and

$$
[u] := b_{1(n_1)}(b_{2(n_2)}(\cdots(b_{k(n_k)} D^i y)\cdots)).
$$

If $k = 0$, then $[u] = D^i y$. We call $k + 1$ the length of $u$, denoted by $|u|$. Note that $[\ ]$ means right normed bracketing.

Denote $span_{\mathbf{k}}[U]$ the $\mathbf{k}$-linear space with a $\mathbf{k}$-basis $[U] := \{[u] \mid u \in U\}$. We will make $span_{\mathbf{k}}[U]$ to be a $C(B, N)$-module.

Note that the set $[T] := \{[a] \mid a \in T\}$ a $\mathbf{k}$-basis of the free associative conformal algebra $C(B, N)$.

We define a scheme of *algorithm* for $C(B, N)$-module $span_{\mathbf{k}}[U]$. For any $a \in T$, $b \in B$, $y \in Y$, $n \geq 0$, $u \in U$, we define $[a]_{(n)}[u]$ as follows.

(i)  $b_{(n)}[u] =$

$$\begin{cases} 0, & \text{if } [u] = y, \ n \geq N(b,y), \\ -\sum_{t \geq 1}(-1)^t \binom{i}{t} \frac{n!}{(n-t)!} b_{(n-t)} D^{i-t}y, & \text{if } [u] = D^i y, \\ & i > 0, \ n \geq N(b,y), \\ -\sum_{t \geq 1}(-1)^t \binom{n}{t} b_{(n-t)}(b_{1(m+t)}[u_1]), & \text{if } [u] = b_{1(m)}[u_1], \\ & n \geq N(b,b_1). \end{cases}$$

(ii) $[a]_{(n)}[u] =$

$$\begin{cases} 0, & \text{if } [a] = D^i b, \ i > n, \\ (-1)^i \frac{n!}{(n-i)!} b_{(n-i)}[u], & \text{if } [a] = D^i b, \ 0 < i \leq n, \\ \sum_{t \geq 0}(-1)^t \binom{m}{t} b_{(m-t)}([c]_{(n+t)}[u]), & \text{if } [a] = b_{(m)}[c]. \end{cases}$$

(iii) For $u = b_{1(n_1)} b_{2(n_2)} \cdots b_{k(n_k)} D^i y \in U,$

$$D([u]) = \sum_{j=1}^{k} [b_{1(n_1)} \cdots b_{j-1(n_{j-1})} D b_{j(n_j)} b_{j+1(n_{j+1})} \cdots b_{k(n_k)} D^i y]$$
$$+ [b_{1(n_1)} b_{2(n_2)} \cdots b_{k(n_k)} D^{i+1} y].$$

**Lemma 3.1.** *For any* $b \in B$, $y \in Y$, $n \geq 0$, $i \geq 0$, *we have*

$$D(b_{(n)} D^i y) = D b_{(n)} D^i y + b_{(n)} D^{i+1} y \tag{1}$$

**Proof.** If $0 \leq n < N(b,y)$ or $i = 1$, the result follows from the definition. We may assume that $n \geq N(b,y)$ and $i > 1$.

Induction on $n$. Let $N := N(b,y)$. The left hand side of (1) is

$$D(b_{(N)} D^i y) = D(-\sum_{t \geq 1}(-1)^t \binom{i}{t} \frac{N!}{(N-t)!} b_{(N-t)} D^{i-t}y)$$
$$= -\sum_{t \geq 1}(-1)^t \binom{i}{t} \frac{N!}{(N-t)!} D(b_{(N-t)} D^{i-t}y)$$
$$= A_1 + A_2,$$

where

$$A_1 := \sum_{t \geq 1}(-1)^t \binom{i}{t} \frac{N!}{(N-t-1)!} b_{(N-t-1)} D^{i-t}y$$
$$= \sum_{p \geq 2}(-1)^{p-1} \binom{i}{p-1} \frac{N!}{(N-p)!} b_{(N-p)} D^{i-p+1}y$$

and $A_2 := -\sum_{t\geq 1}(-1)^t \binom{i}{t}\frac{N!}{(N-t)!}b_{(N-t)}D^{i-t+1}y$. Hence

$$A_1 + A_2 = -\sum_{t\geq 2}(-1)^t \binom{i+1}{t}\frac{N!}{(N-t)!}b_{(N-t)}D^{i-t+1}y + iNb_{(N-1)}D^iy$$

$$= -\sum_{t\geq 1}(-1)^t \binom{i+1}{t}\frac{N!}{(N-t)!}b_{(N-t)}D^{i-t+1}y - Nb_{(N-1)}D^iy.$$

The right hand side of (1) is

$$Db_{(N)}D^iy + b_{(N)}D^{i+1}y$$

$$= -Nb_{(N-1)}D^iy - \sum_{t\geq 1}(-1)^t \binom{i+1}{t}\frac{N!}{(N-t)!}b_{(N-t)}D^{i-t+1}y.$$

So the result holds for $n = N$. Assume that $n > N$. Then the left hand side of (1) is

$$D(b_{(n)}D^iy) = -\sum_{t\geq 1}(-1)^t \binom{i}{t}\frac{n!}{(n-t)!}D(b_{(n-t)}D^{i-t}y)$$

and the right hand side of (1) is

$$Db_{(n)}D^iy + b_{(n)}D^{i+1}y$$

$$= -nb_{(n-1)}D^iy - \sum_{t\geq 1}(-1)^t \binom{i+1}{t}\frac{n!}{(n-t)!}b_{(n-t)}D^{i-t+1}y.$$

By induction, we complete the proof.    □

**Lemma 3.2.** *For any $a \in T$, $u \in U$, there exists $N(a, [u]) \in \mathbb{Z}_{\geq 0}$ such that $[a]_{(n)}[u] = 0$ for all $n \geq N(a, [u])$.*

**Proof.** Induction on $|a|$. If $|a| = 1$, say, $[a] = b$, then $[a]_{(n)}[u] = b_{(n)}[u]$. Now, we find, by induction on $|u|$, $N(b, [u]) \in \mathbb{Z}_{\geq 0}$ such that $b_{(n)}[u] = 0$ for all $n \geq N(b, [u])$. Let $[u] = D^iy$. Then by Lemma 3.1,

$$b_{(n)}[u] = b_{(n)}D^iy = \sum_{t\geq 0}\binom{n}{t}\frac{i!}{(i-t)!}D^{i-t}(b_{(n-t)}y).$$

So $b_{(n)}D^iy = 0$ if $n \geq N(b, y) + i$. Hence, take $N(b, D^iy) = N(b, y) + i$ as required.

Assume that $|u| > 1$ and $[u] = b_{1(m)}[u_1]$. Let $N_0 = N(b, b_1) + N(b_1, [u_1]) - m - 1$. Therefore, for $n \geq N_0 \geq N(b, b_1)$, we have

$$b_{(n)}(b_{1(m)}[u_1]) = -\sum_{t_1\geq 1}(-1)^{t_1}\binom{n}{t_1}b_{(n-t_1)}(b_{1(m+t_1)}[u_1]).$$

Since

$$m + t_1 < N(b_1, [u_1]) \Leftrightarrow t_1 < N(b_1, [u_1]) - m$$
$$\Leftrightarrow n - t_1 > n - N(b_1, [u_1]) + m \Leftrightarrow n - t_1 > n - N_0 + N(b, b_1) - 1$$
$$\Leftrightarrow n - t_1 \geq N(b, b_1),$$

we have

$$b_{(n)}(b_{1(m)}[u_1]) = -\sum_{t_1 \geq 1}(-1)^{t_1}\binom{n}{t_1}b_{(n-t_1)}(b_{1(m+t_1)}[u_1])$$
$$= \sum_{t_1, t_2 \geq 1}(-1)^{t_1+t_2}\binom{n}{t_1}\binom{n-t_1}{t_2}b_{(n-t_1-t_2)}(b_{1(m+t_1+t_2)}[u_1])$$
$$= \cdots$$
$$= \sum_{t_j \geq 1}(-1)^{t+k}\binom{n}{t_1}\binom{n-t_1}{t_2}\cdots\binom{n-t+t_k}{t_k}b_{(n-t)}(b_{1(m+t)}[u_1]),$$

where $t = t_1 + \cdots + t_k$, $n - t < N(b, b_1)$. Thus $m + t \geq N(b_1, [u_1])$. So $b_{(n)}[u] = 0$ for each $n \geq N_0$.

Assume that $|a| > 1$ and $[a] = b_{(m)}[c]$. Since

$$(b_{(m)}[c])_{(n)}[u] = \sum_{p \geq 0}(-1)^p\binom{m}{p}b_{(m-p)}([c]_{(n+p)}[u]),$$

we can assume $n \geq N([c], [u])$. Therefore, $(b_{(m)}[c])_{(n)}[u] = 0$. $\quad\square$

**Lemma 3.3.** *For any $a \in T$, $u \in U$, $m \geq 0$, we have*

$$D[a]_{(m)}[u] = -m[a]_{(m-1)}[u] \tag{2}$$

**Proof.** Assume first $|a| = 1$ and $[a] = D^i b$. By definition, the left hand side of (2) is

$$D[a]_{(m)}[u] = D^{i+1}b_{(m)}[u] = \begin{cases} 0, & i + 1 > m, \\ \dfrac{(-1)^{i+1}m!}{(m-i-1)!}b_{(m-i-1)}[u], & i + 1 \leq m. \end{cases}$$

While the right hand side of (2) is

$$-m[a]_{(m-1)}[u] = (-m)D^i b_{(m-1)}[u]$$
$$= \begin{cases} 0, & i > m - 1, \\ m\dfrac{(-1)^{i+1}(m-1)!}{(m-i-1)!}b_{(m-i-1)}[u], & i \leq m - 1. \end{cases}$$

So $D(D^i b)_{(m)}[u] = -m(D^i b)_{(m-1)}[u]$.

Assume $|a| > 1$, $[a] = b_{(n)}[c]$. Then

$$
\begin{aligned}
D[a]_{(m)}[u] &= D(b_{(n)}[c])_{(m)}[u] \\
&= -n(b_{(n-1)}[c])_{(m)}[u] + (b_{(n)}D[c])_{(m)}[u] \\
&= -\sum_{t \geq 0}(-1)^t n\binom{n-1}{t} b_{(n-1-t)}([c]_{(m+t)}[u]) + \\
&\quad \sum_{t \geq 0}(-1)^t \binom{n}{t} b_{(n-t)}(D[c]_{(m+t)}[u]) \\
&= \sum_{p \geq 1}(-1)^p n\binom{n-1}{p-1} b_{(n-p)}([c]_{(m+p-1)}[u]) - \\
&\quad \sum_{t \geq 0}(-1)^t(m+t)\binom{n}{t} b_{(n-t)}([c]_{(m+t-1)}[u]) \\
&= \sum_{t \geq 1}(-1)^t t\binom{n}{t} b_{(n-t)}([c]_{(m+t-1)}[u]) - \\
&\quad \sum_{t \geq 0}(-1)^t(m+t)\binom{n}{t} b_{(n-t)}([c]_{(m+t-1)}[u]) \\
&= -m\sum_{t \geq 0}(-1)^t\binom{n}{t} b_{(n-t)}([c]_{(m+t-1)}[u]) \\
&= -m(b_{(n)}[c])_{(m-1)}[u] = -m[a]_{(m-1)}[u].
\end{aligned}
$$

Hence (2) is true.    □

**Lemma 3.4.**

For any $a \in T$, $u \in U$, $m \geq 0$, we have

$$
D([a]_{(m)}[u]) = D[a]_{(m)}[u] + [a]_{(m)}D[u] \tag{3}
$$

**Proof.** Case 1. Let $[a] = b$, $[u] = D^i y$. By Lemma 3.1, the identity (3) holds for $m \geq 0$, $i \geq 0$.

Case 2. Let $[a] = b$ and $[u] = b_{1(n)}D^i y$. If $0 \leq m < N(b, b_1)$, the identity (3) is true by the definition. Assume that $[u] = b_{1(n)}D^i y$, $m \geq N(b, b_1)$. Induction on $m$. Let $N := N(b, b_1)$. Then

$$
b_{(N)}(b_{1(n)}D^i y) = -\sum_{t \geq 1}(-1)^t\binom{N}{t} b_{(N-t)}(b_{1(n+t)}D^i y).
$$

So the left hand side of (3) is

$$D(b_{(N)}[u]) = -\sum_{t \geq 1}(-1)^t \binom{N}{t} Db_{(N-t)}(b_{1(n+t)}D^i y)$$

$$- \sum_{t \geq 1}(-1)^t \binom{N}{t} b_{(N-t)} D(b_{1(n+t)}D^i y)$$

$$= A_1 + A_2 + A_3,$$

where

$$A_1 := \sum_{t \geq 1}(-1)^t(N-t)\binom{N}{t} b_{(N-t-1)}(b_{1(n+t)}D^i y),$$

$$A_2 := \sum_{t \geq 1}(-1)^t(n+t)\binom{N}{t} b_{(N-t)}(b_{1(n+t-1)}D^i y),$$

$$A_3 := -\sum_{t \geq 1}(-1)^t \binom{N}{t} b_{(N-t)}(b_{1(n+t)}D^{i+1}y) = b_{(N)}(b_{1(n)}D^{i+1}y).$$

The right hand side of (3) is

$$Db_{(N)}[u] + b_{(N)}D[u] = -Nb_{(N-1)}(b_{1(n)}D^i y)+$$

$$b_{(N)}(Db_{1(n)}D^i y) + b_{(N)}(b_{1(n)}D^{i+1}y)$$

$$= -Nb_{(N-1)}(b_{1(n)}D^i y)+$$

$$\sum_{t \geq 1}(-1)^t n \binom{N}{t} b_{(N-t)}(b_{1(n+t-1)}D^i y) + A_3.$$

Since

$$A_1 + A_2 = \sum_{t \geq 1}(-1)^t(N-t)\binom{N}{t} b_{(N-t-1)}(b_{1(n+t)}D^i y)$$

$$+ \sum_{t \geq 1}(-1)^t(n+t)\binom{N}{t} b_{(N-t)}(b_{1(n+t-1)}D^i y)$$

$$= \sum_{p \geq 2}(-1)^{p-1}(N-p+1)\binom{N}{p-1} b_{(N-p)}(b_{1(n+p-1)}D^i y)$$

$$+ \sum_{t \geq 1}(-1)^t(n+t)\binom{N}{t} b_{(N-t)}(b_{1(n+s-1)}D^i y)$$

$$= \sum_{t \geq 1}(-1)^{t-1}t\binom{N}{t} b_{(N-t)}(b_{1(n+t-1)}D^i y) - mNb_{(N-1)}(b_{1(n)}D^i y)$$

$$+ \sum_{t \geq 1}(-1)^t n \binom{N}{t} b_{(N-t)}(b_{1(n+t-1)}D^i y)$$

$$+ \sum_{t \geq 1} (-1)^t t \binom{N}{t} b_{(N-t)} (b_{1(n+t-1)} D^i y)$$

$$= -N b_{(N-1)} (b_{1(n)} D^i y) + \sum_{t \geq 1} (-1)^t n \binom{N}{t} b_{(N-t)} (b_{1(n+t-1)} D^i y),$$

we have $D(b_{(N)}[u]) = D b_{(N)}[u] + b_{(N)} D[u]$. By induction on $m$, we can get $D(b_{(m)}[u]) = D b_{(m)}[u] + b_{(m)} D[u]$ for any $|u| = 2$, $m \geq 0$.

Next, we use induction on $|u|$. Assume that $|u| > 2$ and the result is true for $|u| < l$. Let $|u| = l$. Then, we can repeat the argument by induction on $m$, and get the identity $D(b_{(m)}[u]) = D b_{(m)}[u] + b_{(m)} D[u]$ for any $[u] \in U$, $m \geq 0$. Hence, the identity (3) is true when $|a| = 1$, $|u| \geq 1$.

Case 3. Suppose $|a| \geq 1$, $[a] = b_{(p)}[c]$. By induction on $|a| + |u|$, the left hand side of (3) is

$$D([a]_{(m)}[u]) = D((b_{(p)}[c])_{(m)}[u])$$

$$= \sum_{t \geq 0} (-1)^t \binom{p}{t} D(b_{(p-t)}([c]_{(m+t)}[u])) = A_1 + A_2 + A_3,$$

where

$$A_1 := -\sum_{t \geq 0} (-1)^t (p-t) \binom{p}{t} b_{(p-t-1)}([c]_{(m+t)}[u]),$$

$$A_2 := -\sum_{t \geq 0} (-1)^t (m+t) \binom{p}{t} b_{(p-t)}([c]_{(m+t-1)}[u]),$$

$$A_3 := \sum_{t \geq 0} (-1)^t \binom{p}{t} b_{(p-t)}([c]_{(m+t)} D[u]) = (b_{(p)}[c])_{(m)} D[u] = [a]_{(m)} D[u].$$

Since

$$D[a]_{(m)}[u] = D(b_{(p)}[c])_{(m)}[u] = (D b_{(p)}[c])_{(m)}[u] + (b_{(p)} D[c])_{(m)}[u]$$

$$= -\sum_{t \geq 0} (-1)^t p \binom{p-1}{t} b_{(p-t-1)}([c]_{(m+t)}[u])$$

$$- \sum_{t \geq 0} (-1)^t (m+t) \binom{p}{t} b_{(p-t)}([c]_{(m-1+t)}[u])$$

$$= A_1 + A_2,$$

the identity (3) is true.    □

**Lemma 3.5.** *Let $n, m \geq 0$, $a, c \in T$, $u \in U$. Then*

$$([a]_{(n)}[c])_{(m)}[u] = \sum_{k \geq 0} (-1)^k \binom{n}{k} [a]_{(n-k)}([c]_{(m+k)}[u]) \qquad (4)$$

**Proof.** (i) We prove that identity (4) is true when $[a] = b$, $|c| = 1$.

Let $[c] = D^i b'$. If $0 \leq n < N(b, b')$, identity (4) follows from the definition.

Let $n \geq N(b, b')$. When $i = 0$, the left hand side of identity (4) is equal to 0, while the right hand side of identity (4) contains the summand

$$b_{(n)}(b'_{(m)}[u]) = -\sum_{k \geq 1}(-1)^s \binom{n}{k} b_{(n-k)}(b'_{(m+k)}[u]).$$

Hence the right hand side of identity (4) is 0 as well.

Induction on $i$. Suppose that $i \geq 1$. The left hand side of identity (4) is equal to

$$(b_{(n)}D^i b')_{(m)}[u] = D(b_{(n)}D^{i-1}b')_{(m)}[u] + n(b_{(n-1)}D^{i-1}b')_{(m)}[u]$$
$$= -m(b_{(n)}D^{i-1}b')_{(m-1)}[u] + n(b_{(n-1)}D^{i-1}b')_{(m)}[u].$$

The right hand side of identity (4) is equal to

$$\sum_{k \geq 0}(-1)^k \binom{n}{k} b_{(n-k)}(D^i b'_{(m+k)}[u])$$

$$= -\sum_{k \geq 0}(-1)^k(m+k)\binom{n}{k} b_{(n-k)}(D^{i-1}b'_{(m+k-1)}[u])$$

$$= -m\sum_{k \geq 0}(-1)^k \binom{n}{k} b_{(n-k)}(D^{i-1}b'_{(m-1+k)}[u])$$

$$\quad + \sum_{k \geq 1}(-1)^{k+1}k \binom{n}{k} b_{(n-k)}(D^{i-1}b'_{(m+k-1)}[u])$$

$$= -m(b_{(n)}D^{i-1}b')_{(m-1)}[u]$$

$$\quad + \sum_{k \geq 0}(-1)^k(k+1)\binom{n}{k+1} b_{(n-k-1)}(D^{i-1}b'_{(m+k)}[u])$$

$$= -m(b_{(n)}D^{i-1}b')_{(m-1)}[u]$$

$$\quad + n\sum_{k \geq 0}(-1)^k \binom{n-1}{k} b_{(n-1-k)}(D^{i-1}b'_{(m+k)}[u])$$

$$= -m(b_{(n)}D^{i-1}b')_{(m-1)}[u] + n(b_{(n-1)}D^{i-1}b')_{(m)}[u].$$

Hence, identity (4) is true for $[a] = b$, $|c| = 1$.

(ii) We prove that identity (4) is true for any $c \in T$, $[a] = b$.

Induction on $|c|$. We have showed the result for $|c| = 1$. Suppose $|c| > 1$ and $[c] = b_{1(p)}[c_1]$. We only need to consider $n \geq N(b, b_1)$. Following the definition, we have

$$b_{(n)}(b_{1(p)}[c_1]) = -\sum_{k\geq 1}(-1)^k\binom{n}{k}b_{(n-k)}(b_{1(p+k)}[c_1]).$$

Now induction on $n$. Let $n = N(b, b_1) =: N$. Then

$$A := (b_{(N)}[c])_{(m)}[u] = -\sum_{k\geq 1}(-1)^k\binom{N}{k}(b_{(N-k)}(b_{1(p+k)}[c_1]))_{(m)}[u]$$

$$= -\sum_{k\geq 1,\ t\geq 0}(-1)^{k+t}\binom{N}{k}\binom{N-k}{t}b_{(N-k-t)}((b_{1(p+k)}[c_1])_{(m+t)}[u])$$

$$= \sum_{k\geq 1,\ t,r\geq 0}(-1)^{k+t+r+1}\binom{N}{k}\binom{N-k}{t}\binom{p+k}{r}\times$$

$$b_{(N-k-t)}(b_{1(p+k-r)}([c_1]_{(m+t+r)}[u])).$$

We denote the right hand side of identity (4) is $A_1 + A_2$ where $A_1 = b_{(N)}((b_{1(p)}[c_1])_{(m)}[u])$, $A_2 = \sum_{k\geq 1}(-1)^k\binom{N}{k}b_{(N-k)}((b_{1(p)}[c_1])_{(m+k)}[u])$. Then

$$A_1 = \sum_{t\geq 0}(-1)^t\binom{p}{t}b_{(N)}(b_{1(p-t)}([c_1]_{(m+t)}[u]))$$

$$= -\sum_{t\geq 0,r\geq 1}(-1)^{t+r}\binom{p}{t}\binom{N}{r}b_{(N-r)}(b_{1(p-t+r)}([c_1]_{(m+t)}[u])),$$

$$A_2 = \sum_{k\geq 1,t\geq 0}(-1)^{k+t}\binom{N}{k}\binom{p}{t}b_{(N-k)}(b_{1(p-t)}([c_1]_{(m+k+t)}[u])).$$

We make a transformation

$$(i,\ j,\ l) = (N-k-t,\ p+k-r,\ m+t+r),$$

where $i$, $j$, $l$ are nonnegative integers, and so $i+j+l = N+m+p$. Then $A$ becomes a sum of the expressions

$$A = \sum_{\substack{i,j,l\geq 0,\\ k\geq 1}}(-1)^{k+l+m+1}\binom{N}{k}\binom{N-k}{N-k-i}\binom{p+k}{p+k-j}b_{(i)}(b_{1(j)}([c_1]_{(l)}[u]))$$

$$= \sum_{\substack{i,j,l\geq 0,\\ k\geq 1}}(-1)^{k+l-m+1}\binom{N}{k}\binom{N-k}{i}\binom{p+k}{j}b_{(i)}(b_{1(j)}([c_1]_{(l)}[u]))$$

$$= \sum_{\substack{i,j,l\geq 0,\\ k\geq 1}}(-1)^{k+N+p-i-j+1}\binom{N}{i}\binom{N-i}{k}\binom{p+k}{j}b_{(i)}(b_{1(j)}([c_1]_{(l)}[u]))$$

$$= \sum_{i,j,l \geq 0} (-1)^{p+j+1} \binom{N}{i}$$

$$\times \left( \sum_{k \geq 1} (-1)^{N-i-k} \binom{N-i}{k} \binom{p+k}{j} \right) b_{(i)} (b_{1(j)} ([c_1]_{(l)} [u]))$$

$$= \sum_{i,j,l \geq 0} (-1)^{p+j+1} \binom{N}{i}$$

$$\times \left( \binom{p}{j-N+i} - (-i)^{N-i} \binom{p}{j} \right) b_{(i)} (b_{1(j)} ([c_1]_{(l)} [u]))$$

$$= \sum_{i,j,l \geq 0} \binom{N}{i} \left( (-1)^{p+j+1} \binom{p}{l-m} + (-1)^{l-m} \binom{p}{j} \right) b_{(i)} (b_{1(j)} ([c_1]_{(l)} [u])).$$

Next, do a similar transformation

$$(i, \ j, \ l) = (N - r, \ p + r - t, \ m + t),$$

where $i$, $j$, $l$ are nonnegative integers, and so $i + j + l = N + m + p$. Then

$$A_1 = \sum_{i,j,l \geq 0} (-1)^{p+j+1} \binom{N}{N-i} \binom{p}{l-m} b_{(i)} (b_{1(j)} ([c_1]_{(l)} [u]))$$

$$= \sum_{i,j,l \geq 0} (-1)^{p+j+1} \binom{N}{i} \binom{p}{l-m} b_{(i)} (b_{1(j)} ([c_1]_{(l)} [u])).$$

Do another transformation

$$(i, \ j, \ l) = (N - s, \ p - t, \ m + s + t),$$

where $i$, $j$, $l$ are nonnegative integers, and so $i + j + l = N + m + p$. Then

$$A_2 = \sum_{i,j,l \geq 0} (-1)^{l+m} \binom{N}{N-i} \binom{p}{p-j} b_{(i)} (b_{1(j)} ([c_1]_{(l)} [u]))$$

$$= \sum_{i,j,l \geq 0} (-1)^{l-m} \binom{N}{i} \binom{p}{j} b_{(i)} (b_{1(j)} ([c_1]_{(l)} [u])).$$

Thus $A = A_1 + A_2$. Assume that $n > N(b, b_1)$, we just repeat the argument of $n = N(b, b_1)$. So identity (4) is true for $[a] = b$.

(iii) We prove that identity (4) holds for $[a] = D^j b$, $j \geq 1$.

By definition, the left hand side of identity (4) is equal to

$$(D^j b_{(n)} [c])_{(m)} [u] = (-1)^j \frac{n!}{(n-j)!} (b_{(n-j)} [c])_{(m)} [u].$$

The right hand side of identity (4) is equal to

$$\sum_{k\geq 0}(-1)^k\binom{n}{k}D^j b_{(n-k)}([c]_{(m+k)}[u])$$

$$=-\sum_{k\geq 0}(-1)^{k+j}\binom{n}{k}\frac{(n-k)!}{(n-k-j)!}b_{(n-k-j)}([c]_{(m+k)}[u])$$

$$=(-1)^j\frac{n!}{(n-j)!}\sum_{k\geq 0}(-1)^k\binom{n-j}{k}b_{(n-j-k)}([c]_{(m+k)}[u]).$$

Therefore, identity (4) holds for $|a|=1$.

(iv) Assume that $|a|>1$. In this case, we write $[a]=b_{(q)}[a_1]$. Then the left hand side of identity (4) is equal to

$$((b_{(q)}[a_1])_{(n)}[c])_{(m)}[u]=\sum_{k\geq 0}(-1)^k\binom{q}{k}(b_{(q-k)}([a_1]_{(n+k)}[c]))_{(m)}[u]$$

$$=\sum_{k,t\geq 0}(-1)^{k+t}\binom{q}{k}\binom{q-k}{t}b_{(q-k-t)}(([a_1]_{(n+k)}[c])_{(m+t)}[u])$$

$$=\sum_{k,t,r\geq 0}(-1)^{k+t+r}\binom{q}{k}\binom{q-k}{t}\binom{n+k}{r}b_{(q-k-t)}([a_1]_{(n+k-r)}([c]_{(m+t+r)}[u]))$$

$$=:A.$$

Do a transformation on the indices

$$(i,\,j,\,l)=(q-k-t,\,n+k-r,\,m+t+r),$$

where $i,\,j,\,l$ are nonnegative integers, and so $i+j+l=n+m+q$. Then

$$A=\sum_{i,j,l,k\geq 0}(-1)^{k+l+m}\binom{q}{k}\binom{q-k}{q-k-i}\binom{n+k}{n+k-j}b_{(i)}([a_1]_{(j)}([c]_{(l)}[u]))$$

$$=\sum_{i,j,l,k\geq 0}(-1)^{k+n+q-i-j}\binom{q}{k}\binom{q-k}{i}\binom{n+k}{j}b_{(i)}([a_1]_{(j)}([c]_{(l)}[u]))$$

$$=\sum_{i,j,l\geq 0}(-1)^{n-j}\binom{q}{i}(\sum_{k\geq 0}(-1)^{q-i-k}\binom{q-i}{k}\binom{n+k}{j})b_{(i)}([a_1]_{(j)}([c]_{(l)}[u]))$$

$$=\sum_{i,j,l\geq 0}(-1)^{n-j}\binom{q}{i}\binom{n}{j-q+i}b_{(i)}([a_1]_{(j)}([c]_{(l)}[u]))$$

$$=\sum_{i,j,l\geq 0}(-1)^{n-j}\binom{q}{i}\binom{n}{n+m-l}b_{(i)}([a_1]_{(j)}([c]_{(l)}[u]))$$

$$= \sum_{i,j,l \geq 0} (-1)^{n-j} \binom{q}{i} \binom{n}{l-m} b_{(i)}([a_1]_{(j)}([c]_{(l)}[u])).$$

Denote the right hand side of identity (4) as $A'$. Then

$$A' = \sum_{k \geq 0} (-1)^k \binom{n}{k} (b_{(q)}[a_1])_{(n-k)}([c]_{(m+k)}[u])$$

$$= \sum_{k,t \geq 0} (-1)^{k+t} \binom{n}{k} \binom{q}{t} b_{(q-t)}([a_1]_{(n-k+t)}([c]_{(m+k)}[u]))$$

$$= \sum_{i,j,l \geq 0} (-1)^{n-j} \binom{n}{l-m} \binom{q}{q-i} b_{(i)}([a_1]_{(j)}([c]_{(l)}[u])) = A,$$

where $(i, j, l) = (q-t, n-k+t, m+k)$. Therefore, we complete the proof of identity (4). $\square$

**Theorem 3.1.** *Let the notation be as above. Then* $span_{\mathbf{k}}[U]$ *is the free module generated by $Y$ over the free associative conformal algebra $C(B, N)$.*

*We denote* $span_{\mathbf{k}}[U]$ *by* $mod_{C(B,N)}(Y)$, *the double free associative conformal module.*

**Proof.** By Lemmas 3.2–3.5, $span_{\mathbf{k}}[U]$ is a module over free associative conformal algebra $C(B, N)$.

For any $C(B, N)$-module $M$ and any map $\varepsilon$ satisfying $[a]_{(n)}\varepsilon(y) = 0$ for all $a \in T, y \in Y, n \geq N([a], y)$, we define the $C(B, N)$-module homomorphism

$$\varphi : mod_{C(B,N)}(Y) \longrightarrow {}_{C(B,N)}M, \quad [u] \longmapsto [u]|_{y \mapsto \varepsilon(y)}$$

such that the following diagram is commutative:

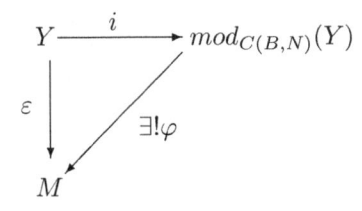

Thus, the result is true. $\square$

### 3.2. *Free associative conformal C-modules*

Let $C = (C, N(-,-), (n), n \in \mathbb{Z}_{\geq 0})$ be an arbitrary associative conformal algebra. Then $C$ is an epimorphic image of some free associative conformal algebra $C(B, N)$. Thus, $C$ has an expression

$$C = C(B, N|S) := C(B,\ N)/Id(S)$$

generated by $B$ with defining relations $S$, where $Id(S)$ is the ideal of $C(B, N)$ generated by $S$.

Let $S \subset C(B, N)$, $C = C(B, N|S)$ and $mod_{C(B,N)}(Y)$ be the $C(B, N)$-module constructed as above. Denote

$$R := \{s_{(m)}[u] \mid s \in S, u \in U, m \geq 0\},$$
$$D^\omega(Y) := \{D^i y \mid i \geq 0,\ y \in Y\}.$$

Then $subm(R) = \sum_{m \geq 0} Id(S)_{(m)} D^\omega(Y)$. Thus, $mod_{C(B,N)}(Y|R)$ is also a $C$-module if we define: for any $f + Id(S) \in C$, $n \geq 0$, $h + subm(R) \in mod_{C(B,N)}(Y|R)$,

$$(f + Id(S))_{(n)}(h + subm(R)) := f_{(n)}h + subm(R).$$

For any left $C$-module $_C M$, we also can regard $_C M$ as a $C(B, N)$-module in a natural way: for any $f \in C(B, N)$, $n \geq 0$, $v \in M$,

$$f_{(n)}v := (f + Id(S))_{(n)}v.$$

**Theorem 3.2.** *Let $C = C(B, N|S)$ be an arbitrary associative conformal algebra, $Y$ a set and $R = \{s_{(m)}[u] \mid s \in S, u \in U, m \geq 0\}$. Then $mod_{C(B,N)}(Y|R)$ is a free $C$-module generated by $Y$.*

*We denote $mod_{C(B,N)}(Y|R)$ by $mod_{C(B,N|S)}(Y)$.*

**Proof.**    Let $_C M$ be a $C$-module and $\varepsilon : Y \to M$ be a map such that $c_{(n)}\varepsilon(y) = 0$ for any $n \geq N(c, y)$, $c \in C$, $y \in Y$. By Theorem 3.1, there exists a unique $C(B, N)$-module homomorphism $\phi : mod_{C(B,N)}(Y) \to {}_C M$, such that $\phi i = \varepsilon$. Let $\pi$ be the natural $C(B, N)$-module homomorphism from $mod_{C(B,N)}(Y)$ to $mod_{C(B,N)}(Y|R)$. Obviously, for any $s_{(m)}[u] \in R$,

$$\phi(s_{(m)}[u]) = s_{(m)}\phi([u]) = (s + Id(S))_{(m)}\phi([u]) = 0.$$

Hence, there exists a unique $C(B, N)$-module homomorphism $\varphi : mod_{C(B,N)}(Y|R) \to {}_C M$ satisfying $\varphi\pi = \phi$.

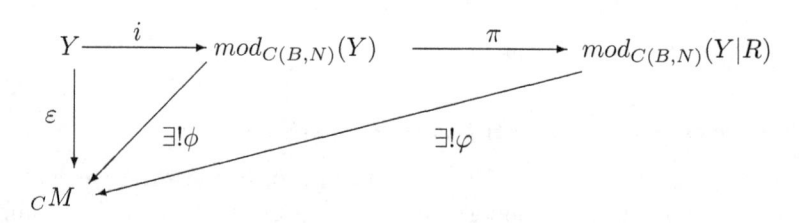

Noting that $\varphi$ is also a $C$-module homomorphism, we complete the proof. $\square$

Let $_CM$ be a $C$-module. Then $_CM$ has an expression $_CM = mod_{C(B,N|S)}(Y|Q)$.

By noting that $mod_{C(B,N|S)}(Y) = mod_{C(B,N)}(Y|R)$, we may assume that $Q \subseteq mod_{C(B,N)}(Y|R)$. Let $Q_1 = \{f \in mod_{C(B,N)}(Y) \mid f + subm(R) \in Q\}$. Then we have

**Proposition 3.1.** *Let the notation as above. Then*

$$mod_{C(B,N|S)}(Y|Q) \cong mod_{C(B,N)}(Y|R \cup Q_1)$$

*as $C(B,N)$-modules and as $C(B,N|S)$-modules.*

## 4. Composition-Diamond lemma for associative conformal modules

We give Composition-Diamond lemma for double free associative conformal module $mod_{C(B,N)}(Y)$ step by step.

In this section, we fix the free $C(B,N)$-module $mod_{C(B,N)}(Y)$ generated by $Y$ with a uniform bounded locality function $N$. Recall that

$$T = \{b_{1(n_1)} \cdots b_{k(n_k)} D^i b_{k+1} \mid b_j, b_{k+1} \in B, 0 \leq n_j < N, 1 \leq j \leq k, i, k \geq 0\},$$

$$U = \{b_{1(n_1)} b_{2(n_2)} \cdots b_{k(n_k)} D^i y \mid b_j \in B,\ y \in Y,$$
$$0 \leq n_j < N,\ 1 \leq j \leq k,\ i, k \geq 0\}$$

and $[U] = \{[u] \mid u \in U\}$ is a **k**-basis of $mod_{C(B,N)}(Y)$, where $[\ ]$ means right normed bracketing.

For $j \geq 0$, $u \in U$, $uD^j$ means that we apply $D^j$ to the last letter of $u$.

A word in $mod_{C(B,N)}(Y)$ is a polynomial of the form $(u)$ (with some bracketing of $u$), where $u = b_{1(n_1)} \cdots b_{k(n_k)} D^i y$, $b_j \in B$, $y \in Y$, $n_j, i, k \geq 0$, $1 \leq j \leq k$. Since $[U]$ is a **k**-basis of $mod_{C(B,N)}(Y)$, each word $(u)$ in $mod_{C(B,N)}(Y)$ is a linear combination of some elements in $[U]$.

### 4.1. *A monomial ordering*

Let $Y, B$ be well-ordered sets. We order elements of $U$ according to the lexicographical ordering of their weights. For any

$$u = b_{1(n_1)} b_{2(n_2)} \cdots b_{k(n_k)} D^i y \in U,$$

denote by

$$wt(u) := (|u|, b_1, n_1, \cdots, b_k, n_k, y, i),$$

where $|u| = k + 1$ is the length of $u$. Then for any $u, v \in U$, define

$$u > v \Leftrightarrow wt(u) > wt(v) \qquad \text{lexicographically.}$$

It is clear that such an ordering is a well ordering on $U$. We will use this ordering in the sequel.

For $f \in mod_{C(B,N)}(Y)$, the leading term of $f$ is denoted by $\bar{f}$ and $\bar{f} \in U$. So $f = \alpha_{\bar{f}}[\bar{f}] + \sum_i \alpha_i[u_i]$, $u_i < \bar{f}$. We will call $f$ monic if $\alpha_{\bar{f}} = 1$.

**Lemma 4.1.** *Suppose that $a \in T$ is $D$-free, $0 \le n < N$, $u, v \in U$. Then*

> (i)    $\overline{[a]_{(n)}[u]} = a_{(n)}u$;
> (ii)   $u < v \Longrightarrow \overline{[a]_{(n)}[u]} < \overline{[a]_{(n)}[v]}$;
> (iii)  $\overline{D^i([u])} = uD^i$, $i \in \mathbb{Z}_{\ge 0}$;
> (iv)   $u < v \Longrightarrow \overline{D^i([u])} < \overline{D^i([v])}$, $i \in \mathbb{Z}_{\ge 0}$.

**Proof.** (i) Induction on $|a|$. If $|a| = 1$, then $[a] = b$ for some $b \in B$ and

$$\overline{[a]_{(n)}[u]} = \overline{b_{(n)}[u]} = b_{(n)}u.$$

Assume that $|a| > 1$ and $[a] = b_{(m)}[a_1]$, where $a_1 \in T$ is $D$-free. Then

$$[a]_{(n)}[u] = (b_{(m)}[a_1])_{(n)}[u] = \sum_{k=0}^{m}(-1)^k \binom{m}{k} b_{(m-k)}([a_1]_{(n+k)}[u]).$$

Since $m < N$,

$$\overline{[a]_{(n)}[u]} = \overline{b_{(m)}([a_1]_{(n)}[u])} = b_{(m)}(\overline{[a_1]_{(n)}[u]}) = b_{(m)}a_{1(n)}u = a_{(n)}u.$$

(ii)  This part follows from (i).

(iii)  Induction on $|u|$. The result is obvious for $|u| = 1$. Let $|u| > 1$, $[u] = b_{(m)}[u_1]$. Then $D^i([u]) = \sum_{k \ge 0}(-1)^k \binom{i}{k} \frac{m!}{(m-k)!} b_{(m-k)} D^{i-k}([u_1])$. Therefore,

$$\overline{D^i([u])} = \overline{b_{(m)}D^i([u_1])} = b_{(m)}\overline{D^i([u_1])} = b_{(m)}u_1 D^i = uD^i.$$

(iv)  This part follows from (iii).    □

Thus, the ordering $>$ on $U$ is a monomial ordering in a sense of (ii) and (iv) in Lemma 4.1.

### 4.2. *S-words and normal S-words*

Let $S \subset mod_{C(B,N)}(Y)$ be a set of monic polynomials, $(u)$ be a word in $mod_{C(B,N)}(Y)$, where $u = b_{1(n_1)} \cdots b_{k(n_k)} D^i y$, $b_j \in B$, $y \in Y$, $n_j, i, k \geq 0$, $1 \leq j \leq k$. We define $S$-word $(u)_{D^i s}$ for any $s \in S$ by induction.

(i) $(D^i s)_{D^i s} = D^i s$ is an $S$-word of $S$-length 1;

(ii) If $(u)_{D^i s}$ is an $S$-word of $S$-length $k$, and $(a)$ is any word in $C(B, N)$ of length $l$, then $(a)_{(n)}(u)_{D^i s}$ is an $S$-words of $S$-length $k + l$.

The $S$-length of an $S$-word $(u)_{D^i s}$ will be denoted by $|u|_{D^i s}$. An $S$-word $\lfloor u \rfloor_{D^i s} := b_{1(n_1)}(b_{2(n_2)}(\cdots (b_{k(n_k)} D^i s) \cdots))$ is called a right normed $S$-word, and it is a normal $S$-word, denoted by $[u]_{D^i s}$, if each $n_j < N$.

**Lemma 4.2.** *Let* $[u]_{D^i s}$ *be a normal S-word. Then* $\overline{[u]_{D^i s}} = u|_{D^i s \mapsto \bar{s} D^i}$.

**Proof.** Let $s = [\bar{s}] + \sum_j \alpha_j [v_j]$, where $v_j < \bar{s}$. Then $D^i s = D^i([\bar{s}]) + \sum_j \alpha_j D^i([v_j])$, and applying Lemma 4.1, we get $\overline{D^i([v_j])} < \overline{D^i([\bar{s}])}$. Hence $\overline{D^i s} = \overline{D^i([\bar{s}])} = \bar{s} D^i$.

If $|u|_{D^i s} = 1$, then $[u]_{D^i s} = D^i s$ and we are done. So assume that $|u|_{D^i s} > 1$. Then $[u]_{D^i s} = b_{(n)}[v]_{D^i s}$, so

$$\overline{[u]_{D^i s}} = \overline{b_{(n)}[v]_{D^i s}} = b_{(n)}\overline{[v]_{D^i s}} = b_{(n)} v\big|_{D^i s \mapsto \bar{s} D^i} = u\big|_{D^i s \mapsto \bar{s} D^i}.$$

We complete the proof by induction on $|u|_{D^i s}$. $\quad\square$

**Lemma 4.3.** *Any S-word* $(u)_{D^i s}$ *can be presented as a linear combination of right normed S-word such that the length of the leading term of each term is less or equal to* $|\overline{(u)_{D^i s}}|$.

**Proof.** Induction on $|u|_{D^i s}$. The result holds trivially if $|u|_{D^i s} = 1$. Let $|u|_{D^i s} > 1$. We assume that

$$(u)_{D^i s} = (a)_{(m)} \lfloor v \rfloor_{D^i s} \tag{5}$$

where $\lfloor v \rfloor_{D^i s}$ is a right normed $S$-word. Next, induction on $|a|$. If $|a| = 1$, then $(a) = D^j b$ for some $b \in B$ and $j \geq 0$, so

$$(a)_{(m)} \lfloor v \rfloor_{D^i s} = D^j b_{(m)} \lfloor v \rfloor_{D^i s} = \begin{cases} 0, & j > m, \\ (-1)^j \dfrac{m!}{(m-j)!} b_{(m-j)} \lfloor v \rfloor_{D^i s}, & j \leq m. \end{cases}$$

Thus, (5) is right normed and we are done. Let $|a| > 1$ and $(a) = (a_1)_{(n)}(a_2)$. Then

$$((a_1)_{(n)}(a_2))_{(m)} \lfloor v \rfloor_{D^i s} = \sum_{k=0}^{n} (-1)^k \binom{n}{k} (a_1)_{(n-k)}((a_2)_{(m+k)} \lfloor v \rfloor_{D^i s}).$$

Now, the result follows from the induction on $|a|$.    $\square$

### 4.3.  *Compositions*

Let $S \subset mod_{C(B,N)}(Y)$ with each polynomial in $S$ monic, $w \in U$ and $f, g \in S$.

We have three kinds of compositions.

- If $w = \bar{f} = \overline{[u]_{D^i g}}$ and $i \geq 0$, then define

$$(f, g)_w = f - [u]_{D^i g},$$

which is a composition of inclusion.

- If $w = \bar{f}D^i = a_{(n)}\bar{g}, i > 0$ and $a \in T$ with $a$ $D$-free, then define

$$(f, g)_w = D^i f - [a_{(n)}g],$$

which is a composition of intersection.

- If $b \in B$ and $n \geq N$, then $b_{(n)}f$ is referred to as a composition of left multiplication.

Let $S \subset mod_{C(B,N)}(Y)$ be a set of monic polynomials and $h \in mod_{C(B,N)}(Y)$. Then $h$ is said to be trivial modulo $S$, denoted by

$$h \equiv 0 \ mod(S) \tag{6}$$

if $h = \sum_i \alpha_i [u_i]_{D^{l_i} s_i}$, where each $[u_i]_{D^{l_i} s_i}$ is a normal $S$-word and $\overline{[u_i]_{D^{l_i} s_i}} \leq \bar{h}$. For $h_1, h_2 \in mod_{C(B,N)}(Y)$, $h_1 \equiv h_2 \ mod(S)$ means $h_1 - h_2 \equiv 0 \ mod(S)$.

The set $S$ is called a Gröbner-Shirshov basis in $mod_{C(B,N)}(Y)$ if all compositions of elements of $S$ are trivial modulo $S$. In particular, if $S$ is $D$-free, then we call $S$ a $D$-free Gröbner-Shirshov basis.

The set $S$ is said to be closed under the composition of left multiplication if any composition of left multiplication of $S$ is trivial modulo $S$. That $S$ is closed under the composition of inclusion and intersection is similarly defined.

**Lemma 4.4.** *Let $S \subset mod_{C(B,N)}(Y)$ with each polynomial in $S$ monic and $[u]_{D^i s}$ a normal $S$-word. Then*

(i)  $D^j([u]_{D^i s}) = [u]_{D^{i+j} s} + \sum_t \beta_t [w_t]_{D^{l_t} s_t}$, *where each* $[w_t]_{D^{l_t} s_t}$ *is a normal $S$-word and* $\overline{[w_t]_{D^{l_t} s_t}} < \overline{[u]_{D^{i+j} s}}$.

(ii)  *If $S$ is closed under the composition of left multiplication, then for any $b \in B$, $n \geq N$, we have $b_{(n)}[u]_{D^i s} \equiv 0 \ mod(S)$.*

**Proof.** (i) Induction on $|u|_{D^i s}$. Let $|u|_{D^i s} = 1$ and $[u]_{D^i s} = D^i s$. Then

$$D^j([u]_{D^i s}) = D^j(D^i s) = D^{i+j}s = [u]_{D^{i+j}s}.$$

Assume that $|u|_{D^i s} > 1$ and $[u]_{D^i s} = b_{(n)}[v]_{D^i s}$. So

$$D^j([u]_{D^i s}) = D^j(b_{(n)}[v]_{D^i s}) = \sum_{p \geq 0}(-1)^p \binom{j}{p}\frac{n!}{(n-p)!}b_{(n-p)}D^{j-p}([v]_{D^i s})$$

$$= b_{(n)}D^j([v]_{D^i s}) + \sum_{p \geq 1}(-1)^p \binom{j}{p}\frac{n!}{(n-p)!}b_{(n-p)}D^{j-p}([v]_{D^i s})$$

$$= b_{(n)}[v]_{D^{i+j}s} + \sum_t \beta_t b_{(n)}[w_t]_{D^{l_t}s_t} + \sum_{p \geq 1}\alpha_p b_{(n-p)}([v]_{D^{i+j-p}s}$$

$$+ \sum_{t'}\gamma_{t'}[w'_{t'}]_{D^{l_{t'}}s_{t'}})$$

$$= [u]_{D^{i+j}s} + \sum_t \beta_t b_{(n)}[w_t]_{D^{l_t}s_t} + \sum_{p \geq 1}\alpha_p b_{(n-p)}([v]_{D^{i+j-p}s}+$$

$$\sum_{t'}\gamma_{t'}[w'_{t'}]_{D^{l_{t'}}s_{t'}}),$$

where $\overline{[w_t]_{D^{l_t}s_t}} < \overline{[v]_{D^{i+j}s}}$, $\overline{[w'_{t'}]_{D^{l_{t'}}s_{t'}}} < \overline{[v]_{D^{i+j-p}s}}$, $\alpha_p = (-1)^p\binom{j}{p}\frac{n!}{(n-p)!}$. Thus $\overline{b_{(n)}[w_t]_{D^{l_t}s_t}} < \overline{b_{(n)}[v]_{D^{i+j}s}} = \overline{[u]_{D^{i+j}s}}$ and $\overline{b_{(n-p)}D^{j-p}([v]_{D^i s})} = b_{(n-p)}\overline{[v]_{D^{i+j-p}s}} < \overline{b_{(n)}[v]_{D^{i+j}s}} = \overline{[u]_{D^{i+j}s}}$ for $p \geq 1$. By induction, we have $D^j([u]_{D^i s}) \equiv [u]_{D^{i+j}s} \ mod(S, w)$.

(ii) Let $|u|_{D^i s} = 1$ and $n = N$. Then

$$b_{(N)}D^i s = D^i(b_{(N)}s) - \sum_{t \geq 1}(-1)^t \binom{i}{t}\frac{N!}{(N-t)!}b_{(N-t)}D^{i-t}s$$

$$= D^i(b_{(N)}s) - \sum_{t \geq 1}\alpha_t[u_t]_{D^{i-t}s},$$

where $\alpha_t = (-1)^t\binom{i}{t}\frac{N!}{(N-t)!}$, and $[u_t]_{D^{i-t}s} = b_{(N-t)}D^{i-t}s$. We may assume that $b_{(N)}s = \sum_j \alpha_j[v_j]_{D^{l_j}s_j}$, where $\overline{[v_j]_{D^{l_j}s_j}} \leq \overline{b_{(N)}s}$. By (i), we get

$$D^i(b_{(N)}s) = \sum_j \alpha_j D^i([v_j]_{D^{l_j}s_j}) = \sum_j \alpha_j([v_j]_{D^{l_j+i}s_j} + \sum_q \beta_{j_q}[w_{j_q}]_{D^{l_{j_q}}s_{j_q}})$$

$$= \sum_j \alpha_j[v_j]_{D^{l_j+i}s_j} + \sum_{j,j_q}\alpha_j\beta_{j_q}[w_{j_q}]_{D^{l_{j_q}}s_{j_q}},$$

where

$$\overline{[w_{j_q}]_{D^{l_{j_q}}s_{j_q}}} < \overline{[v_j]_{D^{l_j+i}s_j}} \quad \text{and} \quad \overline{[v_j]_{D^{l_j+i}s_j}} = \overline{D^i([v_j]_{D^{l_j}s_j})} \leq \overline{D^i(b_{(N)}s)}.$$

By induction on $n$, the result is true for $|u|_{D^i s} = 1$.

Assume that $|u|_{D^i s} > 1$ and the result holds for any $n \geq N$ when the $S$-length of a normal $S$-word is less than $|u|_{D^i s}$. Let $[u]_{D^i s} = b_{1(n_1)}[v]_{D^i s}$. Then

$$b_{(n)}[u]_{D^i s} = b_{(n)}(b_{1(n_1)}[v]_{D^i s}) = -\sum_{t \geq 1}(-1)^t \binom{n}{t} b_{(n-t)}(b_{1(n_1+t)}[v]_{D^i s}).$$

Hence, the result follows from the induction on $n$.    □

**Lemma 4.5.** *Let $S$ be a subset of monic polynomials, closed under the composition of left multiplication. Then any $S$-word $(u)_{D^i s}$ has a presentation $(u)_{D^i s} = \sum_j \alpha_j [u_j]_{D^{l_j} s_j}$, where $[u_j]_{D^{l_j} s_j}$ is a normal $S$-word for each $j$.*

**Proof.** Due to Lemma 4.3, we may assume that $(u)_{D^i s}$ is right normed, i.e. $(u)_{D^i s} = \lfloor u \rfloor_{D^i s}$. We prove the result by induction on $|u|_{D^i s}$. If $|u|_{D^i s} = 1$, then the result holds. Assume that $|u|_{D^i s} > 1$, $\lfloor u \rfloor_{D^i s} = b_{(n)} \lfloor u_1 \rfloor_{D^i s}$, $b \in B$, $i, n \geq 0$. By induction, $\lfloor u_1 \rfloor_{D^i s} = \sum_j \beta_j [v_j]_{D^{l_j} s_j}$. By Lemma 4.4, we can get the result.    □

### 4.4. Key lemmas

The following lemmas play a key role in the proof of the Composition-Diamond lemma, see Theorem 4.1.

**Lemma 4.6.** *Let $S$ be a set of monic polynomials, closed under the compositions of inclusion and intersection, and $s_1, s_2 \in S$. If $w = \overline{[u_1]_{D^{i_1} s_1}} = \overline{[u_2]_{D^{i_2} s_2}}$, then*

$$h := [u_1]_{D^{i_1} s_1} - [u_2]_{D^{i_2} s_2} = \sum_t \beta_t [v_t]_{D^{l_t} s_t} \tag{7}$$

*where each $[v_t]_{D^{l_t} s_t}$ is a normal $S$-word and $\overline{[v_t]_{D^{l_t} s_t}} < w$.*

**Proof.** If $|u_1|_{D^{i_1} s_1} = 1$ or $|u_2|_{D^{i_2} s_2} = 1$, the result follows from Lemma 4.4. We may assume that $[u_t]_{D^{i_t} s_t} = [a_{t(n_t)} D^{i_t} s_t]$, where $a_t \in T$ is $D$-free, $t = 1, 2$. There are two cases to consider.

Case 1. Suppose that the subword $\overline{s_1}$ of $w$ contains $\overline{s_2}$ as a subword. Then $\overline{s_1} = a_{(n_2)} \overline{s_2} D^l$, $a_{1(n_1)} a = a_2$, $l = i_2 - i_1$, $a \in T$ is $D$-free, We have

$$h = [a_{1(n_1)} D^{i_1} s_1] - [a_{2(n_2)} D^{i_2} s_2] = [a_{1(n_1)} D^{i_1} s_1] - [a_{1(n_1)} a_{(n_2)} D^{i_2} s_2]$$

$$= [a_{1(n_1)} D^{i_1} s_1] - [a_{1(n_1)} D^{i_1} [a_{(n_2)} D^l s_2]] + \sum_k \alpha_k [q_k]_{D^{j_k} s_k}$$

$$= [a_{1(n_1)} D^{i_1} (s_1, s_2)_{w_1}] + \sum_k \alpha_k [q_k]_{D^{j_k} s_k},$$

where $\overline{[q_k]_{D^{j_k} s_k}} < \overline{[a_{1(n_1)} a_{(n_2)} D^{i_2} s_2]} = w$, $(s_1, s_2)_{w_1}$ is the composition of inclusion. So $\overline{D^{i_1} (s_1, s_2)_{w_1}} < w_1 D^{i_1}$. By Lemma 4.4 (ii), we can get the result.

Case 2. Suppose $\overline{s_1}$ and $\overline{s_2}$ have a nonempty intersection as a subword of $w$. We may assume that $\overline{s_1} D^l = a_{(n_2)} \overline{s_2}$, $a_{1(n_1)} a = a_2$, $l = i_1 - i_2$, $a \in T$ is $D$-free. Then, similar to Case 1, we can get the result. $\quad\square$

Let $S$ be a set of monic polynomials. Denote

$$Irr(S) = \{[u] \mid u \in U, u \neq \overline{[v]_{D^i s}} \text{ for any normal } S\text{-word } [v]_{D^i s}\}.$$

**Lemma 4.7.** *For any* $0 \neq f \in mod_{C(B,N)}(Y)$, $f$ *has a presentation*

$$f = \sum_i \alpha_i [u_i] + \sum_j \beta_j [v_j]_{D^{l_j} s_j},$$

*where each* $[u_i] \in Irr(S)$, $\overline{[v_j]_{D^{l_j} s_j}} \leq \bar{f}$, $\overline{[u_i]} \leq \bar{f}$ *and* $s_j \in S$.

**Proof.** If $\bar{f} = \overline{[u]_{D^i s}}$ for some normal $S$-word $[u]_{D^i s}$, then let $f_1 = f - \alpha[u]_{D^i s}$, where $\alpha$ is the leading coefficient of $f$. If $[\bar{f}] \in Irr(S)$, then let $f_1 = f - \alpha[\bar{f}]$. In both cases, we have $\bar{f}_1 < \bar{f}$. Thus, the result follows by induction on $\bar{f}$. $\quad\square$

### 4.5. *Composition-Diamond lemma for associative conformal modules*

**Theorem 4.1.** *(Composition-Diamond lemma for associative conformal modules) Let* $S \subset mod_{C(B,N)}(Y)$ *be a set of monic polynomials,* $<$ *the ordering on* $U$ *as before and* $subm(S)$ *the submodule of* $mod_{C(B,N)}(Y)$ *generated by* $S$. *Then the following statements are equivalent.*

(i) *$S$ is a Gröbner-Shirshov basis in* $mod_{C(B,N)}(Y)$.

(ii) *If* $0 \neq f \in subm(S)$, *then* $\bar{f} = \overline{[u]_{D^i s}}$ *for some normal $S$-word* $[u]_{D^i s}$.

(iii) *$Irr(S) = \{[u] \mid u \in U, u \neq \overline{[v]_{D^i s}} \text{ for any normal } S\text{-word } [v]_{D^i s}\}$ is a **k**-basis of the factor module* $mod_{C(B,N)}(Y|S) := mod_{C(B,N)}(Y)/subm(S)$.

**Proof.** $(i) \Rightarrow (ii)$.   Let $S$ be a Gröbner-Shirshov basis and $0 \neq f \in subm(S)$. Then, by Lemma 4.5, we have $f = \sum_j \alpha_j [u_j]_{D^{l_j} s_j}$ where each $\alpha_j \in \mathbf{k}$, $s_j \in S$, $l_j \geq 0$. Let

$$w_j = \overline{[u_j]_{D^{l_j} s_j}}, \quad w_1 = w_2 = \cdots = w_t > w_{t+1} \geq \cdots.$$

We will use induction on $(w_1, t)$ to prove that $\bar{f} = \overline{[u]_{D^i s}}$ for some $s \in S$, $i \geq 0$, where $(w_1, t) < (w_1', t')$ lexicographically.

If $w_1 = \bar{f}$, then the result holds. Assume that $w_1 > \bar{f}, t \geq 2$. We have $w_1 = \overline{[u_1]_{D^{l_1} s_1}} = \overline{[u_2]_{D^{l_2} s_2}}$. By Lemma 4.6,

$$\alpha_1 [u_1]_{D^{l_1} s_1} + \alpha_2 [u_2]_{D^{l_2} s_2} = (\alpha_1 + \alpha_2)[u_1]_{D^{l_1} s_1} + \alpha_2 ([u_2]_{D^{l_2} s_2} - [u_1]_{D^{l_1} s_1})$$

$$= (\alpha_1 + \alpha_2)[u_1]_{D^{l_1} s_1} + \sum \alpha_2 \beta_t [v_t]_{D^{j_t} s_t},$$

where $[v_t]_{D^{j_t} s_t}$ is a normal $S$-word for each $t$ and $\overline{[v_t]_{D^{j_t} s_t}} < w_1$. Now the result follows from the induction whenever $\alpha_1 + \alpha_2 \neq 0$, or $t > 2$, or both $\alpha_1 + \alpha_2 = 0$ and $t = 2$. This shows (ii).

$(ii) \Rightarrow (iii)$.   By Lemma 4.7, $Irr(S)$ generates $mod_{C(B,N)}(Y|S)$ as a linear space. Suppose that $\sum_i \alpha_i [u_i] = 0$ in $mod_{C(B,N)}(Y|S)$, where $0 \neq \alpha_i \in \mathbf{k}$, $[u_i] \in Irr(S)$. It means that $\sum_i \alpha_i [u_i] \in subm(S)$ in $mod_{C(B,N)}(Y)$. Then $\overline{\sum_i \alpha_i [u_i]} = u_j$ for some $j$. Since $[u_j] \in Irr(S)$, it contradicts (ii).

$(iii) \Rightarrow (i)$.   Suppose that $h$ is any composition of polynomial in $S$. Then $h \in subm(S)$. Since $Irr(S)$ is a $\mathbf{k}$-basis for $mod_{C(B,N)}(Y)/subm(S)$ and by Lemma 4.7, we have $h \equiv 0 \; mod(S)$, i.e. $S$ is a Gröbner-Shirshov basis.   $\square$

Let $C = C(B, N|S)$ be an arbitrary associative conformal algebra. We may assume that $S$ is a Gröbner-Shirshov basis in $C(B, N)$ (see Ref. 34). Then by Theorem 3.2, $mod_{C(B,N)}(Y|R) = mod_{C(B,N|S)}(Y)$ is the free $C$-module generated by $Y$, where $R = \{s_{(m)}[u] \mid s \in S, u \in U, m \geq 0\} \subseteq mod_{C(B,N)}(Y)$. Then it is important to find a Gröbner-Shirshov basis for $subm(R)$ in $mod_{C(B,N)}(Y)$.

Note that for a $D$-free monic subset $S$ in $C(B, N)$, $S$ is a Gröbner-Shirshov basis in $C(B, N)$ if $S$ is closed under the compositions of inclusion, intersection and left multiplication in $C(B, N)$, see Ref. 34.

**Lemma 4.8.** *Let $S$ be a $D$-free subset of $C(B, N)$, $R = \{s_{(m)}[u] \mid s \in S, u \in U, m \geq 0\} \subseteq mod_{C(B,N)}(Y)$ and $R_1 = \{s_{(m)}[u] \mid s \in S, u \in U, 0 \leq m < N\}$. Then in $mod_{C(B,N)}(Y)$, $subm(R) = subm(R_1)$.*

**Proof.** We only need to show that $R \subseteq subm(R_1)$. Since $S$ is $D$-free, it is clearly that $s_{(p)}b = 0$, $s_{(p)}y = 0$ for any $s \in S$, $b \in B$, $y \in Y$, $p \geq N$. We will prove $s_{(m)}[u] \in span_{\mathbf{k}}R_1$ for any $m \geq N$.

If $[u] = D^i y$, we have

$$s_{(N)}D^i y = -\sum_{t \geq 1} (-1)^t \binom{i}{t} \frac{N!}{(N-t)!} s_{(N-t)} D^{i-t} y,$$

so $s_{(N)}D^i y \in span_{\mathbf{k}}R_1$.

Let $|u| > 1$ and $[u] = b_{1(n_1)}[u_1]$. Then

$$s_{(m)}[u] = s_{(m)}(b_{1(n_1)}[u_1]) = -\sum_{t \geq 1}(-1)^t \binom{m}{t} s_{(m-t)}(b_{1(n_1+t)}[u_1]).$$

By induction on $m$, we complete the proof. $\square$

**Proposition 4.1.** *Let the notation be as in Lemma 4.8 and $S$ a $D$-free Gröbner-Shirshov basis in $C(B, N)$. Then $R_1$ is a Gröbner-Shirshov basis in $mod_{C(B,N)}(Y)$. Moreover,*

$$Irr(R_1) = \{[a_{(n)}D^i y] \mid [a] \in Irr(S), \ a \in T \text{ with } a \ D\text{-free},$$
$$0 \leq n < N, \ i \geq 0, y \in Y\},$$

*is a $\mathbf{k}$-basis for the module $mod_{C(B,N)}(Y|R)$, where*

$$Irr(S) = \{[u] \mid u \in T, \ u \text{ is } D\text{-free and } u \neq \overline{[v]_{D^i s}}$$
$$\text{for any normal } s\text{-word } [v]_{D^i s}\}.$$

**Proof.** By Lemma 4.8, we have $subm(R) = subm(R_1)$.

For any $f = s_{1(m_1)}[u_1]$, $g = s_{2(m_2)}[u_2] \in R_1$, we assume that $[u_j] = [a_{j(n_j)}D^{i_j}y_j]$, $j = 1, 2$. Hence,

$$f - [s_{1(m_1)}a_1]_{(n_1)}D^{i_1}y_1$$

$$= f - \sum_{t \geq 0}(-1)^t \binom{m_1}{t} s_{1(m_1-t)}([a_1]_{(n_1+t)}D^{i_1}y_1)$$

$$= f - s_{1(m_1)}([a_1]_{(n_1)}D^{i_1}y_1) - \sum_{t \geq 1}(-1)^t \binom{m_1}{t} s_{1(m_1-t)}([a_1]_{(n_1+t)}D^{i_1}y_1)$$

$$= \sum_k \alpha_k h_k,$$

where $h_k \in R_1$ and $\overline{h_k} < \bar{f}$.

Similarly, we have $g - [s_{2(m_2)}a_2]_{(n_2)}D^{i_2}y_2 = \sum_t \gamma_t q_t$, where $q_t \in R_1$ and $\overline{q_t} < \bar{f}$.

If $f, g$ have composition, then $n_1 = n_2 =: n$, $y_1 = y_2 =: y$. There are three cases to consider.

Case 1. $w = \bar{f} = a_{(p)}\bar{g}D^k$ where $a \in T$ is $D$-free and $k \geq 0$. Then $\bar{s}_{1(m_1)}a_1 = a_{(p)}\bar{s}_{2(m_2)}a_2 =: c$, and $i_1 = i_2 + k$. Since $S$ is a $D$-free Gröbner-Shirshov basis in $C(B, N)$, we have

$$[s_{1(m_1)}a_1] - [a_{(p)}s_{2(m_2)}a_2] = \sum_{j'} \beta'_j [v_{j'}]_{D^{l_{j'}} s_{j'}},$$

where each $[v_{j'}]_{D^{l_{j'}} s_{j'}}$ is a normal $S$-word and $\overline{[v_{j'}]_{D^{l_{j'}} s_{j'}}} < c$. Hence,

$$\begin{aligned}
f - [a_{(p)}D^l g] &= [s_{1(m_1)}a_1]_{(n)}D^{i_1}y - [a_{(p)}D^l([s_{2(m_2)}a_2]_{(n)}D^{i_2}y)] \\
&\quad + \sum_k \alpha_k h_k - \sum_t \gamma_t[a_{(p)}D^l q_t] \\
&= [s_{1(m_1)}a_1]_{(n)}D^{i_1}y - [a_{(p)}([s_{2(m_2)}a_2]_{(n)}D^{l+i_2}y)] \\
&\quad - \alpha[a_{(p)}([s_{2(m_2)}a_2]_{(n-l)}D^{i_2}y)] + \sum_k \alpha_k h_k - \sum_t \gamma_t[a_{(p)}D^l q_t] \\
&= ([s_{1(m_1)}a_1] - [a_{(p)}s_{2(m_2)}a_2])_{(n)}D^{i_1}y \\
&\quad - \alpha[a_{(p)}([s_{2(m_2)}a_2]_{(n-l)}D^{i_2}y)] \\
&\quad + \sum_k \alpha_k h_k - \sum_t \gamma_t[a_{(p)}D^l q_t] \\
&\equiv 0 \ mod(R_1),
\end{aligned}$$

where $\alpha = (-1)^l \frac{n!}{(n-l)!}$.

Case 2. $w = \bar{f}D^i = a_{(n)}\bar{g}$ where $a \in T$ is $D$-free and $i \geq 0$. Similar to Case 1, we have $D^i f - [a_{(n)}g] \equiv 0 \ mod(R_1)$.

Case 3. For any $b \in B$, $n \geq N$, we will prove $b_{(n)}f \equiv 0 \ mod(R_1)$. If $n = N$, then by Lemma 4.8,

$$\begin{aligned}
b_{(N)}f &= b_{(N)}(s_{1(m_1)}[u_1]) \\
&= (b_{(N)}s_1)_{(m_1)}[u_1] - \sum_{t \geq 1}(-1)^t \binom{N}{t}b_{(N-t)}(s_{1(m_1+t)}[u_1]).
\end{aligned}$$

Since $S$ is a $D$-free Gröbner-Shirshov basis, we have $b_{(N)}s_1 = \sum_j \beta_j[a_{j(n_j)}s_{j(q_j)}c_j]$, where $a_j$, $c_j \in T$ are $D$-free. So

$$\begin{aligned}
b_{(N)}f &\equiv \sum_j \beta_j[a_{j(n_j)}s_{j(q_j)}c_j]_{(m_1)}[u_1] \\
&\equiv \sum_j \beta_j[a_{j(n_j)}s_{j(q_j)}c_{j(m_1)}u_1] \equiv 0 \ mod(R_1).
\end{aligned}$$

Assume that $n > N$. By induction, we have $b_{(n)}f \equiv (b_{(n)}s_1)_{(m_1)}[u_1] \equiv 0 \bmod(R_1)$.

Therefore, $R_1$ is a Gröbner-Shirshov basis in $mod_{C(B,N)}(Y)$. Now, by Theorem 4.1, the set $R_1$ is a **k**-basis for the module $mod_{C(B,N)}(Y|R)$. $\square$

**Remark:** The condition that $S$ is $D$-free in Proposition 4.1 is essential. For example, let $C = C(a, N = 2|S)$ be an associative conformal algebra, where

$$S = \{a_{(1)}a - a_{(0)}Da, \ [a_{(0)}a_{(1)}a], \ [a_{(1)}a_{(0)}a], \ [a_{(0)}a_{(0)}a], \ [a_{(1)}a_{(1)}a]\}.$$

It is easy to check that $S$ is a Gröbner-Shirshov basis in $C(a, N = 2)$. Let $Y$ be a well-ordered set, $mod_{C(B,N=2)}(Y)$ the double free conformal module and $R = \{s_{(m)}[u] \mid s \in S, u \in U, m \geq 0\}$, where $U = \{a_{(n_1)} \cdots a_{(n_k)}D^i y \mid y \in Y, 0 \leq n_j < 2, 1 \leq j \leq k, i, k \geq 0\}$. Since $(a_{(1)}a - a_{(0)}Da)_{(2)}y = a_{(0)}(Da_{(2)}y) = -2a_{(0)}(a_{(1)}y), a_{(0)}(a_{(1)}y) \in subm(R)$. Thus $(a_{(1)}a - a_{(0)}Da)_{(2)}y = 0 \in mod_{C(B,N=2)}(Y|R)$. Noting that $a_{(0)}a \in Irr(S)$, the set $\{[a_{(n)}D^i y] \mid [a] \in Irr(S)\}$ isn't a **k**-basis of $mod_{C(B,N=2)}(Y|R)$.

## 5. Applications

### 5.1. *Conformal modules over universal enveloping conformal algebra*

Let $L$ be a Lie conformal algebra which is a free **k**$[D]$-module with a well-ordered **k**$[D]$-basis $B = \{a_i \mid i \in I\}$ and a uniform bounded locality $N(a_i, a_j) \leq N$ for all $i, j \in I$. Let the multiplication table of $L$ on $B$ be

$$a_{i[n]}a_j = \Sigma_{t \in I}\alpha_{nijt}a_t, \quad \alpha_{nijt} \in \mathbf{k}[D], \ i \geq j, \ i, j \in I, \ n < N.$$

Then by $\mathcal{U}(L)$, the universal enveloping associative conformal algebra of $L$ with respective to $B$ and $N$, one means the following associative conformal algebra, see Ref. 15,

$$\mathcal{U}(L) = C(B, N \mid a_{i(n)}a_j - \{a_{j(n)}a_i\} - \lfloor a_{i[n]}a_j \rfloor, \ i \geq j, \ i, j \in I, n < N)$$

where

$$\{a_{j(n)}a_i\} = \sum_{k \geq 0}(-1)^{n+k}\frac{1}{k!}D^k(a_{j(n+k)}a_i) \quad \text{and} \quad \lfloor a_{i[n]}a_j \rfloor = \Sigma_{t \in I}\alpha_{nijt}a_t.$$

Let $\mathbb{C}$ be the complex field,

$$Vir = C_{Lie}(v, \ N = 2 \mid v_{[0]}v - Dv, v_{[1]}v - 2v)$$

be the Lie conformal algebra over $\mathbb{C}$ ($Vir$ is called the Virasoro conformal algebra), see Ref. 15, and

$$\mathcal{U}(Vir) = C(v, \ N = 2 \mid v_{(1)}v - v)$$

the universal enveloping associative conformal algebra of $Vir$.

**Example 5.1.** Let $\Delta \in \{0,1\}$ and $\alpha \in \mathbb{C}$. Let

$$M(\Delta, \alpha) = mod_{C(v;N=2)}(y|R \cup Q),$$

where

$$R = \{(v_{(1)}v-v)_{(m)}[v_{(n_1)}v\cdots v_{(n_k)}D^iy] \mid n_j \in \{0,1\}, \ 1 \le j \le k, \ m,k,i \ge 0\}$$

and $Q = \{f_1, f_2\}$ where $f_1 = v_{(0)}y - (D+\alpha)y, \ f_2 = v_{(1)}y - \Delta y$.

Then $Q = \{f_1, f_2\}$ is a Gröbner-Shirshov basis for $M(\Delta, \alpha)$. So, by Theorem 4.1, the set $Irr(Q) = \{D^iy \mid i \ge 0\}$ is a $\mathbb{C}$-basis of $M(\Delta, \alpha)$. It follows that $M(\Delta, \alpha) = \mathbb{C}[D]y$ is a $\mathcal{U}(Vir)$-module, called the Virasoro conformal module, see Ref. 20,27,35.

**Proof.** Let $R_1 = \{(v_{(1)}v - v)_{(m)}[v_{(n_1)}v\cdots v_{(n_k)}D^iy] \mid m, n_j \in \{0,1\}, \ 1 \le j \le k, \ k,i \ge 0\}$. By Lemma 4.2 in Ref. 15, $\{v_{(1)}v - v\}$ is a $D$-free Gröbner-Shirshov basis in $C(v; N = 2)$. Then by Lemma 4.8, $subm(R) = subm(R_1)$ in $mod_{C(v;N)}(y)$. So $M(\Delta, \alpha) = mod_{C(v;N)}(y|R \cup Q) = mod_{C(v;N)}(y|R_1 \cup Q)$.

If $i > 0$, then

$$v_{(m)}D^iy = \sum_{t \ge 0} \binom{m}{t} \frac{i!}{(i-t)!} D^{i-t}(v_{(m-t)}y)$$

$$= \begin{cases} D^i(v_{(0)}y), \ m = 0, \\ D^i(v_{(1)}y) + iD^{i-1}(v_{(0)}y), \ m = 1 \end{cases}$$

$$\equiv \begin{cases} (D+\alpha)D^iy, \ m = 0, \\ \Delta D^iy + i(D+\alpha)D^{i-1}y, \ m = 1 \end{cases}$$

$$\equiv \begin{cases} D^{i+1}y + \alpha D^iy \ mod(Q), \ m = 0, \\ (\Delta + i)D^iy + i\alpha D^{i-1}y \ mod(Q), \ m = 1. \end{cases}$$

So $[v_{(n_1)}v\cdots v_{(n_k)}D^iy] \equiv \sum_{l \ge 0} \beta_l D^l y \ mod(Q)$, where $n_j \in \{0,1\}, \ 1 \le j \le k, \ k,i \ge 0$ and $\beta_l \in \mathbb{C}$. Denote $s = v_{(1)}v - v$. Let $h = s_{(m)}[v_{(n_1)}v\cdots v_{(n_k)}D^iy] \in R_1$. Then we can get $h \equiv 0 \ mod(Q)$ following $s_{(n)}D^ky \equiv 0 \ mod(Q)$ for any $k \ge 0, n \in \{0,1\}$.

Now,

$$v_{(2)}Dy = \sum_{t\geq 0} \binom{2}{t} \frac{1}{(1-t)!} D^{1-t}(v_{(2-t)}y) \equiv 2\Delta y \; mod(Q),$$

$$v_{(2)}D^i y = \sum_{t\geq 0} \binom{2}{t} \frac{i!}{(i-t)!} D^{i-t}(v_{(2-t)}y)$$

$$= 2iD^{i-1}(v_{(1)}y) + i(i-1)D^{i-2}(v_{(0)}y)$$

$$\equiv (i^2 - i + 2\Delta i)D^{i-1}y + \alpha(i^2 - i)D^{i-2}y \; mod(Q), \; i \geq 2,$$

$$s_{(0)}y = (v_{(1)}v)_{(0)}y - v_{(0)}y = v_{(1)}(v_{(0)}y) - v_{(0)}(v_{(1)}y) - v_{(0)}y$$

$$\equiv v_{(1)}(D + \alpha)y - v_{(0)}\Delta y - (D + \alpha)y$$

$$\equiv v_{(1)}Dy - (\Delta + 1)Dy - \alpha y$$

$$\equiv 0 \; mod(Q),$$

$$s_{(0)}D^i y = (v_{(1)}v)_{(0)}D^i y - v_{(0)}D^i y$$

$$= v_{(1)}(v_{(0)}D^i y) - v_{(0)}(v_{(1)}D^i y) - v_{(0)}D^i y$$

$$\equiv v_{(1)}(D^{i+1}y + \alpha D^i y) - v_{(0)}((\Delta + i)D^i y + i\alpha D^{i-1}y)$$
$$\quad - D^{i+1}y - \alpha D^i y$$

$$\equiv (\Delta + i + 1)D^{i+1}y + (i+1)\alpha D^i y + \alpha(\Delta + i)D^i y + i\alpha^2 D^{i-1}y$$
$$\quad - (\Delta + i)(D^{i+1}y + \alpha D^i y) - i\alpha D^i y - i\alpha^2 D^{i-1}y$$
$$\quad - D^{i+1}y - \alpha D^i y$$

$$\equiv 0 \; mod(Q), \quad i > 0,$$

$$s_{(1)}D^i y = \sum_{t\geq 0} \binom{1}{t} \frac{i!}{(i-t)!} D^{i-t}(s_{(1-t)}y)$$

$$= \sum_{t\geq 0} \binom{1}{t} \frac{i!}{(i-t)!} D^{i-t}((v_{(1)}v)_{(1-t)}y - v_{(1-t)}y)$$

$$= \sum_{t\geq 0} \binom{1}{t} \frac{i!}{(i-t)!} D^{i-t}(v_{(1)}(v_{(1-t)}y) - v_{(0)}(v_{(2-t)}y) - v_{(1-t)}y)$$

$$\equiv \begin{cases} v_{(1)}(v_{(1)}y) - v_{(1)}y, \; i = 0, \\ iD^{i-1}(v_{(1)}(v_{(0)}y) - v_{(0)}(v_{(1)}y) - v_{(0)}y) \\ \quad + D^i(v_{(1)}(v_{(1)}y) - v_{(1)}y), \; i > 0 \end{cases}$$

$$\equiv \begin{cases} (\Delta^2 - \Delta)y, \; i = 0, \\ iD^{i-1}(v_{(1)}(D + \alpha)y - v_{(0)}\Delta y - (D + \alpha)y) \\ \quad + (\Delta^2 - \Delta)D^i y, \; i > 0 \end{cases}$$

$$
\equiv \begin{cases} (\Delta^2 - \Delta)y, & i = 0, \\ iD^{i-1}(v_{(1)}Dy - (\Delta+1)Dy - \alpha y) + (\Delta^2 - \Delta)D^i y, & i > 0 \end{cases}
$$
$$
\equiv (\Delta^2 - \Delta)D^i y
$$
$$
\equiv 0 \; mod(Q).
$$

For any $n \geq 2$, we have

$$
v_{(n)}f_1 = v_{(n)}(v_{(0)}y) - v_{(n)}(D + \alpha)y
$$
$$
= -\sum_{t \geq 1}(-1)^t \binom{n}{t}v_{(n-t)}(v_{(t)}y) - v_{(n)}Dy - \alpha v_{(n)}y
$$
$$
\equiv nv_{(n-1)}(v_{(1)}y) - nv_{(n-1)}y
$$
$$
\equiv \begin{cases} 0, & n > 2, \\ 2(\Delta^2 - \Delta)y, & n = 2, \end{cases}
$$
$$
\equiv 0 \; mod(Q),
$$
$$
v_{(n)}f_2 = v_{(n)}(v_{(1)}y) - v_{(n)}\Delta y = -\sum_{t \geq 1}(-1)^t \binom{n}{t}v_{(n-t)}(v_{(1+t)}y) - \Delta v_{(n)}y
$$
$$
\equiv 0 \; mod(Q).
$$

From this it follows that $subm(Q) = subm(R_1 \cup Q)$ and all left multiplication compositions in $Q$ are trivial modulo $Q$.

Then $Q$ is closed under the left multiplication composition. Since $Q$ has no composition of inclusion and intersection, $Q$ is a Gröbner-Shirshov basis of $M(\Delta, \alpha)$.

Now, by Theorem 4.1 and Proposition 3.1, the results follow.    $\square$

**Example 5.2.** Module over the semidirect product of Virasoro conformal algebra and current algebra.

Let $(\mathfrak{g}, [\;])$ be a Lie algebra over $\mathbb{C}$ with a well-ordered $\mathbb{C}$-basis $\{a_i\}_{i \in I}$ and $Cur(\mathfrak{g})$ be the current algebra over $\mathfrak{g}$, where

$$
Cur(\mathfrak{g}) = C_{Lie}(\{a_i\}_{i \in I}, \; N = 1 \mid a_{i[0]}a_j = [a_i a_j], \; i, j \in I).
$$

The semidirect product of $Vir$ and $Cur(\mathfrak{g})$ is

$$
Vir \oplus Cur(\mathfrak{g}) = C_{Lie}(\{v\} \cup \{a_i\}_{i \in I}, \; N = 2 \mid S),
$$

where

$$
S = \{v_{[0]}v - Dv, \; v_{[1]}v - 2v, \; v_{[0]}a_i - Da_i, \; v_{[1]}a_i - a_i,
$$
$$
a_{i[0]}a_j - [a_i a_j], \; a_{i[1]}a_j, \; i > j, \; i, j \in I\}.
$$

Then, see section 4.4 in Ref. 15,

$$\mathcal{U}(Vir \oplus Cur(\mathfrak{g})) = C(\{v\} \cup \{a_i\}_{i\in I}, \ N = 2 \mid S^{(-)}),$$

where $S^{(-)}$ consists of

$$s_1 = a_{i(0)}a_j - a_{j(0)}a_i - [a_i a_j], \ i > j,$$
$$s_2 = a_{i(1)}a_j, \ i > j,$$
$$s_3 = v_{(1)}v - v,$$
$$s_4 = v_{(0)}a_i + a_{i(1)}Dv - 2a_{i(0)}v - Da_i,$$
$$s_5 = v_{(1)}a_i + a_{i(1)}v - a_i,$$
$$s_6 = v_{(0)}(a_{j(0)}a_i) - a_{j(0)}(v_{(0)}a_i), \ i > j,$$
$$s_7 = v_{(0)}(a_{i(0)}v) - a_{i(0)}(v_{(0)}v),$$
$$s_8 = v_{(0)}(a_{i(1)}v) - a_{i(1)}(v_{(0)}v) + a_{i(0)}v.$$

Let $V$ be a $\mathfrak{g}$-module with a $\mathbb{C}$-basis $Y$ and $\Delta \in \{0,1\}$, $\alpha \in \mathbb{C}$. Let

$$M(\Delta, \alpha, V) = mod_{C(\{v\}\cup\{a_i\}_{i\in I}, N)}(Y \mid R \cup Q),$$

where

$$R = \{s_{(m)}[c_{1(n_1)} \cdots c_{k(n_k)} D^t y] \mid s \in S^{(-)}, \ n_j \in \{0,1\}, \ c_j \in \{v\} \cup \{a_i\}_{i\in I},$$
$$y \in Y, \ 1 \le j \le k, \ k, m, t \ge 0\}$$

and $Q = \{f_{1y}, f_{2y}, f_{3iy}, f_{4iy} \mid i \in I, y \in Y\}$ where

$$f_{1y} = v_{(0)}y - (D + \alpha)y,$$
$$f_{2y} = v_{(1)}y - \Delta y,$$
$$f_{3iy} = a_{i(0)}y - a_i y,$$
$$f_{4iy} = a_{i(1)}y, \ i \in I, y \in Y.$$

Then $Q$ is a Gröbner-Shirshov basis for $M(\Delta, \alpha, V)$. Moreover, $M(\Delta, \alpha, V) = \mathbb{C}[D]Y$ is a $Vir \oplus Cur(\mathfrak{g})$-module, see Ref. 20,27.

**Proof.** Let $m \ge 0, k \ge 1$. Then we have, $mod(Q)$,

$$v_{(m)}D^k y = \sum_{t \ge 0} \binom{m}{t} \frac{k!}{(k-t)!} D^{k-t}(v_{(m-t)}y)$$

$$\equiv \begin{cases} \dfrac{k!}{(k-m)!} D^{k-m}(v_{(0)}y) + m\dfrac{k!}{(k-m+1)!} D^{k-m+1}(v_{(1)}y), & k \ge m, \\[2ex] \dbinom{m}{k} v_{(m-k)}y, & k < m \end{cases}$$

$$\equiv \begin{cases} \dfrac{k!}{(k-m)!}(D+\alpha)D^{k-m}y + m\Delta\dfrac{k!}{(k-m+1)!}D^{k-m+1}y, & m \le k, \\ (k+1)\Delta y, & m = k+1, \\ 0, & m > k+1, \end{cases}$$

$$a_{i(m)}D^k y = \sum_{t\ge 0}\binom{m}{t}\frac{k!}{(k-t)!}D^{k-t}(a_{i(m-t)}y)$$

$$\equiv \begin{cases} 0, & 1 \le k < m, \\ \dfrac{k!}{(k-m)!}D^{k-m}(a_i y), & k \ge m \ge 0. \end{cases}$$

So $[c_{1(n_1)}\cdots c_{k(n_k)}D^i y] \equiv \sum_{l\ge 0}\beta_l D^l y \ mod(Q)$, where $n_j \in \{0,1\}$, $c_j \in \{v\} \cup \{a_i\}_{i\in I}$, $1 \le j \le k$, $i,k \ge 0$ and $\beta_l \in \mathbb{C}$. Let $h = s_{(m)}[c_{1(n_1)}\cdots c_{k(n_k)}D^i y] \in R$. Then we can get $h \equiv 0 \ mod(Q)$ following $s_{(n)}D^k y \equiv 0 \ mod(Q)$ for any $s \in S^{(-)}, n, k \ge 0$.

Let $m \ge 0$. For $s_1, s_2, s_3, s_6$, we have, $mod(Q)$,

$$s_{1(m)}D^k y = (a_{i(0)}a_j - a_{j(0)}a_i - [a_i a_j])_{(m)}D^k y$$
$$= a_{i(0)}(a_{j(m)}D^k y) - a_{j(0)}(a_{i(m)}D^k y) - [a_i a_j]_{(m)}D^k y$$
$$\equiv \begin{cases} 0, & 1 \le k < m, \\ \dfrac{k!}{(k-m)!}D^{k-m}(a_i(a_j y)) - (a_j(a_i y)) - [a_i a_j]y), & k \ge m \ge 0, \end{cases}$$
$$\equiv 0,$$

$$s_{2(m)}D^k y = \sum_{t\ge 0}\binom{m}{t}\frac{k!}{(k-t)!}D^{k-t}(s_{2(m-t)}y)$$
$$= \sum_{t\ge 0}\binom{m}{t}\frac{k!}{(k-t)!}D^{k-t}((a_{i(1)}a_j)_{(m-t)}y)$$
$$= \sum_{t\ge 0}\binom{m}{t}\frac{k!}{(k-t)!}D^{k-t}((a_{i(1)}(a_{j(m-t)}y) - (a_{i(0)}(a_{j(m-t+1)}y))$$
$$\equiv 0,$$

$$s_{3(0)}D^k y = D^k((v_{(1)}v - v)_{(0)}y) = D^k(v_{(1)}(v_{(0)}y) - v_{(0)}(v_{(1)}y) - v_{(0)}y)$$
$$\equiv D^k(v_{(1)}(D+\alpha)y - v_{(0)}\Delta y - (D+\alpha)y)$$
$$\equiv 0,$$

$$s_{3(m)}D^k y = \sum_{t\ge 0}\binom{m}{t}\frac{k!}{(k-t)!}D^{k-t}(s_{3(m-t)}y)$$

$$= \sum_{t \geq 0} \binom{m}{t} \frac{k!}{(k-t)!} D^{k-t}((v_{(1)}v)_{(m-t)}y - v_{(m-t)}y)$$

$$= \sum_{t \geq 0} \binom{m}{t} \frac{k!}{(k-t)!} D^{k-t}(v_{(1)}(v_{(m-t)}y) - v_{(0)}(v_{(m-t+1)}y) - v_{(m-t)}y)$$

$$\equiv \begin{cases} 0, & m > k+1 \geq 1, \\ m\dfrac{k!}{(k+1-m)!} D^{k+1-m}(\Delta^2 - \Delta)y, & 1 \leq m \leq k+1 \end{cases}$$

$$\equiv 0,$$

$$s_{6(m)}D^k y = \sum_{t \geq 0} \binom{m}{t} \frac{k!}{(k-t)!} D^{k-t}((v_{(0)}(a_{j(0)}a_i) - a_{j(0)}(v_{(0)}a_i))_{(m-t)}y)$$

$$= \sum_{t \geq 0} \binom{m}{t} \frac{k!}{(k-t)!} D^{k-t}(v_{(0)}(a_{j(0)}(a_{i(m-t)}y)) - a_{j(0)}(v_{(0)}(a_{i(m-t)}y)))$$

$$\equiv \begin{cases} 0, & m > k \geq 0, \\ \dfrac{k!}{(k-m)!} D^{k-m}(v_{(0)}(a_{j(0)}\{a_i y\}) - a_{j(0)}(v_{(0)}\{a_i y\})), & 0 \leq m \leq k \end{cases}$$

$$\equiv 0.$$

For $s_4, s_5, s_7, s_8$ and $m \geq 0$, we have

$$s_{4(m)}D^k y = \sum_{t \geq 0} \binom{m}{t} \frac{k!}{(k-t)!} D^{k-t}((v_{(0)}a_i + a_{i(1)}Dv - 2a_{i(0)}v - Da_i)_{(m-t)}y)$$

$$= \sum_{t \geq 0} \binom{m}{t} \frac{k!}{(k-t)!} D^{k-t}(v_{(0)}(a_{i(m-t)}y) + a_{i(1)}(Dv_{(m-t)}y)$$

$$\qquad - a_{i(0)}(Dv_{(m-t+1)}y) - 2a_{i(0)}(v_{(m-t)}y) - Da_{i(m-t)}y)$$

$$= \sum_{t \geq 0} \binom{m}{t} \frac{k!}{(k-t)!} D^{k-t}(v_{(0)}(a_{i(m-t)}y) + a_{i(1)}(Dv_{(m-t)}y)$$

$$\qquad + (m-t+1)a_{i(0)}(v_{(m-t)}y) - Da_{i(m-t)}y),$$

$$s_{5(m)}D^k y = \sum_{t \geq 0} \binom{m}{t} \frac{k!}{(k-t)!} D^{k-t}((v_{(1)}a_i + a_{i(1)}v - a_i)_{(m-t)}y)$$

$$= \sum_{t \geq 0} \binom{m}{t} \frac{k!}{(k-t)!} D^{k-t}(v_{(1)}(a_{i(m-t)}y) - v_{(0)}(a_{i(m-t+1)}y)$$

$$\qquad + a_{i(1)}(v_{(m-t)}y) - a_{i(0)}(v_{(m-t+1)}y) - a_{i(m-t)}y),$$

$$s_{7(m)}D^k y = \sum_{t \geq 0} \binom{m}{t} \frac{k!}{(k-t)!} D^{k-t}((v_{(0)}(a_{i(0)}v) - a_{i(0)}(v_{(0)}v))_{(m-t)}y)$$

$$= \sum_{t \geq 0} \binom{m}{t} \frac{k!}{(k-t)!} D^{k-t}(v_{(0)}(a_{i(0)}(v_{(m-t)}y)) - a_{i(0)}(v_{(0)}(v_{(m-t)}y))),$$

$$s_{8(m)}D^k y = \sum_{t \geq 0} \binom{m}{t} \frac{k!}{(k-t)!} D^{k-t}((v_{(0)}(a_{i(1)}v) - a_{i(1)}(v_{(0)}v) + a_{i(0)}v)_{(m-t)}y)$$

$$= \sum_{t \geq 0} \binom{m}{t} \frac{k!}{(k-t)!} D^{k-t}(v_{(0)}(a_{i(1)}(v_{(m-t)}y)) - v_{(0)}(a_{i(0)}(v_{(m-t+1)}y))$$

$$- a_{i(1)}(v_{(0)}(v_{(m-t)}y)) + a_{i(0)}(v_{(0)}(v_{(m-t+1)}y)) + a_{i(0)}(v_{(m-t)}y)).$$

There are two cases to consider.

Case 1. $m > k \geq 0$. Then, $mod(Q)$,

$$s_{4(m)}D^k y \equiv \binom{m}{k} k!(a_{i(1)}(Dv_{(m-k)}y) + (m-k-1)a_{i(0)}(v_{(m-k)}y) - Da_{i(m-k)}y$$

$$\equiv \begin{cases} 0, & m-k > 1, \\ m!(-a_{i(1)}(v_{(0)}y) + a_{i(0)}y), & m-k = 1, \end{cases}$$

$$\equiv 0,$$

$$s_{5(m)}D^k y \equiv \binom{m}{k}(v_{(1)}(a_{i(m-k)}y) - v_{(0)}(a_{i(m-k+1)}y) + a_{i(1)}(v_{(m-k)}y)$$

$$\equiv 0,$$

$$s_{7(m)}D^k y = \binom{m}{k}(v_{(0)}(a_{i(0)}(v_{(m-k)}y)) - a_{i(0)}(v_{(0)}(v_{(m-k)}y)))$$

$$\equiv \begin{cases} 0, & m-k > 1, \\ m(v_{(0)}(a_{i(0)}\Delta y) - a_{i(0)}(v_{(0)}\Delta y)), & m-k = 1, \end{cases}$$

$$\equiv \begin{cases} 0, & m-k > 1, \\ m(v_{(0)}\Delta(a_i y) - a_{i(0)}\Delta(D+\alpha)y), & m-k = 1, \end{cases}$$

$$\equiv 0,$$

$$s_{8(m)}D^k y \equiv \binom{m}{k}(v_{(0)}(a_{i(1)}(v_{(m-k)}y)) - a_{i(1)}(v_{(0)}(v_{(m-k)}y)) + a_{i(0)}(v_{(m-k)}y))$$

$$\equiv \begin{cases} 0, & m-k > 1, \\ m(v_{(0)}(a_{i(1)}\Delta y) - a_{i(1)}(v_{(0)}\Delta y) + a_{i(0)}\Delta y), & m-k = 1, \end{cases}$$

$$\equiv \begin{cases} 0, & m-k > 1, \\ m(v_{(0)}(-a_{i(1)}\Delta(D+\alpha)y + \Delta(a_i y), & m-k = 1, \end{cases}$$

$$\equiv 0.$$

Case 2. $0 \leq m \leq k$. Then, $mod(Q)$,

$$s_{4(m)}D^k y = \frac{k!}{(k-m)!}D^{k-m}(v_{(0)}(a_{i(0)}y) - a_{i(0)}(v_{(0)}y))$$

$$+m\frac{k!}{(k+1-m)!}D^{k+1-m}(a_{i(1)}(Dv_{(1)}y) + a_{i(0)}y)$$

$$\equiv \frac{k!}{(k-m)!}D^{k-m}(v_{(0)}(a_i y) - a_{i(0)}(D+\alpha)y)$$

$$+m\frac{k!}{(k+1-m)!}D^{k+1-m}(-a_{i(1)}(D+\alpha)y + a_{i(0)}y)$$

$$\equiv 0,$$

$$s_{5(m)}D^k y \equiv \frac{k!}{(k-m)!}D^{k-m}(v_{(1)}(a_{i(0)}y) + a_{i(1)}(v_{(0)}y) - a_{i(0)}(v_{(1)}y) - a_{i(0)}y)$$

$$\equiv 0,$$

$$s_{7(m)}D^k y = \frac{k!}{(k-m)!}D^{k-m}(v_{(0)}(a_{i(0)}(v_{(0)}y)) - a_{i(0)}(v_{(0)}(v_{(0)}y)))$$

$$+m\frac{k!}{(k-m+1)!}D^{k-m+1}(v_{(0)}(a_{i(0)}(v_{(1)}y)) - a_{i(0)}(v_{(0)}(v_{(1)}y)))$$

$$\equiv \frac{k!}{(k-m)!}D^{k-m}(v_{(0)}(a_{i(0)}(D+\alpha)y) - a_{i(0)}(v_{(0)}(D+\alpha)y))$$

$$+m\frac{k!}{(k-m+1)!}D^{k-m+1}(v_{(0)}(a_{i(0)}\Delta y) - a_{i(0)}(v_{(0)}\Delta y))$$

$$\equiv \frac{k!}{(k-m)!}D^{k-m}(v_{(0)}(D+\alpha)(a_i y)) - a_{i(0)}(D+\alpha)^2 y)$$

$$+m\frac{k!}{(k-m+1)!}D^{k-m+1}(v_{(0)}\Delta(a_i y) - a_{i(0)}\Delta(D+\alpha)y)$$

$$\equiv 0,$$

$$s_{8(m)}D^k y \equiv \frac{k!}{(k-m)!}D^{k-m}(v_{(0)}(a_{i(1)}(v_{(0)}y)) - v_{(0)}(a_{i(0)}(v_{(1)}y))$$

$$-a_{i(1)}(v_{(0)}(v_{(0)}y)) + a_{i(0)}(v_{(0)}(v_{(1)}y)) + a_{i(0)}(v_{(0)}y))$$

$$+m\frac{k!}{(k-m+1)!}\times$$

$$D^{k-m+1}(v_{(0)}(a_{i(1)}(v_{(1)}y)) - a_{i(1)}(v_{(0)}(v_{(1)}y)) + a_{i(0)}(v_{(1)}y))$$

$$\equiv 0.$$

Let $n \geq 2$ and $i, j \in I$. Then we have, $mod(Q)$,

$$v_{(n)}f_{1y} \equiv \begin{cases} 0, & n > 2, \\ 2(\Delta^2 - \Delta)y, & n = 2, \end{cases} \equiv 0,$$

$$v_{(n)}f_{2y} = v_{(n)}(v_{(1)}y) - v_{(n)}\Delta y \equiv 0,$$

$$v_{(n)}f_{3iy} = v_{(n)}(a_{i(0)}y - a_i y) \equiv \sum_{k \geq 1}(-1)^{k+1}\binom{n}{k}v_{(n-k)}(a_{i(k)}y) \equiv 0,$$

$$v_{(n)}f_{4iy} = v_{(n)}(a_{i(1)}y) = \sum_{k \geq 1}(-1)^{k+1}\binom{n}{k}v_{(n-k)}(a_{i(k+1)}y) \equiv 0,$$

$$a_{j(n)}f_{1y} = a_{j(n)}(v_{(0)}y - (D+\alpha)y) = a_{j(n)}(v_{(0)}y) - a_{j(n)}(D+\alpha)y)$$
$$\equiv \sum_{k \geq 1}(-1)^{k+1}\binom{n}{k}a_{j(n-k)}(v_{(k)}y) - na_{j(n-1)}y$$
$$\equiv na_{j(n-1)}(v_{(1)}y) \equiv 0,$$

$$a_{j(n)}f_{2y} = a_{j(n)}(v_{(1)}y - \Delta y) \equiv \sum_{k \geq 1}(-1)^{k+1}\binom{n}{k}a_{j(n-k)}(v_{(1+k)}y) \equiv 0,$$

$$a_{j(n)}f_{3iy} = a_{j(n)}(a_{i(0)}y - a_i y) \equiv \sum_{k \geq 1}(-1)^{k+1}\binom{n}{k}a_{j(n-k)}(a_{i(k)}y) \equiv 0,$$

$$a_{j(n)}f_{4iy} = a_{j(n)}(a_{i(1)}y) \equiv 0.$$

This shows that $subm(Q) = subm(R \cup Q)$ and all left multiplication compositions in $Q$ are trivial modulo $Q$. Since $Q$ has no composition of inclusion and intersection, $Q$ is a Gröbner-Shirshov basis of $M(\Delta, \alpha, V)$.

Now, the results follow from Theorem 4.1 and Proposition 3.1.  $\square$

### Acknowledgement

We wish to express our thanks to Prof. L.A. Bokut and Mr. Zerui Zhang for helpful suggestions and comments.

### References

1. A.A. Belavin, A.M. Polyakov, A.B. Zamolodchikov, Infinite conformal symmetry in two-dimensional quantum field theory, Nuclear Phys. B241(1984), 333-380.
2. G.M. Bergman, The diamond lemma for ring theory, Adv. Math., 29(1978), 178-218.
3. R.E. Borcherds, Vertex algebras, Kac-Moody algebras, and the monster, Proc. Nat. Acda. Sci. U.S.A., 83(1986), 3068-3071.

4. R.E. Borcherds, Monstrous moonshine and monstrous Lie superalgebras, Invent. Math., 109(1992), 405-444.

5. L.A. Bokut, Imbeddings into simple associative algebras, Algebra i Logika, 15(1976), 117-142.

6. L.A. Bokut, Yuqun Chen, Yongshan Chen, Composition-Diamond lemma for tensor product of free algebras, Journal of Algebra, 323(2010), 2520-2537.

7. L.A. Bokut, Yuqun Chen, Yongshan Chen, Groebner-Shirshov bases for Lie algebras over a commutative algebra, Journal of Algebra, 337(2011), 82-102.

8. L.A. Bokut, Yuqun Chen, Xueming Deng, Gröbner-Shirshov bases for Rota-Baxter algebras, Siberian Math. J., 51(6)(2010), 978-988.

9. L.A. Bokut, Yuqun Chen, Cihua Liu, Gröbner-Shirshov bases for dialgebras, International Journal of Algebra and Computation, 20(3)(2010), 391-415.

10. L.A. Bokut, Yuqun Chen, Yu Li, Gröbner-Shirshov bases for Vinberg-Koszul-Gerstenhaber right-symmetric algebras, Fundamental and Applied Mathematics, 14(8)(2008), 55-67(in Russian). J. Math. Sci., 166(2010), 603-612.

11. L.A. Bokut, Yuqun Chen, Qiuhui Mo, Gröbner-Shirshov bases for semirings, Journal of Algebra, 385(2013), 47-63.

12. L.A. Bokut, Yuqun Chen, Jianjun Qiu, Gröbner-Shirshov bases for associative algebras with multiple operators and free Rota-Baxter algebras, Journal of Pure and Applied Algebra, 214(2010), 89-100.

13. L.A. Bokut, Y. Fong, W.-F. Ke, Free associative conformal algebras, Proceedings of the 2nd Tainan-Moscow Algebra and Combinatorics Workshop, Tainan, Springer-Verlag, Hong Kong, 1997, pp.13-25.

14. L.A. Bokut, Y. Fong, W.-F. Ke, Gröbner-Shirshov bases and composition lemma for associative conformal algebras: an example, Contemp. Math., 264(2000), 63-90.

15. L.A. Bokut, Y. Fong, W.-F. Ke, Composition Diamond Lemma for associative conformal algebras, Journal of Algebra, 272(2004), 739-774.

16. L. A. Bokut, Y. Fong, W.-F. Ke, P. Kolesnikov, Gröbner and Gröbner-Shirshov bases in Algebra and Conformal algebras, Fundamental and Applied Mathematics, 6(2000), N3, 669-706.

17. B. Buchberger, An algorithmical criteria for the solvability of algebraic systems of equations, Aequationes Math., 4(1970), 374-383.

18. Yuqun Chen, Yongshan Chen, Chanyan Zhong, Composition-Diamond lemma for modules, Czechoslovak Math. J., 60(135)(2010), 59-76.

19. Yongshan Chen, Yuqun Chen, Groebner-Shirshov bases for matabelian Lie algebras, Journal of Algebra, 358(2012), 143-161.

20. S.-J Cheng, V. Kac, Conformal modules, Asian J. Math., 1(1997), 181-193.

21. E.S. Chibrikov, On free Lie conformal algebras, Vestnik Novosibirsk State University, 4(1)(2004), 65-83.

22. P.M. Cohn, Universal Algebra, 2nd Edition, in: Mathematics and its Applications, vol. 6, Reidel, Dordrecht- Boston, MA, 1981.

23. V. Dotsenko, A. Khoroshkin, Gröbner bases for operads, Duke Mathematical Journal, 153(2)(2010), 363-396.

24. I. Frenkel, J. Lepowsky, A. Meurman, Vertex Operator Algebras and the Monster, in: Pure Appl. Math., vol. 134, Academic Press, Boston, MA,

1988.

25. M. Gao, Y. Xu, X. Yue, The Lie conformal algebra of a Block type Lie algebra, Algebra Colloquium, 3(2015), 367-382.

26. H. Hironaka, Resolution of singularities of an algebraic variety over a field if characteristic zero, I, II, Ann. of Math., 79(1964), 109-203, 205-326.

27. V. Kac, Vertex algebras for beginners, University Lecture Series, Vol. 10., AMS, Providence, RI, 1996.

28. V. Kac, The idea of locality, in: M.-D. Doebner, et al. (Eds.), Physical Applications and Mathematical Aspects of Geometry, Groups and Algebras, World Scientific, Singapore, (1997), 16-32.

29. S.-J. Kang, K.-H. Lee, Gröbner-Shirshov bases for irreducible $sl_{n+1}$-modules, Journal of Algebra, 232(2000), 1-20.

30. A.A. Mikhalev, The junction lemma and the equality problem for color Lie superalgebras, Vestnik Moskov. Univ. Ser. I Mat. Mekh., 5(1989), 88-91. English translation: Moscow Univ. Math. Bull., 44(1989), 87-90.

31. A.A. Mikhalev, The composition lemma for color Lie superalgebras and for Lie $p$-superalgebras, Contemp. Math., 131(2)(1992), 91-104.

32. A.A. Mikhalev, Shirshov's composition techniques in Lie superalgebra (noncommutative Gröbner bases). Trudy. Sem. Petrovsk., 18(1995), 277-289. English translation: J. Math. Sci., 80(1996), 2153-2160.

33. A.A. Mikhalev, A.A. Zolotykh, Standard Gröbner-Shirshov bases of free algebras over rings, I. Free associative algebras, International Journal of Algebra and Computation, 8(6)(1998), 689-726.

34. Lili Ni, Yuqun Chen, A new Composition-Diamond lemma for associative conformal algebras, Journal of Algebra and its Applications, 16(5)(2017), 1750094(28pages).

35. Alexander Retakh, Structure and representations of conformal algebras, rings and their representations, World Sci. Publ., Hackensack, NJ, (2006), 289-311.

36. M. Roitman, On free conformal and vertex algebras, Journal of Algebra, 217(2)(1999), 496-527.

37. A.I. Shirshov, Some algorithmic problem for $\varepsilon$-algebras, Sibirsk. Mat. Z., 3(1962), 132-137.

38. A.S. Wightman, Quantum field theory in terms of vacuum expectation values, Phys. Rev., 101(1956), 860-866.

39. Selected works of A.I. Shirshov, Eds L.A. Bokut, V. Latyshev, I. Shestakov, E. Zelmanov, Trs M. Bremner, M. Kochetov, Birkhäuser, Basel, Boston, Berlin, 2009.

# De Morgan Semi-Heyting and Heyting Algebras

Hanamantagouda P. Sankappanavar

*Department of Mathematics*
*State University of New York*
*New Paltz, NY 12561*

*sankapph@newpaltz.edu*

Dedicated to Professor Leonid A. Bokut on his 80th birthday

The variety **DMSH** of semi-Heyting algebras with a De Morgan negation was introduced in Ref. 12, and an increasing sequence $\mathbf{DMSH_n}$ of level $n$, for $n \in \omega$, of its subvarieties was investigated in the series Ref. 12–17, of which the present paper is a sequel. It should also be mentioned that the variety **DMSH** is an equivalent algebraic semantics for the propositional logic, called "De Morgan semi-Heyting logic" which is recently introduced in Ref. 3. Since the lattice of subvarieties of **DMSH** is dually isomorphic to the lattice of extensions of the De Morgan semi-Heyting logic, the results of this paper and the earlier papers in this series have logical counterparts in the corresponding (axiomatic) extensions of the De Morgan semi-Heyting logic; and, in particular, of De Morgan Heyting logic.

In this paper, we prove two main results: Firstly, we prove that $\mathbf{DMSH_1}$-algebras of leve 1 satisfy Stone identity, generalizing an earlier result that regular $\mathbf{DMSH_1}$-algebras of level 1 satisfy Stone identity. Secondly, we prove that the variety **DmsStSH** of dually ms, Stone semi-Heyting algebras is at level 2. As an application, it is derived that the variety of De Morgan semi-Heyting algebras is also at level 2. It is also shown that these results are sharp.

*2010 Mathematics subject classification:* 03G25, 06D20, 08B15, 06D15, 03C05, 03B50,08B26, 06D30, 06E75.

*Keywords*: Blended dually quasi-De Morgan semi-Heyting algebra of level $n$, dually ms Stone semi-Heyting algebra, De Morgan semi-Heyting algebra, De Morgan Heyting algebra, discriminator variety.

## 1. Introduction

The variety **DQDSH** of semi-Heyting algebras with a dually quasi-De Morgan negation was introduced and investigated in Ref. 12. Several important subvarieties of **DQDSH** were also introduced in the same paper, some of which are: Subvarieties of **DQDSH** of level $n$, for $n \in \omega$, the variety **DMSH** of De Morgan (symmetric) semi-Heyting algebras, and **DmsStSH** of dually ms, Stone semi-Heyting algebras. The work of Ref. 12 was continued in Ref. 13–17. We also note that the variety **DMSH** is an equivalent algebraic semantics for the propositional logic, called "De Morgan semi-Heyting logic" which is recently introduced in Ref. 3. Since the lattice of subvarieties of **DMSH** is dually isomorphic to the lattice of extensions of the De Morgan semi-Heyting logic, the results of this paper and the earlier papers in this series have logical counterparts in the corresponding (axiomatic) extensions of the De Morgan semi-Heyting logic; and, in particular, of De Morgan Heyting logic.

In this paper, we present two main results: Firstly, we prove that the variety **DMSH₁** of De Morgan semi-Heyting algebras of level 1 satisfies Stone identity, generalizing an earlier result that regular **DMSH₁**-algebras of level 1 satisfy Stone identity, proved in Ref. 15. Secondly, we prove that the variety **DmsStSH** of dually ms, Stone semi-Heyting algebras is at level 2. As an application, it is derived that the variety of De Morgan Stone semi-Heyting algebras is at level 2. Finally, it is shown that these results are sharp.

## 2. Preliminaries

In this section we recall some definitions and results needed in this paper. For other relevant information, we refer the reader to the textbooks listed in Refs. 1, 2, 7.

An algebra $\mathbf{L} = \langle L, \vee, \wedge, \rightarrow, 0, 1 \rangle$ is a *semi-Heyting algebra* (Ref. 11) if $\langle L, \vee, \wedge, 0, 1 \rangle$ is a bounded lattice and $\mathbf{L}$ satisfies:

(SH1) $x \wedge (x \rightarrow y) \approx x \wedge y$,

(SH2) $x \wedge (y \rightarrow z) \approx x \wedge [(x \wedge y) \rightarrow (x \wedge z)]$,

(SH3) $x \rightarrow x \approx 1$.

Semi-Heyting algebras are distributive and pseudocomplemented, with $a^* := a \rightarrow 0$ as the pseudocomplement of an element $a$. These and other properties (see Ref. 11) of semi-Heyting algebras are frequently used without explicit mention, throughout this paper.

Let $\mathbf{L}$ be a semi-Heyting algebra. $\mathbf{L}$ is a *Heyting algebra* if $\mathbf{L}$ satisfies:

(H) $(x \wedge y) \rightarrow y \approx 1$.

$\mathbf{L}$ is a *Stone semi-Heyting algebra* if $\mathbf{L}$ satisfies:

(St) $x^* \vee x^{**} \approx 1$.

The variety of Stone semi-Heyting algebras is denoted by **StSH** or just by **St**.

The following definition, taken from Ref. 12, is central to this paper.

**Definition 2.1.** An algebra $\mathbf{L} = \langle L, \vee, \wedge, \rightarrow, ', 0, 1 \rangle$ is a *semi-Heyting algebra with a dual quasi-De Morgan operation* or *dually quasi-De Morgan semi-Heyting algebra* (**DQD**-algebra, for short) if $\langle L, \vee, \wedge, \rightarrow, 0, 1 \rangle$ is a semi-Heyting algebra, and $\mathbf{L}$ satisfies:

(a) $0' \approx 1$ and $1' \approx 0$,
(b) $(x \wedge y)' \approx x' \vee y'$,
(c) $(x \vee y)'' \approx x'' \vee y''$,
(d) $x'' \leq x$.

Let $\mathbf{L}$ be a **DQD**-algebra. $\mathbf{L}$ is a *De Morgan semi-Heyting algebra* (**DM**-algebra) if $\mathbf{L}$ satisfies:

(DM) $x'' \approx x$.

$\mathbf{L}$ is a *dually ms semi-Heyting algebra* (**Dms**-algebra) if $\mathbf{L}$ satisfies:

(JDM) $(x \vee y)' \approx x' \wedge y'$ ($\vee$-De Morgan law).

$\mathbf{L}$ is a *blended dually quasi-De Morgan semi-Heyting algebra* (**BDQD**-algebra) if $\mathbf{L}$ satisfies the following identity:

(BDM) $(x \vee x^*)' \approx x' \wedge x^{*'}$ (Blended $\vee$-De Morgan law).

$\mathbf{L}$ is regular if $\mathbf{L}$ satisfies the following inequality:

(R) $x \wedge x^+ \leq y \vee y^*$, where $x^+ := x'^{*'}$.

*The reader should be cautioned that this notion of regularity is totally different from the one used in Ref. 12.*

The varieties of **DQD**-algebras, **Dms**-algebras, and **DM**-algebras are denoted, respectively, by **DQD**, **Dms**, and **DM**. The variety of regular **DQD**-algebras will be denoted by **RDQD**, while **DQDSt** denotes the subvariety of **DQD** with Stone semi-Heyting reducts. If the underlying semi-Heyting algebra of a **DQD**-algebra is a Heyting algebra, then we add "**H**" at the end of the names of the varieties considered in the sequel; for example, **DQDH** denotes the variety of dually quasi-De Morgan Heyting algebras.

**Lemma 2.1.** *Let* $\mathbf{L} \in \mathbf{DQD}$ *and let* $x, y, z \in L$. *Then*

(i) $1'^* = 1$, *and* $1 \to x = x$,

(ii) $x \leq y$ *implies* $x' \geq y'$,

(iii) $(x \wedge y)'^* = x'^* \wedge y'^*$,

(iv) $x''' = x'$,

(v) $x \vee x^+ = 1$.

The following definition is from Ref. 12; it helps us to classify subvarieties of **DQD** by means of "levels". It also plays a crucial role in describing an increasing sequence of discriminator subvarieties of **BDQD** (see Ref. 12 for more details).

**Definition 2.2.**

Let $\mathbf{L} \in \mathbf{DQD}$ and $x \in \mathbf{L}$. For $n \in \omega$, we define $x^{n('*)}$ and $t_n(x)$ recursively as follows:

$$x^{0('*)} := x;$$
$$x^{(n+1)('*)} := (x^{n('*)})'^*, \text{ for } n \geq 0;$$
$$t_0(x) := x,$$
$$t_{n+1}(x) := t_n(x) \wedge x^{(n+1)('*)}, \text{ for } n \geq 0.$$

Let $n \in \omega$. The subvariety $\mathbf{DQD_n}$ *of level* $n$ of **DQD** is defined by the identity:

$$t_n(x) \approx t_{(n+1)}(x); \tag{$L_n$}$$

For a subvariety **V** of **DQD**, we let $\mathbf{V_n} := \mathbf{V} \cap \mathbf{DQD_n}$. We say that $\mathbf{V_n}$ is "at level $n$" or "of level $n$".

It should be noted that, for $n \in \omega$, the variety $\mathbf{BDQD_n}$, is a discriminator variety (Ref. 12).

## 3. The variety $\mathbf{DM_1}$ of De Morgan semi-Heyting algebras of level 1

It was proved in Ref. 15 that regular De Morgan semi-Heyting algebras of level 1 satisfy the Stone identity. The purpose of this section is to prove a more general theorem which says that De Morgan semi-Heyting algebras of level 1 satisfy the Stone identity.

The following Lemma, whose proof is immediate since $x^{2('*)} \leq x$ in a **DM**-algebra, gives an alternate definition of "level $n$", for $n \in \omega$ such that $n \geq 1$.

**Lemma 3.1.** *Let* $\mathbf{L} \in \mathbf{DM}$ *and let* $n \in \omega$ *such that* $n \geq 1$. *Then* $\mathbf{L}$ *is at level* $n$ *iff* $\mathbf{L}$ *satisfies the identity:*

$(\mathrm{L}'_n)$ $\quad (x \wedge x'^*)^{(n-1)(^{'*})} \approx (x \wedge x'^*)^{n(^{'*})}.$

Recall that $\mathbf{DM_1}$ denotes $\mathbf{DM} \cap \mathbf{DQD_1}$.

**Lemma 3.2.** *Let* $\mathbf{L} \in \mathbf{DM_1}$ *and* $x \in L$. *Then,* $x^{*'} \wedge x' \wedge x^* = 0$.

**Proof.**

$$
\begin{aligned}
x^{*'} \wedge x' \wedge x^* &= x^{*'} \wedge (x' \wedge x''^*)'^* \quad \text{by Lev 1 and (DM)} \\
&= x^{*'} \wedge (x'' \vee x''^{*'})^* \\
&= x^{*'} \wedge (x \vee x^{*'})^* \quad \text{by (DM)} \\
&\leq x^{*'} \wedge x'^{*} \\
&= 0.
\end{aligned}
$$

$\square$

We are ready to present our main theorem of this section.

**Theorem 3.1.** *Let* $\mathbf{L} \in \mathbf{DM_1}$. *Then* $\mathbf{L}$ *satisfies Stone identity.*

**Proof.** We have $x^{*'} \wedge x^{**'} \wedge x^* = x^{*'} \wedge x^{**'} \wedge (x^{**} \rightarrow 0) = x^{*'} \wedge x^{**'} \wedge [(x^{**} \wedge x^{*'} \wedge x^{**'}) \rightarrow 0]$, which, by Lemma 3.2, implies $x^{*'} \wedge x^{**'} \wedge x^* = x^{*'} \wedge x^{**'}$, from which, in view of Lemma 3.2, we get $x^{*'} \wedge x^{**'} = 0$. Hence, $x^{*''} \vee x^{**''} = 1$. which, by (DM), implies $x^* \vee x^{**} = 1$. $\square$

**Corollary 3.1.** $\mathbf{DM_1} = \mathbf{DMSt_1}$. *In particular,* $\mathbf{DMH_1} = \mathbf{DMStH_1}$.

Recall that $\mathbf{RDM_1}$ denotes the variety of regular De Morgan semi-Heyting algebras of level 1. The following result, proved in Ref. 15 (Theorem 3.8), is now a special case of the preceding corollary.

**Corollary 3.2.** (Ref. 15, Theorem 3.8) $\mathbf{RDM_1} = \mathbf{RDMSt_1}$.

It is natural to wonder if Theorem 3.1 can be further generalized. There are two possible directions to try to generalize, which led us to ask if the theorem holds in $\mathbf{DM_2}$ or in $\mathbf{Dms_1}$. However, the theorem fails in both cases, as shown by the following two examples:

**Example 3.1.** Theorem 3.1 fails in the variety $\mathbf{DM_2}$ as witnessed by the following algebra:

Let $\mathbf{A} = \langle A, \vee, \wedge, \to, ', 0, 1 \rangle$ be an algebra, with $A = \{0, a, b, c, d, e, 1\}$, whose lattice reduct and the operatons $\to$ and $'$ are defined in Figure 1. It is routine to verify that $\mathbf{A} \in \mathbf{DM_2}$, but fails to satisfy the Stone identity (at $b$).

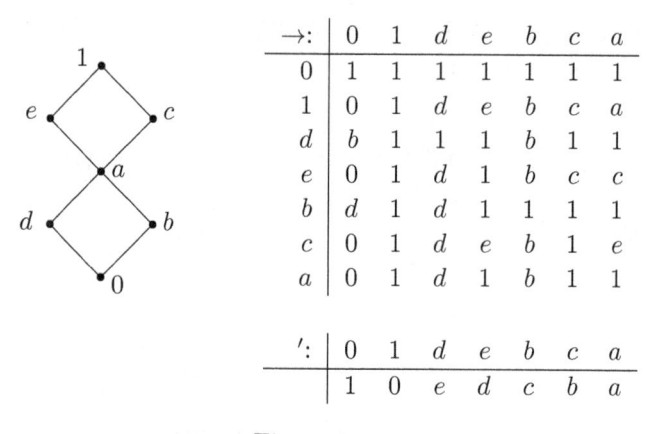

| $\to$: | 0 | 1 | $d$ | $e$ | $b$ | $c$ | $a$ |
|---|---|---|---|---|---|---|---|
| 0 | 1 | 1 | 1 | 1 | 1 | 1 | 1 |
| 1 | 0 | 1 | $d$ | $e$ | $b$ | $c$ | $a$ |
| $d$ | $b$ | 1 | 1 | 1 | $b$ | 1 | 1 |
| $e$ | 0 | 1 | $d$ | 1 | $b$ | $c$ | $c$ |
| $b$ | $d$ | 1 | $d$ | 1 | 1 | 1 | 1 |
| $c$ | 0 | 1 | $d$ | $e$ | $b$ | 1 | $e$ |
| $a$ | 0 | 1 | $d$ | 1 | $b$ | 1 | 1 |

| $'$: | 0 | 1 | $d$ | $e$ | $b$ | $c$ | $a$ |
|---|---|---|---|---|---|---|---|
| | 1 | 0 | $e$ | $d$ | $c$ | $b$ | $a$ |

Figure 1

Thus, $\mathbf{DMSt_2} \subset \mathbf{DM_2}$. Also, since the algebra in Figure 1 is actually in $\mathbf{RDM}$, we also have $\mathbf{RDMSt_2} \subset \mathbf{RDM_2}$.

**Example 3.2.** The following algebra is in $\mathbf{RDms_1}$, but fails to satisfy the Stone identity.

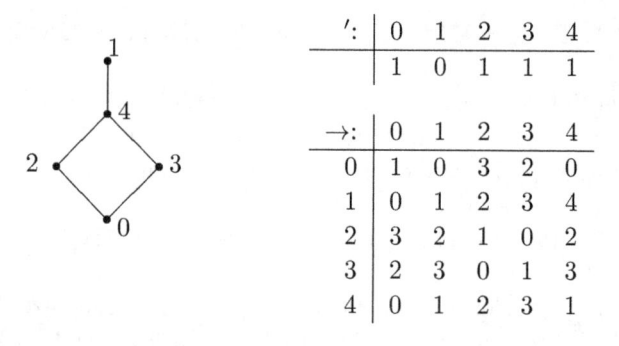

| $'$: | 0 | 1 | 2 | 3 | 4 |
|---|---|---|---|---|---|
| | 1 | 0 | 1 | 1 | 1 |

| $\to$: | 0 | 1 | 2 | 3 | 4 |
|---|---|---|---|---|---|
| 0 | 1 | 0 | 3 | 2 | 0 |
| 1 | 0 | 1 | 2 | 3 | 4 |
| 2 | 3 | 2 | 1 | 0 | 2 |
| 3 | 2 | 3 | 0 | 1 | 3 |
| 4 | 0 | 1 | 2 | 3 | 1 |

Figure 2

Thus, $\mathbf{DmsSt_1} \subset \mathbf{Dms_1}$.

## 4. The level of the variety DmsSt of dually ms, Stone semi-Heyting algebras

Recall that the variety of dually ms Stone semi-Heyting algebras, **DmsSt**, is defined, relative to **DQD**, by the following identities:

(St)  $x^* \vee x^{**} \approx 1$,

(JDM)  $(x \vee y)' \approx x' \wedge y'$    ($\vee$-De Morgan law).

Since **DQD**-algebras satisfy $x'' \leq x$, it is clear that the $\{\vee, \wedge, '\}$-reduct of a **DQD**-algebra satisfying (JDM) is indeed a dual ms algebra.

In this section our goal is to show that the variety **DmsSt** lies at level 2, but not at level 1. The following theorem, proved in Ref. 13 (Theorem 2.5), gives an alternative definition of **DQDSt$_n$**.

**Theorem 4.1.** (Ref. 13, Theorem 2.5) *For* $n \in \omega \setminus \{0\}$, **DQDSt$_n$** *is defined by the identity:*

$(\mathrm{L}'_n)$    $(x \wedge x'^*)^{(n-1)\,(\prime *)} \approx (x \wedge x'^*)^{n\,(\prime *)},$

*relative to* **DQDSt**.

*In particular, the varieties* **DQDSt$_1$** *and* **DQDSt$_2$** *are defined, respectively, by the identities:* $x \wedge x'^* \approx (x \wedge x'^*)'^*$ *and* $(x \wedge x'^*)'^* \approx (x \wedge x'^*)'^{*\prime *}$, *relative to* **DQDSt**.

Throughout this section, **L** denotes an algebra in **DmsSt**, and $x \in \mathbf{L}$. Let $x^+ := x'^{*\prime}$.

**Lemma 4.1.** $x' \vee x^{*\prime **} = 1$.

**Proof.** Since $x' \vee x^{*\prime *} = (x' \vee x^{*\prime *}) \wedge (x \wedge x^*)' = (x' \vee x^{*\prime *}) \wedge (x' \vee x^{*\prime}) = x' \vee (x^{*\prime *} \wedge x^{*\prime}) = x'$, we get

$$x^{*\prime *} \leq x'. \tag{1}$$

Hence, $x' \vee x^{*\prime **} = x' \vee x^{*\prime *} \vee x^{*\prime **} = 1$, in view of (1) and (St).    □

**Lemma 4.2.** $x^{*\prime} \leq x^{**\prime *}$.

**Proof.** $x^{*\prime} \wedge x^{**\prime *} = x^{*\prime} \wedge (x^{**\prime} \to 0) = x^{*\prime} \wedge [(x^{**\prime} \wedge x^{**\prime}) \to (x^{**\prime} \wedge 0)] = x^{*\prime} \wedge [(x^{*\prime} \wedge x^{**\prime}) \to 0] = x^{*\prime} \wedge [(x^* \vee x^{**})' \to 0] = x^{*\prime} \wedge (0 \to 0) = x^{*\prime}$, in view of the identities (JDM) and (St).    □

**Lemma 4.3.** $x^{**\prime} = x^{*\prime *}$. *Hence,* $x^{*+} \leq x^{**}$.

**Proof.** From the equation (1) of Lemma 4.1, and Lemma 4.2, we have $x^{*'} = x^{**'*}$, from which, replacing $x$ by $x^*$, we get $x^{**'} = x^{***'*}$, leading to $x^{**'} = x^{*'*}$. Hence, $x^{*+} = x^{**''} \leq x^{**}$. $\square$

**Lemma 4.4.** $x^{*''} = x^*$.

**Proof.** From $x^* \wedge x^{**''} = 0$, we have $(x^{*''} \vee x^*) \wedge (x^{*''} \vee x^{**''}) = x^{*''}$, which implies $(x^{*''} \vee x^*) \wedge (x^* \vee x^{**})'' = x^{*''}$, whence, by (St), we get $x^{*''} \vee x^* = x^{*''}$. Thus we can conclude that $x^{*''} = x^*$. $\square$

**Lemma 4.5.** $x^{*''} \leq x'^{*'}$.

**Proof.** From $x \wedge x^{*''} = 0$ we have $(x \vee x'^{*'}) \wedge (x^{*''} \vee x'^{*'}) = x'^{*'}$, implying that $x^{*''} \leq x'^{*'}$, in view of Lemma 2.1 (v). $\square$

**Lemma 4.6.** $x^{*+} = x^{**}$.

**Proof.** Applying Lemma 4.4 and Lemma 4.5, we get $x^* \leq x^+$. Then, from Lemma 4.3, we have $x^{**} \leq x^{*+} \leq x^{**}$. $\square$

We are now ready to prove our main theorem of this section.

**Theorem 4.2.** *The variety* **DmsSt** *is at level 2.*

**Proof.** Let $\mathbf{L} \in$ **DmsSt** and let $a \in L$. Then, $(a \wedge a'^*)'^{*'*} = (a'^* \wedge a'^{*'*})'^* = (a'^{*'} \vee a'^{*+})^* = (a^+ \vee a'^{**})^*$ by Lemma 4.6. Hence, $(a \wedge a'^*)'^{*'*} = a^{+*} \wedge a'^{***} = a^{+*} \wedge a'^* = (a^+ \vee a')^* = (a'^* \wedge a)'^*$. $\square$

Recall that **DmsStH** denotes the variety of dually ms, Stone Heyting algebras.

**Corollary 4.1.** **DmsSt** = **DmsSt₂**. *In particular,* **DmsStH** = **DmsStH₂**.

The following corollary is now immediate.

**Corollary 4.2.** *The variety* **DMSt** *of De Morgan Stone semi-Heyting algebras is at level 2.*

**Proof.** Observe that **DMSt** $\subset$ **DmsSt** and apply Theorem 4.2. $\square$

We note that Theorem 4.2 is sharp in the sense that it fails to satisfy the level 1 identity, as shown by the following example.

**Example 4.1.** It is easy to see that the following algebra is in **DmsSt**; but fails to satisfy the level 1 identity: $x \wedge x'^* = (x \wedge x'^*)'^*$ at $a$.

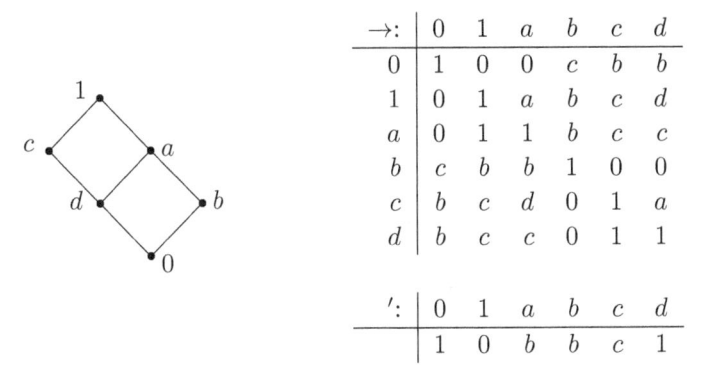

| $\rightarrow$: | 0 | 1 | $a$ | $b$ | $c$ | $d$ |
|---|---|---|---|---|---|---|
| 0 | 1 | 0 | 0 | $c$ | $b$ | $b$ |
| 1 | 0 | 1 | $a$ | $b$ | $c$ | $d$ |
| $a$ | 0 | 1 | 1 | $b$ | $c$ | $c$ |
| $b$ | $c$ | $b$ | $b$ | 1 | 0 | 0 |
| $c$ | $b$ | $c$ | $d$ | 0 | 1 | $a$ |
| $d$ | $b$ | $c$ | $c$ | 0 | 1 | 1 |

| $'$: | 0 | 1 | $a$ | $b$ | $c$ | $d$ |
|---|---|---|---|---|---|---|
| | 1 | 0 | $b$ | $b$ | $c$ | 1 |

Figure 3

Let **DmsE2** denote the subvariety of **Dms** defined by the following Lee's identity (Ref. 4):

(E2) $\quad (x \wedge y)^* \vee (x^* \wedge y)^* \vee (x \wedge y^*)^* \approx 1.$

**Example 4.2.** The 15-element algebra whose universe is: $\{0, 1, 2, 3, 4, 5, 6, 7, 8, 9, 10, 11, 12, 13, 14\}$, whose lattice reduct and the operations $\rightarrow$, $'$ are given below in Figure 4, is in **DmsE2** \ **DmsSt** and is at level $\geq 3$ (the level 2 identity fails at 2). Thus, the variety **DmsE2** is at a level $\geq 3$. (We suspect that its level is 4.)

Here is yet another direction to consider for a possible generalization of Theorem 4.2. Recall that **BDQDSt** denotes the subvariety of **DQD** defined by

(B) $(x \vee x^*)' \approx x' \wedge x^*{}',$
(St) $x^* \vee x^{**} \approx 1.$

In this context, we have the following theorem, with M. Kinyon, which will be published in a future paper.

**Theorem 4.3.** *The variety* **BDQDSt** *is at level 2.*

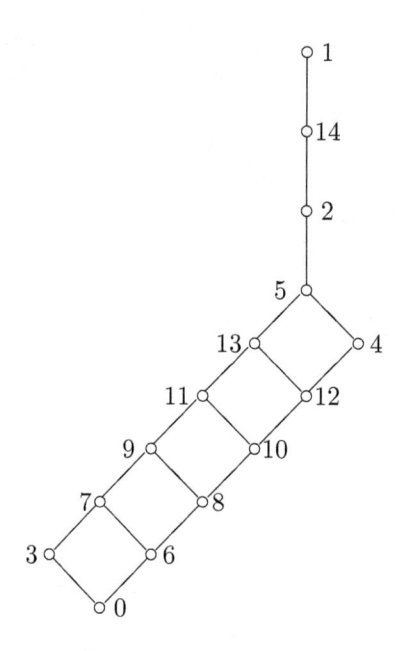

| →: | 0 | 1 | 2 | 3 | 4 | 5 | 6 | 7 | 8 | 9 | 10 | 11 | 12 | 13 | 14 |
|---|---|---|---|---|---|---|---|---|---|---|---|---|---|---|---|
| 0 | 1 | 1 | 2 | 1 | 1 | 5 | 7 | 7 | 9 | 9 | 11 | 11 | 13 | 13 | 14 |
| 1 | 0 | 1 | 2 | 3 | 4 | 5 | 6 | 7 | 8 | 9 | 10 | 11 | 12 | 13 | 14 |
| 2 | 0 | 1 | 1 | 3 | 4 | 5 | 6 | 7 | 8 | 9 | 10 | 11 | 12 | 13 | 14 |
| 3 | 4 | 1 | 2 | 1 | 4 | 5 | 6 | 7 | 8 | 9 | 10 | 11 | 12 | 13 | 14 |
| 4 | 3 | 1 | 2 | 3 | 1 | 5 | 7 | 7 | 9 | 9 | 11 | 11 | 13 | 13 | 14 |
| 5 | 0 | 1 | 2 | 3 | 4 | 1 | 6 | 7 | 8 | 9 | 10 | 11 | 12 | 13 | 14 |
| 6 | 3 | 1 | 2 | 3 | 1 | 5 | 1 | 1 | 9 | 9 | 11 | 11 | 13 | 13 | 14 |
| 7 | 0 | 1 | 2 | 3 | 4 | 5 | 4 | 1 | 8 | 9 | 10 | 11 | 12 | 13 | 14 |
| 8 | 3 | 1 | 2 | 3 | 1 | 5 | 7 | 7 | 1 | 1 | 11 | 11 | 13 | 13 | 14 |
| 9 | 0 | 1 | 2 | 3 | 4 | 5 | 6 | 7 | 4 | 1 | 10 | 11 | 12 | 13 | 14 |
| 10 | 3 | 1 | 2 | 3 | 1 | 5 | 7 | 7 | 9 | 9 | 1 | 1 | 13 | 13 | 14 |
| 11 | 0 | 1 | 2 | 3 | 4 | 5 | 6 | 7 | 8 | 9 | 4 | 1 | 12 | 13 | 14 |
| 12 | 3 | 1 | 2 | 3 | 1 | 5 | 7 | 7 | 9 | 9 | 11 | 11 | 1 | 1 | 14 |
| 13 | 0 | 1 | 2 | 3 | 4 | 5 | 6 | 7 | 8 | 9 | 10 | 11 | 4 | 1 | 14 |
| 14 | 0 | 1 | 2 | 3 | 4 | 5 | 6 | 7 | 8 | 9 | 10 | 11 | 12 | 13 | 1 |

| ′: | 0 | 1 | 2 | 3 | 4 | 5 | 6 | 7 | 8 | 9 | 10 | 11 | 12 | 13 | 14 |
|---|---|---|---|---|---|---|---|---|---|---|---|---|---|---|---|
|  | 1 | 0 | 3 | 2 | 1 | 2 | 1 | 2 | 1 | 2 | 1 | 2 | 1 | 2 | 3 |

Figure 4

# References

1. R. Balbes and PH. Dwinger, Distributive lattices, Univ. of Missouri Press, Columbia, 1974.
2. S. Burris and H.P. Sankappanavar, A course in universal algebra, Springer–Verlag, New York, 1981. The free, corrected version (2012) is available online as a PDF file at math.uwaterloo.ca/∼snburris
3. J. M. Cornejo and H. P. Sankappanavar, *A logic for dually hemimorphic semi-Heyting algebras and axiomatic extensions*, (2020). Preprint.
4. K.B. Lee, Equational classes of distributive pseudo-complemented lattices, Canad. J. Math. 22 (1970), 881–891.
5. A. Monteiro, *Sur les algebres de Heyting symetriques*, Portugaliae Mathemaica 39 (1980), 1–237.
6. W. McCune, *Prover9 and Mace 4*, http://www.cs.unm.edu/mccune/prover9/
7. H. Rasiowa, An algebraic approach to non-classical logics, North–Holland Publ.Comp., Amsterdam, (1974).
8. H.P. Sankappanavar, *Heyting algebras with dual pseudocomplementation*, Pacific J. Math. 117 (1985), 405–415.
9. H.P. Sankappanavar, *Heyting algebras with a dual lattice endomorphism*, Zeitschr. f. math. Logik und Grundlagen d. Math. 33 (1987), 565–573.
10. H.P. Sankappanavar, *Semi-De Morgan algebras*, J. Symbolic. Logic 52 (1987), 712–724.
11. H.P. Sankappanavar, *Semi–Heyting algebras: An abstraction from Heyting algebras*, Actas del IX Congreso Dr. A. Monteiro (2007), 33-66.
12. H.P. Sankappanavar, *Expansions of Semi-Heyting algebras. I: Discriminator varieties*, Studia Logica 98 (1-2) (2011), 27-81.
13. H.P. Sankappanavar, *Dually quasi-De Morgan Stone Semi-Heyting algebras I. Regularity*, Categories and General Algebraic Structures with Applications, 2 (2014), 55-75.
14. H.P. Sankappanavar, *Dually quasi-De Morgan Stone Semi-Heyting algebras II*, Categories and General Algebraic Structures with Applications, 2 (2014), 77-99.
15. H.P. Sankappanavar, *A note on regular De Morgan semi-Heyting algebras*, Demonstratio Mathematica 49 (2016), 252-265.
16. H.P. Sankappanavar, *JI-distributive dually quasi-De Morgan semi-Heyting and Heyting algebras*, Scientia Mathematica Japonica 82, No. 3 (2019). (To appear).
17. H.P. Sankappanavar, *Regular dually Stone semi-Heyting algebras*. In Preparation.

458

# Restriction semigroups * †

Yanhui Wang

*College of Mathematics and Systems Science,*
*Shandong University of Science and Technology,*
*Qingdao, 266590, P.R. China*

*yanhuiwang@sdust.edu.cn*

Xueming Ren

*Department of Mathematics, Xi'an University of Architecture and Technology,*
*Xi'an, 710055, P.R. China*

*xmren@xauat.edu.cn*

K. P. Shum

*Institute of Mathematics, Yunnan University,*
*Kunming, Yunnan 650091, P.R. China*

*kpshum@ynu.edu.cn*

This article is dedicated to celebrate the 80th birthday of
Professor Leonid A. Bokut

As a generalization of inverse semigroups, restriction semigroups have arisen in a number of contexts, one being by using relations $\widetilde{\mathcal{R}}_E$ and $\widetilde{\mathcal{L}}_E$ which may be regard as the generalized Green relations, the other being as varieties of (bi)-unary semigroups. The purpose of this article is to give a survey of certain techniques used in the study of restriction semigroups. The main objects that we introduce and discuss in this article are $W$-products, Bruck–Reilly extensions and Szendrei expansions.

*Keywords*:   Restriction semigroups; W-products; Bruck–Reilly extensions; Szendrei expansions

---

*The research was supported by NSFC (Grant No:11471255, 11501331), Shandong Province Natural Science Foundation(Grant No: BS2015SF002), SDUST Research Fund (No. 2014TDJH102), and Joint Innovative Center for Safe and Effective Mining Technology and Equipment of Coal Resources, Shandong Province.

†The authors would like to thank Professor Victoria Gould for her suggestions and useful comments towards to this manuscript.

## Introduction

It is well known in the literature that the restriction semigroups are the weakly $E$-ample semigroups which are non-regular generalizations of the inverse semigroups. These special semigroups are the semigroups with two additional unary operations. Various aspects of restriction semigroups and their one-sided analogues have been extensively studied and investigated by many authors in thee last two decades, see, e.g., Refs. 4,11–13,15 and the references therein.

In this article , we give a survey to the techniques used in the study of restriction semigroups. The study of the restriction semigroups has in a large part been motivated by the establishment of the structure theory for inverse semigroups. For instance, we adopt the following approaches:

(a) The applications of Ehresmann-Schein-Nambooripad characterizations of inverse semigroups in terms of inductive groupoids;

(b) The usage of the Munn's fundamental theory of inverse semigroups and its construction of the semigroup $T_E$ from a semilattice;

(c) To cite the core results of McAlister to show on the one hand that every inverse semigroup has a proper ($E$-unitary) cover.

On the other hand, in order to determine the structure of proper inverse semigroups in terms of groups, semilattices and partially ordered sets, Gould[12] has written a note to discuss the above techniques for studying the restriction semigroups, and hence, in this article we do not mention again the above three approaches in details.

In Section 1, we first introduce the basic definitions of restriction semigroups. For the basic properties of these semigroups determined by the relations that may be thought of as analogues of the well known Green's relations. We always recall the morphisms and congruences on the restriction semigroups and also in various proper restriction semigroups.

In Section 2 and Section 3, we focus on the (generalized) $W$-products. The construction of $W(T,Y)$ with $Y$ being a semilattice and $T$ a right cancellative monoid was introduced in Ref. 7 as a construction of a left ample semigroup. In Ref. 15, it was a generalized version for any unipotent monoid $T$, and it was noticed that there is a natural unary operation $*$ on the semigroup $W(T,Y)$ so that it becomes the so-called weakly ample semigroup. A weakly ample semigroup is a restriction semigroup in which every idempotent of the semigroup is a projection. The arguments in Ref. 15 can be easily applied for any monoid $T$, so if we drop the requirement for $T$ being unipotent, then $W(T,Y)$ is immediately seen a restriction semigroup.

In Section 2, we recall the construction of $W$-products for the left restriction semigroups, free restriction semigroups, $F$-restriction semigroups and (almost) factorizable restriction semigroups. In Section 3,we consider the construction of generalized $W$-products for almost perfect restriction semigroups. These construction theorems are analogous to the known results of inverse semigroup[21–25].

In fact, the method of construction of the semigroups $W(T,Y)$ and the constructions used in the covering theorems are all instances of a very general construction $S_{T,R}$ which, despite its somewhat technical hypotheses, has a simple verification since the calculations are performed within a direct product $S \times T$. In Section 4 , we recall the results of $S_{T,R}$ for almost $T$-proper restriction semigroups, proper restriction semigroups.

In Section 5, we will investigate the restriction $\omega$-semigroups. In fact, a restriction $\omega$-semigroup is a generalization of an inverse $\omega$-semigroup. Based on the Bruck–Reilly extensions, we give here a description of a class of restriction $\omega$-semigroups , namely, the restriction $\omega$-semigroups with an inverse skeleton.

An *expansion* is a functor $F$ from one category of semigroups to a larger one such that there exists a natural transformation $\nu$ from $F$ to the identity functor, with each arrow $\nu_S$ surjective. Birget and Rhodes[3] first introduced the concept of semigroup expansions. In Ref. 9, the original Szendrei expansion of Refs. 5,31 has already been generalized to the weakly left $E$-ample case. Gould and Hollings[13] further generalized it to the left restriction semigroups. Notice that these generalized Szendrei expansions can be applied to semigroups; the original Szendrei expansion could only be applied to monoids. In Section 6, we recall the construction of Szendrei expansion of a left restriction semigroup.

## 1. Preliminaries

### 1.1. *Restriction semigroups*

In this section, we first introduce the restriction semigroups in two rather distinct view points. One is by using the relations $\widetilde{\mathcal{R}}_E$ and $\widetilde{\mathcal{L}}_E$, which may be thought of as the generalized Green's relations. The other is as the varieties of the (bi)-unary semigroups. For the sake of convenience, we give both descriptions beginning with the former. We refer the reader to Refs. 6,8,11,30 and 34– 36 for more details.

Let $S$ be a semigroup and $E \subseteq E(S)$. Then, we define the relation

$$\widetilde{\mathcal{L}}_E = \{(a,b) \in S \times S : ae = a \Leftrightarrow be = b \text{ for all } e \in E\}.$$

The relation $\widetilde{\mathcal{R}}_E$ can be defined dually. We define

$$\widetilde{\mathcal{H}}_E = \widetilde{\mathcal{R}}_E \cap \widetilde{\mathcal{L}}_E \text{ and } \widetilde{\mathcal{D}}_E = \widetilde{\mathcal{R}}_E \vee \widetilde{\mathcal{L}}_E.$$

We remark here that $\mathcal{L} \subseteq \mathcal{L}^* \subseteq \widetilde{\mathcal{L}}_E$ and $\mathcal{R} \subseteq \mathcal{R}^* \subseteq \widetilde{\mathcal{R}}_E$, where $\mathcal{L}$ and $\mathcal{R}$ are the usual Green's relations on a semigroup $S$ and $\mathcal{L}^*$ and $\mathcal{R}^*$ are the generalized Green's $*$-relations on $S$. Let $S$ be a semigroup and $E \subseteq E(S)$. If $e, f \in E$, then $e \mathrel{\mathcal{R}} f$ if and only if $e \mathrel{\widetilde{\mathcal{R}}_E} f$ , and dually, $e \mathrel{\mathcal{L}} f$ if and only if $e \mathrel{\widetilde{\mathcal{L}}_E} f$. In particular, if $E$ is a semilattice, then, each $\widetilde{\mathcal{L}}_E$-class contains at most one idempotent in $E$ (in this case we denote the unique idempotent of $E$ in the $\widetilde{\mathcal{L}}_E$-class of $a$, if it existby $a^*$) and also each $\widetilde{\mathcal{R}}_E$-class contains at most one idempotent from $E$ (in this case we denote the unique idempotent of $E$ in the $\widetilde{\mathcal{R}}_E$-class of $a$, if it exists, by $a^+$). Observe that $\mathcal{L}$ and $\mathcal{L}^*$ are always right congruences but the same fact need not be true for $\widetilde{\mathcal{L}}_E$. Dually, $\widetilde{\mathcal{R}}_E$ is not necessarily a left congruence. Another fact is that, as is the case with $\mathcal{R}^*$ and $\mathcal{L}^*$, the relations $\widetilde{\mathcal{R}}_E$ and $\widetilde{\mathcal{L}}_E$ need not commute.So it is not usually the case that $\widetilde{\mathcal{D}}_E = \widetilde{\mathcal{L}}_E \circ \widetilde{\mathcal{R}}_E = \widetilde{\mathcal{R}}_E \circ \widetilde{\mathcal{L}}_E$. For the sake of convenience, we denote the $\widetilde{\mathcal{R}}_E$-class ($\widetilde{\mathcal{L}}_E$-class, $\widetilde{\mathcal{H}}_E$-class,$\widetilde{\mathcal{D}}_E$-class) for any $a \in S$ by $\widetilde{R}_E^a$ ($\widetilde{L}_E^a$, $\widetilde{H}_E^a$, $\widetilde{D}_E^a$). If we do not stress on the subset $E$ of $E(S)$, we simply write $\widetilde{\mathcal{K}}_E$ as $\widetilde{\mathcal{K}}$, where $\mathcal{K} \in \{\mathcal{L}, \mathcal{R}, \mathcal{D}, \mathcal{H}\}$.

A semigroup $S$ with $E \subseteq E(S)$ is said to be *left restriction*. If $E$ is a semilattice in $S$, then every $\widetilde{\mathcal{R}}_E$-class contains exactly one idempotent of $E$, moreover, $\widetilde{\mathcal{R}}_E$ is a left congruence and it satisfies the *ample condition*

$$ae = (ae)^+ a$$

for all $a \in S$, $e \in E$.

**Lemma 1.1.** [11] *A semigroup is left restriction such that for any $a \in S$, we have $a \, \widetilde{\mathcal{R}}_E \, a^+$, where $a^+ \in E$, if and only if it satisfies the identities*

$$x^+ x = x, x^+ y^+ = y^+ x^+, (x^+ y)^+ = x^+ y^+, xy^+ = (xy)^+ x$$

*and*

$$E = \{a^+ : a \in S\}.$$

The *Right restriction semigroups* can be defined dually. A semigroup $S$ is said to be *restriction* if it is both a left and a right restriction semigroup with respect to the same semilattice $E \subseteq E(S)$, that is, $E$ is a semilattice in $S$, every $\widetilde{\mathcal{R}}_E$-class and every $\widetilde{\mathcal{L}}_E$-class of $S$, contains exactly one idempotent of $E$, $\widetilde{\mathcal{R}}_E$ is a left congruence and $\widetilde{\mathcal{L}}_E$ is a right congruence on $S$, and it satisfies the *ample conditions*:

$$ae = (ae)^+ a, \quad ea = a(ea)^*$$

for all $a \in S$, $e \in E$.

It is known that the restriction semigroups form a variety of algebras of type $(2, 1, 1)$. The defining identities (other than associativity) are those of Lemma 1.1 and their dual, together with

$$(x^+)^* = x, (x^*)^+ = x.$$

Let $S$ be a restriction semigroup. It follows from the axioms $(x^+)^* = x$ and $(x^*)^+ = x$ that

$$\{x^* : x \in S\} = \{x^+ : x \in S\}.$$

We denote this set by $E$. It is easy to see that the set $E$ is closed with respect to the multiplication is a semilattice and also $x^* = x^+ = x$ for all $x \in E$. This implies that $E$ is a $(2, 1, 1)$-subalgebra of $S$. It is called the *semilattice of distinguished idempotents* of $S$. Sometimes the set $E$ is called the *semilattice of projections*. In this paper, we prefer to call the later and denote it by $E_S$ (or $E$). Notice that a projection is necessarily an idempotent, but a restriction semigroup may contain idempotents which are not projections.

It is easy to see that an inverse semigroup $S$ is a restriction semigroup with respect to the semilattice $E(S)$. For any $a \in S$, we have $a^* = a^{-1}a$ and $a^+ = aa^{-1}$, where $a^{-1}$ is the unique inverse of $a$ in $S$. In addition, a restriction semigroup $S$ is *ample* if for all $a, b, c \in S$,

$$ac = bc \Rightarrow ac^+ = bc^+ \text{ and } ca = cb \Rightarrow c^*a = c^*b.$$

Fruitful results related to the restriction semigroups and their one sided versions have been obtained from various points of view and under different names since 1960s. To emphasize that the class of restriction semigroups is a natural generalization of the class of ample semigroups, restriction semigroups were formerly called weakly $E$-ample semigroups ($E$-ample semigroups in Ref. 29). For detailed studies of a historical overview of restriction semigroups and their different versions, the reader is referred to Ref. 11.

Let $S$ be a restriction semigroup. An ideal $I$ of $S$ is said to be a $\sim$-*ideal* if it is the union of $\widetilde{\mathcal{R}}_E$- and $\widetilde{\mathcal{L}}_E$-classes, that is, if $a \in I$ then $\widetilde{R}_E^a, \widetilde{L}_E^a \subseteq I$. The principal $\sim$-ideal generated by $a$ is denoted by $\widetilde{J}(a)$. Thus $\widetilde{J}(a)$ is the smallest $\sim$-ideal which is the union of $\widetilde{\mathcal{D}}_E$-classes. We now define the relation $\widetilde{\mathcal{J}}_E$ on $S$ by

$$a \; \widetilde{\mathcal{J}}_E \; b \Leftrightarrow \widetilde{J}(a) = \widetilde{J}(b).$$

### 1.2. *Morphisms and congruences*

A morphism of restriction semigroups is required to preserve the multiplication and the operations $*$ and $^+$, that is to be a morphism of (2, 1, 1)-algebras, and not only a semigroup morphism. For emphasis, we will often use the term (2, 1, 1)-morphisms, (2, 1, 1)-subalgebra, etc. A bi-unary subsemigroup of a restriction semigroup $S$ with respect to $E \subseteq E(S)$ is *full* if it contains all of the projections of $S$. We call a morphism $S \to T$ of restriction semigroups with respect to $E_S$ and $E_T$, respectively *projection-separating* if it is injective on $E_S$.

Let $S$ be a restriction semigroup with respect to $E \subseteq E(S)$. We denote the biggest projection-separating congruence by $\mu$ and observe that $\mu \subseteq \widetilde{\mathcal{H}}_E$. In the standard terminology, a congruence $\rho$ on a semigroup is *perfect* if $(a\rho)(b\rho) = (ab)\rho$. Denote by $\sigma_E$ the least congruence $\rho$ such that $e\rho f$ for all $e, f \in E$. Thus $a \; \sigma_E \; b$ if and only if $a = b$ or there exists a sequence

$$a = c_1 e_1 d_1, c_1 f_2 d_1 = c_2 f_2 d_2, \ldots, c_n f_n d_n = b,$$

where $c_1, d_1, \ldots, c_n, d_n \in S^1$ and $(e_1, f_1), \ldots, (e_n, f_n) \in E \times E$.

It is clear that the relation $\sigma_E$ in a restriction semigroup $S$ is the least congruence identifying all the projections of $E$. If we do not emphasize the set of projections, we just write $\sigma_E$ as $\sigma$ on a restriction semigroup. It is well known that if $S$ is an inverse semigroup and $\sigma = \sigma_{E(S)}$, then $a \; \sigma \; b$ if and only if $ea = eb$ for some $e \in E(S)$ [26], and further, we see that $\sigma$ is the least congruence such that $S/\sigma$ is a group.

The *natural partial order* on a restriction semigroup $S$ with respect to $E \subseteq E(S)$ is defined by $a \leq b$ if $a = eb$ for some $e \in E$; equivalently if $a = a^+ b$. It is self-dual, compatible with the operations on $S$ and extends the usual order on $E$. Put $a \downarrow = \{b \in S : b \leq a\}$ be the principal order ideal generated by $a$. An order ideal of $S$ is then a nonempty subset of $S$, that is , closed under $\downarrow$. The order ideals of a semilattice are just its ideals.

A mapping $\alpha : S \to T$ of restriction semigroups is a *subhomomorphism* if $a\alpha b\alpha \leq (ab)\alpha$ for all $a, b \in S$ (and the unary operations are respected). Notice that for inverse semigroups, Lawson [19] calls such maps *dual prehomomorphisms*, while for Petrich[27] a prehomomorphism is a subhomomorphism that, in addition, respects inversion.

### 1.3. *Various proper restriction semigroups*

In this subsection, we recall basic definitions and properties about various proper restriction semigroups. We beginning with proper inverse semigroups.

An inverse semigroup is *proper* if $\mathcal{R} \cap \sigma = \iota$, where $\iota$ denotes the trivial congruence. It is ease to see that this one sided definition is equivalent to $\mathcal{L} \cap \sigma = \iota$. Moreover, an inverse semigroup is proper if and only if it is $E$-unitary, that is, $E(S)$ forms a $\sigma$-class.

A restriction semigroup $S$ is *proper* if $\widetilde{\mathcal{R}}_E \cap \sigma = \widetilde{\mathcal{L}}_E \cap \sigma = \iota$ (where $\iota$ is the identical relation). From this definition it is immediate that $\sigma\mu = \iota$, that is, $S$ is a subdirect product of $S/\sigma$ and $S/\mu$. Given a restriction semigroup $U$, a *cover* for $U$ [over a monoid $T$] is a proper restriction semigroup $S$ [such that $S/\sigma \simeq T$] having $U$ as a projection-separating homomorphic image[17].

A restriction semigroup $S$ is *almost perfect* if it is proper and $\sigma$ is perfect. A restriction monoid is *perfect* if it is proper, $\sigma$ is perfect and each $\sigma$-class has a greatest element.

The relationship between the perfect and almost perfect restriction monoids is described in the following lemma.

### Lemma 1.2.[17]

(i) *Every perfect restriction monoid is almost perfect; the converse need not hold;*

(ii) *if $S$ is a proper restriction semigroup without identity and $S^1$ is almost perfect, then $S$ is almost perfect; the converse holds if $S/\sigma$ has trivial group of units.*

As a generalization of the inverse semigroup, a restriction semigroup is *F-restriction* if it is proper and each $\sigma$-class has a greatest element. Since in that case the projections form a $\sigma$-class, such a semigroup is of necessity a monoid. In any such monoid, let $m_a$ denote the greatest element of $a\sigma$, so that $a\sigma = m_a \downarrow$.

### Lemma 1.3.[17] Let $M$ be an $F$-restriction monoid and put $T = M/\sigma$. The relation $m_a m_b \leq m_{ab}$ always holds. Thus the map $\kappa_M : T \to M$, $(a\sigma)\kappa_M = m_a$, is a subhomomorphism. Furthermore, $M$ is perfect if and only if $m_a m_b = m_{ab}$ for all $a, b \in M$ and, therefore, if and only if $\kappa_M$ is a homomorphism.

The motivation for the term "T-proper" comes from the consideration of generators. Suppose that a restriction semigroup $S$ is generated as such by a subset $X$. LeWE use $T$ to denote the restriction monoid of $S^1$ generated by $X$. Then we have $S = ET$. In this case, we say that $S^1$ is $E$-generated by $T$.

A restriction monoid $M$ with identity $1_M$ and submonoid $T$, is *strongly T-proper* if $M = ET$ and $\sigma$ separates $T$.

Let $S$ be a restriction semigroup. A nonempty subset $A \subseteq S$ is said to be *permissible* if $A$ is an order ideal of $S$ with respect to the partial order $\leq$, and the equalities $a^+b = b^+a$ and $ab^* = ba^*$ are valid for every $a, b \in A$. Denote by $C(S)$ the set of all permissible subsets of $S$, consider the usual set multiplication on $C(S)$, and define, for any $A \in C(S)$,

$$A^+ = \{a^+ : a \in A\} \text{ and } A^* = \{a^* : a \in A\}$$

**Theorem 1.1.** [33] *The algebra $C(S) = (C(S), \cdot, ^+, ^*)$ is a restriction monoid with identity element $E_S$, where the set of projections is*

$$E_{C(S)} = \{A \subseteq P(S) : A \text{ is an order ideal in } P_S\},$$

*and the natural partial order is the set inclusion. The mapping $\tau_S : S \to C(S)$, $a \mapsto (a]$, where $(a]$ stands for the principal order ideal of $S$ generated by $a$, is a (2, 1, 1) embedding of $S$ into $C(S)$. If $S = S^1$ then $\tau_S$ is also a monoid embedding. Moreover, $C(S)$ is (left) ample if and only if $S$ is (left) ample.*

**Theorem 1.2.** [33] *The restriction monoid $C(S)$ is proper if and only if $S$ is proper. If this is the case then each $\sigma_{C(S)}$-class has a maximum element, and the monoids $S/\sigma_S$ and $C(S)/\sigma_{C(S)}$ are isomorphic.*

**Corollary 1.1.** [33] *Every proper restriction semigroup $S$ can be (2, 1, 1)-embedded in a proper restriction semigroup $T$ where $T/\sigma_T$ is isomorphic to $S/\sigma_S$ and each $\sigma_T$-class has a maximum element.*

A restriction semigroup $S$ is *almost T-proper* if $C(S)$ is strongly T-proper (with respect to some submonoid $T$). By an abuse of terminology, given any monoid $T$, we may call $M$ strongly T-proper when it is strongly $T'$-proper with respect to some submonoid $T'$ isomorphic to $T$.

**Lemma 1.4.** [17] *Let $M$ be a restriction monoid and $T$ a submonoid. Then $M$ is strongly T-proper if and only if for all $m \in M$, $m \leq t$ for some unique $t \in T$; or equivalently, $m = m^+t$ for some unique $t \in T$; and equivalently if and only if for all $m \in M$, $m = tm^*$ for a unique $t \in T$. In that event, $M$ is necessarily proper and $T \simeq M/\sigma$.*

**Lemma 1.5.** [17] *The following are equivalent for a proper restriction monoid $M$:*

   *(i) $M$ is strongly T-proper for some monoid $T$;*

   *(ii) $M$ is strongly $M/\sigma$-proper;*

   *(iii) $M$ is perfect.*

By Lemma 1.4, that a semigroup be almost T-proper is equivalent to the property that every $A \in C(S)$ be contained in a unique member of $T$. Also by the above lemma, in that case $C(S)$ and, therefore, $S$ itself is necessarily proper. Hence, we have the following lemma.

**Lemma 1.6.** [17] *The following conditions are equivalent for a proper restriction monoid $M$:*

   *(i) $M$ is almost T-proper for some monoid $T$;*

   *(ii) $M$ is almost $M/\sigma$-proper;*

   *(iii) $M$ is almost perfect.*

## 2. W-products

Let $T$ be a monoid and $Y$ be a semilattice, with binary operation of meet , denoted by $\wedge$. Then $T$ acts on $Y$ (on the left) by morphisms if there is a map $T \times Y \to Y$, $(t, a) \mapsto {}^t a$, such that for all $a, b \in Y$, $s, t \in T$ we have

$$^1 a = a, \quad {}^{st} a = {}^s ({}^t a) \text{ and } {}^s (a \wedge b) = {}^s a \wedge {}^s b.$$

Now, we suppose that the monoid $T$ acts by morphisms on the left of a semilattice $Y$. We denote it by $Y \rtimes T$ the *semidirect product* of $Y$ by $T$, so that

$$Y \rtimes T = Y \times T \text{ and } (e, s)(f, t) = (e \wedge {}^s f, st),$$

for all $(e, s), (f, t) \in Y \times T$.

We remark here that the right action of a monoid $T$ on a semilattice $Y$ by morphisms can be defined dually to the left, where $a^t$ for the right action of $t \in T$ on $a \in Y$. The reverse semidirect product $T \ltimes Y$ is then dual to the construction above.

Let $T$ be a monoid and $Y$ a semilattice. We say that $T$ acts on $Y$ on the right (resp. left) if a monoid morphism is given from $T$ into the endomorphism monoid End $Y$ of $Y$ (resp. into the dual $\text{End}^d Y$ of the endomorphism monoid of $Y$). For the sake of brevity, we use $a^t$ (resp. ${}^t a$) to denote the image of the element $a \in Y$ under the endomorphism assigned to the element $t \in T$. It is well known that $T$ acts on $Y$ on the right is equivalent to requiring that the equalities

$$(ab)^t = a^t b^t, \quad (a^t)^u = a^{tu}, \quad a^1 = a$$

are valid for every $a, b \in Y$ and $t, u \in T$. Suppose that $T$ acts on the right on $Y$ by injective endomorphisms such that the range of each endomorphism corresponding to an element of $T$ forms an order ideal in $Y$ . Consider the set

$$W(T, Y) = \{(t, a^t) \in T \times Y : a \in Y, t \in T\},$$

and define a multiplication and two unary operations on it by the following rules: for any $(t, a^t), (u, b^u) \in W(T, Y)$, put

$$(t, a^t)(u, b^u) = (tu, a^{tu} \cdot b^u),$$

$$(t, a^t)^+ = (1, a), \quad (t, a^t)^* = (1, a^t).$$

It is straightforward to see that $W(T, Y)$ is a subsemigroup in the reverse semidirect product $T \ltimes Y$.

We now state the following theorem concerning a monoid acting on a semilattice.

**Theorem 2.1.** [33] *Let $Y$ be a semilattice and $T$ a monoid acting on $Y$ on the right. Then*

*(i) $W(T, Y) = (W(T, Y), \cdot, {}^+, {}^*)$ is a restriction semigroup, and its set of projections is $E = \{(1, a) : a \in Y\}$, which is isomorphic to $Y$;*

(ii) *the first projection* $\pi : W(T,Y) \to T$ *is a surjective homomorphism whose kernel is* $\sigma$. *Consequently,* $W(T,Y)/\sigma$ *is isomorphic to* $T$;

(iii) $W(T,Y)$ *is proper;*

(iv) $W(T,Y)$ *is a monoid if and only if* $Y$ *has an identity.*

Notice that if $Y$ is a semilattice and $T$ a monoid acting on $Y$ on the right by automorphisms then $W(T,Y) = T \ltimes Y$. If $Y$ is a semilattice and $T$ a monoid acting on $Y$ on the right, then the restriction semigroup $W(T,Y)$ is called a $W$-*product* of $Y$ by $T$.

We say that a monoid $T$ acts *doubly* on a semilattice $Y$ with identity $\varepsilon$, if $T$ acts by morphisms on the left and right of $Y$ and the compatibility conditions hold, that is

$$({}^t e)^t = \varepsilon^t \wedge e \text{ and } {}^t(e^t) = e \wedge {}^t \varepsilon.$$

for all $t \in T$, $e \in Y$.

Suppose that $T$ acts doubly on $Y$ as above, with the additional property that ${}^t\varepsilon = \varepsilon$ for all $t \in T$. Then

$$Y *_m T = \{(e,t) : e \leq^t \varepsilon\} = Y \rtimes T$$

is proper restriction with $(e,t)^+ = (e,1)$ and $(e,t)^* = (e^t, 1)$ [15].

If $\mathcal{P}$ is a partially ordered set, then we denote the smallest order ideal containing $P \subseteq \mathcal{P}$ by $\langle P \rangle$, abbreviated to $\langle p \rangle$ where $P = \{p\}$ is a singleton.

**Proposition 2.1.** [15] *Let* $W(T,Y)$ *be a* $W$-*product and denote by* $\overline{Y}$ *the semilattice of order ideals of* $Y$ *under intersection* ($\phi$ *is regarded as an order ideal). Then* $T$ *acts doubly on* $\overline{Y}$ *such that* ${}^t Y = Y$ *for all* $t \in T$, *and* $W(T,Y)$ *embeds (as a bi-unary semigroup) into* $\overline{Y} \rtimes T$.

## 2.1. *Left restriction semigroups.*

Let $S$ be a left restriction semigroup. Then, we define a sequence $\omega_1, \omega_2, \cdots$ of congruences on $S$ such that $\omega_1 \subseteq \omega_2 \subseteq \cdots$ and $\cup_{i \in \mathbb{N}} \omega_i = \omega_S$. Put

$$\tau_S = \{(a,b) \in S \times S : a^+ = b^+ \text{ and } a \ \omega_S \ b\}.$$

For later purposes, we define two conditions on a left restriction semigroup $S$:

(C) for any $a, b \in S$, if $a^+ = b^+$ and $a \ \omega_S \ b$, then $(ae)^+ = (be)^+$, for all $e \in E$;

(D) for any $r, s, t, x \in S$, if $x^+ = r^+, rs^+ = r, xt^+ = x$ and $rs \ \omega_S \ x$, then $r(st)^+ = r$.

**Theorem 2.2.** [15] *The following statements are equivalent for a left restriction semigroup* $S$:

(i) $S$ *is embeddable into a* $W$-*product;*

(ii) $S$ *is proper, and the relation* $\tau_S$ *is a congruence on* $S$;

(iii) $S$ *is proper, and* $\rho_S$, *the least proper left ample congruence on* $S$, *is projection-separating;*

*(iv)  S is proper and satisfies Condition (C);*

*(v)  S is proper and satisfies Condition (D).*

If $S$ is proper left ample, then $\rho_S$ is the identity relation and so certainly is projection separating. From (i)$\Leftrightarrow$ (iii) in Theorem 2.2, we deduce the following:

**Corollary 2.1.**[7] *Each proper left ample semigroup is embeddable into a W-product.*

### 2.2. Free restriction semigroups.

In the following, we obtain a free restriction semigroup on $X$ as a subsemigroup in a $W$-product.

Let $X$ be a set, and consider the free monoid $X^*$ and the free group $F\mathcal{G}(X)$ on $X$. The elements of $X^*$ are said to be *words* in $X$. The multiplication in $X^*$ is juxtaposition. The identity element of $X^*$ is the empty word which we denote by 1. The subset $X^+ = X^* \backslash \{1\}$ forms a subsemigroup in $X^*$, and $X^+$ is the free semigroup on $X$.

The elements of $F\mathcal{G}(X)$ are supposed to be the reduced words in $X \cup X^{-1}$. For any word $w \in X \cup X^{-1}$, the reduced form of $w$ is denoted by red$(w)$. The prefix order $\leq_p$ is a partial order defined on $F\mathcal{G}(X)$ by $u \leq_p v$ if $u$ is a prefix of $v$, that is, $v = uw(= \text{red}(uw))$ for some $w \in F\mathcal{G}(X)$.

Denote by $\mathcal{Y}$ the set of all finite order ideals of $(F\mathcal{G}(X), \leq_p)$ with at least two elements. For any $v \in F\mathcal{G}(X)$ and any subset $S \in F\mathcal{G}(X)$, define

$$^v S = \{\text{red}(vs) : s \in S\},$$

and let $\mathcal{X} = ^{F\mathcal{G}(X)} \mathcal{Y}$. In Ref. 33, $\mathcal{X} = (\mathcal{X}, \vee)$ is a semilattice. The left action of the group $F\mathcal{G}(X)$ on the semilattice $\mathcal{X}$ naturally defines a right action of $F\mathcal{G}(X)$ on $\mathcal{X}$ by the rule

$$S^u = {}^{u^{-1}} S \quad (u \in F\mathcal{G}(X), S \in \mathcal{X}).$$

Clearly, Its restriction to $X^*$ is a right action of the monoid $X^*$ on the semilattice $\mathcal{X}$ by automorphisms.

Now we consider the set

$$F\mathcal{RS} = \{(A, u) \in \mathcal{Y} \times F\mathcal{G}(X) : u \in X^* \cap A\},$$

and define a multiplication by

$$(A, u)(B, v) = (A \cup {}^u B, uv),$$

and unary operations by

$$(A, u)^+ = (A, 1) \text{ and } (A, u)^* = ({}^{u^{-1}} a, 1).$$

We now state a theorem of proper restriction semigroups.

The following statements on a proper restriction semigroups are equivalent:

**Theorem 2.3.** [32]

(i) *$FRS(X)$ is a proper restriction semigroup, and it is ample;*

(ii) *the set of projections of $FRS(X)$ is*

$$E_{FRS(X)} = \{(A, 1) : A \in \mathcal{Y}\}$$

*which is isomorphic to $\mathcal{Y}$;*

(iii) *the restriction semigroup $FRS(X)$ together with the injective mapping*

$$X \to FRS(X), x \mapsto (\{1, x\}, x)$$

*is a free restriction semigroup on $X$.*

Consider the subset

$$\mathcal{Q} = \mathcal{Y}^{X^*} = \{Q \in \mathcal{X} : Q \cap (X^*)^{-1} \neq \phi\}$$

in $\mathcal{X}$, where $T^{-1}$ is used to denote the subset $\{t^{-1} : t \in T\}$ in $FG(X)$ for any $T \subseteq X^*$. Then $\mathcal{Q}$ forms a subsemilattice in $\mathcal{X}$, and the monoid $X^*$ acts on it on the right by injective endomorphisms. Moreover, for each $t \in X^*$, we have

$$\mathcal{Q}^t = \{Q \in \mathcal{X} : Q \cap (X^*t)^{-1} \neq \phi\},$$

and so it is a dual order ideal in $(\mathcal{Q}, \supseteq)$.

**Theorem 2.4.** [32] *The free restriction semigroup $FRS(X)$ is (2, 1, 1)-embeddable in the $W$-product $W(X^*, Q)$, where*

$$W(X^*, Q) = \{(t, Q) \in X^* \times Q : Q \cap (X^*t)^{-1} \neq \phi\}.$$

In Theorem 2.4, the (2, 1, 1)-embedding $\iota : FRS(X) \to W(X^*, Q)$ is given by $(A, t) \mapsto (t, A^t)$. It immediately follows from this theorem that $FRS(X)\iota$ is a (2, 1, 1)-subsemigroup in $W(X^*, Q)$. Furthermore, we have:

**Theorem 2.5.** [33] *The subset*

$$\begin{aligned} F_W RS(X) &= \{(t, A^t) \in W(X^*, Q) : A \in \mathcal{Y} \text{ and } t \in A\} \\ &= \{(t, A^t) \in X^* \times \mathcal{Y} : A \in \mathcal{Y}\} \end{aligned}$$

*forms a (2, 1, 1)-subsemigroup in $W(X^*, Q)$. Furthermore, $F_W RS(X)$ together with the injective mapping*

$$X \to F_W RS(X), x \mapsto (x, \{1, x^{-1}\})$$

*is a free restriction semigroup on $X$.*

Next we focus on that each restriction semigroup has a proper cover (actually, a proper ample cover) which is embeddable in a $W$-product of a semilattice by a monoid. we first characterize the (2, 1, 1)-congruences of $F_W RS(X)$ contained in $\sigma$ as follows:

**Lemma 2.1.** [32] *Let $\kappa$ be a congruence on $Y$ such that the following condition is fulfilled:*

(P) *for every $A, B \in \mathcal{Y}$ and $t \in X^* \cap A \cap B$, we have $A \kappa B$ if and only if $a^T \kappa b^T$.*

*Then the relation $\rho$ defined by the rule*

$$(t, A^t) \rho (u, B^u) \text{ if and only if } t = u \text{ and } A \kappa B$$

*is a (2, 1, 1)-congruence on $F_W \mathcal{RS}(X)$ contained in $\sigma$. Conversely, each (2, 1, 1)- congruence on $F_W \mathcal{RS}(X)$ contained in $\sigma$ can be obtained in this way.*

**Theorem 2.6.** [32] *Suppose that $\rho$ is a (2, 1, 1)-congruence on $F_W \mathcal{RS}(X)$ contained in $\sigma$. Denote by $\kappa$ the congruence on $Y$ corresponding to $\rho$ given in Lemma 2.1, and consider the congruence $\nu$ on the semilattice $\mathcal{Q}$ generated by the relation*

$$\{(P, Q) \in \mathcal{Q} \times \mathcal{Q} : {}^t P, {}^t Q \in \mathcal{Y} \text{ with } {}^t P \kappa {}^t Q \text{ for some } t \in X^*\}.$$

*Then the right action of $X^*$ on $\mathcal{Q}$ induces a right action of $X^*$ on the factor semilattice $\mathcal{Q}/\nu$ by putting $(Q\nu)^{\nu} = (Q^{\nu})\nu$ for every $Q \in \mathcal{Q}, v \in X^*$, such that $W(X^*, \mathcal{Q}/\nu)$ is defined, and the mapping*

$$F_W \mathcal{RS}(X)/\rho \to W(X^*, \mathcal{Q}/\nu), (t, A^t)\rho \to (t, (A\nu)^t)$$

*is a (2, 1, 1)-embedding.*

The above result can be formulated, in a less technical way, as follows.

**Theorem 2.7.** [32] *Each (2, 1, 1)-factor semigroup of a free restriction semigroup over a (2, 1, 1)-congruence contained in the least unary trivial (2, 1, 1)-congruence (equivalently, in the least cancellative congruence) is (2, 1, 1)-embeddable in a $W$-product of a semilattice by a monoid.*

**Theorem 2.8.** [32] *Each restriction semigroup has a proper (ample) cover which is (2, 1, 1)-embeddable into a $W$-product of a semilattice by a monoid.*

### 2.3. F-restriction semigroups.

A restriction semigroup is an $F$-restriction semigroup if every $\sigma$-class has a maximal element.

**Proposition 2.2.** [18] *A $W$-product $W(T, Y)$ is $F$-restriction if and only if $Y$ has an identity if and only if $W(T, Y)$ is a monoid.*

We observe that Cornock and Gould[4] have provided a structure theorem for the proper restriction semigroups in general, based on pairs of partial actions of a monoid $T$ on a semilattice $Y$.

Let $T$ be a monoid and let $X$ be a set, Then $T$ acts partially on $X$ (on the left) if there is a partial map $T \times X \to X, (t, x) \mapsto t \cdot x$, such that for all $s, t \in T$ and $x \in X$,

$$\exists 1 \cdot x \text{ and } 1 \cdot x = x$$

and
$$\text{if } \exists t \cdot x \text{ and } \exists s \cdot (t \cdot x) \text{ then } \exists st \cdot x \text{ and } s \cdot (t \cdot x) = st \cdot x,$$
where we write $\exists u \cdot y$ to indicate that $u \cdot y$ is defined. Dually, we may define the (partial) right action of $T$ on $X$, using the symbol "∘" to replace "·".

Let $T$ be a monoid, acting partially on the left and right of a semilattice $Y$, via · and ∘ respectively. Suppose that both actions preserve the partial order and the domains of each $t \in T$ are order ideals. Suppose in addition that for $e \in Y$ and $t \in T$, the following and their duals hold:

(a) if $\exists e \circ t$, then $\exists t \cdot (e \circ t)$ and $t \cdot (e \circ t) = e$;

(b) for all $t \in R$, there exists $e \in Y$ such that $\exists e \circ t$. Then $(T, Y)$ is a *strong M-pair*.

For a strong $M$-pair $(T, Y)$ we put
$$\mathcal{M}(T, Y) = \{(y, t) \in Y \times T : \exists y \circ t\}$$
and define the multiplication on $\mathcal{M}(T, Y)$ by
$$(x, s)(y, t) = (s \cdot (x \circ s \wedge f), st), (x, s)^+ = (x, 1) \text{ and } (x, s)^* = (x \circ s, 1).$$

**Theorem 2.9.**[4] *A semigroup is proper restriction if and only if it is isomorphic to some $\mathcal{M}(T, Y)$.*

It is natural to look for conditions on the semigroup $\mathcal{M}(T, Y)$, under which the poset $X$ constructed in the previous section would be a semilattice and the action $*$ would have nice properties. This can be achieved if $T$ is a free monoid and $\mathcal{M}(T, Y)$ is perfect restriction, as follows:

**Theorem 2.10.**[18] *Let $T = A^*$ be the free $A$-generated monoid and assume that a left partially defined action · of $T$ on a semilattice $Y$ is given such that the semigroup $\mathcal{M}(T, Y)$ can be formed and is a perfect restriction monoid. Let $X$ and $*$ be the poset and the left action of $T$ on $X$ constructed as above. Then*

*(i)  $X$ is semilattice;*

*(ii)  the $W$-product $W(T, X)$ may be formed;*

*(iii)  $\mathcal{M}(T, Y)$ embeds into $W(T, X)$.*

**Theorem 2.11.**[18] *Every restriction monoid $S = \langle A \rangle$ has a perfect restriction cover $\mathcal{M}(A^*, E_S)$ which (2, 1, 1)-embeds into a $W$-product $W(A^*, X)$. Every restriction semigroup $S = \langle A \rangle$ has a almost perfect restriction cover $\mathcal{M}(A^*, E_S)$ which (2, 1, 1)-embeds into a $W$-product $W(A^*, X')$.*

### 2.4. *(Almost) factorizable restriction semigroups*

In Refs.[10,33] a restriction semigroup S is *almost left factorizable* if every element of $S$ belongs to some member of the $\mathcal{R}$-class of the identity in $C(S)$. Again there are naturally right and two-sided versions of this definition. They extend to restriction semigroups the definition for inverse semigroups of Lawson[19]. Note that in Ref. 33 it proved that each restriction semigroup is (2, 1, 1)-embeddable into a so-called almost left factorizable restriction semigroup. A restriction semigroup is almost left factorizable if and only if it has a proper cover isomorphic to a $W$-product of a semilattice by a monoid.

**Lemma 2.2.**[17] *A proper restriction semigroup $S$ is almost [left, right] factorizable if and only if $C(S)$ is [left, right] factorizable, and thus if and only if $S$ is almost $T$-proper, where $T = [\widetilde{\mathcal{R}}_1, \widetilde{\mathcal{L}}_1] \widetilde{\mathcal{H}}_1$ of $C(S)$.*

**Theorem 2.12.**[33] *For every restriction semigroup $S$, the following conditions are equivalent:*

  (i) *$S$ is almost left factorizable;*

  (ii) *$S$ is a projection separating (2, 1, 1)-homomorphic image of a $W$-product of a semilattice by a monoid;*

  (iii) *$S$ is a (2, 1, 1)-homomorphic image of a $W$-product of a semilattice by a monoid.*

**Theorem 2.13.**[33] *A restriction semigroup is proper and almost left factorizable if and only if it is (2, 1, 1)-isomorphic to a $W$-product of a semilattice by a monoid.*

In Ref. 33 a restriction monoid is factorizable if and only if it is both left and right factorizable. A restriction semigroup $S$ is said to be *almost factorizable* if, for any $a \in S$, there exists $A \in \widetilde{\mathcal{H}}_{P(S)}(C(S))$ such that $a \in A$. A proper restriction semigroup is almost factorizable if and only if it is both almost left and almost right factorizable. A restriction monoid is almost factorizable if and only if it is factorizable.

**Theorem 2.14.**[33] *If $M$ is a factorizable restriction monoid then $M \setminus \widetilde{\mathcal{H}}_1(M)$ is an almost factorizable restriction semigroup. Conversely, each almost factorizable restriction semigroup is (2, 1,1)-isomorphic to a restriction semigroup of the form $M \setminus \widetilde{\mathcal{H}}_1(M)$, where $M$ is a factorizable restriction monoid.*

**Theorem 2.15.**[33] *For every restriction semigroup $S$, the following conditions are equivalent:*

  (i) *$S$ is almost factorizable;*

  (ii) *$S$ is a projection separating (2, 1, 1)-homomorphic image of a reverse semidirect product $T \ltimes Y$ where $Y$ is a semilattice and $T$ is a monoid acting on $Y$ on the right by automorphisms;*

  (iii) *$S$ is a (2, 1, 1)-homomorphic image of a reverse semidirect product $T \ltimes Y$ where $Y$ is a semilattice and $T$ is a monoid acting on $Y$ on the right by automorphisms.*

**Theorem 2.16.**[33] *A restriction semigroup is proper and almost factorizable if and only if it is (2, 1, 1)-isomorphic to a reverse semidirect product $T \ltimes Y$ where $Y$ is a semilattice and $T$ is a monoid acting on $Y$ on the right by automorphisms.*

**Theorem 2.17.**[33] *Each restriction semigroup is (2, 1, 1)-embeddable into an almost left factorizable restriction semigroup.*

## 3. Generalized W-product

Let $T$ be a monoid, $Y$ a semilattice and suppose that there is a $(2, 1, 1)$-homomorphism $\alpha : T \to TI_Y$, where $TI_Y$ is the inverse semigroup of isomorphisms between ideals of $Y$. Adopting the usual language of actions, we say that $T$ acts on $Y$ (on the right) by isomorphisms between ideals. If the image lies in $T_Y$, then we say that the action is by isomorphisms between the principal ideals.

For the element $t \in T$, write $\alpha_t$ instead of $t\alpha$ and denote it by $\triangle t$ and $\nabla t$, respectively, its domain and range. Expressed in the notation of actions, for $e \in \triangle t$, write $e^t$ instead of $e\alpha_t$.

Consider the set

$$GW(T, Y) = \{(t, f) \in T \times Y : f \in \nabla t\} = \{(t, e^t) \in T \times Y : e \in \triangle t\}$$

The alternative form $(t, e^t)$ for $(t, f)$ results from the bijectivity of $\alpha_t$. The product is defined by:

$$(t, e^t)(u, f^u) = (tu, (e^t f)^u).$$

The Unary operations are defined by:

$$(t, e^t)^+ = (1, e) \text{ and } (t, f)^* = (1, f).$$

**Theorem 3.1.**[17] *Let $T$ be a monoid, $Y$ a semilattice, and $\alpha : T \to TI_Y$ (2, 1, 1)-homomorphism, that is, $T$ acts on $Y$ by isomorphisms between ideals. Then $W = GW(T, Y)$ is isomorphic to the semigroup $(T_Y)_{T, TI_Y}$ and is therefore an almost perfect restriction semigroup, with $E_W \simeq Y$ and $W/\sigma \simeq T$.*

*Furthermore, if $Y$ is also a monoid and $\alpha$ is a homomorphism into $T_Y$, that is, $T$ acts on $Y$ by isomorphisms between principal ideals, then $W$ is isomorphic to $(T_Y)_T$ and is therefore a perfect restriction monoid.*

It is noted that the original $W$-semigroup construction[33] mentioned in Section 2 corresponds precisely to the special case whereby $\triangle tY$ for all $t \in T$, that is, the action is by endomorphisms of $Y$. In that case, $GW(T, Y) = T \times Y$ and so is a "reverse" semidirect product. Observe that the action is then not simply by endomorphisms, however, since these endomorphisms must be injective and their images must be ideals of $Y$. The original construction of course includes the case that the action be by automorphisms of $Y$.

In our general situation, the dual to that just considered would be the special case whereby $\nabla t = Y$ for all $t \in T$, that is, the representation is by isomorphisms from ideals of $Y$ onto $Y$ itself.

The converse of Theorem 3.1 is stated as follows:

**Theorem 3.2.**[17] *Let $S$ be an almost perfect restriction semigroup. Putting $T = S/\sigma$ and $Y = P(S)$. Then $S \simeq GW(T, Y)$, where $Y = E_S$ and the action of $T$ on $Y$, by isomorphisms between ideals, is induced by the homomorphism $\kappa\hat{\theta}$,* *the composition of the injection of $T$ in $C(S)$ with the extension of the Munn representation of $S$ to $C(S)$.*

*If $S$ is a perfect restriction monoid, then $Y = Y^1$ and the action of $T$ on $Y$ is by isomorphisms between principal ideals, induced by the Munn representation of $S$ itself.*

**Corollary 3.1.** [17] *The free restriction monoid $FRM_X$ on $X$ is isomorphic to $GW(X^*, Y)$, where $X^*$ is the free monoid on $X$, acting on the semilattice of projections $Y$ of $FRM_X$ according to the Munn representation.*

**Proposition 3.1.** [17] *The following are equivalent for a proper restriction semigroup $S$:*

    *(i) $S$ is almost left factorizable [almost right factorizable, almost factorizable];*

    *(ii) $S$ is almost $T$-proper, where $T$ is the $\widetilde{\mathcal{R}}$-class [$\widetilde{\mathcal{L}}$-class, $\widetilde{\mathcal{H}}$-class] of the identity in $C(S)$;*

    *(iii) the action that is induced by the Munn representation of $S$, by isomorphisms between ideals of $E_S$, is by endomorphisms [onto mappings, automorphisms];*

    *(iv) $S \simeq GW(T, Y)$ for such an action of a monoid $T$ upon a semilattice $Y$ (that is, in the left and two-sided cases, the 'original' W-product).*

If there is a subhomomorphism $\alpha : T \to TI_Y$, then $T$ is said to *sub-act* on the semilattice $Y$.

**Theorem 3.3.** [17] *Let $T$ be a monoid, $Y$ a semilattice, and $\alpha : T \to TI_Y$ a $(2, 1, 1)$-subhomomorphism, that is, $T$ sub-acts on $Y$ by isomorphisms between ideals. Then $W = GW(T, Y)$ is isomorphic to the semigroup $(T_Y)_{T,TI_Y}$, and so is a proper restriction semigroup, with $E_W \simeq Y$ and $W/\sigma \simeq T$; it is almost $T$-proper if and only if $\alpha$ is a homomorphism. Furthermore, if $Y$ is a monoid and $\alpha$ is a subhomomorphism into $T_Y$, that is, $T$ sub-acts on $Y$ by isomorphisms between principal ideals, then $W$ is isomorphic to the monoid $(T_Y)_T$ and so is an $F$-restriction monoid; it is strongly $T$-proper if and only if $\alpha$ is a homomorphism.*

The converse part of the above theorem is stated as follows:

**Theorem 3.4.** [17] *Let $S$ be a proper restriction semigroup. Put $T = S/\sigma$ and $Y = P(S)$. Then $S \simeq GW(T, Y)$, where the sub-action of $T$ on $Y$, by isomorphisms between ideals, is induced by the subhomomorphism $\kappa\bar{\theta}$, the composition of the injection of $T$ in $C(S)$ with the extension of the Munn representation of $S$ to $C(S)$.*

*If $S$ is an $F$-restriction monoid then $Y = Y^1$ and the sub-action of $T$ on $Y$ is by isomorphisms between principal ideals, induced by the Munn representation of $S$ itself.*

It is a good place to remark that in Theorem 2.9, the construction of $\mathcal{M}(T, Y)$ describes the proper restriction semigroups. Clearly, there must be a correspondence between $\mathcal{M}(T, Y)$ constructed by Cornock and Gould[4] and that in this section, but so far, there is no any result can be found in the literature.

At the end of this section, we turn to the inverse case. The proper inverse semigroups are usually termed $E$-unitary. The specializations of the general parts of Theorem 3.3 and Theorem 3.4 to inverse semigroups are easily obtained. In Ref. 27, a prehomomorphism of inverse semigroups is a subhomomorphism that respects inverses.

We stsate the folowing corollary.

**Corollary 3.2.**[28] *Let $T$ be a group, $Y$ a semilattice, and $\alpha : T \to TI_Y$ a (2, 1, 1)-prehomomorphism. Then $W = GW(T, Y)$ is an $E$-unitary inverse semigroup, isomorphic to $(T_Y)_{T,TI_Y}$ with $E(W) \simeq Y$ and $W/\sigma \simeq T$. Conversely, for any $E$-unitary inverse semigroup $S$, let $T$ be the group $S/\sigma$ and $Y = E(S)$. Then $S \simeq GW(T, Y)$, where the sub-action of $T$ on $Y$, by isomorphisms between ideals, is induced by the prehomomorphism $\kappa\bar{\theta}$.*

The point that we wish to emphasize here is that the specialization of almost perfection and perfection to inverse semigroups does not yield a general theory as it does for the restriction semigroups. They are some well-studied classes:

**Proposition 3.2.**[17] *The almost perfect inverse semigroups are the semidirect products of semilattices and groups. The perfect inverse monoids are the monoidal such products.*

## 4. The construction $S_{T,R}$

Let $R$ be a restriction monoid and $S$ a restriction subsemigroup such that for each $r \in R$ the following conditions and their duals are satisfied: (i) there exists $e \in E$, $e \leq r^+$, and (ii) for any such $e, er \in S$. Let $T$ be a monoid and $\alpha : T \to R$ a (2, 1, 1)-homomorphism. Let

$$S_{T,R} = \{(a, t) \in S \times T : a \leq t\alpha \text{ in } R\}.$$

Then, Write $S_T$ in case $R = S^1$. The following theeorem follows:

**Theorem 4.1.**[17] *The set $S_{T,R}$ is an almost $T$-proper restriction subsemigroup of $S \times T$, with $S_{T,R}/\sigma \simeq T$, now regarding $T$ as a (reduced) restriction semigroup, and the first projection is a projection-separating homomorphism onto a full subsemigroup of $S$. Further, we have the following results:*

- *(i) if $S \subseteq (T\alpha) \downarrow$ then $S_{T,R}$ is a subdirect product of $S$ and $T$ and is therefore a cover of $S$ over $T$; in particular, if $R = S^1$ and $R$ is $E$-generated by $T\alpha$, then $S_T$ is a cover of $S$;*

- *(ii) if $S$ is a monoid and $\alpha$ is (2, 1,1)-morphism, then $S_T$ is strongly $T$-proper.*

**Corollary 4.1.** [17] *Let $S$ be a restriction semigroup. If $T$ is a monoid and $\alpha :$
$T \to S^1$ is a (2, 1, 1)-homomorphism, the image of which $P$-generates $S^1$, then
$S_T = \{(s,t) \in S \times T : s \leq t\alpha \text{ in } S^1\}$ is an almost $T$-proper cover of $S$ that is
a subdirect product of $S$ and $T$. In particular, if $S$ is generated, as a restriction
semigroup, by a subset $X$ and $T$ is the submonoid of $S^1$ generated by $X$, then $S_T$
is such a cover.*

*If $S$ is a monoid to begin with, then the above covers are strongly $T$-proper
restriction monoid.*

In the following Corollary, we shall identify a restriction semigroup $S$ with
its image in $C(S)$ under the embedding $\tau_S : a \mapsto a \downarrow$ in $C(S)$. Let $Y$ is a
semilattice, $T_Y$ denotes the Munn semigroup on $Y$: the inverse subsemigroup of
the symmetric inverse semigroup $\mathcal{I}_Y$ consisting of the isomorphisms between the
principal ideals of $Y$. We denote the inverse subsemigroup of $\mathcal{I}_Y$ by $TI_Y$, which
consists of the isomorphisms between arbitrary ideals of $Y$.

**Corollary 4.2.** [17] *Let $S$ be a restriction semigroup, $R$ a submonoid of $C(S)$ that
contains $S$, $T$ a restriction monoid and $\alpha : T \to R$ a (2, 1, 1)-homomorphism.
The almost $T$-proper semigroup*

$$S_{T,R} = \{(a,t) \in S \times T : a \in t\alpha\}$$

*is well defined. In particular, $S_{T,C(S)}$ is well defined.*

*For any semilattice $Y$, a restriction monoid $T$ and (2, 1, 1)-homomorphism
$\alpha : T \to TI_Y$, the semigroup $(T_Y)_{T,TI_Y}$ is a well defined, almost $T$-proper restriction semigroup. If $Y$ has an identity element and $\alpha : T \to T_Y$, then $(T_Y)_T$
is a strongly $T$-proper monoid.*

The following theorem may be regarded as a converse of Theorem 4.1. Its
generality allows two distinct important applications. Again we identify $S$ with
its image in $C(S)$ under $\tau_S$, except where additional clarity is required.

**Theorem 4.2.** [17] *Let $N$ and $S$ be restriction semigroups, with $T = N/\sigma$. If $N$
is almost perfect and $\beta : N \to S$ is a projection-separating homomorphism whose
image is full in $S$, then $N \simeq S_{T,C(S)}$, with respect to $\alpha = \kappa\hat{\beta} : T \to C(T)$.*

*Let $N$ and $M$ be restriction monoids. If $N$ is perfect and $\beta : N \to S$ is a
projection-separating homomorphism whose image is full in $M$, then $N \simeq M_T$,
where $\alpha = \kappa\beta : T \to M$.*

The first application of this theorem is a description of the almost perfect
covers of a restriction semigroup.

**Corollary 4.3.** [17] *Let $N$ be an almost perfect cover of a restriction semigroup $S$,
via the homomorphism $\beta$. Put $T = N/\sigma$. Then $N \simeq S_{T,C(T)}$, where $\alpha = \kappa\hat{\beta} :$
$T \to C(T)$ and $S \subseteq (T\alpha) \downarrow$ is satisfied.*

*If $N$ is a perfect, restriction monoid cover of the restriction monoid $M$, again
via $\beta$, then $N \simeq M_T$, where $\alpha = \kappa\beta : T \to M$ and $M$ is projection-generated by
$T\alpha$.*

In the case that $S$ itself is almost perfect, regarded as its own cover, that is, $\beta$ is the identity map, put $T = S/\sigma$. Then $S_{T,C(S)} \simeq S$, since if $(s,t) \in S_{T,C(s)}$ then $t$ must be $s\sigma$. Likewisely, if $M$ is a perfect monoid, then $M \simeq M_T$.

**Corollary 4.4.** [17] *Let $S$ be an almost perfect restriction semigroup, with $T = S/\sigma$ and $Y = P_S$. Then $S \simeq F_{T,C(F)}$, where $F \simeq S/\mu$ is the image of $S$ in $T_Y$ under the Munn representation.*

Finally, we have the following theorem.

**Theorem 4.3.** [17]

(i) *If $\alpha : T \to R$ is a subhomomorphism, then $S_{T,R}$ is a proper restriction semigroup;*

(ii) *if $S \leq R \leq C(S)$ then $S_{T,R}$ is almost perfect if and only if $\alpha$ is a homomorphism. In particular, this is true for $S_T$;*

(iii) *if $S$ is a monoid and $\alpha$ is a (2, 1,1)-morphism, then $S_T$ is an F-restriction monoid; thus $S_T$ is strongly $T$-proper (that is, perfect) if and only if $\alpha$ is also a homomorphism.*

## 5. Bruck–Reilly extensions

In this section, our aim is to investigate a Bruck–Reilly semigroup which is a $\widetilde{\mathcal{J}}$-simple restriction $\omega$-semigroup with an inverse skeleton. We recall some definitions and results which are closely related to inverse skeletons.

Let $S$ be a semigroup with $E \subseteq E(S)$. An element $a \in S$ is $E$-*regular* if $a$ has an inverse $a^\circ$ such that $aa^\circ, a^\circ a \in E$.

**Lemma 5.1.** [14] *If $S$ is a restriction semigroup such that every $\widetilde{\mathcal{H}}_E$-class contains an $E$-regular element, then $\widetilde{\mathcal{R}}_E \circ \widetilde{\mathcal{L}}_E = \widetilde{\mathcal{L}}_E \circ \widetilde{\mathcal{R}}_E$, so that $\widetilde{\mathcal{D}}_E = \widetilde{\mathcal{R}}_E \circ \widetilde{\mathcal{L}}_E$, and if $a, b \in S$ with $a \, \widetilde{\mathcal{D}}_E \, b$, then $|\widetilde{\mathcal{H}}_E^a| = |\widetilde{\mathcal{H}}_E^b|$.*

A subset $V \subseteq S$ is an $\widetilde{\mathcal{H}}_E$-*transversal* of $S$ if $|V \cap \widetilde{\mathcal{H}}_E^a| = 1$ for all $a \in S$. Let $W$ be an inverse subsemigroup of $S$ consisting of $E$-regular elements such that $E \subseteq W$. If $W$ is an $\widetilde{\mathcal{H}}_E$-transversal of $S$, then $W$ is an *inverse skeleton* of $S$.

If $S$ is a restriction semigroup with an inverse skeleton then every $\widetilde{\mathcal{H}}_E$-class contains an $E$-regular element. According to Lemma 5.1, the relations $\widetilde{\mathcal{R}}_E$ and $\widetilde{\mathcal{L}}_E$ commute in a restriction semigroup with an inverse skeleton.

A restriction semigroup $S$ with semilattice of projections $E$ is an $\omega$-*semigroup* if $E$ is isomorphic to $(\mathbb{N}^0, \geq)$. Hence, we can write

$$E = \{f_i : i \in \mathbb{N}^0\},$$

where $f_i \leq f_j$ if and only if $i \geq j$ for all $i, j \in \mathbb{N}^0$. Hence, $E$ is a descending chain $f_0 > f_1 > f_2 > \ldots$.

For a restriction $\omega$-semigroup $S$, we have $\widetilde{\mathcal{H}}_E = \mu$, that is, $\widetilde{\mathcal{H}}_E$ is a congruence on $S$.

Now, let $T$ be a monoid and $\theta : T \to T$ be a monoid morphism. We define a multiplication on the set $\mathbb{N}^0 \times T \times \mathbb{N}^0$ by

$$(m, a, n)(p, b, q) = (m - n + t, a\theta^{t-m} b\theta^{t-p}, q - p + t),$$

where $t = \max\{n, p\}$, and we define $x\theta^0 = x$ for all $x \in T$. This multiplication is associative [16], and we call the resulting semigroup a *Bruck–Reilly semigroup* and we denote it by $BR(T, \theta)$.

Let $T$ be an arbitrary monoid with identity $e$ and $\theta : T \to T$ be a monoid morphism. Let $E = \{(m, e, m) : m \in \mathbb{N}^0\}$ be a subset of idempotents of $BR(T, \theta)$. In Ref. 29, $BR(T, \theta)$ is also shown to be a restriction semigroup.

**Lemma 5.2.** [29] *Let $T$ be an arbitrary monoid and $\theta : T \to T$ be a monoid morphism. Let $E = \{(m, e, m) : m \in \mathbb{N}^0\}$ be a subset of idempotents of $BR(T, \theta)$. Then $BR(T, \theta)$ is a restriction semigroup with respect to $E$ and having an inverse skeleton.*

The Bruck–Reilly semigroup in Lemma 5.2 are $\widetilde{\mathcal{D}}_E$-simple. In order to characterize the $\widetilde{\mathcal{J}}$-simple restriction semigroups, we need a slightly more complicated generalization which is related to a strong semilattice of monoids.

Let $T = \cup_{i=0}^{d-1} M_i$ be a strong semilattice of the monoids $M_i$, where $d \in \mathbb{N}^0$, the indices $i$ form a chain $0 > 1 > \ldots > d-1$ and the structure morphisms are all monoid morphisms. Let $\theta : T \to M_0$ be a monoid morphism. Let $S = BR(T, \theta)$. Furthermore, let $E = \{(m, e_i, m) : m \in \mathbb{N}^0, 0 \leq i \leq d - 1\}$, where $e_i$ is the identity of $M_i$. The Bruck–Reilly semigroup $BR(T, \theta)$ constructed under the above assumptions is called a $\widetilde{\mathcal{J}}_E$-*Bruck–Reilly semigroup*.

Let $A$ be an ideal of a semigroup $S$. Then we call $S$ an ideal extension of the semigroup $A$ by the semigroup $T$, where the Rees semigroup $S/A$ is isomorphic to $T$.

Let $m$ be a natural number. The set $E_m$ is defined to be the empty set when $m = 0$ and for $m \geq 1$, we define

$$E_m = \{(0, 0), \cdots, (m - 1, m - 1)\}.$$

Then $E_m$ consists of the top $m$ idempotents of the bicyclic monoid. Let $d$ be a non-zero natural number. Define the set

$$I_{(m,d)} = \{(a, b) \in B : m \leq a, b \text{ and } a \equiv b \bmod (d)\}.$$

We put $B_{(m,d)} = E_m \cup I_{(m,d)}$ and $B_d = B_{(0,d)} = \{(a, b) \in B : a \equiv b \bmod(d)\}$. Then we have the following lemma:

**Lemma 5.3.** [19] $B_{(m,d)}$ *is a full inverse subsemigroup of the bicyclic monoid such that*

(i)   $B_1$ *is the bicyclic monoid and bisimple;*

(ii)   *for $d \geq 2$, $B_d$ is a simple inverse monoid with $d$ $\mathcal{D}$-classes;*

(iii)   *for $m \geq 1$, $B_{(m,d)}$ is non-simple having $I_{(m,d)}$ as a proper ideal. Furthermore, $I_{(m,d)}$ is isomorphic to $B_d$.*

A restriction semigroup $S$ with semilattice of projections $E$ is said to be a *super-restriction* semigroup if each $\widetilde{\mathcal{H}}_E$-class contains a projection.

Finally, we give a description of a restriction $\omega$-semigroup having an inverse skeleton.

**Theorem 5.1.** [1,2] *Let $S$ be a restriction $\omega$-semigroup with semilattice of projections $E$ and an inverse skeleton. Then the folwoing statements hold:*

(i) *the semigroup $S$ is a super-restriction semigroup if and only if $S/\mu$ is a semilattice which is isomorphic to $E$;*

(ii) *the semigroup $S$ is $\sim$-bisimple if and only if $S/\mu$ is isomorphic to the bicyclic monoid;*

(iii) *the semigroup $S$ is $\widetilde{\mathcal{J}}_E$-simple if and only if $S/\mu$ is isomorphic to $B_d$, where $d \geq 2$.*

(iv) *if the semigroup $S$ is none of the above then it is an ideal extension of a $\widetilde{\mathcal{J}}_E$-simple $\omega$-restriction semigroup by a super-restriction semigroup with a finite chain of distinguished idempotents with a zero adjoined.*

In closing this section, we turn to the adequate semigroups. We replace the set of projections $E$ by the whole set of idempotents $E(S)$ and use the relations $\mathcal{R}^*$, $\mathcal{L}^*$ and $\mathcal{H}^*$ instead of $\widetilde{\mathcal{R}}_E$, $\widetilde{\mathcal{L}}_E$ and $\widetilde{\mathcal{H}}_E$ in the definition of $\widetilde{\mathcal{J}}_E$-simple restriction $\omega$-semigroups with an inverse skeleton. We thus obtain $\mathcal{J}^*$-simple ample $\omega$-semigroups with an inverse skeleton. We define a $\widetilde{\mathcal{J}}$-Bruck–Reilly semigroup $S = BR(T, \theta)$ to be a $\mathcal{J}^*$-*Bruck–Reilly semigroup* if the monoids $M_i$ ($i = 0, 1, \ldots, d-1$) are cancellative, where $T = \cup_{i=0}^{d-1} M_i$ is a strong semilattice of the cancellative monoids $M_i$.

**Corollary 5.1.** [2] *Every $\mathcal{J}^*$-Bruck–Reilly semigroup is a $\mathcal{J}^*$-simple ample $\omega$-semigroup $S$ with an inverse skeleton $W = \{(m, e_i, n) : m, n \in \mathbb{N}^0, 0 \leq i \leq d-1\}$, where $e_i$ is the identity of $M_i$. Conversely, every $\mathcal{J}^*$-simple inverse $\omega$-semigroup $S$ with an inverse skeleton is isomorphic to a $\mathcal{J}^*$-Bruck–Reilly semigroup.*

Now we focus on the inverse semigroups. We replace the distinguished set of idempotents $E$ by the whole set of idempotents $E(S)$ and use the relations $\mathcal{R}$, $\mathcal{L}$ and $\mathcal{H}$ instead of $\widetilde{\mathcal{R}}_E$, $\widetilde{\mathcal{L}}_E$ and $\widetilde{\mathcal{H}}_E$ in the definition of $\widetilde{\mathcal{J}}_E$-simple restriction $\omega$-semigroups with an inverse skeleton. We thus obtain $\mathcal{J}$-simple inverse $\omega$-semigroups with an inverse skeleton. We define a $\widetilde{\mathcal{J}}$-Bruck–Reilly semigroup $S = BR(T, \theta)$ to be a $\mathcal{J}$-*Bruck–Reilly semigroup* if the monoids $M_i$ ($i = 0, 1, \ldots, d - 1$) are groups, where $T = \cup_{i=0}^{d-1} M_i$ is a strong semilattice of the groups $M_i$.

**Corollary 5.2.** [2] *Every $\mathcal{J}$-Bruck–Reilly semigroup is a $\mathcal{J}$-simple inverse $\omega$-semigroup $S$ with an inverse skeleton $W = \{(m, e_i, n) : m, n \in \mathbb{N}^0, 0 \leq i \leq d-1\}$, where $e_i$ is the identity of $M_i$. Conversely, every $\mathcal{J}$-simple inverse $\omega$-semigroup $S$ with an inverse skeleton is isomorphic to $\mathcal{J}$-Bruck–Reilly semigroup.*

## 6. Szendrei expansions

Let $S$ be a left restriction semigroup with respect to $E$. Then, we define the *Szendrei expansion* of $S$ to be the set

$$Sz(S) = \{(A, a) \in \mathcal{P}^f(S) \times S : a, a^+ \in A \text{ and } A \subseteq \widetilde{\mathcal{R}}^a_E\},$$

together with the operation

$$(A, a)(B, b) = ((ab)^+ A \cup aB, ab),$$

where $\mathcal{P}^f(S)$ denotes the collection of all finite subsets of $S$. It is easy to verify that

$$E(Sz(S)) = \{(F, f) \in Sz(S) : f \in E(S) \text{ and } fF \subseteq F\}.$$

Let $S$ and $T$ be the left restriction semigroups with respect to $E \subseteq E(S)$ and $F \subseteq E(T)$, respectively. Then the mapping $\theta : S \to T$ is a *strong premorphism* if $(s\theta)(t\theta) = (s\theta)^+(st)\theta$ and $(s\theta)^+ \leq s^+\theta$. In Ref. 9, the mapping $\iota : S \to Sz(S)$ given by $s\iota = (\{s^+, s\}, s)$, is a strong premorphism. We mention here that in Ref. 9, $\iota$ is denoted by $\eta^*_S$.

**Theorem 6.1.** [13] *Let $S$ and $T$ be left restriction semigroups with respect to $E \subseteq E(S)$ and $F \subseteq E(T)$, respectively. If $\theta : S \to T$ is a strong premorphism, then there exists a unique (2, 1)-morphism $\bar{\theta} : Sz(S) \to T$ such that $\iota\bar{\theta} = \theta$.*

*Conversely, if $\bar{\theta} : Sz(S) \to T$ is a (2, 1)-morphism, then $\theta = \iota\bar{\theta}$ is a strong premorphism.*

In Ref. 13, a map between inverse semigroups is an order-preserving premorphism if, and only if, it is a strong premorphism. Then we obtain the following corollary.

**Corollary 6.1.** [20] *Let $S$ and $T$ be inverse semigroups. Then all order-preserving inverse semigroup premorphisms are of the form $\iota\varphi$, where $\varphi : Sz(S) \to T$ is a morphism. Consequently, all partial actions of inverse semigroups on a set $X$ are of the form $\iota\varphi$, where $\varphi : Sz(S) \to \mathcal{I}_X$ is a morphism.*

## References

1. D. Abdulkadir, '$\widetilde{\mathcal{J}}$-restriction $\omega$-semigroups', PhD thesis, The University of York (2014).
2. D. Abdulkadir, Y. H. Wang, 'Restriction $\omega$-semigroups', to appear.
3. J. C. Birget, J. Rhodes, 'Almost finite expansions of arbitrary semigroups', *J. Pure Appl. Algebra* 32 (1984), 239–287.
4. C. Cornock, V. Gould, 'Proper two-sided restriction semigroups and partial actions', *J. Pure Appl. Algebra* 216 (2012) 935–949.
5. J. Fountain, G. M. S. Gomes, 'The Szendrei expansion of a semigroup', *Mathematika* 37 (1990), 251–260.
6. J. Fountain, G. M. S. Gomes, G. Goule, 'A Munn type representation for a class of of $E$-semiadequate semigroups', *J. Algebra* 218 (1999) 693–714.

7. J. Fountain, G. M. S. Gomes, 'Proper left type-A monoids revisited', *Glasg. Math. J.* 35 (1993), 293–306.
8. J. Fountain, G. M. S. Gomes, G. Goule,'The free ample monoid', *Internat. J. Algebra Comput.* 19 (2009) 527–554.
9. G. M. S. Gomes, 'The generalised prefix expansion of a weakly left ample semigroup', *Semigroup Forum* 72(2006), 387–403.
10. G. M. S. Gomes, M. B. Szendrei,' Almost factorizable weakly ample semigroups', *Comm. Algebra* 35 (2007), 3503–3523.
11. V. Gould, 'Notes on restriction semigroups and related structures', `http://www-users.york.ac.uk/varg1/restriction.pdf`.
12. V. Gould, 'Restriction and Ehresmann semigroups', *Proceedings of the International Conference on Algebra 2010*, Page 265–288.
13. V. Gould, C. Hollings, 'Partial actions of inverse and weakly left $E$-ample semigroups', *J. Aust. Math. Soc.* 86 (2009), 355–377.
14. V. Gould, Rida E-Zenab, 'Semigroups with inverse skeletons and Zappa–Szep products', *CGASA* 1 (2013), 59–89.
15. V. Gould, M. B. Szendrei, 'Proper restriction semigroups semidirect products and $W$-products', *Acta Math. Hungar.* 141 (1-2) (2013), 36057.
16. J. M. Howie, Fundamentals of semigroup theory, London Mathematical Society Monographs. New Series vol. 12, The Clarendon Press Oxford University Press, New York, 1995. Oxford Science Publications.
17. P. R. Jones, 'Almost perfect restriction semigroups', *J. Algebra* 445 (2016), 193–220.
18. G. Kudryavtseva, 'Partial monoid actions and a class of restrictiion semigroups', *J. Algebra* 429 (2015) 342–370.
19. M. V. Lawson, Inverse semigroups,World Scientific Publishing Co. Pte. Ltd, 1998.
20. M. V. Lawson, S. W. Margolis, B. Steinberg, 'Expansions of inverse semigroups', *J. Aust. Math, Soc.* 80 (2006), 205–228.
21. M.V. Lawson, 'Almost factorisable inverse semigroups', *Glasgow Math. J.* 36 (1994), 97–111.
22. D. B. McAlister, 'Groups semilattices and inverse semigroups', *Trans. Amer. Math. Soc.* 192 (1974), 227–244.
23. D. B. McAlister, 'Groups semilattices and inverse semigroups II',*rans. Amer. Math. Soc.* 196 (1974), 351–370.
24. D. B. McAlister, 'Some covering and embedding theorems for inverse semigroups', *J. Austral. Math. Soc.* 22 (1976), 188–211.
25. D. B. McAlister, N. R. Reilly, '$E$-unitary covers for inverse semigroups', *Pacific J. Math.* 68 (1977), 161–174.
26. W. D. Munn, 'A class of irreducible matrix representations of an arbitrary inverse semigroup', *Proc. Glasgow Math. Ass.* 5(1961), 41–48.
27. M. Petrich, 'Inverse Semigroups', Wiley, New York, 1984.
28. M. Petrich, N. Reilly, 'A representation of E-unitary inverse semigroups', *Quart. J. Math. Oxford* 30 (1979) 339–350.
29. S. Ma, X. Ren, Y. Yuan, 'On $U$-ample $\omega$-semigroups', *Front. Math. China* 8 (2013), 1391–1405.

30. X. M. Ren, Y. H. Wang, K. P. Shum, 'On $U$-orthodox semigroups', *Sci. China Ser A* 52 (2009), 329–350.

31. M. B. Szendrei, 'A note on Birget-Rhodes expansion of groups', *J. Pure Appl. Algebra* 58 (1989), 93–99.

32. M. B. Szendrei, 'Proper covers of restriction semigroups and $W$-products', *J. Algebra Comput.* 22 (2012), 1250024(16 pages).

33. M. B. Szendrei, 'Embedding into almost left factorizable restriction semigroups', *Communications in Algebra* 41 (2013), 1458–1483.

34. Y. H. Wang, 'Weakly B-orthodox semigroups', *Period. Math. Hung.* (68) 1 (2014), 13–38.

35. Y. H. Wang, 'Hall-type representations for generalised orthogroups', *Semigroup Forum* (89) 3 (2014), 518–545.

36. Y. H. Wang, 'Beyond regular semigroups', *Semigroup Forum* (92) 2 (2016), 414–448.

# Author index